Plastic Optical Fiber Sensors

Series in Fiber Optic Sensors

Series Editor: Dr Alexis Méndez, MCH Engineering, LLC

This series of practical, concise, and modern guidebooks encompasses all types of fiber optic sensors, including fiber Bragg grating sensors, Fabry–Perot sensors, interferometric sensors, distributed sensors, and biomedical sensors. The aim of the series is to give a broadly approachable, essential overview of the fundamental science, core technologies, design principles, and key implementation challenges in applications, such as oil, gas, and mining; renewable energy; defense/security; biomedical sciences; civil and structural engineering; and industrial process monitoring. Scientists, engineers, technicians, and students in any relevant field of practice or research will benefit from these unique titles.

Titles in the series

Fiber-Optic Fabry-Perot Sensors
An Introduction
Yun-Jiang Rao, Zeng-Ling Ran, Yuan Gong

An Introduction to Distributed Optical Fibre Sensors
Arthur H. Hartog

Plastic Optical Fiber Sensors
Science, Technology and Applications
Edited by Marcelo Martins Werneck, Regina Célia da Silva Barros Allil

For more information about this series, please visit
https://www.crcpress.com/Series-in-Fiber-Optic-Sensors/book-series/CRCSERINFIB

Plastic Optical Fiber Sensors

Science, Technology and Applications

Edited by
Marcelo Martins Werneck
Regina Célia da Silva Barros Allil

CRC Press is an imprint of the
Taylor & Francis Group, an **informa** business

CRC Press
Taylor & Francis Group
6000 Broken Sound Parkway NW, Suite 300
Boca Raton, FL 33487-2742

First issued in paperback 2022

ISBN-13: 978-1-138-29853-8 (hbk)
ISBN-13: 978-1-03-233755-5 (pbk)
DOI: 10.1201/b22357

Library of Congress Cataloging-in-Publication Data

Names: Werneck, Marcelo Martins, 1949- editor. | Allil, Regina Célia da Silva Barros, editor.
Title: Plastic optical fiber sensors : science, technology and applications / edited by Marcelo M. Werneck and Regina Celia S. B. Allil.
Description: Boca Raton : CRC Press, 2020. | Series: Series in fiber optic sensors | Includes bibliographical references and index.
Identifiers: LCCN 2019030930 | ISBN 9781138298538 (hardback : acid-free paper) | ISBN 9781315098593 (ebook)
Subjects: LCSH: Optical fiber detectors. | Plastic optical fibers.
Classification: LCC TA1815 .P53 2020 | DDC 681/.25--dc23
LC record available at https://lccn.loc.gov/2019030930

Visit the Taylor & Francis Web site at
http://www.taylorandfrancis.com

and the CRC Press Web site at
http://www.crcpress.com

Contents

Foreword

Congratulations on the publication of *Plastic Optical Fiber Sensors*. The field of sensors has been increasingly important as the sensor technologies have become tremendously useful not only for disaster countermeasures and sensing of constructions, but also for AI and IoT. Needless to say, Professor Marcelo Martins Werneck is the leading expert worldwide in POF sensors. I am lucky to have known him through annually held POF conferences (officially, *International Conference on Plastic Optical Fibers*) since 1992. POF conferences are organized by the International Cooperative of Plastic Optical Fiber (ICPOF), and Marcelo has been the committee member as the delegate of Brazil. He organized the 22nd POF Conference in Rio de Janeiro, Brazil, in 2013, which was a great success.

The first plastic optical fiber (POF) was reported from Du Pont in the United States in 1960s. The main purpose of the POF at that time was for illumination of light, because the light injected into the core region of POF can transmit through the POF and illuminate from the end of the POF. However, the attenuation (transmission loss) was as large as 1,000 dB/km, and the length of POF was quite limited to a few meters. The main application was only decoration or toys. The attenuation of POF is caused by the scattering loss and the absorption loss. In 1970s, rigorous purification of contaminants in POF material was developed by several institutes, mainly in Japan, and the scattering loss was remarkably improved. On the other hand, the absorption loss attributes to the molecular structure of polymer, and is an intrinsic characteristic of polymer itself. The main reason for the absorption loss was the absorption of light by the carbon-hydrogen stretching vibration in polymer structure. Therefore, the dominant idea of eliminating the serious absorption loss was the substitution of hydrogen for a much heavier atom such as fluorine or chlorine. As a result, the absorption loss by the carbon-fluorine vibration shifted to a much longer wavelength, which improved the absorption loss at the wavelength of laser so as to be negligible. In the more than thirty years since then, low-loss POF made of perfluorinated polymer has been developed, and the total attenuation including both scattering and absorption losses achieved only 10 dB/km, which can transmit light up to 1 km.

We learn from history that the innovation of technologies that dramatically changed the world is, in many cases, achieved by the innovation from the material side. For example, in the 1960s, the roll of vacuum tube was ended by the invention of semi-conductor material (transistor) that opened the door to real electronics society. In the 1990s, a CRT television was replaced by a liquid crystal display (LCD) by the invention of liquid crystal molecules. These are examples that "the material changed the system." Now we have a variety of POFs with different materials (acrylate materials and fluorinated polymer, etc.) and different structures (step-index (SI) and graded-index (GI) POFs) based on their inherent characteristics. Such POFs are mainly used for various sensor applications and high-speed data transmission, utilizing the inherent characteristics of polymer, such as easy handling and flexibility. For example, compared to silica optical fiber, the thermal expansion and the isothermal compressibility of POFs are two orders of magnitude greater, which means that such a POF can become an excellent sensitive sensor to external information, such as temperature change or slight amount of stress.

POFs have additional advantages of high strain limit, high durability, and negative thermo-optic coefficients. These unique properties of POFs have been utilized in various applications, such as chemical/biological and radiation sensing as well as those of strain, temperature, and displacement. For localized sensors that can be multiplexed, fiber Bragg grating (FBG) has been implemented in POFs. Moreover, microstructured POFs (mPOFs) allow for single mode operation for a wide range of wavelengths, improving performances of POF sensors.

By reading this book, readers will have an in-depth understanding of POF in sensors. I hope that this book opens up opportunity for many researchers, scientists, engineers, and students to formulate new ideas for POF sensors.

Yasuhiro Koike
Keio University
Keio Photonics Research Institute
International Cooperative of Plastic Optical Fiber (ICPOF)

Series Foreword

Optical fibers are considered among the top innovations of the twentieth century, and Sir Charles Kao, a visionary proponent who championed their use as a medium for communication, received the 2009 Nobel Prize in Physics. Optical fiber communications have become an essential backbone of today's digital world and internet infrastructure, making it possible to transmit vast amounts of data over long distances with high integrity and low loss. In effect, most of the world's data flows nowadays as light photons in a global mesh of optical fiber conduits. As the optical fiber industry turns fifty in 2016, the field might be middle aged, but many more advances and societal benefits are expected of it.

What has made optical fibers and fiber-based telecommunications so effective and pervasive in the modern world? Its intrinsic features and capabilities make it so versatile and very powerful as an enabling and transformative technology. Among their characteristics we have their electromagnetic (EM) immunity, intrinsic safety, small size and weight, capability to perform multi-point and multi-parameter sensing remotely, and so on. Optical fiber sensors stem from these same characteristics. Initially, fiber sensors were lab curiosities and simple proof-of-concept demonstrations. Nowadays, however, optical fiber sensors are making an impact and serious commercial inroads in industrial sensing, biomedical applications, as well as in military and defense systems, and have spanned applications as diverse as oil well downhole pressure sensors to intra-aortic catheters.

This transition has taken the better part of thirty years and has now reached the point where fiber sensor operation and instrumentation are well understood and developed, and a variety of diverse variety of commercial sensors and instruments are readily available. However, fiber sensor technology is not as widely known or deeply understood today as other more conventional sensors and sensing technologies such as electronic, piezoelectric, and MEMS devices. In part this is due to the broad set of different types of fiber sensors and techniques available. On the other hand, although there are several excellent textbooks reviewing optical fiber sensors, their coverage tends to be limited and does not provide sufficiently in-depth review of each sensor technology type. Our book series aims to remedy this by providing a collection of individual tomes, each focused exclusively on a specific type of optical fiber sensor.

The goal of this series has been from the onset to develop a set of titles that feature an important type of sensor, offering up-to-date advances as well as practical and concise information. The series encompasses the most relevant and popular fiber sensor types in common use in the field, including fiber Bragg grating sensors, Fabry-Perot sensors, interferometric sensors, distributed fiber sensors, polarimetric sensors, polymer fiber sensors, structural health monitoring (SHM) using fiber sensors, biomedical fiber sensors, and several others.

This series is directed at a broad readership of scientists, engineers, technicians, and students involved in relevant areas of research and study of fiber sensors, specialty optical fibers, instrumentation, optics and photonics. Together, these titles will fill the need for concise, widely accessible introductory overviews of the core technologies, fundamental design principles, and challenges to implementation of optical fiber-based sensors and sensing techniques.

Series Foreword

This series has been made possible due to the tenacity and enthusiasm of the series manager, Mr. Lou Han, to whom I owe a debt of gratitude for his passion, encouragement, and strong support—from the initial formulation of the book project and throughout the full series development. Lou has been a tremendous resource and facilitator, and a delight to work within the various stages of development for each of the series volumes.

Information, as the saying goes, is knowledge. And thanks to the dedication and hard work of the individual volume authors as well as chapter co-authors, the readers have enriched their knowledge on the subject of fiber optic sensors. I thank all the authors and extend my deep appreciation for their interest and support for this series and for all the time and effort they poured into its writing.

To the reader, I hope that this series is informative, fresh, and of aid in his/her ongoing research, and wish much enjoyment and success!

Alexis Méndez
Series Editor
MCH Engineering, LLC
Alameda, CA

Preface

About the Book

This book has been written by top experts in the field of optical sensors, with many developments, with real field applications and patents. This book will be an invaluable resource for physicists, electronic engineers, and those working in the sensing area.

There are many books available on fiber optic sensors; however, none deals specifically with plastic optical fiber (POF) sensors, maybe because POF sensors are relatively new in the optical fiber sensing area. POF sensors are much easier to design, particularly because the unique POF characteristics, such as easy handling, large diameter, cheap peripheral components, and simple tools. These unique properties will be clearly evident to the readers of this book.

When I was invited by Taylor & Francis Group to write this book, I believed it would be a simple task, just another book, I thought. But, despite our expertise in many sensor technologies and the expertise of the other authors, it took two years to write it and even so, it went out incomplete. This is so because only during the research for new sensor technologies and authors to help them did we realize how wide the area of plastic optical fiber sensor is. It will probably need another book to include everything we left out.

However, making a comparison and of course respecting the proportions, I decided to publish this book unfinished because I remembered that Charles Darwin is said to have experienced the same dilemma. He had so much information to include in his famous book and at the same time he was aware that someone else could publish something ahead of him. Having this in mind and also compelled by the other authors to publish their chapters right away, I present to you *Plastic Optical Fiber Sensors—Theory and Applications.*

In this preface I have included a description and a short history on the Photonics and Instrumentation Laboratory. It is a research laboratory dedicated to sensor technology, it is a real cauldron of ideas that accumulated a lot of material and motivated me to write this book.

I have also included a few paragraphs to acknowledge students and the contributors. Without them this book would have been even more unfinished than it is now.

Finally, I want to share with you some of the many ideas that occurred in my mind during the writing of this book. Many of the ideas had absolutely nothing to do with the book subject, but are indirectly in related to it, and I thought was worth mentioning.

The Photonics and Instrumentation Laboratory

The Photonics and Instrumentation Laboratory is a multidisciplinary laboratory devoted to researching projects in photonic technology, optoelectronic instrumentation, and optical fiber sensing. Our research is concentrated in the areas of the electric power industry, biotechnologies, and the oil and gas industry.

We are linked to the largest high-technology complex in Latin America, the Institute Alberto Luiz Coimbra of Post-Graduation Studies and Research at the Federal University of Rio de Janeiro (COPPE/UFRJ).

Our laboratory occupies an area of 650 square meters joining together several internal modules such as the thin film lab, the FBG lab, the high voltage lab, and the biosensor lab.

In the academic area we have students in BSc, MSc, DSc, and post-doctorate, all involved in researching Optical Fiber Sensors, Optoelectronic Instrumentation, Fiber Bragg Grating, and POF technologies applied in the Electric Energy, Biotechnology, and Oil & Gas sectors.

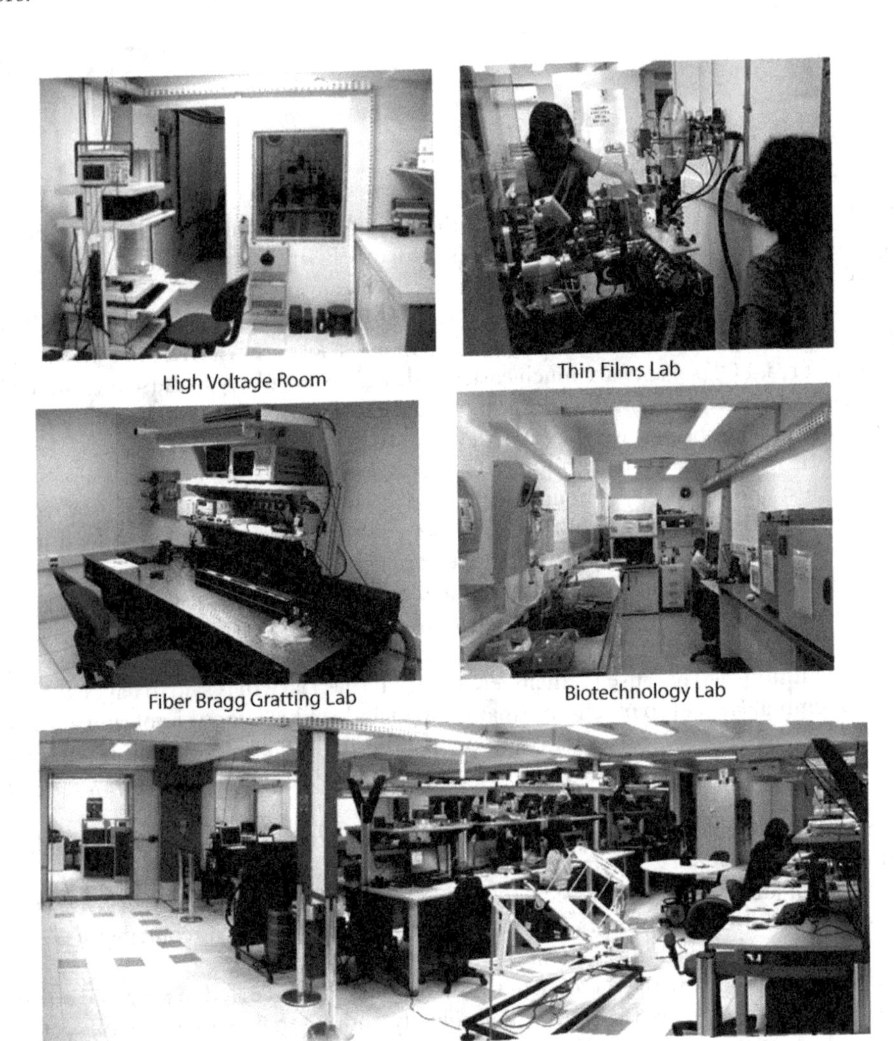

The infrastructure of the Photonics and Instrumentation Laboratory.

The Authors

The real authors of this book are our students who trusted us, following our supervising without question (sometimes). As much as they learned with us, they have been an endless source of new ideas and challenges. To them all, I am grateful and acknowledge their importance in the development of this book.

2017 end of year celebration with some of our students.

No less important are the other authors of this book, who contributed to one or more chapters from many research laboratories all over the word. The authors list can be found at the end of the book, but I do not have words to thank you all for the effort you made to widen the range of this book; without you, the book would not be complete. So, thank you Portugal, England, Australia, Spain, Germany, United States, Japan, and Brazil for your collaboration.

A Tribute to Charles Kao

By the end of the 60s all telecom channels, either transmitting through copper cables or microwaves, were already saturated and a solution had to be found to supply the world's need for bandwidth. Optical fibers were not a possible solution, as its optical attenuation was about 1000 dB/km.

However, as per Charles Kuen Kao's famous paper (Kao and Hockham, 1966), silica optical fiber presented its large attenuation simply due to impurities, without which optical fibers would be a perfect wave guide "for optical frequencies." According to Kao, an optical fiber with 20 dB/km of attenuation could be a solution for long distance telecom, with repeaters. Remember that 20 dB/km means that only 1% of the injected light will reach the end of a 1-km long optical fiber.

It may have been difficult to accept Kao's concept because by that time, optical glasses were already perfectly transparent and used for precision optical equipment such as telescopes, microscopes and all kinds of lenses, for cameras, spectacles, etc. How could we have imagined possible improvement for those lenses? On the other hand, it is easy to understand that, even with good transparency and with impurity concentrations on the order of magnitude of ppb, the light travels inside a lens only a few millimeters, whereas inside an optical fiber, light travels several kilometers, and therefore the probability of a photon bumping together with an impurity is much higher.

Optical lenses were made of fused quartz crystal, that is silica or silicon dioxide (SiO_2), either found in nature or artificially grown, and the impurities come with the raw material. In order to improve transparency, new techniques of producing silica certainly would have to be devised.

Four years later, by 1970, Corning Glass produced the first optical fiber in the world with an attenuation of 20 dB/km. Such attenuation still looks too high; however, this is the attenuation of a copper coaxial cable running at 100 Mbits/s, which demands a signal regeneration every now and then. This is probably why Kao used this figure to suggest that a optical fiber could be used in telecom if it possesses such attenuation. For this discovery, Kao was awarded the Nobel Prize of Physics in 2009.

Sir Charles Kuen Kao, 2009 Nobel of Physics
(Born in Shanghai in November 1933, died in Hong Kong in September 2018).

Corning technology that produces a low attenuation optical fiber was continuously improved today reaching an astonishing 0.15 dB/km. The rest of this story we all know: Optical fibers were installed all over the world and are not saturated yet, despite the never-ending need for speed.

From then on, there has been a meteoric rise in the use of optical fibers for data transmission on medium- to long-length transmission paths. For this reason, optical fiber costs have decreased a lot, as much as the peripheral components, making the use of this technology possible for the development of sensors of all kinds in hundreds of applications.

Notice that optical fiber links can be expensive; however, due to their high bandwidth, the high cost is divided by the thousands of users so that the cost per single user is reasonable. On the other hand, sensors do not attend many applications per unit and therefore they have to be as cheap as the competing technologies, otherwise they will not be useful in the industry. For this reason, optical fiber sensors took a long time to appear in the market, as they had to wait for the decrease in the prices of the optical technologies due to the great demand. This happened to the silica fiber sensors and then, a few decades ago, to the POF sensors.

Evolution of POFs and Charles Kao in the *14th International Conference in Plastic Optical Sensors*

POFs were first developed by DuPont in 1963, made of polystyrene, with losses in the range of 500 to 1000 dB/km. Their first use was for short-distance illumination. Nowadays we use the conventional POF made of polymethyl methacrylate (PMMA). They were also introduced by DuPont in 1968, with losses in the range of 300 dB/km.

Contrary to the silica fibers, since their first applications, POFs did not evolve much in transmission losses. Just for comparison, the ESKA® PMMA fiber from Mitsubishi Rayon presents an attenuation value of 180 dB/km at 650 nm. POFs found applications mainly for short-distance telecom, such as 100 m, and at the same time POFs attracted attention for sensor development. The reason for this is that POFs can be connected to transmission components at low cost by using simple tools as will be seen throughout this book.

Due to its low fabrication cost and many other advantages as compared to silica fiber, POFs would be perfect for long-distance datacom if the manufacturers could improve the attenuation values. This challenge has been paramount for many manufacturers all over the world, trying to follow the silica path fabrication that dropped its attenuation value and making silica fibers ideal for long-distance telecom industry.

To recap on the past and open this discussion in this field, Professor Pak Chu of the City University of Hong Kong, then chairman of the *14th International Conference on Polymer Optical Fiber* (*ICPOF*) held in Hong Kong in September 2005, invited Charles Kao for the opening speech.

Charles Kao presenting a talk at the *14th International Conference on Polymer Optical Fiber* held in 2005, Hong Kong.

For those involved with optical fibers, whether related to material research, sensors, or telecom, e.g., all participants of the conference, the presence of Charles Kao was exciting and fascinating.

There we were, Pak Chu from City University of Hong Kong, the conference chairman, Yasuhiro Koike from Keio University, Japan, involved in telecom research, Charles Kao the Nobel Prize laureate, myself, involved in POF sensors research, and Masaki Naritomi, general manager of the Lucina Division of Asahi Glass Co., POF manufacturer and interested in material research for POF fabrication. This was really a unique, and fascinating meeting.

Pak Chu, Yasuhiro Koike, Charles Kao, Marcelo Martins Werneck, and Masaki Naritomi, 2005.

The Werneck Factor

This book took two years to be written. During the writing of the first chapters, I realized that if I did not distribute key subjects to some specialized colleagues, this book would have never been ready. So, I looked for colleagues that participated in recent ICPOF to help create these chapters whose subject would have been easier to write about. However, writing a book chapter is not as simple as writing a paper. For instance, it is necessary to have more elaborated introduction and bibliography review. This makes the writing and the time dedicated to it longer. And since we are all busy with many other projects, the writing of this book took longer than I anticipated.

This situation happened to me, and of course to all other authors, so that I had to hurry them (and myself) every now and then, trying to convince everybody the importance of this book and how good it will be when it's published.

But we all know that, if we have to rush someone, we have to moderate adequately the rush frequency, otherwise the effect could be disastrously inversed.

A few years ago, I modeled the rush process when I gave an assignment to a group of students. They had to design a device and send the project to the workshop to be fabricated. Of course, there was a line of other projects to be done by the technician and additionally, every other project supposedly more important than ours would be done first. Since we were getting close to the end of the academic semester, the chances were that the prototype would not been completed in time.

In order to try to have the prototype finished in time, I asked the students to visit the workshop every day to check on how the it was progressing. The group decided that each day a student would go to check on the progress. This schedule went on for a few days until, one day, the technician got fed up with the charging and finished the prototype in record time. This is the end of the story, but then, I decided to model the system by a very simple equation:

$$T = T_0 + \frac{W}{n} \quad (1)$$

where:

T is the total time taken to do an assignment;

T_0 is the minimal time it would take to do the assignment with a continuous non-stop dedication;

W is the Werneck factor, a factor that depends on interaction of the environment with the person responsible for the job;

n is the number of times you call the person asking for the job done.

Note that W is completely unpredictable, since it can change without notice with the weather conditions and different seasons of the year, with indispositions, yesterday's result of the football match, family problems, bank account problems, a joyful day—that is, anything can change W.

Therefore, if $n = 1$ in **Eq. 1**, that is, if you contact the responsible person for the project just once, you will never know when the job will be ready, since W changes from day to day. On the other hand, if you make $n \to \infty$, then W ceases to be important in the equation since $T \to T_0$ and you will have the project done in a minimum time.

This method worked well for the technician from whom we asked for the student's prototype; however, it could not work for everybody. The reason for this is the existence of another factor, neglected in **Eq. 1** that, unlike the W factor, can increase T with n. By including this factor, **Eq. 1** becomes:

$$T = T_0 + \frac{W}{n} + FU(n-1) \quad \text{(2)}$$

where FU is the Fed-Up factor.

FU occurs in several levels in people that get annoyed when one rushes him or her. For this kind of personality, the more you rush, the more the person gets fed-up and decreases the speed of the project.

So it is not certain that an indefinite increase of n will speed the project, as T can increase instead of decrease due to the FU factor. In these cases, the solution is to find the right n to get a minimum in **Eq. 2**. But since you do not know either W or FU, one has to test the system and monitor the response in order to correctly model the system.

The graph below shows two examples of **Eq. 2**. The red line represents T with a minimum time T_0 equal to two days, W factor equal to 10 days, and $FU = 0$.

For $n = 1$ the total time T will be 12 days; however, you can increase n up to 10 and will have a total time approaching 2 days.

The blue line contains the same parameters as the red line, except for a small value of the Fed-Up factor, equal to 1. For $n = 1$ the total time is still 12 days. Notice that the minimum of this combination in **Eq. 2** occurs for $n = 3$. For larger n, the total time tends to increase and there is no advantage in annoying the person responsible for the project any longer. If you ask him or her 10 times, the total time will return to 12 days!

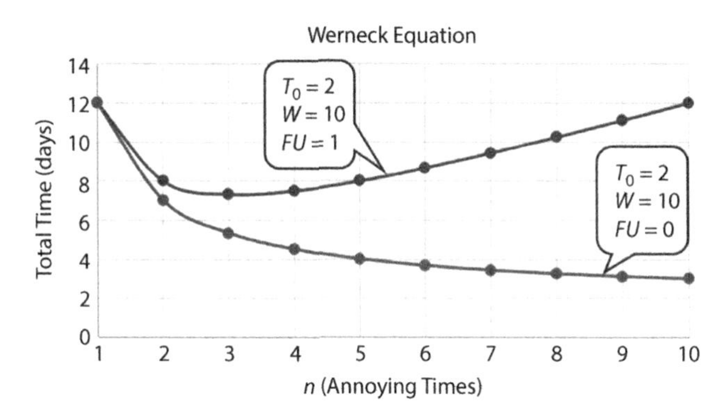

The graph of the Werneck equation for two sets of parameters. For $FU = 0$ (red line), the more one increases n, the more the total time tends to T_0. For an FU as small as 1 (blue line), there is no point in trying to increase n above 3, as the total time tends to increase from then on.

I described the model above to justify the lack of important sensor technologies that should be in this book but are not. This is because during these two years it took to write this book, I have applied the Werneck equation to most of the authors with success. However, in one or two cases, factor *FU* dominated the model and I did not have the chapters in time.

Reference: Dielectric-Fibre Surface Waveguides for Optical Frequencies

K. C. Kao, B.Sc. (Eng.), Ph.D., A.M.I.E.E., and G. A. Hockham, B.Sc. (Eng.), PROC. IEE, Vol. 113, No. 7, July 1966.

Synopsis

A dielectric fiber with a refractive index higher than its surrounding region is a form of dielectric waveguide which represents a possible medium for the guided transmission of energy at optical frequencies. The particular type of dielectric-fiber waveguide discussed is one with a circular cross-section. The choice of the mode of propagation for a fiber waveguide used for communication purposes is governed by consideration of loss characteristics and information capacity. Dielectric loss, bending loss, and radiation loss are discussed, and mode stability, dispersion, and power handling are examined with respect to information capacity. Physical-realization aspects are also discussed. Experimental investigations at both optical and microwave wavelengths are included.

Marcelo Martins Werneck

Editors

Marcelo Martins Werneck has been leading a team of researches and students at the Photonics and Instrumentation Laboratory (LIF) at the Universidade Federal do Rio de Janeiro for the last 20 years. The research lines at LIF center on instrumentation and sensors, particularly fiber optic sensors.

He and his team started research in plastic optical fibers (POF) around the year of 2000 with applications of POF sensors in the high voltage sector for measuring several electrical parameters such as voltage, current, temperature, leakage currents, switching contact resistance, contact position, gas, oil leakage, etc. Today they have already applied these technologies in areas such as biotechnology, electrical energy, and oil and gas with innumerous projects and industrial solutions, many of them reverted in several patents.

Prof. Werneck was born in Petrópolis, Brazil. He received a degree in electronic engineering from the Pontifícia Universidade Católica of Rio de Janeiro, Rio de Janeiro, Brazil, in 1975 and an MSc degree from the Biomedical Engineering Program, Federal University of Rio de Janeiro (UFRJ), Rio de Janeiro, in 1977. His PhD degree was earned from the University of Sussex, Brighton, UK, in 1985. He has been with UFRJ since 1978, where he is currently a lecturer and researcher. He is also the Coordinator of the Instrumentation and Photonics Laboratory at the Electrical Engineering Program, UFRJ. His research interests include fiber optics sensors, nano-biosensors, transducers, and instrumentation.

Regina Célia da Silva Barros Allil worked 30 years with the Biological, Chemistry and Nuclear Defense Division of the Brazilian Army Technological Center (CTEx). She joined LIF as an MSc student, then she earned a DSc degree and followed a post-doctoral research program. By 2006 she joined the LIF team as Research Assistant where she supervises students and coordinate projects, particularly the Bacteriosensor Project in the development of a rapid biosensor for *Escherichia coli* detection based in the inmunocapture effect over a POF nanosensor.

Dr. Allil was born in Rio de Janeiro, Brazil. She received her BSc Degree in electronic engineering from the Faculdade Nuno Lisboa, Rio de Janeiro, in 1988, and the MSc degree from the Biomedical Engineering Program, Federal University of Rio de Janeiro (UFRJ), Brazil, in 2004. Her DSc degree was earned from the Electronic Engineering Program, Instrumentation and Photonics Laboratory, UFRJ, in 2010. She is currently retired from the Brazilian Army Technological Center (CTEx), Rio de Janeiro and is a researcher at the Instrumentation and Photonics Laboratory, UFRJ. Her research interest lies in fiber optics sensors, optoelectronic instrumentation, and biosensors.

Contributors

Alexandre Silva Allil got his Electrical Engineering bachelor's degree in 2018. He is presently a MSc student at the Federal University of Rio de Janeiro (UFRJ), Brazil and researcher at the Photonics and Instrumentation Laboratory from Alberto Luiz Coimbra Institute for Graduate Studies and Research in Engineering (LIF/PEE/COPPE), Brazil. Currently his research interests are on low-cost renewable illumination systems through POFs attached to solar trackers and FBG-based gas flow sensors for the oil and gas Industry.

Lúcia Bilro is presently a researcher at Telecommunications Institute (IT) in Portugal, co-founder and CTO of Watgrid, Lda and board member of the Portuguese Optical Society (SPO). Prior to that she was Post-doc research fellow at IT. She received a PhD degree from the Aveiro University in 2011. Her currently research interests are Bragg sensors, plastic optical fiber sensors, physics medicine and rehabilitation, and environmental monitoring.

Daniel André Pires Duarte is presently a PhD candidate at the Department of Physics in Aveiro University, Portugal, and researcher at Telecommunications Institute Aveiro, Portugal. He received the MSc degree in engineering physics from the same university in 2013. His current research interests are the application of machine learning and data fusion algorithm to optical sensor data analysis and development of low-cost fiber optic-based sensors.

Meysam M. Keley is a Brazilian post–doctorate research fellow at the Photonics and Instrumentation Laboratory at Federal University of Rio de Janeiro (UFRJ). He is currently running a study dealing with applications of sensitive metal oxide thin films as gas sensors coupled to plastic optical fibers. In 2017 he received the title of Doctor of Philosophy with a thesis regarding hydrophobic surfaces. Dr. Keley has vast experience in the area of fabrication and characterization of thin films and functionally graded materials, which is the theme of his Master of Science thesis.

Maryanne Large studied at the University of Sydney and was awarded a PhD from Trinity College Dublin. After graduating she won a Marie Curie Fellowship to do postdoctoral work in Paris, and then worked as a lecturer at the Dublin Institute of Technology. Upon her return to Australia in 2000, she joined the Optical Fibre Technology Centre, as a research fellow and Deputy Director. She led the team that developed the first microstructured polymer fibers (mPOF) and has numerous publications exploring the modeling, fabrication, and applications of mPOF. She has published widely on materials science and interdisciplinary research, including optical biomimetics.

Rogério N. Nogueira is presently a Principal Research Scientist at Telecommunications Institute (IT) in Portugal, co-founder and CEO of Watgrid, Lda and co-founder and vice-president of the Portuguese Optical Society. Prior to that he was an invited professor at Aveiro University and Innovation Manager at Nokia Siemens Networks. He received the PhD degrees from the Aveiro University in 2005 and an Executive Certificate on Management and Leadership from MIT. He coordinates a group at IT in the topics of Fiber Optical Components and Devices for Optical Communications and Sensors.

Ricardo Oliveira received a PhD degree in Physics Engineering from the University of Aveiro, Portugal, in 2017. He is now a researcher at Instituto de Telecomunicações, Aveiro, Portugal. Ricardo has worked in several research projects involving fiber optic devices and components. His main interests include optical fiber devices and components for communications and sensors using polymer optical fibers.

Hui Pan is Chief Analyst with Information Gatekeepers (IGI) and the Plastic Optical Fiber Trade Organization (POFTO). IGI is an internationally recognized market research, publishing, and consulting firm focusing on the fiber optics and telecommunications markets worldwide. POFTO is a trade association that has been established to promote the common business interests of its members and the plastic optical fiber industry in general.

As Chief Analyst, Dr. Pan tracks the global telecom, fiber optics, and plastic optical fiber (POF) market development and provides overall direction to the market and technology analysis of IGI. He has done research on the U.S. and global policies relating to telecom, fiber topics, POF, and ICT services and markets. He has focused on issues such as network convergence, competition, next generation Internet (IPv6), and optical networking. Dr. Pan has authored numerous market and technology reports on the telecommunications, fiber optics, and POF markets worldwide. He has undertaken many major consulting projects as lead consultant for companies wishing to enter the U.S. and international markets.

Dr. Pan has also chaired and organized many conferences and forums, both for IGI and for other industry associations such as the Optical Society and the International Cooperative of Plastic Optical Fibers.

Dr. Pan has a PhD in economics from Northeastern University in the United States.

Filipa Sequeira is presently a PhD candidate at the Physics Department in Aveiro University, Portugal, and researcher at the Telecommunications Institute (IT-Aveiro), Portugal. She received the MSc degree in Applied Physics from the same university in 2008. Her current research interests are the development of low-cost POF sensors for water quality assessment, namely turbidity, refractive index, and the detection of contaminants through the deposition of selective layers.

Pavol Stajanca studied physics and received a PhD degree in Quantum Electronics and Optics from Comenius University in Bratislava, Slovakia, in 2016 for the work on nonlinear propagation of ultrafast pulses in specialty photonic crystal fibers. Since 2014 he has been with Federal Institute for Materials Research and Testing (BAM) in Berlin, Germany. He is a member of BAM's Fibre Optic Sensors division, where he has worked on several European and national projects related to various applications of fiber optic sensors. His research interests include optical fibers, fiber optic sensors, lasers, and nonlinear and ultrafast optics.

David Webb holds the 50th Anniversary Chair in Photonics in the Aston Institute of Photonic Technologies (AIPT) at Aston University. He is a Deputy Director of AIPT and Associate Dean for Research in the School of Engineering and Applied Science. His main interests cover optical fiber sensing using in-fiber Bragg gratings and fiber interferometers. Current research involves medical and biochemical applications of optical sensing technology and the development of polymer optical fiber-based grating devices. He has published over 400 journal and conference papers on these subjects.

Aleksander Wosniok is a postdoctoral research fellow at Federal Institute for Materials Research and Testing (BAM) based in Berlin, Germany. He has many years of experience in the development and application of distributed fiber optic sensor systems for structural health monitoring. He received his PhD degree from the Technical University of Berlin in the field of distributed Brillouin sensing in 2013. His current research interests include geotechnical monitoring, special fibers for high-temperature applications, and optical fiber communications.

Joseba Zubia received a degree in solid-state physics in 1988 and a PhD degree in physics from the University of the Basque Country, in 1993. His PhD work focused on optical properties of ferroelectric liquid crystals. He is a full professor at the Telecommunications Engineering School (University of the Basque Country, Bilbao, Spain). He has more than 25 years of experience doing basic research in the field of plastic optical fibers. At present, he is involved in research projects in collaboration with universities and companies from Spain and other countries in the field of plastic optical fibers, fiber-optic sensors, and liquid crystals. Prof. Zubia won a special award for best thesis in 1995 and the Euskoiker prize in 2010.

Introduction

Why Plastic Optical Fibers?

Hui Pan

Contents

1.1 Why Plastic Optical Fibers?

Plastic optical fibers (POFs) compete with copper wires, coaxial cables, glass optical fibers, and wireless, and they require a transmitter, receiver, cables, and connectors similar to those used in glass optical-fiber links. Manufacturers form POFs out of plastic materials such as polystyrene, polycarbonates, and polymethyl methacrylate (PMMA). These materials have transmission windows in the visible range (520–780 nm). However, the loss of light transmitted at these wavelengths is high, ranging from 150 dB/km for PMMA to 1,000 dB/km for polystyrene and polycarbonates. These losses often handicap plastic fibers in competing against high-quality glass fibers, which have losses of 0.2 dB/km for a single-mode fiber and less than 3 dB/km for multimode fibers. Hence, plastic fibers have been relegated to short-distance applications, typically of a few hundred meters or less, compared with the hundreds of kilometers for glass. Nonetheless, POFs have found many applications in areas such as automobiles, airplanes, consumer electronics, industrial controls, short data links, sensors, signs, and illuminations. Basically, POF applications are divided into data communication and non-data applications such as sensors and signs. Today, the growth in POF production stems from its use in data transmission.

There are both advantages and disadvantages of plastic optical fibers versus glass fiber and copper wire [1]. Certain users find POF systems provide benefits compared to glass fiber or copper wire, which include:

- Simpler and less expensive components
- Lighter weight
- Operation in the visible spectrum
- Greater flexibility, and resilience to bending, shock, and vibration
- Immunity to electromagnetic interference (EMI)
- Ease in handling and connecting (POF diameters are 1 mm compared with 8–100 microns for glass)
- Use of simple and inexpensive test equipment
- Greater safety than glass fibers as glass requires a laser light source
- Transceivers require less power than copper transceivers

With these advantages come disadvantages that researchers and manufacturers are working to overcome. They include:

- High loss during transmission
- Small number of systems and suppliers
- A lack of awareness among users of how to install and design with POFs

- Lack of major funding for research and education
- Certification programs from POF installers are incomplete
- Lack of high temperature fibers (125°C)

As the market for POF develops further, progress is being made by researchers and POF companies to overcome these challenges.

1.2 History and Evolution of Polymer Optical Fibers

Plastic optical fibers were first developed by DuPont in the early 1960s. The first fibers were made of polystyrene and acrylics which had high loss—on the order of 500–1,000 dB/km—and were used mainly for light guides in short-distance illumination applications. Some of the first applications were in automobiles for the lighting of dashboards. In 1968, DuPont reported the first results of plastic fibers manufactured using polymethyl methacrylate (PMMA) base material. After many years of development, Du Pont decided in 1978 to sell the POF business to Mitsubishi Rayon of Japan. The losses of the PMMA POF were 1000 dB/km when Du Pont sold the business to Mitsubishi Rayon. Mitsubishi Rayon, over the next few years, was able to reduce the loss of the PMMA fibers to close to the theoretical limit of 150 dB/km at 650 nm. This was a step index fiber with a bandwidth of 50 Mbps over 100 meters.

The next major development was by Prof. Yasuhiro Koike and colleagues at Keio University who developed a process to manufacture graded index POF (GI-POF) using PMMA material in 1990. Koike and his colleagues reported a bandwidth of 3 GHz-km with losses of 150 dB/km at 650 nm. The development by Koike solved the bandwidth problem of the step index PMMA POF, but losses were still high.

The next major development was a graded index perfluorinated polymer, which was also developed by Koike and colleagues at Keio University in 1995. The perfluorinated polymer had losses of less than 50 dB/km over a range of 650–1300 nm. The theoretical limit of the perfluorinated material was reported as 10 dB/km. Bell Laboratories reported at OFC 1999 that they had achieved 10 Gbps transmission over 100 m of Lucina® fiber (trade name for the perfluorinated polymer fiber developed commercially by Asahi Glass).

The next major development was the simultaneous reporting in 2001 of a microstructured polymer optical fiber (holey fiber) by groups in Australia and Korea.

The first commercially available data link using graded index fiber was announced by Fuji Photo Film in 2005. The link was a 30 meter DVI link operating at 1 Gbps using a 780 nm VCSEL. The link used PMMA GI-POF developed by Fuji Photo under license from Keio University. In early 2007, PMMA GI-POF became commercially available from Optimedia, a company based in South Korea.

Although Asahi Glass was wiring buildings in Japan with its perfluorinated GI-POF Lucina® fiber since 2002, PF-GI-POF fiber cable by itself was not available until fiber cable alone was introduced commercially in 2005 by Chromis Fiberoptics, Inc. of the United States. Chromis Fiberoptics, a spin-off of OFS and Bell Laboratories, had licensed the manufacture of Lucina fiber from Asahi. Chromis has developed a continuous extrusion process for the manufacture of PF-GI-POF compared to the batch process developed by Asahi. The Chromis fiber process produces higher quality fibers at a lower cost.

Although this chapter has focused on the historical development of plastic optical fibers, comparable developments have occurred in low cost, low profile connectors, optical sensors, and detectors.

A complete history of the development of POF for data communications is shown in **Table 1.1**. Some may question or have different views on the dates or the importance of some of the items, or even the omission of some important dates and events, but the list does provide a perspective on the historical development of POF through the present time.

Table 1.1 History of Plastic Optical Fiber Developments for Data Communications

Year	Development
1968	DuPont first to develop PMMA
1977	PMMA—d8 developed by DuPont
1978	DuPont sells all products and patents to Mitsubishi Rayon (MR)
1990	Professor Koike of Keio University announces development of graded index POF with bandwidth of 3 GHz/km
1992	2.5 Gbps over 100 m of POF using red laser reported by Koike et al.
1992	Professor Koike reports GI-POF with bandwidth of more than 19 GHz
1993	Sasaki et al., of Keio University report POF optical amplifier
1994	DARPA announces award for High Speed POF Network (HSPN)
1994	2.5 Gbps 100 m transmission with GI-POF demonstrated using high-speed 650 nm LD
1995	Keio University and KAIST develop fluoropolymer
1996	Yamazaki of NEC demonstrates 155 Mbps ATM LAN
1997	ATM Forum approves POF PMD for 155 Mbps over 50 meters
1997	Yamazaki of NEC reports 400 Mbps POF link for 1394 over 70 m
1997	Asahi Glass reports on perfluorinated POF GI-POF with one-third the loss of conventional PMMA POF
1997	DARPA funds PAVNET, a follow-on to HSPN, with Lucent Technologies added to consortium
1997	Imai of Fujitsu reports 2.5 Gbps over 200 m using 1.3 nm FP-LD, InGaAs and FP laser with GI-POF
1998	Asahi Glass reports at OFC-98 new PF fiber, CYTOP(r), with attenuation of 50 dB/km from 650 to 1300 nm, with further improvements in bus possible. Bandwidth of 300–500 MHz-km. Theory shows the possibility 10 GHz/km. Fiber stable over the temperature range of –40°C–90°C.
1998	XaQti, a Gigabit semiconductor supplier, demonstrates at Interop 1 Gbps over 200 m using PF GI-POF
1998	Nanoptics reports on continuous process for manufacturing GI-POF
1998	MOST Standard for Automobiles started
1999	Lucent reports 11 Gbps over 100 m of Lucina fiber (PF GI-POF)
2000	European Commission starts POF program under name of Optomist, which consists of three programs: Agetha, I/O, and Home Planet
2001	First "POF Applications Center" established in Nuremberg, with grant from the Bavarian Government
2001	Lucina fiber available in limited quantities
2001	The University of Sydney, Australia, and KAIST of Korea announce developments of photonic crystal, or microstructured POF (mPOF)
2001	European Commission I/O Project and Home Planet initiated
2002	IEEE 1394B Standard ratified; IDB-1394 for Automobiles completed

(*Continued*)

Table 1.1 (*Continued*) History of Plastic Optical Fiber Developments for Data Communications

Year	Development
2002	Fuji Film announces availability of GI-POF
2003	Digital Optronics plans to offer graded index POF for commercial sale
2004	Optimedia of Korea offers PMMA GI-POF for sale in the U.S.
2005	First commercially available PF GI-POF announced by Chromis Fiberoptics
2005	The University of Sydney demonstrates first hollow-core POF
2005	Aston University, UK, and University of Sydney, Australia, demonstrate Fiber Bragg Gratings (FBG) in mPOF
2006	Fuji Photo Film announces GI-POF with a bandwidth of 10 Gbps over 50 m
2007	Georgia Tech researchers show 30 meters at 40 Gbps over PF GI-POF
2008	Georgia Tech researchers show 100 meters at 40 Gbps over PF GI-POF
2010	Partially chlorinated GI-POF—Sekisui Chemical
2011	1 Gbps over 50 meters of SI PMMA—POF-PLUS
2011	Southern Photonics, New Zealand, and University of Sydney researchers show 50 meters at 9.5 Gbps over PMMA-mPOF
2012	40 Gbps GI-POF Interface for 4K3D Video Transmission successful at Keio University
2012	Technical University of Eindhoven, and University of Sydney researchers show 50 meters at 7.3 Gbps using DMT modulation over PMMA-mPOF
2012	ITU G.hn (ITU-T G.9960) over POF standard published
2013	Several silicon manufacturers announce G.hn support
2014	First HDMI/DVI cables introduced by Chromis using partially fluorinated POF
2014	First results of using GaN sources
2015	Chromis Fiberoptics announces POF Active Optical Cables (AOC) for commercial installations
2015	Boeing announces first application of POF for a commercial aircraft
2016	Panasonic, working with KAI Photonics (Keio University), develops single POF cable and connector solution for transmission of full spec 8K signals
2016	Mitsubishi Pencil and KPRI demonstrate at OFC 2016 a 4K/8K connector through multiple GI POF micro-collimators based on ball-point pen technology
2017	IEEE 802.3bv Gigabit Ethernet over POF standard published
2017	Keio University presents consumer friendly GI POF interface cable for uncompressed 8K video transmission up to 120 Gbps (12 GI-POFs) using commercial optical modules for 10 Gbps Datacom at POF Symposium at OFC 2017 [3]
2018	Eindhoven University of Technology presents lab results of wired and wireless joint transport of 9 LTE-A bands, 1.66 Gbit/s PAM-4, WLAN, and WSN services over 50 m Ø1mm core GI-POF at POF Symposium at OFC 2018 [4]
2018	Keio University shows optical interface for 8K endoscope using GI-POF at POF 2018 [5]

Source: Polishuk, P., "POF Market and Technology Assessment Report", IGI Consulting, 2016.

The history of POF development has been a long one, without the publicity and glamour of glass optical fiber. However, the time for POF may have come, as the need for low-cost, high-speed, easy-to-use optical data communications is being recognized.

1.3 Silica vs. Polymer Fibers

There are many discussions in the trade literature concerning the pros and cons of using copper cables, glass optical fiber (GOF), and POF. The question is always which is best for what applications.

In its simplest form, a POF data link consists of a transmitter, receiver, cable, and connectors. The transmitters and receivers are electrical-to-optical and optical-to-electrical converters, respectively. More complicated data-link configurations include rings (each receiver on a network responds only to its address), stars (signals go to a hub for relay), and meshes (all receivers are interconnected in a manner similar to the Internet). As when glass-fiber systems were introduced, simple point-to-point POF links were installed first, followed by rings and stars.

The high-loss problem with POF is being addressed with new perfluorinated polymer materials, which have brought losses down to potentially 10 dB/km. **Table 1.2** compares the different transmission media—glass, plastic, and copper.

Table 1.2 Comparison of Plastic Optical Fiber, Glass Optical Fiber, and Copper Wire

Parameter	Plastic	Glass	Copper
Component costs	Potentially low-cost fiber and components	More expensive fiber and components	Low cost
Loss	High-medium loss (short distance)	Medium-low loss (long distance)	High loss
Connectorization	Easy to connectorize, requires little training or special tools	Takes longer, requires special tools and training	Easy
Handling	Easy to handle	Requires training and care	Easy
Flexibility	Flexible	Brittle	Flexible
Wavelength operating range	Operating in visible	Operating in infrared	NA
Numerical aperture	High (0.4 N.A.)	Low (0.1–0.2 N.A.)	NA
Bandwidth	High (40 Gbps over 100 meters)	Large (40 Gbps)	Limited to 100 meters at 100 Mbps
Test equipment	Low-cost	Expensive	High
Systems costs	Low overall	High	Medium

1.4 POF Markets and Applications

Plastic optical fiber applications cover a wide range of industries. Unlike the telecommunications field, if the market goes down in one sector, there are other industries that will pick up the slack. In addition, developments for any one sector could have a dramatic impact on all the others. For example, the ability of automobiles to drive the price down could have a large impact on the rest of the industry. According to research conducted by IGI Consulting/Information Gatekeepers, the potential worldwide POF market is expected to grow from $2.6 billion in 2010 to over $7 billion in 2020.

The following are the key market sectors, but others are developing, such as smart fabrics, medical diagnosis and treatment, and sensors.

- Automotive
- Avionics
- Consumer electronics
- Interconnection
- Industrial controls and IoT
- Home networking
- Medical and sensors
- Homeland security

In addition, the increased interest in homeland security in the United States could have a major impact on the demand for low-cost, mass-produced sensors that can be purchased at Home Depot and other similar home improvement centers.

The following are descriptions of the major market sectors and the market drivers.

1.4.1 Automotive

The number of electronic devices in cars has been increasing over the years in order to enhance safety, entertainment, and comfort to drivers and passengers. In 2000, German auto manufacturer Daimler-Benz recognized that the increasing use of digital devices in automobiles increased the weight, susceptibility to EMI, and complexity of wiring harnesses. Until then, each auto manufacturer developed its own proprietary wiring standards, which hampered them from achieving the economies of scale provided by mass production. Daimler-Benz realized that the way to reduce costs was to develop and buy to a common standard, and its analysis indicated that POF ring networks would meet the needs of future automobiles. The auto maker convinced six other European auto manufacturers, including BMW and Volkswagen, to join it in developing a standard called MOST (Media Oriented Systems Transport) [6] and to agree to purchase against the standard.

The seven companies also formed an organization called the MOST Cooperation to coordinate the standard's development and promotion. The MOST Cooperation now consists of most of the world's major auto manufacturers, including Audi, Daimler, General Motors, Honda, Jaguar, Land Rover, Mazda, Porsche, Toyota, Volkswagen, Volvo, and dozens of POF suppliers worldwide. Late in 2000, sixteen European automakers ratified the "MOST" standard, which specifies POF for data networks in automobiles. By developing a standard that is accepted by 16 European automobile manufacturers, suppliers can produce to the same standard for 16 customers and thus can obtain economics of scale. There is tremendous pressure by the automakers on their suppliers to reduce prices. This had a major impact on all sectors of the POF industry to realize the promise of POF's low cost! [7] The combination of an accepted standard and the agreement by a group of auto manufacturers to purchase against the standard has created the economies of scale needed by the industry.

Daimler-Benz and BMW have led the way in the use of POF in automobiles. Over 1 million POF nodes were installed in "S" line Mercedes-Benz automobiles in 2000. The number increased to 55 million by the end of 2007 in 50 different models of cars from the low end to the high end. In 2013, over 150 models of automobiles had over 133 million POF nodes installed worldwide. In 2007, for the first time, two Korean auto manufacturers—Hyundai and Kia—announced that they will adopt the POF MOST standard. The MOST systems have been installed in more than 200 car models since 2000.

The original MOST system was designed for 25 Mbps. The standard itself has progressed, and now it has specifications covering 50 Mbps and 150 Mbps. Although MOST was developed for non-mission-critical applications in automobiles, BMW also developed a separate, 10-Mbps POF star network, called Byteflight, for critical elements such as airbag sensors. A third auto network, Flexray, used POFs as part of an automotive fly-by-light system.

The IEEE formed a Working Group in 2015—802.3bv "Gigabit Ethernet over Plastic Optical Fiber" that was intended to support Gigabit transmission over POF in key markets such as automotive, industrial, and home networking. In March 2017, The IEEE published a new IEEE 802.3bv standard "Gigabit Ethernet over Plastic Optical Fiber." The publication of the IEEE GEPOF standard has paved the way for the adoption of POF by automobile manufacturers.

The automotive industry is facing unprecedented technological and service model changes. Auto manufacturers are putting their investment in autonomous and all-electric cars. These cars will require high bandwidth for in-vehicle, vehicle to vehicle, and vehicle to infrastructure communications. POF is well positioned to be the ideal solution for vehicle networking in the future.

1.4.2 Avionics

Because of their small size, low weight, resiliency to shock and vibration, and high bandwidth capabilities over short distances, POF has been used for applications in aircraft, tanks, ships, helicopters, missiles, and spacecraft. In the mid-1990s, DARPA (Defense Advanced Research Program Agency) of the United States invested heavily in POF technology to develop high speed links for military applications. Unfortunately, the technology was not at that time ready for commercialization. Now, the technology has evolved and has already been used for military and aircraft applications.

There are potential savings of hundreds of pounds in weight per commercial airplane if fiber optics and POF are adopted. Aircraft have hundreds of data links per plane.

Reduced weight benefits [8]:

- Airplanes: wiring separation for different electromagnetic environment (EME) and critical signals is very complex
- Airlines: half of an airline's direct operating cost is from fuel burn
- Environment: every 5 pounds savings per airplane equals 1 ton of CO_2 reduction per year per airplane
- Up to 150 kilometers of signal wiring including 30 km of data bus wiring per airplane
- Boeing Commercial Airplanes (BCA) delivered about 750 airplanes per year

Currently less than 10% of data wiring uses fiber optics on a commercial airplane, hence, there is great need for more fiber to reduce weight, EME, and wire integration complexity [9].

For commercial airplanes, Boeing is leading the industry in fiber optics usage, particularly POF, for aircraft. Recently, Boeing has also developed a passive optical network using plastic optical fiber for the 777X flagship, which went into production in 2018. It is the first passive

POF network in aviation history. Boeing is now evaluating graded index perfluorinated POF for multi-Gbps data links. Avionics represents a great opportunity for POF suppliers.

1.4.3 Consumer Electronics

One of the first uses of plastic optical fibers for data communications was in connecting audio components. Because of its low cost and ease of use, POF was a natural fiber for consumer electronics. As the number of consumer devices increases and the trend continues toward digital electronics, POF is becoming an even greater choice for interconnecting consumer electronics.

The completion of the 1394b standard—which increases the distance between nodes of the existing 1394 standard from 4.5 to 100 meters, and increases speeds from 400 Mbps and eventually to 3.2 Gbps—specifies a low-cost high increase of usage of POF as one of the transmission media (50 meters at speeds of 200 Mbps). Sony and other Japanese firms are leading the charge to increase the usage of POF for video games and video cameras using 1394 interfaces. In the United States, Apple is promoting 1394 through its trade name "FireWire." The number of 1394-capable devices is expected to grow. This could offer a good opportunity for POF.

1.4.4 Interconnection

The interconnect market includes a wide range of interconnect products, including circuit boards, backplanes, chassis-to-chassis, and rack-to-rack applications.

Interconnects for large terabit routers, optical cross-connects, and optical switches are becoming a major issue as these devices are reaching terabit speeds. Today 10 Gbps and 40 Gbps interfaces are a reality. The pressure is on reducing the cost and complexity and improving reliability of backplanes for component-to-component, board-to-board, chassis-to-chassis, and rack-to-rack wiring.

At the same time, more and more systems are being packed into central offices, data centers, and co-location facilities requiring the interconnection of different types of equipment. In addition, some digital cross-connect products require 70 bays of equipment that have to be interconnected. Speeds between bays require 40 Gbps, making this one of the first markets for 40G components. These interconnections could range from a few inches to tens of meters, well within the range of existing or new plastic optical fibers.

An evolving market is that of "server optical interconnects," which are essentially supercomputers. These have tremendous data transfer requirements in the tens of Gbps over 10–50 meters.

An aspect favoring POF is the cooling required for these large machines and servers. The dissipation of heat is a major problem in costs of cooling required as well as the electricity costs. Transceivers for 10G copper require 15 watts of power whereas a 10G POF link requires 1.5 watts.

1.4.5 Industrial Controls and IoT

The industrial controls market has always been one of the most stable markets for POF, mainly due to the need for protection of the data links from radio frequency interference caused by radiation from motors, laser welders, high voltage devices, etc. Standards such as "SERCOS," "Profibus," and Interbus have been available for a number of years and specify POF as a transmission medium.

Another driver is the recent adoption by the industrial control industries of Ethernet data buses. Ethernet has been used in the plant operations and management but is now being pushed further down onto the factory floor. This has put pressure on the need for maintenance and repair on the factory floor. The ease of connectorization with or without connectors makes it easier for craftspeople in tight spaces.

The rapidly growing Internet of Things (IoT) market represents another great opportunity to POF as there will be a tremendous need for a lower bandwidth, flexible, and inexpensive connectivity solution.

1.4.6 Home Networking

Home networks are starting to emerge as a major market segment for POF, driven by the need to network computers and audio/video devices in the home. Until recently, there was little interest in home wiring, mainly due to the complexity and cost of rewiring existing homes. With the trend toward digital TV and ultra-high definition TV (4K/8K), and the fact that there are now multiple PCs and TVs per home, there is an increasing desire by the consumer to be able to interconnect these devices throughout the home.

There are several competing technologies that are vying for this market, including coaxial cable, powerline, Ethernet, and wireless, but no one system now dominates. As home electronics become digital and the need for high speeds increases, POF becomes an ideal network solution to interconnect clusters of home network appliances.

The market adoption of 4K/8K ultra HD TV has gained momentum in the last two years. IGI estimates that shipments of 4K/8K UHD sets are expected to reach 135 million units in 2020, equivalent to almost half of all sets sold worldwide. In the age of 4K/8K TV, home networks are becoming the bandwidth bottleneck. Fiber in the Home (FITH) requires optical fibers with easy termination and high enough bandwidth to support up to 10 Gbps with at least 50-meter reach. FITH and 4K/8K could be one of the killer applications for POF.

1.4.7 Medical and Sensors

The POF medical market is a relatively small but growing market for POF. The main applications are POF sensors and POF for data links in major medical machines such as X-ray, MRI, and CAT scanners, etc., when there are internally large electromagnetic fields. POF provides immunity to this source of radio frequency interference (RFI) and provides isolation.

1.4.8 Homeland Security

Another market driver that is just emerging is that of "Homeland Security," which could require POF technology for secure communications such as fiber to the desktops, POF sensors and systems for monitoring, and data collection systems for ships, aircraft, and other vehicles.

1.5 POF Activities, Organizations, and Global Centers of Research and Education

1.5.1 POF in Japan

Plastic optical fiber developments in Japan have been ongoing since Mitsubishi Rayon purchased the rights to POF from DuPont in the late 1970s. A complete industry has developed that produces POF fibers, components, cables, and systems. Multiple suppliers now provide products on a commercial basis. In order to promote sales of POF, the Japan Institute of Standards (JIS) has developed standards for cables, connectors, and test methods.

The industry organized itself into a "POF Consortium" in 1994 with the objective of "constructing a POF based multimedia society." The POF Consortium now consists of over 60 corporate and university members. It organizes conferences and seminars on the subject.

A major force in the development of POF technology and especially plastic optical fiber has been Professor Yasuhiro Koike of Keio University. A new Keio Photonics Research Institute (KPRI) has been formed. In 2015, Keio University established KAI Photonics Co. Ltd., which is a venture company with a purpose to commercialize research results from KPRI. In 2018, Nitto Denko Corporation licensed the POF technology from Keio University to produce high speed interconnect products.

Prof. Koike has been successful in obtaining government funding for a multimedia society. Recent major research of KPRI includes:

- High speed graded index polymer optical fibers
- Highly scattering optical transmission (HSOT) polymer
- Zero birefringence polymer
- High power optical fiber amplifier and laser

The research is supported by the Japan Society for the Promotion of Sciences (JSPS) through its "Funding Program for World-leading Innovative R&D on Science and Technology" (FIRST Program).

There are several other universities and research centers in Japan that are active in POF research and education. They include the University of Tokyo, Tokyo Institute of Technology, Utsunomiya University, and Toyota Central R&D Labs.

1.5.2 POF in the United States

With the departure of DuPont from the POF field, there was a period when POF for data communications was at a virtual standstill. In the 1990s, one of the major early developers of optical fiber communication, Hewlett-Packard, seeing the need for low-cost production, shifted its POF manufacturing operations to Singapore while continuing to develop the product in the United States. POF cable was still purchased from the Japanese suppliers.

The developments in the POF field had attracted the attention of the Defense Advanced Research Program Agency (DARPA), a research arm of the U.S. Department of Defense that is involved in developing low-cost interconnection links for dual commercial and military uses. In 1994, DARPA awarded a three-year, US$60 million contract to a consortium of Packard Hughes Interconnect, Boeing, Honeywell, and Boston Optical Fiber to develop cost-effective, high-bandwidth, plastic fiber optic solutions for short- to medium-distance applications that would traditionally use copper. The program, known as the "High Speed Plastic Network (HSPN)," was focused on developing graded index plastic optical fiber (GI-POF)-based systems using VCSEL sources. Packard Hughes would develop the connectors; Boston Optical Fiber, the fiber and cables; Honeywell, the VCSELs; and Boeing, the applications.

In 1997, DARPA funded another project, OMNET (Optical Micro-Network). This DARPA program planned to build on advances in optoelectronic device technology, device integration, and fabrication resources to develop and demonstrate revolutionary new system concepts and resulting capabilities. This program also was to investigate the use of graded index perfluorinated POF (GI-PF POF). The OMNET project was funded in fiscal year 1997 to the total of approximately US$30 million for three years. In 2001, the projects were completed and the consortium disbanded.

Since the early 2000s, with mergers and acquisitions on the rise, new companies entered the POF business while others sought exit. One of the companies that entered the POF market is Chromis Fiberoptics, Inc. In 2004, Chromis Fiberoptics bought the entire POF business from OFS and began to manufacture PF GI-POF under license from Asahi Glass. PF GI-POF fiber and cable became available in June 2005. In recent years, the company has expanded its product lines by producing GI POF active optical cables, which has gained considerable success in the market.

There are several other companies that have been aggressively pursuing the POF business for many years in the United States. They include Mitsubishi International PolymerTrade Corporation, FiberFin, Industrial Fiber Optics, Micronor, and the Canadian firm BL Lighting. They are either producing and selling their own products or are major distributors for other POF suppliers. In addition, MCH Engineering is a leading consulting firm on fiber-optic and POF sensors.

In 1995, the POF Interest Group (POFIG) was founded in the United States by a group of POF manufacturers and interested parties with the purpose of promoting POF business. The name was changed to POF Trade Organization (POFTO) in 1999 (www.pofto.org). The mission of POFTO is to actively promote the proliferation of low cost, low and high bandwidth POF systems serving the data communications and non-datacom markets, coordinate with other organizations, and seek out new market opportunities.

Information Gatekeepers, Inc. (IGI) serves as the Secretariat of POFTO, which maintains the POFTO website and organizes POF events at major trade shows such as OFC, CEDIA, SAE, ECOC, etc. IGI also organizes a yearly POF tradeshow in the United States called POF World, which has become the POF Symposium, held in conjunction with OFC in recent years. IGI hosts the International Conference on Optical Fibers conference (ICPOF) when the event rotates back in the United States. In addition, IGI conducts POF market research and consulting, and publishes several POF studies and newsletters (see www.igigroup.com).

Several universities in the United States have professors and researchers working in the POF area. They include: University of Washington, Washington State University, Georgia Tech, City University of New York—the College of Staten Island, Polytechnic Institute of NYU, and University of Alabama—Birmingham.

1.5.3 POF in France

POF activity in Europe started in 1982—mainly for lighting, illumination, and sensing—by the French "POF Club," an informal organization of French POF companies supported by the French government. This club has not been active in recent years.

1.5.4 POF in Germany

In the 1990s, several companies in Germany such as Hirschmann, Siemens, and AMP were developing data links for industrial applications. Hoechst of Germany entered the POF market in the early 1990s both in Europe and the United States, with research facilities in Japan. In 1995, because of a downturn in the chemical business, Hoechst decided to take its products off the U.S. market. Siemens was one of the few companies in Europe producing components for POF. Siemens spun off its active components business (POF transceivers) units into Infineon, which was subsequently sold to Avago Technologies. Avago was later purchased by Broadcom.

In 1998, a group of German and European automobile manufacturers led by Daimler Benz formed the MOST Consortium to develop a POF data bus for automobiles (see section on automotive applications), which was ratified in 2000.

In 2000, a POF Application Center (POFAC) was inaugurated at the University of Applied Science "Georg Simon Ohm" in Nuremberg, Germany, funded by a grant from the Bavarian government [10]. The objective of the center was to develop applications and standards for POF. Even though government funding has ceased, POFAC has been able to continue with industry support. The center has focused its research on POF applications in data communication, illumination, optical slip rings, and sensors. The POFAC offers a wide range of research and development services, measurements, trainings, and simulations of active and passive components. The center has been a major contributor to POF publications over the years. In 2015, POFAC successfully hosted the *24th International Conference on Plastic Optical Fibers* (*POF 2015*), which was chaired by Prof. Olaf Ziemann.

The Federal Institute for Materials Research and Testing (BAM) is a key German organization on POF research and development. BAM is a senior scientific and technical federal institute with responsibility to the Federal Ministry for Economic Affairs and Energy of Germany. In POF related areas, BAM is engaged in the development, validation, and testing of fiber optic sensors, fiber optic components, and systems.

1.5.5 POF in European Union (EU)/European Commission (EC)

The European Union (EU)/European Commission (EC) funded POF projects early in its advanced technology programs (RACE). The RACE program was changed to the "IST-Optomist" program, and the EC initiated three POF development programs in 2000 and 2001.

In 2006, the IST program funded a group of European companies to develop a "POF-All" program to develop POF applications for 1 mm step index fibers. Funding was EUR1.6 million, matched by the member companies. The program lasted until 2009.

POF-PLUS—"Plastic Optical Fibre for Pervasive Low-cost Ultra-high capacity Systems"—began in May 2008, running three years until April 2011. Total cost was EUR3.6 million, with EUR2.6 million provided by the EC.

The ALPHA Project—"Architectures for fLexible Photonic Home and Access networks"—was another three-year project, from January 1, 2008 to December 31, 2010. Total cost was EUR16.5 million, with a contribution of EUR11.2 million by the EC (see www.ict-alpha.eu). The general focus of the ALPHA project is the integration of access, home, and in-building networks.

In the last few years, funding on POF research from the EU has slowed down. Hopefully, more funds will be available from individual governments and from the corporate sectors in the future.

1.5.6 POF in the United Kingdom

Professor Demetri Kalymnios at the London Metropolitan University was actively involved in POF applications for sensor and in research on methods for increasing the bandwidth of standard SI-POF links. He also organized the UK POF Consortium.

Aston Institute of Photonics Technology (AIPT) at Aston University, Birmingham, UK, has an active POF research program. Prof. David Webb, deputy director of AIPT, hosted and chaired POF 2016 on the Aston Campus. The Center for Photonic Systems at the University of Cambridge is conducting research on POF optical interconnects and polymer optical waveguide. University of Manchester is also engaged in POF research.

1.5.7 POF in Spain

The Applied Photonics Group at the University of the Basque Country (UPV/EHU) is a leading research center on POF in Spain. Led by Prof. Joseba Zubia, the core of its research team, which mainly works at the School of Engineering of Bilbao, includes over a dozen professors and research associates and PhD students. Its recent research activities have focused on fabrication of microstructured optical fiber, POF sensors for structural health monitoring, POF optical amplifiers, and lasers. POF 2011 was chaired by Prof. Zubia and hosted by the group. The conference has received rave reviews.

The following universities are also active in POF research and education in Spain: Universidad Carlos III de Madrid, Universidad de Zaragoza, Universidad de Cantabria, and Universidad de Alcala.

One leading POF company in Spain is the fabless semiconductor supplier Knowledge Development for POF S.L. (KDPOF). KDPOF provides chipset for gigabit communications over POF. KDPOF's technology enables 1 Gbps POF links for automotive, industrial, and home networks. Founded in 2010 in Madrid, KDPOF offers their technology as either Application-Specific Integrated Circuit (ASIC)/chip or as Intellectual Property (IP) to be integrated in third party Systems on Chips (SoC).

1.5.8 POF in Portugal

Instituto de Telecomunicações (IT) was established in 1992 in Aveiro, Portugal. It is a private, non-profit association of six Portuguese universities, one polytechnic, one public telecom operator, and one telecom equipment manufacturer. Its mission is to create and share scientific knowledge in telecommunications worldwide and to host and tutor graduate and post-graduate students. IT has just completed a two-year project—*Paving the way for high capacity POF based communication systems (HiPOF project)*. A team of researchers is working on POF sensors and fiber Bragg gratings. IT organized and hosted the POF 2017 conference chaired by Prof. Rogerio Nogueira.

University of Aveiro, University of Lisboa, and the Technical University of Lisbon are also engaged in POF research and education.

1.5.9 POF in the Netherlands

The Institute for Photonic Integration, Eindhoven University of Technology in the Netherlands, led by Prof. A.M.J. (Ton) Koonen, has engaged in the study of POF system techniques for wired and wireless indoor delivery of broadband services for several years. The group has shown that a single in-home fiber network can be the universal converged backbone for both wirebound and wireless delivery of services. Its research shows that large-core POF is very suited due to its ease of handling and installation, provided appropriate spectrum-efficient modulation methods are used. In-building fiber solutions are more future proof and energy efficient than copper and radio solution, and more cost efficient at higher data capacities, in particular the POF solutions.

Prof. Koonen presented the lab results of wired and wireless joint transport of 9 LTE-A bands, 1.66 Gbit/s PAM-4, WLAN and WSN services over 50 m Ø1mm core GI-POF at the POF Symposium at OFC 2018. The group also conducted research on POF for 2D position sensing [4].

1.5.10 POF in Brazil

POF activities in Brazil are concentrated in two institutions: the Instrumentation and Photonics Laboratory (LIF) at the Federal University of Rio de Janeiro (UFRJ) and the Specialty Optical Fiber Laboratory (LaFE) at the State University of Campinas (UNICAMP) located in the state of São Paulo [11].

The Instrumentation and Photonics Laboratory (Laboratório de Instrumentação e Fotônica—LIF)—working in collaboration with the Institute of Chemical, Biological, Radiological and Nuclear Defense, Brazilian Army Technological Center (CTEx), Rio de Janeiro—started the study of POF in the mid-1990s with applications in the electric power industry. From then on, they diversified research areas into fiber-optic sensors, with applications in many areas of the industry, such as electrical power industry, renewable energy sources, oil and gas, environmental monitoring, and biological defense.

The team, headed by Prof. Marcelo M. Werneck, has also hosted two main events in Brazil in the POF area—*the International Workshop on Polymer Optical Fibers and Micro-Structured POF* in 2006 and the *22nd International Conference on Plastic Optical Fibers* (*POF 2013*), chaired by Prof. Werneck and held at the city of Armação dos Búzios in the state of Rio de Janeiro in September 2013. In recent years, the Instrumentation and Photonics Laboratory (LIF) and the Federal University of Rio de Janeiro (UFRJ) have expanded their POF research and education activities. They have become a leading center for POF research in Brazil.

The Specialty Optical Fiber Laboratory, working in collaboration with the Institute of Advanced Studies (IEAv) at the Aerospace Technical Center (CTA), has been working during the last several years with silica and softglass microstructured optical fibers. Fibers are being fabricated, modeled, characterized, and used in applications from telecom to sensing. Attracted by polymers' characteristics such as flexibility, low processing temperature, interesting mechanical properties, low cost, and functionalization possibilities, microstructured polymers fibers are also being studied.

By the end of 2012 the group infrastructure was ready to produce structured preforms and to draw from them plastic microstructured optical fibers using a state-of-the-art dedicated optical drawing tower. Solid polymeric fibers made by extrusion can also be produced using raw or processed pellets.

Using this infrastructure, single-mode, multimode, spun, and Hi-Bi microstructured PMMA optical fiber are currently being fabricated and characterized. Cyclo-olefin fibers (Zeonex) are also being studied, aimed at THz applications.

1.5.11 POF in Korea

The Korean government was instrumental in 2001 in starting a "Korean POF Consortium" of manufacturers, research organizations, and government agencies to develop a POF industry in Korea. There are several Korean companies that are producing POF products.

Optimedia, Inc., founded in 1999, is a GI-POF manufacturing company in Korea. One of their products, OM-Giga, is a PMMA-based GI-POF for high-speed short-distance data communication applications. It does not contain any refractive-index modifying dopant, and has excellent mechanical properties and thermal stability.

A new company, Optomind Inc., is producing advanced optical engines for POF active optical cables or embedded optics for both data center and non-data center applications.

Another company, Opto Marine Co., Ltd., founded in 2011, is producing 650 nm fiber optic transmitter and receiver modules for 1 mm SI POF with data rates of 500 Mbps, 5 Gbps, and 10 Gbps at 100 meters.

1.5.12 POF in Australia

POF-related activities in Australia started at the University of New South Wales in the 1990s, with research on the fabrication of step-index, single-mode POF for the infrared, intended to be combined with gratings and utilized in sensing [12].

In 2001, the first microstructured POF (mPOF) was reported by the University of Sydney Optical Fibre Technology Centre (OFTC) at the ICPOF of that year, held in Amsterdam. This work sought to combine the flexibility of microstructured optical fiber designs, which had recently begun to be explored in silica, with all the advantages of using polymer fibers. The work was supported briefly by Redfern Polymer Optics, and subsequently by their spin-off, Cactus Fibres. In 2004, a grant from the Australian Research Council funded the purchase of a specialized draw tower, around which an mPOF fabrication facility was established. MPOF research focused on the traditional datacoms applications, as well as sensing.

In 2008, the fabrication facility was subsumed under the Australian National Fabrication Facility (ANFF), an initiative of the Australian Commonwealth and New South Wales governments, and in 2009, with the closure of the OFTC, the facility was relocated to the School of Physics, University of Sydney. From January 2009, a spin-off company, Kiriama, was formed to make mPOF commercially available, predominantly to other researchers; however, operations ceased in July 2013. The fabrication facility continues to operate at the University of Sydney, under ANFF, as a research and training facility.

One leading manufacturer of POF test and measurement equipment in Australia is Kingfisher International Pty Ltd. Found in 1986, Kingfisher manufactures quality handheld fiber optic test solutions including for POF applications.

1.5.13 POF in Greater China

The study of plastic optical fiber in mainland China began in the 1980s at Chinese universities and government research institutions. The initial applications of POF in China were for signage and illumination. The researchers started to focus on decreasing the transmission loss of POF and its application in short distance data communications in the 1990s. Efforts were also made to develop heat resistant POF and to produce Graded Index POF and its testing techniques.

The following universities and institutes are some of the leading institutions that are engaged in POF research and education in mainland China:

- University of Science and Technology of China (USTC)
- Institute of Chemistry of Chinese Academy of Sciences (CAS)
- Technical Institute of Physics and Chemistry of CAS
- Xi'an Institute of Optics and Precision Mechanics of CAS
- Nanjing Fiberglass Research and Design Institute
- Wuhan Research Institute of Post and Telecommunications
- Huazhong University of Science and Technology
- Yanshan University
- Shanghai University
- Southeastern University
- Xi'an Jiaotong University
- Shanghai Jiaotong University
- Zhejiang University
- Chongqing University of Technology

These government research institutions and universities have collaborated with Chinese companies on POF research and development. Together, they have made China one of the major POF manufacturing centers in the world. In order to promote POF, several POF companies in China established the China POF Industry Alliance in 2011. The alliance invited its member companies to jointly exhibit at the PT/EXPO COMM CHINA that year. At the trade show, several POF-related applications were demonstrated including lighting, data communications, and consumer, industrial, and FTTH solutions.

The following are leading POF companies in greater China:

- Century Opticomm Co., Ltd.
- Sichuan HuiYuan POF Co., Ltd.
- Nanjing Chunhui Science and Technology Industrial Co., Ltd.
- Guangdong IPT Industrial Co., Ltd.
- Firecomms China
- Jiangxi Daishing POF Co., Ltd.
- XiaMen San-U Optronics Co., Ltd.
- COMOSS Electronic Co., Ltd.
- Shenzhen HWAYING

Several leading universities in Hong Kong and Taiwan are also active in POF research. They include:

- City University of Hong Kong
- The Chinese University of Hong Kong

- The Hong Kong Polytechnic University
- National Central University, Taiwan
- National Cheng-Kung University, Tainan

1.5.14 POF in Other Countries

There are growing POF research and education activities in Belgium, Cyprus, Ireland, Israel, Denmark, Italy, Malaysia, Mexico, Poland, Romania, and Switzerland. These seem to be individual companies or research institutes, but no formal government or industry consortia are being formed in these countries.

1.6 POF Resources

1.6.1 Proceedings of International Conference on Plastic Optical Fibers

The International Conference on Plastic Optical Fibers (ICPOF) has been held annually around the world since 1992. The ICPOF conference is widely recognized as one of the most authoritative academic conferences on POF. Papers presented at ICPOF meetings and published in the conference proceedings represent the latest research results at the time. Information on how to obtain copies of the proceedings is available from Information Gatekeepers (www.igigroup.com).

List of International POF Conferences and Their Locations

1992	1993	1994	1995	1996	1997
Paris	The Hague	Yokohama	Boston	Paris	Kauai
1998	**1999**	**2000**	**2001**	**2002**	**2003**
Berlin	Chiba	Boston	Amsterdam	Tokyo	Seattle
2004	**2005**	**2006**	**2007**	**2008**	**2009**
Nuremberg	Hong Kong	Seoul	Turin	Santa Clara	Sydney
2010	**2011**	**2012**	**2013**	**2014**	**2015**
Yokohama	Bilbao	Atlanta	Rio de Janeiro	Yokohama	Nuremberg
2016	**2017**	**2018**	**2019**		
Birmingham	Aveiro	Seattle	Yokohama		

1.6.2 Proceedings of POF World and POF Symposium

Since 1998, Information Gatekeepers (IGI) and the Plastic Optical Fiber Trade Organization (POFTO) have organized the POF World and POF Symposium annually in the United States, sometimes in conjunction with OFC (Optical Fiber Communications conference). These meetings are more industry and applications oriented. There is a product exhibit area called POF Technology and Application Pavilion where companies can demonstrate their products and technologies. The proceedings from POF World and POF Symposium are available from IGI (www.igigroup.com) and POFTO (www.pofto.org).

List of POF World and Their Locations					
1998	1999	2000	2002	2004	2005
Providence	San Jose	San Jose	San Jose	San Jose	Santa Clara
2006	2007	2008			
Santa Clara	Santa Clara	Santa Clara			

List of POF Symposium and Their Locations					
2009	2010	2011	2012	2013	2014
San Diego	San Diego	Los Angeles	Los Angeles	Anaheim	San Francisco
2015	2016	2017	2018	2019	
Los Angeles	Anaheim	Los Angeles	San Diego	San Diego	

1.6.3 POF Market Reports and Newsletters

Information Gatekeepers has published a series of market and technology reports on POF. Here is a list of the published reports:

- Plastic Optical Fiber Market & Technology Assessment Study
- The Market for Plastic Optical Fibers in Automobiles
- The Market for Plastic Optical Fibers in Aircraft
- The Market for Plastic Optical Fibers in Consumer Electronics
- The Market for Plastic Optical Fiber Sensors
- The Market for Plastic Optical Fiber Interconnects
- The Market for Plastic Optical Fibers in Woven Fabrics
- The Market for Plastic Optical Fibers in Home Networks
- Top 40 Actual & Potential Plastic Optical Fiber Markets
- Market Assessment for Graded Index Plastic Optical Fiber (GI-POF)
- Study of the Market for Plastic Clad Silica Fibers
- Plastic Optical Fibers in Industrial Controls
- POF Source Book
- POF Directory
- POF Newsletter

Further information on these reports can be found on the IGI web site, www.igigroup.com.

1.6.4 Select Books on Plastic Optical Fibers

Many books on plastic optical fibers (POF) have been published in recent years. The list below is meant to give the reader some background reading on POF.

Fundamentals of Plastic Optical Fibers
By Yasuhiro Koike
Wiley-VCH, 2015

POF Handbook—Optical Short Range Transmission Systems
By Olaf Ziemann, Juergen Krauser, Peter E. Zamzow, and Werner Daum
Springer, 2008

Microstructured Polymer Optical Fibers
by Maryanne Large, Leon Poladian, Geoff Barton, and Martijn A. van Eijkelenborg
Springer, 2008

MOST—The Automotive Multimedia Network
by Andreas Grzemba
Franzis Verlag GmbH, 2008

Polymer Fiber Optics: Materials, Physics, and Applications
by Mark G. Kuzyk
Taylor & Francis Group, 2007

POF—Polymer Optical Fibers for Data Communication
By Werner Daum, Juergen Krauser, Peter E. Zamzow, and Olaf Ziemann
Springer, 2002

Acknowledgments

This paper is based on the research and publications on POF conducted by the late Dr. Paul Polishuk and published by IGI Consulting /Information Gatekeepers Group. Paul devoted most of his life to the promotion of the POF technology and business. I would like to humbly dedicate this paper to him.

I would also like to thank Prof. Marcelo Werneck of the Universidade Federal do Rio de Janeiro and Prof. Alexander Argyros of the University of Sydney for their contributions to the sections on POF in Brazil and in Australia, respectively.

References

1. Yasuhiro Koike, *Fundamentals of Plastic Optical Fibers*, Wiley-VCH, 2015, pp. 6–8.
2. Paul Polishuk, "POF Market and Technology Assessment Report", IGI Consulting, 2016.
3. Yasuhiro Koike and Azusa Inoue, "Status of GI POF towards Noise-Free 8K Data Transmission", *POF Symposium at OFC 2017*, March 23, 2017, Los Angeles, CA.
4. Ton Koonen, Federico Forni, Henrie van den Boom, and Eduward Tangdiongga, "POF system techniques for wired & wireless indoor delivery of broadband services", *POF Symposium at OFC 2018*, March 15, 2018, San Diego, CA.
5. Tetsuya Toma, Hiroshi Takizuka, Toshio Chiba, and Yasuhiro Koike, "Study on Optical Interface For 8K Endoscope Using GI-POF", *POF 2018 Conference*, September 4–6, 2018, Seattle, WA.
6. MOST Standard for Automobiles www.mostcooperative.com.
7. Harold Schopp, "MOST Standard Applied to Consumer Electronics", *POF World* (*West*), June 21–23, 2005, Santa Clara, CA.
8. Trieu Kien Truong, "Boeing Commercial Airplanes Fiber Optic Evolution and Applications of POF", *POF Symposium at OFC 2016*, March 24, Anaheim, CA.
9. Trieu Kien Truong, "Boeing Commercial Airplanes O-enable Technologies & Application Opportunities", *POF Symposium at OFC 2018*, March 15, San Diego, CA.

10. B. O. Ziemann and H. Poisel, “15 years polymer optical fiber application center—A summary”, *24th International Conference on Plastic Optical Fibers (POF 2015) Proceedings*, September 22–24, 2015, Nuremberg, Germany, pp. 21–25.
11. Marcelo M. Werneck, contributed section on POF in Brazil in “POF Market and Technology Assessment Report”, IGI Consulting, Inc., 2016, pp. 45–46.
12. Alexander Argyros, contributed section on POF on Australia in “POF Market and Technology Assessment Report”, IGI Consulting, Inc., 2016, pp. 44–45.

Principles of Polymer Optical Fibers

Ricardo Oliveira, Lúcia Bilro, and Rogério N. Nogueira

Contents

2.1 Principles of Operation

In fiber optics, light is generally guided through a dielectric medium by total internal reflection. The behavior of light in such medium can be simply explained through the ray theory. When light travels from one medium to another, the speed can decrease or increase, depending on the density of the two media. The density of each medium determines its refractive index (n), which can be expressed as:

$$n = \frac{c}{v} \tag{2.1}$$

where c and v are the velocity of the light in vacuum and in the medium, respectively. Moreover, the ray direction is also changed when the two media have different refractive index. In order to explain such phenomenon, consider two media with refractive index n_1 and n_2, where $n_1 > n_2$ (see **Figure 2.1**).

When a light ray traveling in a medium n_1, with a small angle ϕ_1 with respect to the normal, crosses the interface between the media, a small portion of the light will be reflected back to the

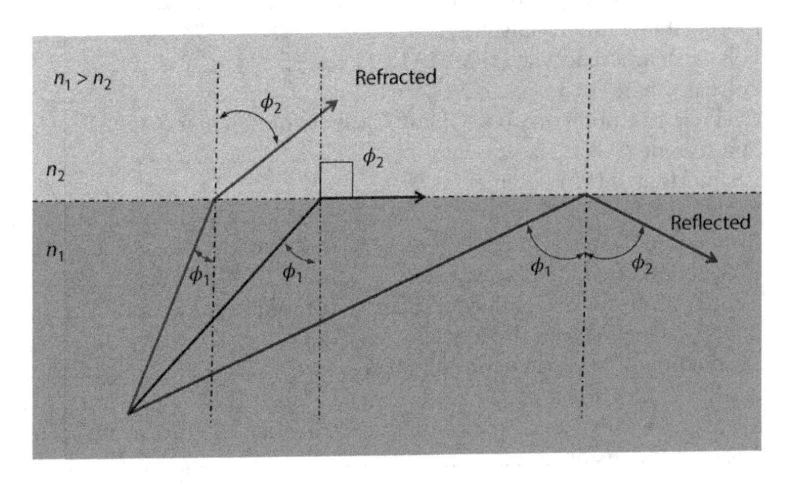

FIGURE 2.1 Refraction and reflection in materials with different refractive index ($n_1 > n_2$).

first medium and the majority of the light will be refracted for the second medium with an angle ϕ_2. On the other hand, when ϕ_1 increases, ϕ_2 will increase in the same proportion. For a certain angle $\phi_1 = \phi_c$, the ray crossing the interface will become 90°. In this specific case, the refracted ray will travel parallel to the interface between the two media. This specific angle is defined as the critical angle, and above this angle, the ray will be reflected to the same medium, where $\phi_1 = \phi_2$. The relations between the refractive index of the two media and the angle of incidence and refraction are governed by Snell's law given in (2.2).

$$n_1 \sin(\phi_1) = n_2 \sin(\phi_2) \tag{2.2}$$

Using the Snell's law, one can find that, when $\phi_2 = 90°$, the critical angle will be given as:

$$\phi_c = \arcsin\left(\frac{n_2}{n_1}\right) \tag{2.3}$$

Thus, the total internal reflection will always occur when $90° > \phi_1 > \phi_c$. This is the case found in optical fibers (see **Figure 2.2a**). In fact, an optical fiber is a cylindrical waveguide composed of two dielectric materials which can be made of silica, polymer, or the combination of both. In such waveguide, a cladding layer with refractive index n_{cl}, surrounds the core waveguide

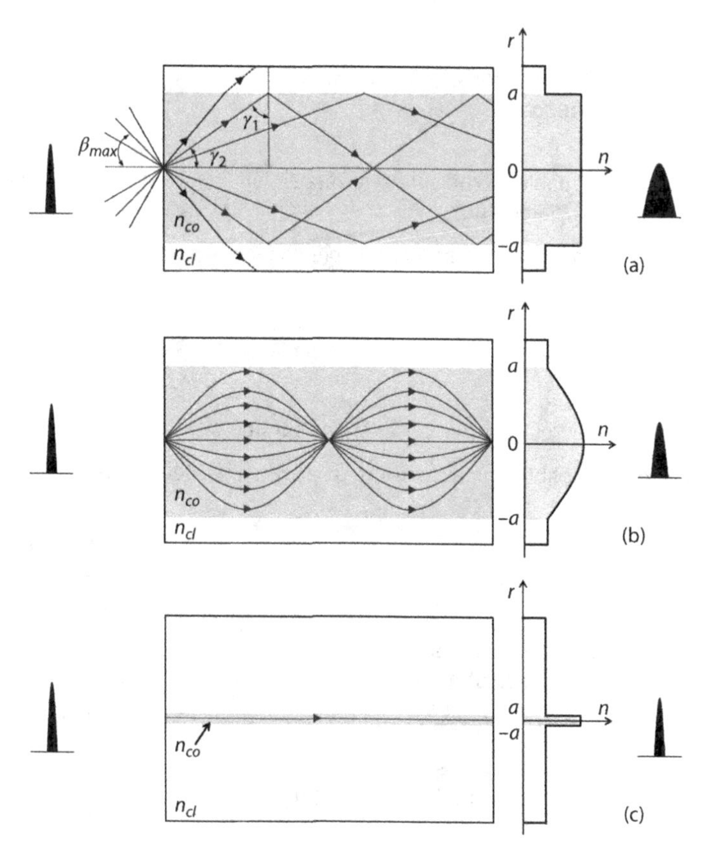

FIGURE 2.2 Refractive index profiles of three different fibers, namely: (a) step index multimode fiber, (b) graded index multimode fiber, and (c) step index single mode fiber. The refractive index profile, together with the pulse shape before and after propagating through the fiber, is also shown.

with refractive index n_{co}. Thus, in order to reach total internal reflection, n_{co} needs to be higher than n_{cl}, and the incidence angle needs to be higher than ϕ_c, but lower than 90° (Ioannides et al., 2014; Ziemann, Krauser, Zamzow, & Daum, 2008).

2.2 Types of Optical Fibers

Nowadays, there are a variety of applications that are employing optical fibers. Some of those require specific properties and for that reason, different fibers have been developed. **Table 2.1** lists some of the most relevant parameters found in literature to define an optical fiber.

Table 2.1 Types of Optical Fibers Depending on Different Parameters

Parameter	Description	Fiber Example
Index profile	Composed of a step index profile Composed of double step-index profile Composed of graded index profile Composed of multi-step index profile	SI-POF DSI-POF GI-POF MSI-POF
Numerical aperture	With high acceptance cone	High numerical aperture (HNA)
Number of modes	Supports only one mode Supports three to ten modes Supports thousands of modes	SMF FM-mPOF MM fiber (OM2 (50/125))
Dispersion	Capable to shift the zero dispersion wavelength Capable to offer a flat dispersion Capable to compensate the dispersion	Dispersion shifted fiber (DSF) Dispersion flattened fiber (DFF) Dispersion compensating fiber (DCF)
Microstructure	Photonic crystal fibers Microstructured polymer optical fibers	PCF mPOF
Polarization behavior	Capable to maintain the polarization state	Polarization maintaining fiber (PMF)
Material	Variants of glass Silica (SiO_2) Polymers	ZBLAN fiber SMF-28e POF
Number of cores	Composed of a single core Composed of multiple cores	SMF Multicore fiber (MCF)
Bending losses	Insensitive to curvature	Bending insensitive fiber (BIF)
Cladding material	Composed of a hybrid structure such as SiO_2 core and polymer cladding	Polymer clad silica fiber (PCS)
Standard	ITU-Standards	(SMF-28e)

2.2.1 Refractive Index Profile

One of the important parameters presented by an optical fiber and described in **Table 2.1** is the refractive index profile. This parameter is particularly important when working with data transmission systems, where the transmission capacity is severely reduced with the modal dispersion. To understand this phenomenon, consider the fiber shown in **Figure 2.2a**.

The fiber shown in **Figure 2.2a** is defined as step index (SI), meaning that the refractive index across the entire cross section of the core and cladding is constant. Additionally, from the same figure, it can be observed that different rays are traveling in different paths with different lengths. Even considering those rays traveling with the same velocity and coincident at the input of the optical fiber, they will disperse at the output fiber due to their different path lengths. Thus, the impulse signal will broaden on the transmission link; this becomes a serious problem since the pulses will overlap and interfere with each other, making it impossible to retrieve the signal at the receiver side (Koike & Koike, 2011). To suppress the problem, one can consider single mode fibers (SMF) (i.e. **Figure 2.2c**), a type of SI fiber that only supports single mode behavior (a mode, in this sense, is a spatial distribution of optical energy in one or more dimensions that remains constant in time). In fact, this type of fiber has been the preferred choice for transmission applications. The choice is undoubtedly due to its higher performance when compared with multimode (MM) fibers. However, this type of fiber has a typical core radius of about 4 μm. This, in fact, is a disadvantage due to the inherent precision needed to couple light into this type of fiber. Therefore, the low tolerances on the coupling process imposed by MM fibers, led this type of fiber to be extensively studied in telecommunication applications up until the mid-1980s (Koike & Koike, 2011). The progress on MM fibers led to the development of a fiber with a special refractive index profile, capable to reduce the modal dispersion presented by the SI profile. This type of fiber was engineered to have a graded index (GI) profile (see **Figure 2.2b**), where the core refractive index is dependent on the radial distance (r), decreasing its value from the core to the periphery, forming a parabolic distribution. The refractive index distribution can be expressed through (Marcuse, 1978):

$$n(r)=\begin{cases} n_{co}\sqrt{1-2\Delta\left(\dfrac{r}{a}\right)^{g}}, r\leq a \\ n_{co}\sqrt{1-2\Delta}, r>a \end{cases} \tag{2.4}$$

a being the core radius and g the profile exponent, where $g = 2$ can be found for fibers with GI profile and in the limit ($g \rightarrow \infty$), for fibers with SI profile. The parameter Δ is defined as the relative difference in refractive index, defined as:

$$\Delta=\frac{n_{co}^{2}-n_{cl}^{2}}{n_{co}^{2}} \tag{2.5}$$

Regarding **Figure 2.2b** and **Equation** (2.4), one can easily understand why the modal dispersion is reduced in fibers presenting a parabolic distribution. In fact, the rays traveling at the periphery of the core (higher order modes) have longer distances to travel; however, the refractive index in the periphery is lower than in the center of the fiber, thus its velocity is increased. Regarding the rays propagating along the fiber axis, their path is shorter, but the refractive index is higher in this region, meaning that those rays will propagate with lower velocities. The control of the refractive index profile can thus reduce the modal dispersion in MM fibers (Koike & Koike, 2011).

2.2.2 Refractive Index Contrast between Core and Cladding

The refractive index contrast between the core and cladding is one of the most relevant properties when designing an optical fiber. Indeed, the maximum angle in which incident light can be guided through the core, commonly known as the angle of acceptance (β_{max}), is dependent on this contrast. Normally, β_{max} is expressed in terms of numerical aperture (NA), which can be calculated through the following relations:

$$\begin{aligned} n_{ext}\sin(\beta_{max}) &= n_{co}\sin(\gamma_2) \\ &= n_{co}\cos(\gamma_1) \\ &= n_{co}\sqrt{1-\sin^2(\gamma_1)} \\ &= n_{co}\sqrt{1-(n_{cl}/n_{co})^2} \\ \leftrightarrow \sin(\beta_{max}) &= \frac{\sqrt{n_{co}^2-n_{cl}^2}}{n_{ext}} \\ \leftrightarrow NA = \sin(\beta_{max}) &= \sqrt{n_{co}^2-n_{cl}^2} \end{aligned} \tag{2.6}$$

n_{ext} being the refractive index of the external medium and γ_1 and γ_2 the angles defined in **Figure 2.2a**. Due to the wide availability of polymer materials, it is possible to create POFs with high refractive index contrast between the core and cladding. In fact, polymers can be found with refractive indices that vary from 1.32 for highly fluorinated materials to around 1.6 for some cast phenolic resins (Emslie, 1988). With such refractive index contrast, POFs have been fabricated with high NA, varying from 0.2 up to 0.7 (Zubia & Arrue, 2001). POFs with such high NA are good for easy light coupling to low cost devices such as LEDs. Therefore, whenever POFs are compared with their silica counterparts, this property gives an obvious opportunity. Conversely, the ease of POF fabrication with high NAs becomes problematic when SM behavior is required. Indeed, the number of modes found in an optical fiber is related to the normalized frequency (V_{SIF}), which for SI fibers is given by:

$$V_{SIF} = \frac{2\pi a}{\lambda} NA \tag{2.7}$$

where λ defines the free space wavelength. Therefore, in order to show SM behavior, V_{SIF} needs to be lower than 2.405 (Marcuse, 1974; Snyder & Love, 1984). From this expression, one can notice that for a specific wavelength, the occurrence of SM behavior depends on the balance between the core radius and the refractive index contrast between the core and cladding. Thus, a higher refractive index contrast, which for POFs is easier to achieve, means the need of a smaller core radius, leading to an increase of the scattering losses at the core–cladding boundary (Large, Poladian, Barton, & van Eijkelenborg, 2008). On the contrary, considering the case where the fiber core is large, it implies a lower refractive index contrast that can be achieved by doping. However, this process can be challenging since the dopant diffusion needs to be carefully controlled in order to have a stable refractive index. Another important feature that can be seen in (2.7) is that at short wavelengths, the balance between the refractive index contrast and core radius is even more pronounced. This takes more relevance when considering POFs, since their low loss region is usually localized at the visible region. For that reason, POFs with a SI profile have only been reported with SM behavior at the near-infrared region (Bosc & Toinen, 1992; Garvey et al., 1996; Kuzyk, Paek, & Dirk, 1991; Peng & Chu, 2000; Yang, Yu, Tao, & Tam, 2004). Such fibers were first reported by Kuzyk and co-workers in the early 1990s, for a POF operating at 1.3 μm and composed of 8 μm core and 125 μm cladding (Kuzyk et al., 1991). The commercialization of POFs of this type was initialized by Paradigm Optics, Inc., for fibers composed of PMMA in the cladding and PMMA copolymerized with PS (<3%) in the core., with cutoff wavelengths ranging from 750 to 1100 nm. POFs with SM behavior have also been fabricated with materials that benefit specific applications. In one of those works, reported by Zhou et al., the fiber employs a graded index perfluorinated core (one of the most transparent materials at the near infrared region) and a PMMA cladding (Zhou et al., 2010). Woyessa et al. have also reported a new SM-POF composed of ZEONEX® 480R cladding and TOPAS® grade 5013S-04 core, providing humidity insensitiveness and high temperature operation.

Considering other configurations than the step index profile, Leon-Saval et al. reported the use of an array of tight and thin ZEONEX® 480R capillarity involved in a PMMA matrix, allowing SM behavior at the visible region (Leon-Saval, Lwin, & Argyros, 2012). However, the manufacturing process is more challenging.

2.2.3 Microstructured Fibers

In conventional optical fibers, the guidance of light occurs through the effect of total internal reflection due to the refractive index contrast between the core and cladding. This contrast is normally obtained by doping one of the regions, or through the utilization of different materials. In fact, this is an advantage for POFs due to the wide range of polymer materials. However, there exists another class of optical fibers which are composed of tiny air holes arranged normally in a regular pattern surrounding the core region, as shown in **Figure 2.3**. This type of fiber is classified as microstructured fiber.

The presence of air holes surrounding the solid core region allows the presence of a depressed index and thus, the condition for total internal reflection, or the equivalent mode confinement is satisfied (Large et al., 2008). The first works reporting this type of fibers were made in the 1970s with the purpose to avoid chemical doping (Kaiser, Marcatili, & Miller, 1973; United States patent No. 3712705, 1973); however, at that time, the possibilities of those fibers were not fully understood. Two decades later, thanks to the work developed by Philip Russell and his colleagues, it was possible to present the microstructured silica fiber (Birks, Knight, & Russell, 1997; Knight, Birks, Russell, & Atkin, 1996), also referred as photonic crystal fiber (PCF). Its unique properties revealed numerous applications, spanning many areas of science. This is because the structure can be easily manipulated by changing parameters like the distance between the air holes (Λ_d) and their dimension (d), allowing special opportunities, such as the ability to remain

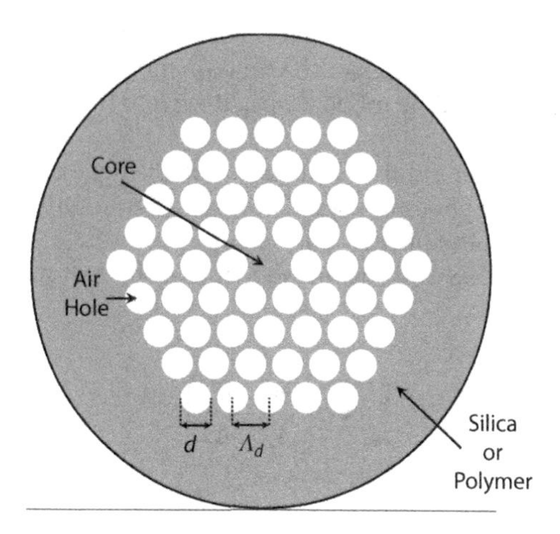

FIGURE 2.3 Microstructured optical fiber.

single mode at all wavelengths for which fused silica is transparent, also known as endless single mode operation (Birks et al., 1997). Other possibilities could be found, such as the ability to control the dispersion by adjusting the air filling fraction (d/Λ_d) (Mortensen, 2005); the possibility to create highly nonlinear effects by concentrating the mode field in a small area, allowing supercontinuum generation (Dudley, Genty, & Coen, 2006); the capability to create highly birefringent fibers through the use of asymmetric structures (Ortigosa-Blanch et al., 2000); etc.

The fabrication of a microstructured silica fibers is normally made through the stack-and-draw-technology, which is considered a time dependent process. Contrary to silica, polymers are easily processed, requiring low processing temperatures, and technologies such as drilling, casting, and extrusion can be employed. Because of that and considering the opportunities of polymer materials, van Eijkelenborg et al. reported in 2001 the first microstructured polymer optical fiber (mPOF) (van Eijkelenborg et al., 2001). Since then, several works have succeeded in the fabrication of fibers with graded index profile, high birefringence, dual core, suspended core, and hollow core (van Eijkelenborg et al., 2003). Doping these type of fibers is also possible (Large, Ponrathnam, Argyros, Pujari, & Cox, 2004), allowing the creation of fiber lasers and amplifiers. Additionally, the ability to use different polymer materials allows the development of fibers capable to be humidity sensitive (Zhang & Webb, 2014) or insensitive (Woyessa et al., 2016), or the ability to sustain high temperatures (above 100°C) (Markos et al., 2013; Woyessa et al., 2016).

As described previously, SM behavior in SI-POFs has only been reported at the infrared window. This is due to the compromise between the core radius and refractive index contrast that need to be balanced in order to have a normalized frequency lower than 2.405, making it particularly difficult when the operating wavelength is shorter. One clear advantage of mPOFs when compared with SI-POFs is the ease of the manipulation of the fiber structure. Therefore, by manipulating the hole dimensions and their arrangement, it is possible to easily produce a fiber with SM behavior at the visible region (van Eijkelenborg et al., 2001; Zagari et al., 2004).

The definition of normalized frequency in microstructured fibers (V_{PCF}) is not straightforward as previously shown for SI fibers. This is because (2.7) fails to describe the cutoff properties and it has the problem of determining an equivalent core radius. For that reason Mortensen et al. defined a mathematical expression shown below (Birks et al., 1997; Mortensen, Folkenberg, Nielsen, & Hansen, 2003):

$$V_{PCF} = \frac{2\pi\Lambda_d}{\lambda}\sqrt{(n_{FM}^2(\lambda) - n_{FSM}^2(\lambda))} \quad (2.8)$$

While this expression takes the same mathematical form as the one shown in (2.7), the unique nature of microstructured fibers is taken into account. In this equation $n_{FM}(\lambda)$ represents the wavelength dependent effective index of the fundamental mode and $n_{FSM}(\lambda)$ corresponds to the effective index of the first cladding mode propagating into the air-hole lattice region (see the example shown in **Figure 2.4**).

Considering the second order cutoff, Mortensen (2002) suggested a phase diagram for the SM-MM operation regime for a photonic crystal fiber with a triangular air-hole lattice cladding*, the details of such boundary subsequently followed up in more detail by Kuhlmey, McPhedran, and de Sterke (2002). From those works it was shown that the best fit equation describing such boundary transition is:

$$\frac{\lambda^*}{\Lambda_d} \approx \alpha\left(\frac{d}{\Lambda_d} - \frac{d^*}{\Lambda_d}\right)^{\gamma} \quad (2.9)$$

* This approach applies also to microstructured fibers in general (Mortensen, 2002).

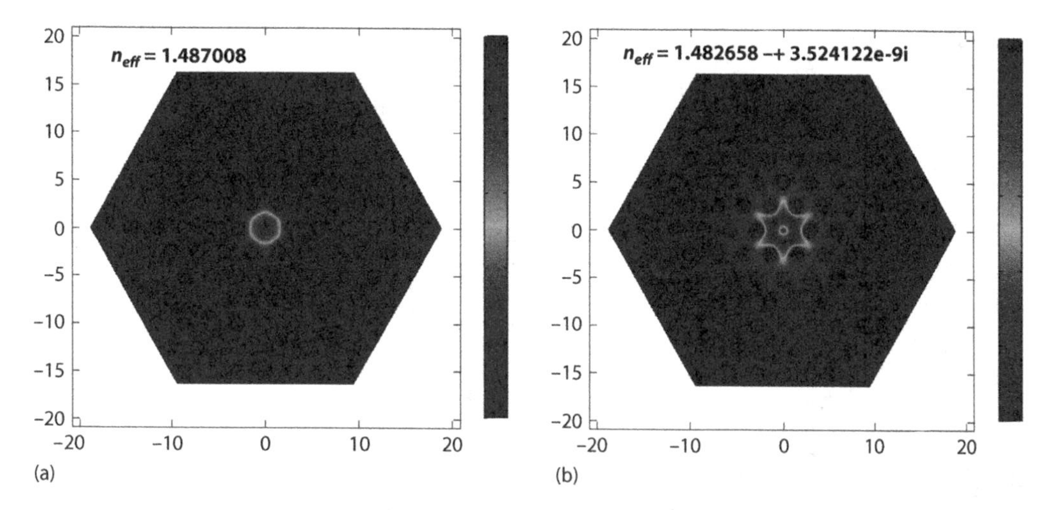

FIGURE 2.4 Two dimensional full vector finite element model simulation showing the two first modes propagating in an endlessly SM-mPOF composed of PMMA, operating at 650 nm region. The fiber is composed of six layers of holes, where $d = 1.4$ μm and $\Lambda_d = 2.9$ μm. (a) Fundamental mode (core region) and (b) first cladding mode propagating into the microstructure. (Parameters taken from an mPOF, produced by Kiriama Pty., Ltd., and sold with trade name SM-125.)

where λ^* is the second order cutoff, Λ_d the separation between two holes, $\alpha = 2.80 \pm 0.12$ and $\gamma = 0.89 \pm 0.02$ the fitting coefficients, and $d^*/\Lambda_d = 0.406$ the air filling fraction (Kuhlmey et al., 2002; Mortensen et al., 2003). Thus, for values of $d/\Lambda_d < d^*/\Lambda_d$, the fiber has SM behavior at all wavelengths, also called endlessly single mode (Birks et al., 1997). On the other hand, when $d/\Lambda_d > d^*/\Lambda_d$ it supports a second-order mode at wavelengths $\lambda/\Lambda_d < \lambda^*/\Lambda_d$, and it is single mode for $\lambda/\Lambda_d > \lambda^*/\Lambda_d$ (Mortensen et al., 2003).

Regarding the cutoff, Mortensen et al. showed that the multimode cutoff can be associated with the value $V_{PCF} = \pi$ (Mortensen et al., 2003).

2.3 Quality Requirements for Optical Polymers

Nowadays, our society is looking for new optical devices and components that have to meet several requirements. A careful choice of an optical material is crucial to fulfill the needs of a specific application. Transparency is obviously the most wanted characteristic; however, other characteristics such as mechanical, thermal, and chemical properties need to be taken into account. Recently, glass materials have been conventionally used in most of the photonic technology; however, the paradigm is changing. Polymer materials are being viewed as a viable alternative to silica-based applications. One example of that can be found in the optical fiber domain. In fact, POFs are being considered as an alternative for several applications, such as the one found in automobile, imaging, lightning, and sensing. Additionally, it is expected to have a market explosion based on the widespread deployment of FTTH applications. What makes this fiber interesting for new emerging technologies is related with the properties of the polymer materials. Many of the properties of some of those polymers, such as PMMA, polycarbonate (PC); cycloolefin polymers (COP), e.g. ZEONEX® 480R; cycloolefin copolymers (COC), e.g. TOPAS® 5013L-10; PS; and perfluorinated polymer (PF), e.g. CYTOP®, may be seen in **Table 2.2**. Nevertheless, a more detailed analysis will be given in the next subsections.

Table 2.2 Properties of Different Optical Polymers

Property	Symbol & Unit	Polymer, Brand Name, Supplier					
		PMMA: ACRYPET™ VH (Mitsubishi Rayon Co., Ltd.)	PC: PANLITE® AD5503 (Teijin Chemicals)	COP: ZEONEX® 480R (Zeon)	COC: TOPAS® 5013L-10 (Topas Advanced Polymer GmbH)	PS: Styron™ 666D (Americas Styrenics)	PF: CYTOP® (AGC Chemicals)
Density	ρ (g/cm^3)	1.19[a]	1.20[b]	1.01[c]	1.02[d]	1.04[e]	2.03[f]
Refractive index@ 589 nm	n_D^{25}	1.49[a]	1.59[b]	1.53[c]	1.53[d]	1.59[e]	1.34[f]
Abbe number	V_D	57[g]	30[h]	56[c]	56[d]	31[g]	90[f]
Water absorption	CME (%)	0.3[a]	0.2[b]	<0.01[c]	<0.01[d]	0.2[g]	<0.01[f]
Young's modulus	E (GPa)	3.3[a]	2.5[b]	2.2[c]	3.2[d]	2.9[e]	1.5[f]
Yield strength	σ_y (MPa)	77[a]	63[b]	59[c]	46[d]	45[e]	40[f]
Yield strain	ε_y (%)	2.3[h]	2.5[i]	2.7[i]	1.4[i]	1.6[i]	2.7[i]
Differential stress-optic coefficient	ΔC (×10^{-12}/Pa)	−6.0[h]	72.0[h]	6.5[h]	−2 to −7[d]	−55.0[h]	6.5[f]
Poisson's ratio	ν	0.35–0.4[j]	0.37[j]	0.40[c]	0.37[d]	0.35[j]	0.42[f]
Glass transition temperature	T_g (°C)	107[a]	143[b]	138[c]	134[d]	99[e]	108[f]
Thermo-optic coefficient	ξ (×10^{-5}/°C)	−8.5[h]	−14.3[h]	−12.6[h]	−10.4[h]	−12.0[h]	−5.0[k]
Thermal expansion coefficient	α (×10^{-5}/°C)	6[a]	7[b]	7[c]	6[d]	9[e]	12[f]

[a] Mitsubishi Rayon Co. Ltd., General Properties of Acrypet™, 2015.
[b] Teijin Limited, Panlite® AD-5503 – Polycarbonate, 2016.
[c] Zeon Chemicals, ZEONEX® Cyclo Olefin Polymer (COP), 2016.
[d] Topas Advanced Polymers GmbH, Cycloolefin Copolymer (COC), 2013.
[e] AmericasStyrenics LLC, STYRON 666D General Purpose Polystyrene Resin, 2008.
[f] Asahi Glass Co. Ltd., Amorphous Fluoropolymer (CYTOP), Tokyo, Japan, 2009.
[g] Khanarian, G. and Celanese, H., *Opt. Eng.*, 40, 1024–1029, 2001.
[h] Minami, K., *Handbook of Plastic Optics*, 2nd ed., Wiley-VCH Verlag GmbH & Co. KGaA., Weinheim, Germany, 2010.
[i] Calculated from the stress – strain relation, using σ_y and E.
[j] Goodfellow Corporation, Standard Price List for all Polymers, 2016.
[k] Lacraz, A. et al., Bragg grating inscription in CYTOP polymer optical fibre using a femtosecond laser, in K. Kalli (Ed.), *Micro-structured and Specialty Optical Fibres IV*, Vol. 9507, p. 95070K, 2015.

2.3.1 Optical Properties

2.3.1.1 Transparency

The most required property needed for optical plastics is to have high transparency at specific wavelengths, for which the optical device will operate. Transparency is intrinsically related to the molecular structure. When a polymer material is exposed to light, various interatomic/intermolecular interactions will occur, causing optical absorption in the UV region due to the electronic transitions, and optical absorption in the infrared region due to vibrational transitions (Minami, 2010).

The random disposition of the molecular chains is also an important requirement for transparency. On the other hand, a material presenting crystalline structures will scatter light at the boundary between the amorphous and crystalline phases, causing haziness, and thus, low transparency. Transparency is also related to refractive index, since reflectance (R), depends on the refractive index as follows (Minami, 2010):

$$R = \frac{(n-1)^2}{(n+1)^2} \tag{2.10}$$

According to (2.10), a material with lower refractive index will present lower reflectance and thus higher transparency.

Finally, the transparency can also be due to extrinsic factors such as contaminations or imperfections, which may absorb or scatter the light, causing lower transparency.

2.3.1.2 Refractive Index

Refractive index as defined previously in (2.1) is a dimensionless physical quantity that reflects how the light propagates in a medium. It varies depending on the molecular polarizability and density, as defined by the Lorentz-Lorentz equation (Minami, 2010):

$$\frac{n^2-1}{n^2+2} = \frac{4\pi}{3} N_m \alpha \tag{2.11}$$

where N_m is the number of molecules per unit volume and α the mean polarizability. The refractive index of a material is dependent on the wavelength of light, and this refractive index dependence is defined as dispersion. One way to quantify the amount of dispersion in a material is through the Abbe number (V_D), described in (2.12):

$$V_D = \frac{n_D - 1}{n_F - n_c} \tag{2.12}$$

where n_D, n_F, and n_C are the refractive indices of the material at the wavelengths of the Fraunhofer D-, F-, and C-lines, which corresponds to 589.3, 486.1, and 656.3 nm, respectively. Therefore, a small V_D value corresponds to larger dispersion in the visible spectrum. For comparison, polycarbonate is a polymer material with low Abbe number, contrary to PF that has the higher value among the optical plastics (see **Table 2.2**).

For a more accurate description of the refractive index – wavelength dependence, the Sellmeier empirical formula can be used:

$$n^2(\lambda) = 1 + \frac{B_1\lambda^2}{\lambda^2 - C_1} + \frac{B_2\lambda^2}{\lambda^2 - C_2} + \frac{B_3\lambda^2}{\lambda^2 - C_3} \tag{2.13}$$

where $B_{1,2,3}$ and $C_{1,2,3}$ are the Sellmeier coefficients that are frequently quoted instead of refractive index in tables.

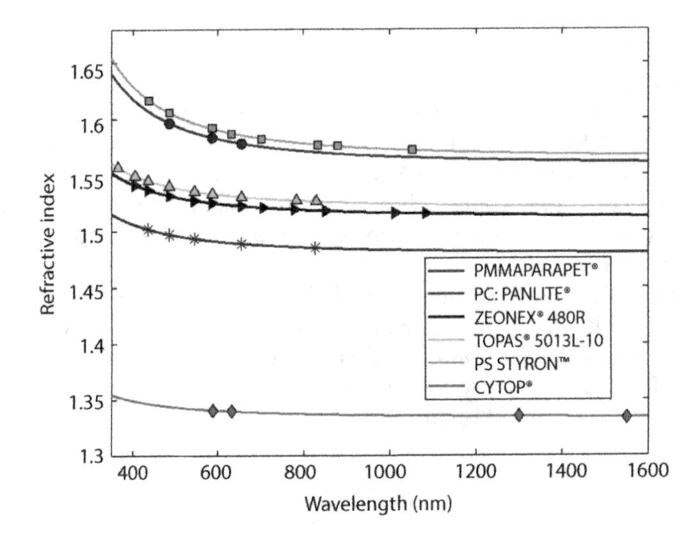

FIGURE 2.5 Refractive index evolution of various materials. The marker points define the experimental values found in literature for PMMA, PC, and ZEONEX®. (Minami, 2010); TOPAS® (Topas Advanced Polymers GmbH, 2013); PS (Sultanova et al., 2009); and CYTOP® (Asahi Glass Co. Ltd., 2009). The correspondent lines are referred to the fitting curves, following the Sellmeier equation.

From the above assumptions and in order to clarify the optical properties of different amorphous polymer materials that have been used for the production of POFs [specifically PMMA, PC, and ZEONEX® (Minami, 2010); TOPAS® (Topas Advanced Polymers GmbH, 2013); PS (Sultanova, Kasarova, & Nikolov, 2009); and CYTOP® (Asahi Glass Co. Ltd., 2009)], which can also be found in **Table 2.2**, their refractive indices are plotted in **Figure 2.5** (marker points). The data points were then adjusted to the Sellmeier equation (i.e. [2.13]), in order to verify the evolution of the refractive index over the visible and near infrared region.

From **Figure 2.5**, it can be observed that materials on the top of the graph have high refractive index, which according to (2.10), poses lower transparency. On the other hand, it can also be observed that those materials also have strong dependency with wavelength and thus, present higher material dispersion. One obvious conclusion from this discussion is that *CYTOP®* has superior optical characteristics among the presented polymers.

2.3.1.3 Birefringence

Birefringence, also referred to as double refraction of light, occurs when a beam of non-polarized light passes through an optical material and splits into two polarization components (see **Figure 2.6**). These two components travel at different velocities through the material and emerge out of phase, the latter defined as retardance.

The numerical difference between the index of refraction of the components within the optical material is called the birefringence number. Birefringence arises from the material anisotropy, which refers to a non-uniform spatial distribution of the material properties. Therefore, birefringence can be an intrinsic property of the optical material (i.e. absorbance, refractive index, density, etc.), or can be induced for instance by the application of external forces to the material. Induced birefringence can be found in two different types: temporary birefringence (i.e. when the material is oscillated or stretched and released) and residual birefringence (i.e. "frozen" stresses in the material after the production process, or by the employment of stress applying elements in the optical material).

In optical fibers, different values of birefringence can be found depending on the material employed. Additionally, special fibers capable of showing high birefringence can also be developed through different techniques. One example is the elliptical core fiber, where the fiber is

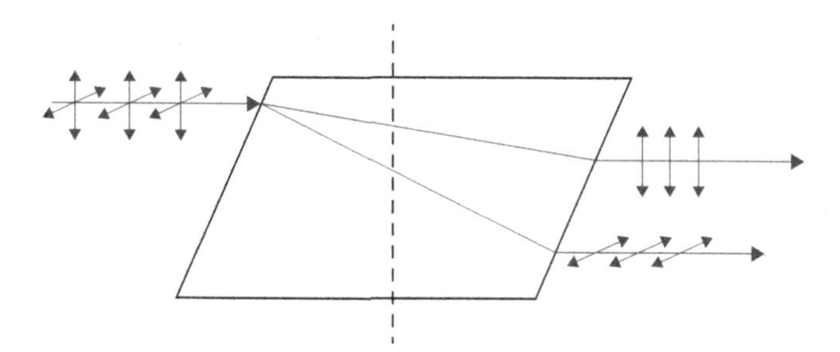

FIGURE 2.6 Light traveling through a birefringent medium will take one of two paths depending on its polarization.

composed of an elliptical core, in which the shape asymmetry creates both geometrical anisotropy and asymmetrical stress across the core. On the other hand, the employment of stress applying parts made out of materials with different thermal expansion coefficients at opposite sides of the core in the cladding region of the fiber can also be used. Typical fibers using this design structure are referred to in literature as elliptical cladding fibers, bow-tie fibers, and panda fibers. Furthermore, for the case of PCF, the symmetry of the hexagonal structure in the microstructured cladding needs to be broken. Here, a great advantage may be found for POFs, related to silica fibers, since they have more flexible technologies for the creation of the structure.

2.3.2 Mechanical Properties

Nowadays, the scientific community is very active and interested in the use of POFs in sensing applications. This has been essentially motivated by the better mechanical properties that POFs may offer when compared with their silica counterparts. Among those mechanical properties is the Young's modulus (E), which for silica can reach values as high as 70 GPa (Antunes, Lima, Monteiro, & Andre, 2008), where for polymers it can range from 1.6 to 3.4 GPa (Jiang, Kuzyk, Ding, Johns, & Welker, 2002; Kiesel, Peters, Hassan, & Kowalsky, 2007; Yang et al., 2004). Such low values allow the use of POFs to measure strain in materials that have a small Young's modulus, providing a better strain transfer from the material to the POF, a condition that cannot be satisfied with silica fibers which can act to locally stiffen the material (Ye et al., 2009). Nevertheless, the lower POFs Young's modulus can be used to get better sensitivities than silica fibers in pressure sensing (Bhowmik, Peng, Luo, et al., 2015). Conversely, regarding the elastic limit, POFs can recover strains up to 4.7%, depending on the strain rate (Kiesel et al., 2007), a value that is much higher than the one found in uncoated silica fibers that is approximately equal to 0.6% (Antunes et al., 2008). Additionally, POFs can survive in elongations that can be up to 100% (Webb et al., 2005), a feature that is much superior to the one found for uncoated (0.6%) or coated (1.6%) silica fibers (Antunes et al., 2008). With such properties, the capability to integrate POFs in large deformation applications such as the ones found in distributed strain measurements or in structural health applications is feasible (Liehr, Lenke, & Wendt, 2009). Nevertheless, it should be noted that above the elastic limit the fiber will deform permanently and thus, the light transmission can be severely affected (Guerrero, Guinea, & Zoido, 1998). Applications in such conditions should employ higher powers at the emitter side and/or higher sensitivities at the receiver side.

2.3.3 Thermal Properties

Polymer optical fibers have several advantages over silica fibers; however, they present a drawback related with their operational temperature, which for the most used POF based material (PMMA) can reach in the best case 90°C (Carroll et al., 2007) (only if the fiber is annealed after the drawing process). This value is, however, well below the one reported for silica fibers, which is around 500°C.

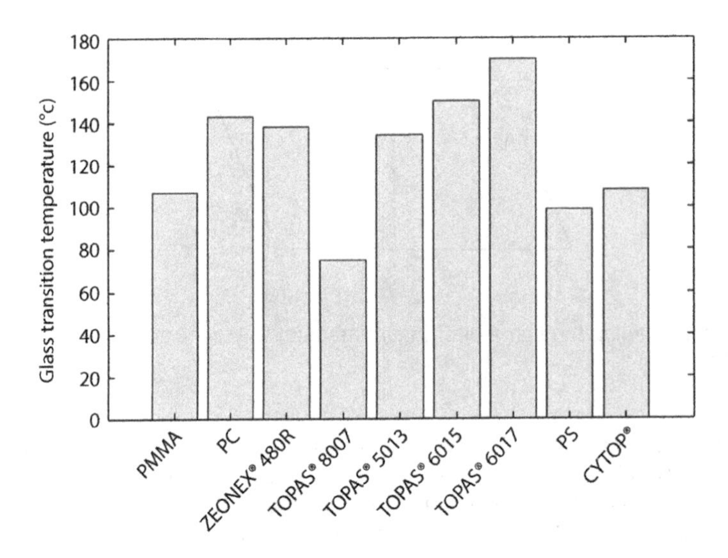

FIGURE 2.7 Glass transition temperature of several polymer materials. (Data collected from Table 2, Polyplastics Co. Ltd., TOPAS®: Thermoplastic olefin polymer of amorphous structure (COC), 2011.)

Nevertheless, other polymers with high operational temperature can be used. For comparison purposes, the glass transition temperature of different polymers is shown in **Figure 2.7**.

Among the polymer materials shown in **Figure 2.7**, it can be observed that polycarbonate offers superior temperature performance. This polymer has proven to have an operational temperature up to 125°C (Fasano et al., 2016; Guerrero, Zoido, Escudero, & Bernabeu, 1993; Tanaka, Sawada, Takoshima, & Wakatsuki, 1988), and for that reason it has been pointed as a candidate for high temperature sensing solutions (Fasano et al., 2016) and for automobile applications (Guerrero et al., 1993). Nevertheless, other polymers such as some TOPAS® grades (Markos et al., 2013; Polyplastics Co. Ltd., 2011) or ZEONEX® (Woyessa et al., 2016) can also be employed in temperatures above the ones achieved by PMMA.

When designing a POF sensor, it is important to know that fibers drawn under low temperature will present a high degree of birefringence, low ductility, high yield strength and tensile strength, and low thermal and strain stability (Carroll et al., 2007; Jiang et al., 2002; Yuan, Stefani, et al., 2011). This occurs due to the high degree of chain alignment along the fiber length. In fact, drawing a POF with low temperature will require a higher drawing force leading to the formation of a high degree of chain alignment, which is directly related to the parameters above mentioned. Fortunately, if a thermal treatment is given to the fiber drawn under such conditions, it can be possible to revert the process and achieve characteristics that are found in fibers drawn under high temperatures/low drawing forces. The thermal treatment also designated by annealing process consists of heating the fiber for a period of time that will depend on the degree of chain alignment present in the polymer. In most of the works reported to date, this can be up to 24 hours (Fasano et al., 2016). The temperature should be higher, but below the glass transition temperature of the polymer employed in the fiber. By employing a proper thermal annealing, the fibers can show lower birefringence, high ductility, and high thermal and strain stability at a cost of fiber length decrease, diameter increase, and lower yield strength and tensile strength (Jiang et al., 2002; Yuan, Stefani, et al., 2011).

2.3.4 Chemical Properties

The chemistry and structure of a polymer material influence its chemical resistance. A material with low chemical resistance can be affected in strength, flexibility, surface appearance, color, and dimension/weight. The chemical attack may occur on the polymer chain causing

Table 2.3 Chemical Resistance of Different Polymer Materials

		PMMA	PC	ZEONEX® 480R	TOPAS®	PS	CYTOP®	PVC
Acid	Hydrochloric acid	o	o	o	o	o	o	o
	Sulfuric acid	o	o	x	o	o	o	o
	Nitric acid	o	o	x	o	o	o	o
Ketones	Ketones	x	x	o	Δ*	x	o	x
	Methyl ethyl ketone	x	x	o	Δ*	x	o	x
Hydrocarbon	Gasoline	Δ	Δ	x	x	x	o	Δ
Alcohols	Isopropyl alcohol	Δ	Δ	o	o	Δ	o	o
Oils		o	o~Δ	x	x	x	o	Δ
Alkalis		o	x	o	o	o	o	Δ

Source: Asahi Glass Co. Ltd., Amorphous Fluoropolymer (CYTOP), Tokyo, Japan, 2009; Polyplastics Co. Ltd., TOPAS®: Thermoplastic olefin polymer of amorphous structure (COC), 2011; Zeon Chemicals, ZEONEX® Cyclo Olefin Polymer (COP), 2016.

Note: **o**: usable, **Δ**: usable with care, **x**: not usable. (*) Grade 8007 is not usable.

a reduction on the physical properties of the polymer. The absorption of the solvent may induce softening and swelling of the polymer material. Additionally, stress cracking may occur as a result of the above phenomena.

POFs have been proposed in sensing applications due to their great advantages when compared with their silica counterparts. For that reason, they have been studied for the use in liquids like the ones found in the automobile industry, such as petrol, oil, and battery liquid (Guerrero et al., 1993).

It is known that the reduction of the diameter of a POF through chemical attack is an easy method that can enhance different properties, such as the Brillouin signal (Nakamura, Mizuno, & Hayashi, 2013), the time response in POFBG humidity sensors (Chen, Zhang, Liu, Hong, & Webb, 2015), and the sensitivity in strain, temperature, and pressure sensors (Bhowmik et al., 2016a, 2016b; Bhowmik, Peng, Ambikairajah, et al., 2015; Bhowmik, Peng, Luo, et al., 2015).

Regarding the above discussion, it is important to know which polymer is most suited for a specific application. For that reason **Table 2.3** presents some of the most known transparent polymers (PMMA, PC, ZEONEX®, TOPAS®, CYTOP®, and polyvinyl chloride (PVC)) and their chemical resistance in terms of being "usable," "usable with care," and "not usable."

As can be seen from **Table 2.3**, CYTOP® presents a much wider resistance to corrosion than any other polymer shown and is therefore one of the most preferred.

2.3.4.1 Moisture Absorption

Most of the polymers present the ability to absorb water from the environment (see **Figure 2.8**). The water uptake can lead to different changes in the polymer properties, such as volumetric changes and also changes in its refractive index (Watanabe, Ooba, Hida, & Hikita, 1998). The capability to monitor those properties can be useful for the development of applications such as food storage, paper manufacturing, pharmaceutical industries, among others, where humidity needs to be controlled to ensure the quality of the product.

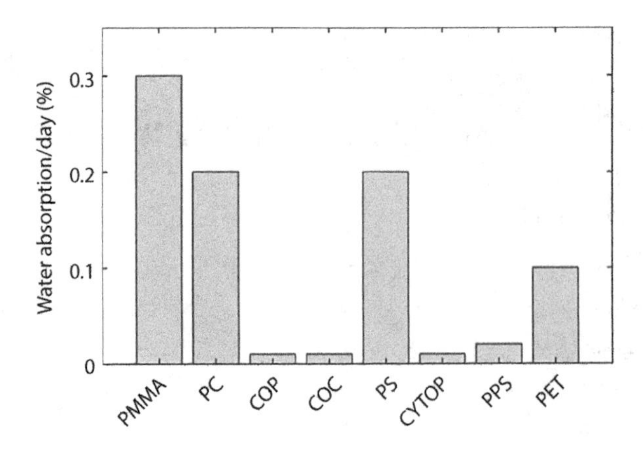

FIGURE 2.8 Water uptake per day at 23°C, for different polymer materials. (Data collected from Table 2, Zeon Chemicals, ZEONEX® Cyclo Olefin Polymer [COP], 2016.)

Despite the capability to measure the amount of humidity, the water absorption by a polymer material is usually an undesired effect. Water absorption leads to the appearance of attenuation bands due to the O–H molecular vibration overtones, compromising the transparency of the polymer in the useful low loss windows. Indeed, at 90% relative humidity (RH) and temperature of 50°C, the water content in the most used polymer (PMMA) rises its attenuation in the interesting low loss regions, such as the 670 and 766 nm, the latter being the most pronounced, where an increase of 250 dB/km may be expected (Kaino, 1985b). For deuterated polymers (i.e. perdeuterated PMMA (PMMA-d8)), increment losses of 800 dB/km and 600 dB/km were observed at 840 and 746 nm, respectively (Kaino, 1985b). Due to the poor stability transmission property, this polymer material should be avoided for near infrared transmission applications. Based on the above discussion, polymers offering low water absorption such as CYTOP®, COCs, or COPs are desirable for transmission applications. An example of the absorption capabilities of different polymers (PMMA, PC, COP, COC, CYTOP, PS, polyphenylene sulfide (PPS), and polyethylene terephthalate (PET)) can be seen in **Figure 2.8**.

2.4 Polymer Optical Fiber Materials

From the early developments of POF-based technology, a variety of polymer materials have been proposed for the production of polymer optical fibers. The reason is related to their special properties that allow their use in a variety of applications. If compared with silica-based optical fibers, POFs reveal higher losses, limiting their use for short range applications. However, polymers have a considerably lower Young's modulus, superior elastic limit, negative and much higher thermooptic coefficient, biocompatibility, flexibility, non-brittle nature, among other features (Cusano, Cutolo, & Albert, 2011) (see **Table 2.4**).

Another important feature presented in polymer optical fibers is the possibility of doping with organic materials including fluorescent or photosensitive materials such as Rhodamine dyes (Arrue, Jiménez, Ayesta, Illarramendi, & Zubia, 2011; Kuriki et al., 2000; Large et al., 2004). Indeed, organic materials can be added during the polymerization stage (prior to the heat and draw technique), or can be diffused into the polymer matrix through solution doping (Large et al., 2004). Moreover, organic materials cannot be incorporated in silica-based fibers, since their glass transition temperature (~2000°C) is much higher than the temperature at which all organic materials decompose (~400°C) (Large et al., 2008).

Table 2.4 Comparison between Silica- and Polymer-Based Materials

Property	Material	
	Silica	Polymer
Density (g/cm³)	~1[a]	>2
Young's modulus (GPa)	68:74	~2:3
Failure strain (%)	<1[b]	>6
Glass transition temperature (°C)	~2000	100:150
Thermo-optic coefficient (×10⁻⁵/°C)	~1	−5:−15
Biocompatibility	No[c]	Yes
Brittle	Yes	No
Flexibility	No	Yes

Source: Large, M. C. J. et al., *Microstructured Polymer Optical Fibres: Microstructured Polymer Optical Fibres*, Springer US, Boston, MA, 2008.

[a] Except for fluorinated polymers where values higher than 2 g/cm³ are expected.

[b] Considering the raw fiber (without the protective coating material).

[c] The choice to consider the material as non-biocompatible was based on its brittle nature.

The advantages of POFs over silica fibers have been explored in a variety of fields such as in short-range transmission systems (Shao, Cao, Huang, Ji, & Zhang, 2012), sensing applications [i.e. biosensing (Markos, Yuan, Vlachos, Town, & Bang, 2011), civil engineering (Kuang, Quek, Koh, Cantwell, & Scully, 2009), medical (Bilro, Oliveira, Pinto, & Nogueira, 2011), etc.], and lighting (Zubia & Arrue, 2001). Nevertheless, even among the polymer materials, there are different features that may benefit some application in particular. For instance, the heat and chemical resistance or the water absorption are different from polymer to polymer. Those characteristics are thus a drawback or an advantage depending on the application.

2.4.1 Polymethylmethacrylate (PMMA)

Among the different polymer materials, the most popular and widely used in polymer optical fibers is the PMMA, commercially known as Plexiglas®. This polymer material is made through the polymerization of the monomer methyl methacrylate (MMA). PMMA is an organic compound forming long chains with typical molecule weights of around 105 (Ziemann et al., 2008). It has a density of 1.19 g/cm³, tensile strength of 77 MPa, glass transition temperature of 107°C, and a refractive index of 1.49 at 589 nm (Mitsubishi Rayon Co. Ltd, 2015) (see **Table 2.2** for more characteristics). PMMA is resistant to water, lyes, diluted acids, petrol, mineral oil, and turpentine oil (Ziemann et al., 2008). This polymer has an amorphous structure, presenting high transparency in the visible region. Additionally, PMMA has the capability to absorb water from the external environment [0.3%/day (Mitsubishi Rayon Co. Ltd, 2015)]. This can be used to develop humidity sensors. However, this parameter is normally unwanted due to the losses caused by the OH overtones (Kaino, 1985b) and the cross sensitivity issues in sensing applications.

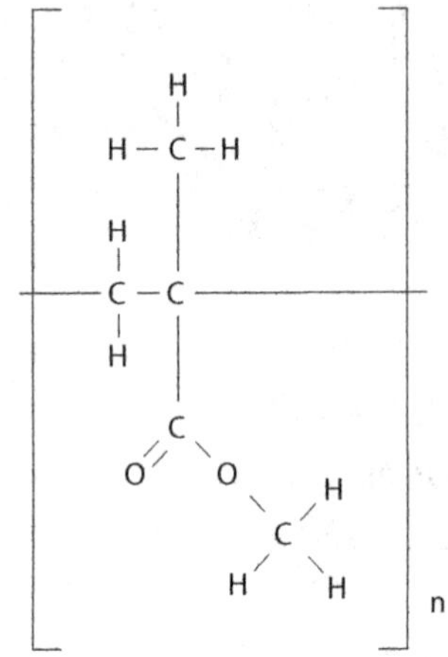

FIGURE 2.9 Chemical structure of PMMA.

The chemical structure of PMMA can be visualized in **Figure 2.9**. Note that each monomer is composed of eight aliphatic C–H bonds, being the overtones of the C–H vibration are the main reason for the losses presented by this polymer (Groh, 1988).

In 1963, DuPont manufactured the first PMMA-based POF (Crofon™) (United Kingdom Patent No. GB 2006790, 1979; Large et al., 2008; Ziemann et al., 2008). The fiber was based on a SI profile with an MM core, with attenuation around 1000 dB/km at 650 nm. Since then, the attenuation has been improved, and PMMA-based POFs with losses around 120 dB/km at 650 nm and less than 70 dB/km at 570 nm are now commercially available (Minami, 1994).

Due to the high qualities presented by this polymer material, such as high transparency, relatively low cost, ease of processing, and inherent resistance, today it is the preferred choice for almost any fiber sensor.

2.4.2 Polycarbonate (PC)

Polycarbonate is another commonly used optical material. It is made from the polymerization of bisphenol-A with carbonyl chloride or diphenylether (Minami, 2010). The PC chemical structure can be seen in **Figure 2.10**.

The first report of the use of polycarbonate in an optical fiber was in 1986 by Fujitsu (Koike, Ishigure, & Nihei, 1995), where the core and cladding materials were made of polycarbonate and polyolefin polymers, respectively. The minimum attenuations of such fiber were located at 656 and 764 nm with theoretical intrinsic losses of 166 and 224 dB/km, respectively (Yamashita & Kamada, 1993). However, the experimental attenuation values achieved for POFs based on this material are two times higher than those theoretically predicted. Compared with PMMA, PC has attenuation four times higher, and for that reason it is used only in applications where high attenuation can be tolerated. Despite the high loss, PC presents high glass transition temperature (143°C (Teijin Limited, 2016)). Thereby, it can be used in high-temperature environments, such as the ones found in the automobile industry, specifically in the engine area, where temperatures can go over 100°C (Ziemann et al., 2008). However, it tends to degrade in high-temperature and humidity environments (Ziemann et al., 2008). PC has a refractive index of 1.59 at 589 nm (Teijin Limited, 2016), which is considerably high when compared with other polymer based materials, this being another motivation for its use. Conversely, PC has high moisture absorption and high birefringence (unless special processing equipment and

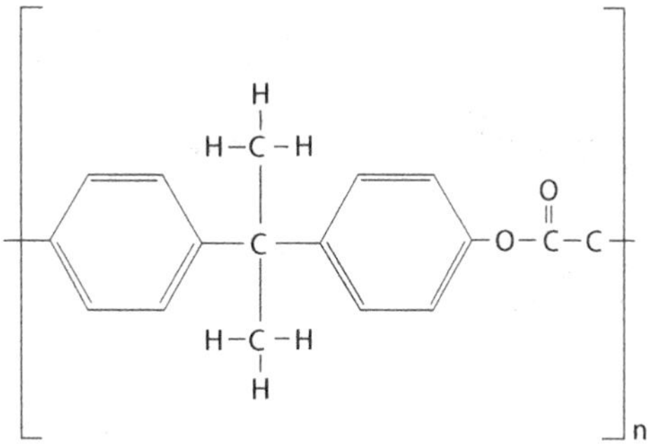

FIGURE 2.10 Chemical structure of polycarbonate.

conditions are used) (Khanarian & Celanese, 2001). Finally, PC yields and breaks at elevated values of strain, being flexible in bending. For other properties not mentioned here, please refer to **Tables 2.2** and **2.3**.

2.4.3 Cycloolefin Copolymer (COC) and Cycloolefin Polymer (COP)

Among the polyolefins, polypropylene (PP) and polyethylene (PE) are the most popular. However, their transparency is quite low, resulting from the light scattering on the interface between the crystalline and amorphous parts in the polymer matrix (Obuchi, Komatsu, & Minami, 2015).

Besides that, cycloolefin copolymers (COCs) and cycloolefin polymers (COPs) are a new class of thermoplastics, which are essentially amorphous, having excellent optical properties as a result of the presence of cyclic structures in their polymer chain.

Overall COCs and COPs have good heat resistance, low moisture absorption, low birefringence, good chemical resistance to common solvents such as acetone (see **Table 2.3**), large Abbe number, high durability under high temperature, good moldability, and a transparency similar to that of PMMA (Cui, Yang, Li, & Li, 2015; Topas Advanced Polymers GmbH, 2013; Zeon Chemicals, 2016). They have a refractive index close to 1.53 at 589 nm (Topas Advanced Polymers GmbH, 2013; Zeon Chemicals, 2016), which is higher than that of PMMA, allowing their use as core material in optical fibers (Leon-Saval et al., 2012). For general properties, **Tables 2.2** and **2.3** can be addressed.

COCs were patented in 1997 to be used in optical waveguides (United States Patent No. *US 5637400A*, 1997) due to their low melt viscosity, high tensile strength, and low attenuation (Beckers, Schlüter, Vad, Gries, & Bunge, 2015). COCs can be processed through copolymerization of cycloolefin, such as norbornene or cyclopentene, with ethylene or alpha-olefin, using a metallocene catalyst (Cui et al., 2015; Khanarian & Celanese, 2001), (see **Figure 2.11**). COCs are being commercialized with the names APEL™ by Mitsui Chemicals Co., Ltd and TOPAS® by Topas Advanced Polymer GmbH (formerly Ticona and Hoechst) (Minami, 2010).

The glass transition temperature of COCs can be enhanced by increasing the cycloolefin content in the polymer chains (Khanarian & Celanese, 2001). An example of such material is TOPAS® grade 6017, which has a glass transition temperature of 178°C (Topas Advanced Polymers GmbH, 2013), clearly outstanding when compared with other polymer materials (see **Figure 2.7**).

In recent years, several authors have reported the use of different grades of TOPAS® to produce microstructured (Emiliyanov et al., 2007; Markos et al., 2013; Nielsen et al., 2009; Yuan, Khan, et al., 2011) and SI (Woyessa et al., 2015, 2016) fibers, with intended uses in the near-infrared (Emiliyanov et al., 2007; Markos et al., 2013; Woyessa et al., 2015, 2016; Yuan, Khan, et al., 2011) and terahertz applications (Nielsen et al., 2009). Those works were intended

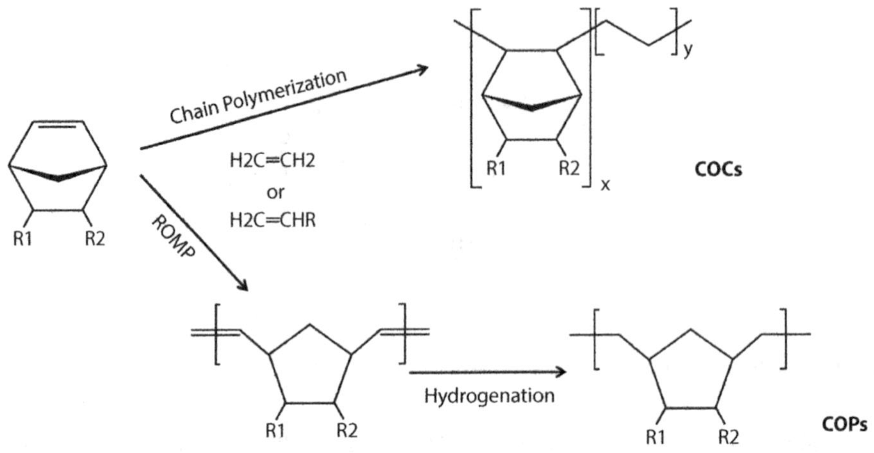

FIGURE 2.11 Typical polymerization routes for the production of COCs and COPs.

to explore the special characteristics offered by COC materials, such as biocompatibility (Emiliyanov et al., 2007), low moisture absorption (Woyessa et al., 2015, 2016; Yuan, Khan, et al., 2011), high temperature resistance (Markos et al., 2013; Woyessa et al., 2016), and high transparency (Nielsen et al., 2009).

COPs are amorphous polyolefins with a cyclic structure in the main chain. The polymerization is done through ring opening metathesis polymerization (ROMP) of norbornene derivatives followed by hydrogenation of double bonds (see **Figure 2.11**), providing more stability in terms of heat and weather resistance (Cui et al., 2015; Obuchi et al., 2015). The commercialization of COPs was initialized in 1991 by Zeon Corporation (Obuchi et al., 2015). Today it is sold under the trade names of ZEONEX® and ZEONOR® by Zeon, and ARTON® by Japan Synthetic Rubber (JSR) (Minami, 2010).

ZEONEX® has been used in microstructured fibers (Anthony, Leonhardt, Argyros, & Large, 2011), SI fibers (Woyessa et al., 2015, 2016) (as cladding material), and multicore composite single mode fibers (Leon-Saval et al., 2012). The material choice for those POFs relies on the material compatibility with COCs (Woyessa et al., 2015, 2016), humidity insensitiveness (Woyessa et al., 2015, 2016), high temperature resistance (Woyessa et al., 2016), and high transparency (Anthony et al., 2011).

2.4.4 Polystyrene (PS)

Polystyrene is a widespread and costless material used in the production of POFs (Makino et al., 2013). It has a refractive index of 1.59 at 589 nm (AmericasStyrenics LLC, 2008), this being a motivation for its use as the core of optical fibers, since most of the polymers can be used as cladding material. PS can be obtained by thermal polymerization without any initiator as is needed in PMMA, which sometimes produces bubbles during the polymerization reaction (Kaino, Fujiki, & Nara, 1981). PS has a glass transition temperature of 99°C (AmericasStyrenics LLC, 2008), which is one of the lowest values among the thermoplastics, where the maximum operation temperature is estimated to be 70°C. The chemical resistance and mechanical properties (see **Tables 2.2** and **2.3**) are inferior to those of PMMA (Kaino et al., 1981).

The first step-index-based POF composed of PS was reported in 1972 by Toray Co., Ltd., presenting an attenuation of 1100 dB/km at 670 nm (Koike et al., 1995; Zubia & Arrue, 2001). In 1979, Oikawa et al. at the NTT (former Nippon Telegraph and Telephone Public Corporation) reported a PS fiber with reduced attenuation of 114 dB/km at the same wavelength region (Oikawa, Fujiki, & Katayama, 1979).

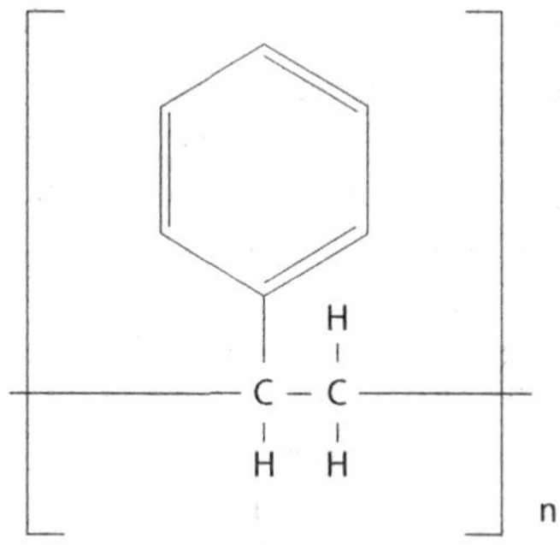

FIGURE 2.12 Chemical structure of polystyrene.

Theoretically, the attenuation of PS can be as low as 70 dB/km at 670 nm (Kaino et al., 1981), this being the intrinsic value related with the overtones of the of three aliphatic and five aromatic carbon-hydrogen (C–H) bonds (Koike & Koike, 2011) (seen in **Figure 2.12**), and also due to the phenyl group present in each monomer unit, that due to the flat physical geometry, gives rise to molecular anisotropy and hence, scattering (Emslie, 1988). Because of that, side-emitting or fluorescent fibers made of PS are being used (Spigulis, 2005).

2.4.5 Perfluorinated (PF) Polymer

Perfluorinated polymers are formed via free radical mechanism of perfluoromonomers. They tend to form partially crystalline structures, and thus are opaque due to the light scattering between the amorphous and crystalline phases. The most efficient way to give transparency to the polymer is through the introduction of aliphatic rings into the main chain, blocking the formation of crystalline structures. The most known examples of such amorphous polymers are Teflon® AF and CYTOP®, developed by DuPont and Asahi Glass Co., Ltd. (AGC), respectively (Koike & Koike, 2011). Teflon® AF is obtained through the co-polymerization of perfluoro-2,2-dimethyl-1,3-dioxole (PDD) (with cyclic structure in its monomer unit), and tetrafluoroethylene (TFE). Contrary to Teflon® AF, CYTOP® is a homopolymer obtained from perfluoro(4-vinyloxyl-1-butene), yielding penta- and hexa-cyclic structures in the polymer chain (Yamamoto & Ogawa, 2005) (see **Figure 2.13**).

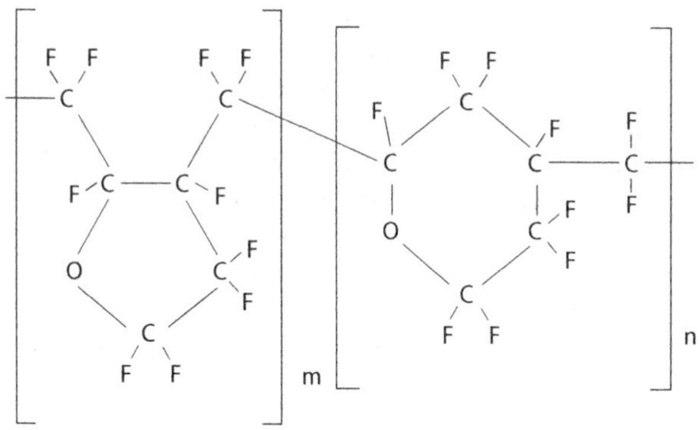

FIGURE 2.13 Chemical structure of CYTOP® based on penta- and hexa-cyclic structures.

Both Teflon® AF and CYTOP® have excellent transparency due to the cyclic structures in the main chain. Additionally, they have excellent solubility in fluorinated solvents, thermal and chemical durability, high electrical isolation, low water absorption, low refractive index, and low dielectric properties (Asahi Glass Co. Ltd., 2009; Yamamoto & Ogawa, 2005).

Despite the good transparency of Teflon® AF, the polymer is being used as the cladding material in fibers and waveguides due to the extremely low refractive index ($n_{589\ \mathrm{nm}} = 1.29$) (Koike & Koike, 2011). On the other hand, in 1995 professor Koike from Keio University introduced a graded-index POF composed of CYTOP®. In 2000, AGC commercialized the first CYTOP®-based GI-POF (Lucina®). Since then, the fiber has been used in different areas with emphasis in home networking, due to the low loss (i.e. 10 dB/km at 1000 nm and 15 dB/km at 1300 nm) (Koike & Ishigure, 2006), which is suitable for home networking applications.

In 2010, AGC released another CYTOP®-based GI-POF called Fontex™. The fiber was an improvement of the Lucina® fiber regarding the bending loss, by employing a double cladding structure with a considerably lower refractive index (Koike & Koike, 2011).

CYTOP® fibers can have a theoretical loss limit at 1000 and 1300 nm of 0.7 and 0.26 dB/km, respectively (Koike & Koike, 2011; Tanio & Koike, 2000), a value that could be comparable with that of silica fibers (Koike & Gaudino, 2013). These lower losses are related to the low refractive index of this polymer ($n_{589\ \mathrm{nm}} = 1.34$), which provides low scattering losses at longer wavelengths (Koike & Koike, 2011). Nevertheless, this polymer has no hydrogen atoms in the main chain, contrary to most POF-based materials. Thus, the molecular absorption is only due to the overtones of C–C, C–F and C–O bonds, which are lower in intensity, since the wavelengths of their fundamental stretching-vibration are found at longer wavelengths (Koike & Koike, 2011). One way to reduce the fiber losses may be realized, for instance, by employing new fabrication methods or reducing the impurities.

2.5 Polymer Optical Fiber Fabrication

Polymer optical fibers, unlike silica fibers, can be fabricated with a wider range of techniques. The reason is mainly associated with the lower processing temperatures (lower T_g) and the ease of machining in the case of microstructured fibers. For the production of a POF, it is necessary to know which fabrication method is more adequate for each fiber structure/profile. Therefore, it is worth knowing the differences between the techniques in order to have better economies, lower processing times, and lower losses, among others.

Overall, the fabrication techniques can be associated with three types of POFs, specifically, SI, GI, and mPOF. Additionally, the techniques can be separated concerning the flow which can be: continuous and discontinuous.

Discontinuous manufacturing techniques always involve a two-step procedure: (i) beginning with the construction of a preform, which is a scaled-up (larger in diameter, sorter in length) version of the fiber that contains the essential features, such as structure and refractive index (Argyros, 2013) and (ii) a second step concerning the drawing process, in which the preform is drawn to fiber using a draw tower. For that, the preform is secured at the top of the draw tower and heated in a furnace to a temperature larger than T_g, allowing the necessary viscosity, such that it can be drawn to fiber. The process can be done in intermediate stages, involving the drawing of the initial preform to an intermediate size called "cane," sleeving it to increase the diameter, or combining several canes to perform a new preform (Argyros, 2013; Barton, van Eijkelenborg, Henry, Large, & Zagari, 2004). The velocity in which the fiber is pulled down and the speed at which the mandrel rotates are adjusted by a regulatory mechanism performed in the main control. For that, the POF diameter readings executed by the laser measuring system are continuously analyzed in order to allow the fine diameter control of the fiber. One drawback of this technique is related with the limited length and diameter of the preform. Therefore, a disruption of the drawing process is inevitable, leading to a discontinuous process (Beckers et al., 2015).

For the continuous manufacturing techniques, the procedures involved are made simultaneously, allowing the fabrication of a fiber with a theoretical infinite length. In conclusion, the main advantage when compared with discontinuous manufacturing techniques is the high production rate that can be obtained.

2.5.1 Production of Step – Index Polymer Optical Fibers

The production of SI-POFs through discontinuous process, can be made through the heat drawing or by the batch extrusion process. For the heat drawing process, the preform composed with core and cladding can be prepared by a wet or dry process. For the wet process, the polymer gets in direct contact with the liquid monomer (in situ polymerization) before it is completely polymerized. In the dry process, the core and cladding are polymerized separately, and finally joined for the creation of the preform (Beckers et al., 2015; Harmon Julie, 2001). After that, the preform is drawn to a fiber using the drawing tower as described before. Another possible way to produce the fiber using this method is basically through the use of a preform only composed of a core polymer. After pulling the core preform to a small diameter, it is passed through a plasticized coating material that is later cured through UV radiation. The coating will act thus as the cladding of the fiber. The simplicity and flexibility of the technique, together with the good quality of the produced fiber, are the key aspects when using this kind of manufacturing process. However, the process can be tedious and expensive for large-scale purposes.

The other discontinuous process used for the production of SI-POF is the batch extrusion technique. This technique involves a two-step procedure, specifically, the polymerization and the extrusion of the polymer melt (Beckers et al., 2015). The monomer, initiator, chain transfer agent, and additives are pumped to a reactor where the polymerization will occur. After that, the temperature is raised to form a polymer melt that will be carried to a spinning nozzle using nitrogen. For the creation of the cladding of the fiber, a secondary extruder containing the molten cladding material is conveyed to a second spinning nozzle (Beckers et al., 2015; Ziemann et al., 2008).

SI-POFs can also be produced through continuous processes. From those, it is possible to find the continuous extrusion, the melt spinning, and the photochemical polymerization technique. The first two are quite similar regarding the manufacturing processes; however, the raw materials are different. For the continuous extrusion, the raw materials that are continually fed into the mixing chamber are the monomer, the initiator, and the chain transfer agent, which need to achieve a combination of 80% polymer and 20% monomer before being extruded by the spinning nozzle. The introduction of the cladding material can be performed through a co-extrusion process, where the same spinning nozzle used for the extrusion of the core polymer is used (Beckers et al., 2015). On the other hand, in the melt spinning technique, the polymerization reaction is no longer needed since the raw materials are already polymer granulates instead of monomers. The material is thus molten in the extruder and conveyed to the spinning nozzle. The cladding material is applied in a co-extrusion process and after passing through the spinning nozzle, the fiber is cooled down and wound up in a reel unit. The complete process can be seen in **Figure 2.14**.

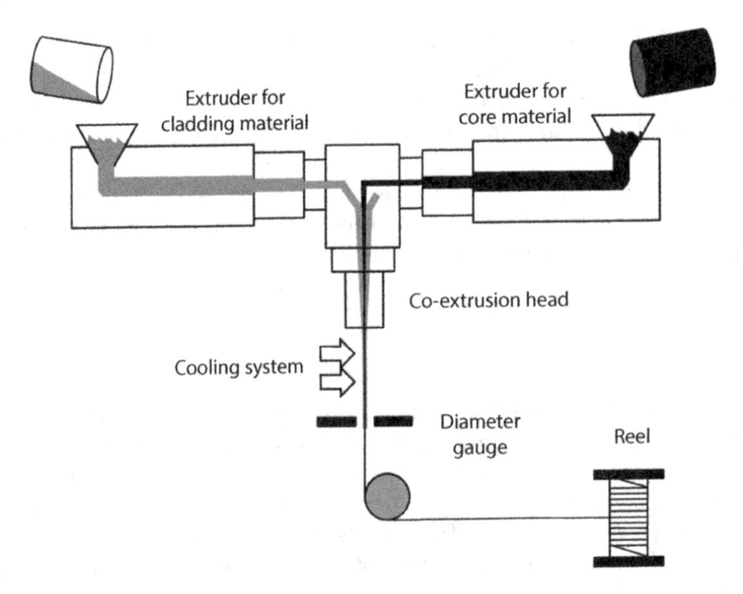

FIGURE 2.14 Schematic of the production of POFs through the melt spinning technique.

2.5.2 Production of Graded Index Fibers

One of the most important factors degrading the bandwidth in multimode step-index fibers is the modal dispersion. However, as mentioned in Section 2.2.1, the use of a refractive index with graded index profile in the core region allows the control of the propagation speed of each fiber mode. The first GI-POF was fabricated in 1982 (Koike, Kimoto, & Ohtsuka, 1982), and since then, several methods of creating a GI profile have been reported (Im et al., 2002; Koike, 1992; Koike et al., 1995; Villegas, Ocampo, Luna-Bárcenas, & Saldívar-Guerra, 2009). Among those, the low molecular doping method is one of the most efficient and simple ways to create a radial distribution of the refractive index.

2.5.2.1 Interfacial Gel Polymerization Technique

The interfacial gel polymerization technique is classified as a discontinuous technique, and it was first reported by professor Koike from Keio University (Koike, 1992). The process involves the preparation of the preform with a GI profile, following heat drawn to the GI-POF.

The different steps of the production process may be seen in **Figure 2.15**. At the beginning of the process, a glass vessel tube is filled with MMA monomer mixtures, initiator and chain

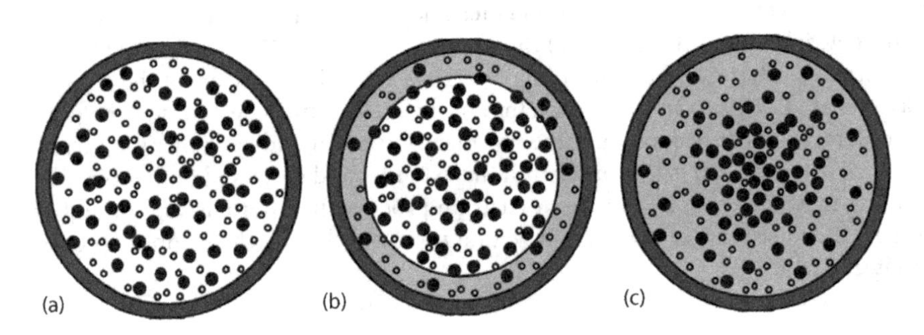

FIGURE 2.15 Gel polymerization technique. (a) PMMA tube filled with different monomers (M1 (•) and M2 (o)); (b) formation of the gel layer; and (c) final preform with GI profile.

transfer agent, which are then rotated at 3000 rpm and heated at 70°C for 3–6 hours. During the process, the MMA monomer coats the inner wall of the glass vessel due to the centrifugal force, and it is then gradually polymerized (Koike & Koike, 2011). After heat treatment at 90°C for 24 hours, a PMMA hollow tube is created and will serve as the cladding layer of the graded index preform. The hollow tube is then filled with a mixture of two different monomers, M1 (high refractive index and large molecules (dopant)) and M2 (smaller refractive index and smaller molecules); initiator; and chain transfer agent. The tube mixture is heated at a temperature of 120°C during 48 hours, with 0.6 MPa nitrogen pressure to avoid bubble formation (Koike & Koike, 2011). At the early stage, the PMMA wall tube is slightly liquefied, resulting in a layer of gel that accelerates the polymerization. During the process, the smaller molecules (M1) are more susceptible to diffuse into this gel layer, leading to a higher concentration of M2 molecules toward the center region (Ziemann et al., 2008). The final result is a 15 to 22 mm thick preform composed of a graded index profile. Finally, the preform is then heat drawn to the fiber at temperatures ranging from 220°C–250°C.

2.5.2.2 Coextrusion Process

The continuous production of GI-POFs (without the use of an initial preform) was first proposed to reduce the fabrication costs (United States patent No. 5783636, 1998). In this process, polymers doped with low-molecular-weight dopants and homogeneous polymers are first prepared as the base materials of the core and cladding layers, respectively (Hirose, Asai, Kondo, & Koike, 2008). The two materials are melted in each extrusion section and conveyed to the spinning nozzle. The cladding material is applied in a co-extrusion process. The fiber leaving the spinning nozzle at this stage presents a SI profile. For that reason, it is passed through a diffusion section that heats the fiber and promotes the diffusion of the low molecular weight dopant outwards, producing a GI profile. The final stage of the production comprises the cooling system and the winding of the fiber in the reel unit (Hirose et al., 2008). Due to the simplicity of the process, GI-POFs can be produced in large scale with relatively low cost. One example of a commercial available fiber produced through this method is the GigaPOF® fiber, a CYTOP® based fiber, produced by Chromis Fiberoptics, Inc.

2.5.3 Production of Microstructured Fibers

The use of tiny holes with a special arrangement around a solid core allows the light guidance in a similar way to the total internal reflection that occurs in SI fibers. This type of fiber can exhibit properties that cannot be attained with conventional step index fibers, such as the ability to remain SM in a wide range of wavelengths (Birks et al., 1997). To achieve such features, it is necessary to adjust the position and diameter of the air holes. However, this is not an easy task to perform with silica fibers due to their brittle nature and high glass transition temperature. Unlike silica, polymers are easier to manipulate due to their non-brittle nature and their lower glass transition temperatures. Additionally, the fabrication of SI-POFs with SM behavior at the low loss region of polymers is a challenging task due to the balance between the refractive index of the core/cladding and the core diameter (see [2.7]). As it happens with silica fibers, this feature is easier to achieve with the formation of a microstructured cladding. For such reasons, mPOFs have been produced with a variety of techniques (all discontinuous), such as stacking (van Eijkelenborg, Argyros, & Leon-saval, 2008), drilling (van Eijkelenborg et al., 2001), extrusion (Ebendorff-Heidepriem, Monro, van Eijkelenborg, & Large, 2007), casting (Zhang et al., 2006), and 3D printing (Cook et al., 2015a). These preform production processes will be described in the next sections and can be seen in **Figure 2.16**.

2.5.3.1 Stacking

Stacking is the main method used for the production of silica photonic crystal fibers (PCFs) (see **Figure 2.16a**), being also employed for the fabrication of mPOFs based on PS (Shin & Park, 2004),

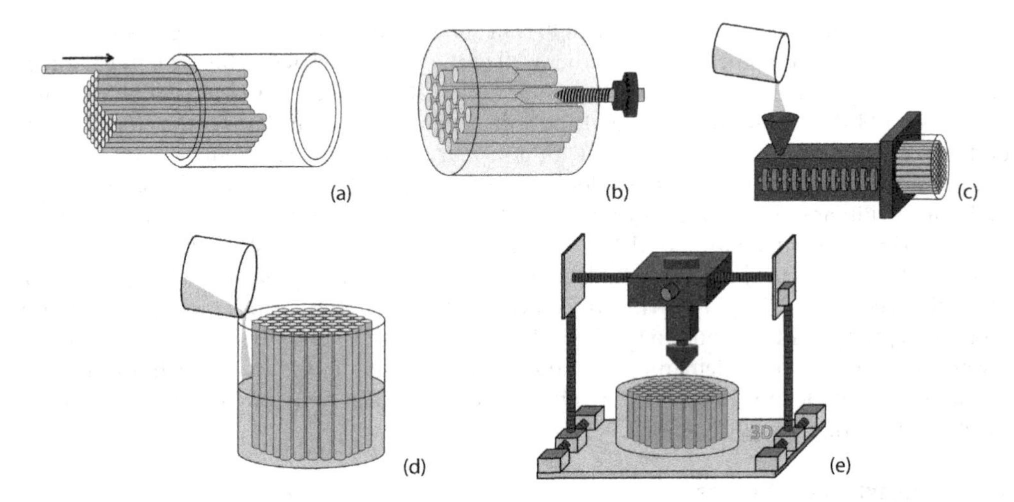

FIGURE 2.16 Discontinuous fabrication methods used for mPOF fabrication. (a) Stacking; (b) drilling; (c) extrusion from resin, billets, or monomer; (d) casting/molding; and (e) 3D printing.

PF (Nagasawa, Kondo, Ishigure, & Koike, 2005), and PMMA (van Eijkelenborg et al., 2008). In this method, tiny capillary tubes are piled up with the predefined arrangement and then filled into a hollow core tube in order to form the preform that later will be heat drawn to fiber. One obvious limitation of this procedure is the limited hole arrangement. Additionally, the process can be time consuming, prone to human errors, and expensive, since it is a discontinuous process and requires pilling up of thousands of capillary tubes to form a preform.

2.5.3.2 Drilling

The drilling technique was the one used to produce the first mPOF in 2001 (van Eijkelenborg et al., 2001). The technique is based on the perforation of a solid rod (**Figure 2.16b**). For that, a computer numerical controlled mill is used. Semi-synthetic cutting fluid is constantly flushed onto the drilling zone, permitting cooling of the drilled part and minimizing the likelihood of surface roughness. Nevertheless, the drill is controlled in an up and down process, allowing the removal of the small plastic bits. Due to the limited drill length, the preform needs to be drilled at both ends in order to have enough length for the heat drawing process. Therefore, the length of the rod is about two times the length of the drill. Due to these two last reasons, a preform composed of tens of holes can take several hours to create. The holes can be arbitrarily located on the rod, being only limited by the technological limits of drilling quality (Barton et al., 2004; Beckers et al., 2015). The technique is very easy and versatile and can be used for several structures. Examples of preforms after being drilled can be seen in **Figure 2.17**.

After the drilling process, the preform is flushed with water for several hours in order to remove any impurities from the cutting fluid and PMMA micro-swarf that adhered to the preform (van Eijkelenborg et al., 2004). A 24-hour annealing process is then applied to the preform, evaporating the water absorbed by the polymer on the cleaning process. The preform is now prepared for the use of the heat drawing process. However, the dimensions required for the holes structure in the final fiber impose more than one step procedure in the heat drawing technique. Generally, the preform is primarily drawn to a "stretched" secondary preform (6–12 mm in diameter and 15 cm in length) that is subsequently sleeved to form a secondary preform. The process may be repeated many times as needed (Barton et al., 2004). In the

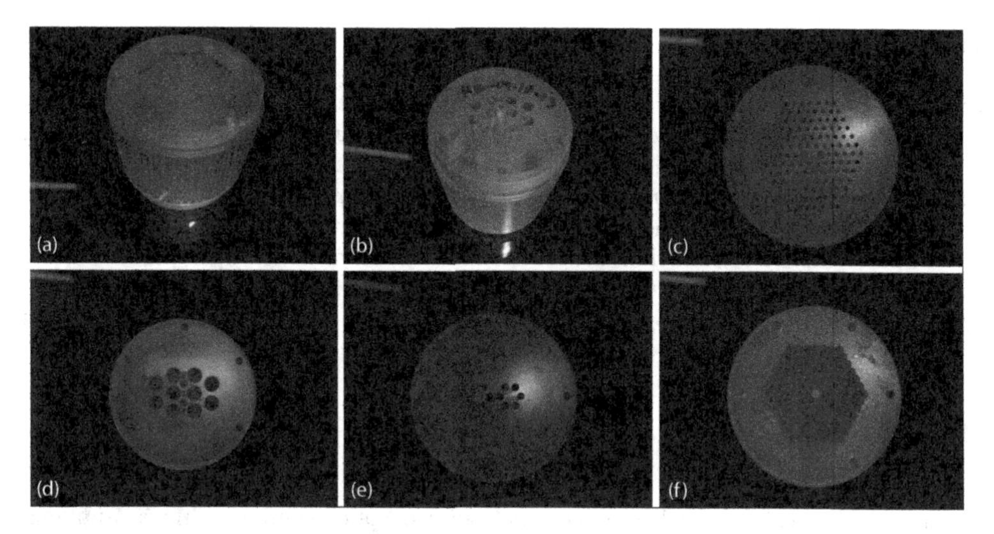

FIGURE 2.17 Side view (a, b) and top view (c, d, e, f) photographs of different PMMA preforms, used for the fabrication of HiBi mPOFs (a, b, c, d), dual core mPOF (e), and SM mPOF (f).

heat drawing process for mPOFs, it is necessary to pressurize the holes in order to avoid their collapse. By doing that, a more precise control of the holes dimension/air filling fraction may be achieved. However, this brings another parameter to control which further increases the complexity of the process. Nevertheless, the precise control of all the variables enhances the quality of the mPOFs prepared.

2.5.3.3 Extrusion

The extrusion method is possibly the most indicated method for large scale production of mPOFs (Kang, Wang, Yang, & Wang, 2008). In this method, the molten polymer, resin, or monomers/polymer mixture are fed into the extruder. The material is then forced to pass through a die that contains the inverted pattern of the required mPOF (Ebendorff-Heidepriem, Monro, van Eijkelenborg, & Large, 2006; Ebendorff-Heidepriem et al., 2007). With this method it can be possible to extrude complex structures (Ebendorff-Heidepriem et al., 2007), with hole shapes that cannot be performed with, for instance, the drilling technique. Additionally, the preforms can be obtained in a single automated step, with a length determined by the drawing tower and handling issues. The correspondent setup can be seen in **Figure 2.16c**.

2.5.3.4 Casting

The preform casting is another technique that can be used for large scale production of mPOFs due to its low cost of production. This technique has been used to produce both glass and plastic preforms (United States Patent No. US 20050286847A1, 2005; Zhang et al., 2006).

Preform casting refers to the technique where a mold that mirrors the shape of pretended microstructured fiber is filled with the necessary chemical precursors (i.e. monomer, initiator, and chain transfer agent). The correspondent setup can be seen in **Figure 2.16d**. During the polymerization, the formation of bubbles will occur, so the reaction is performed in vacuum atmosphere. After the polymerization, the solid preform is removed from the mold and is then ready for the heat draw technique (Zhang et al., 2006).

Compared with stacking methods, casting is more advantageous, since the hole pattern, size, and spacing can be altered independently, and it does not create interstitial holes within the lattice (Zhang et al., 2006).

2.5.3.5 3D Printing

In 2015 Cook et al. presented a new methodology to fabricate mPOFs, the 3D printing (Cook et al., 2015a, 2015b). The technique described in their paper is based on a low-cost, commercially available thermosetting 3D printer that prints a 3D preform fiber. The material used for the preform production is a PS mixture containing styrene - butadiene - copolymer and PS. The injection of the material is done through an extruder, where its temperature is set to 210°C, keeping the bed temperature at 95°C. The process is done by depositing several layers of the filament material (see **Figure 2.16e**), and thus, the process can take several hours to be completed, which is a disadvantage for this technique (Cook et al., 2015a).

Producing optical preforms by 3D printing is a new milestone in optical fiber manufacturing process, and it could be one of the directions to follow in the near future.

2.6 Attenuation in Polymer Optical Fiber Materials

Nowadays the use of fiber optics has its main application in optical communications. Additional, uses such as lightning and sensing are also being employed. However, any of these applications are subject to light attenuation through the waveguide material, this being more pronounced for long distance applications. Fiber attenuation limits how far the light can propagate before being too weak to be detectable. The best known method to measure the fiber attenuation is through the cutback method (Koike & Koike, 2011). For that, the output power (P_{out1}) of the fiber under test is measured. Later, and without disturbing the input conditions, the test fiber is cut back to a shorter length, leaving a few meters from the input source, in order to allow an equilibrium mode distribution. The output power at this near end (P_{out2}) is then measured. The attenuation α, expressed in decibels per kilometer, is defined as follows:

$$\alpha(dB/km) = -\frac{10}{L_{out1} - L_{out2}} \log_{10}\left(\frac{P_{out1}}{P_{out2}}\right) \quad (2.14)$$

where L_{out1} and L_{out2} define the length of the fiber before and after the cleaving process.

Today, POFs are being considered appropriate for short range applications (Koike & Koike, 2011; Zubia & Arrue, 2001). This is due to two main reasons: first, the use of large core permits an easy coupling between the fibers, allowing low skill operators and the implementation of cheap plastic connectors. Secondly, the use of large diameter core fiber allows the use of low cost, high NA light sources. Loss mechanisms affect silica fibers as well as POFs. However, this is much more pronounced for POFs. Indeed, for all optical fibers the Rayleigh scattering is present at short wavelengths, but for polymers, the vibration of the C–H harmonic becomes very significant at wavelengths longer than 600 nm.

The basic attenuation mechanisms in a POF can be classified in two main groups: intrinsic (internal) and extrinsic (external). Intrinsic attenuation results from the physical and chemical structure of the polymer, which are due to absorption and scattering. Both contributions are dependent on the composition of the optical fiber and cannot be eliminated, thus representing the ultimate transmission loss limit (Zubia & Arrue, 2001). Extrinsic attenuation is due to absorption by organic contaminants, transition metals, and water absorbed by the fiber material (Kaino, 1985b). It is also due to the scattering of light in micro-voids and dust, as well as core diameter fluctuations, birefringence, and core cladding imperfections during fiber manufacturing process (Ioannides et al., 2014). Basically, these are the losses that can be reduced by purification of the material and on the optimization of the manufacturing process. All these mechanisms can be seen in **Table 2.5**.

Table 2.5 Classification of Intrinsic and Extrinsic Factors Affecting POF Attenuation

Type	Mechanism	Type	Abbreviation
Intrinsic	Absorption	Electronic transitions	α_{eT}
		Molecular vibration absorption	α_{mv}
	Scattering	Rayleigh scattering	α_{Rs}
Extrinsic	Absorption	Transition metals	α_i
		Organic contaminants	
		Water absorption	
	Dispersion	Dust	
		Voids	
		Microfractures	
	Radiation	Structural imperfections	
		Microbends	
		Macrobends	

Note: From the above classification, the attenuation can be written as the sum of each contribution ($\alpha = \alpha_{eT} + \alpha_{mv} + \alpha_{Rs} + \alpha_i$).

2.6.1 Intrinsic Losses

2.6.1.1 Electronic Transition Absorption (UV Absorption)

Light induced electronic transitions in polymers are responsible for absorptions in the UV region, which may tail into the visible region depending on the extent to which electrons are delocalized in the structure (Harmon Julie, 2001). The light absorption occurs when a photon excites an electron to a higher energy level. Absorption due to electronic transitions occurs due to the transfer of exciting valence electrons from an occupied lower energy orbital to an empty higher energy level, with correspondent absorption of light. The transitions occur in order of energy as $\eta \rightarrow \pi^* < \pi \rightarrow \pi^* < \eta \rightarrow \sigma^* <<< \sigma \rightarrow \sigma^*$ (Harmon Julie, 2001), (see **Figure 2.18**).

When saturated bonds or single bonds exist, electrons can undergo $\sigma \rightarrow \sigma^*$ transitions, occurring in deep UV regions (135 nm), having no effect on the practical transmission windows of polymers (Harmon Julie, 2001). On the other hand, electrons in unsaturated bonds (double or triple bonds) can have $\eta \rightarrow \pi^*$, $\pi \rightarrow \pi^*$, and $\eta \rightarrow \sigma^*$ transitions with absorptions near 220–230 nm (Gupta, Liang, Tsay, & Moacanin, 1980), extending their tails into the visible region. There are two important transitions for organic groups, $\pi \rightarrow \pi^*$ in the C=C bond as well as in the phenyl group, and $\eta \rightarrow \pi^*$ in the C=O bond.

The absorption peak originated from the electronic transitions in the far UV has its tail extended to the visible and near infrared. This absorption tail (α_{eT}) follows the Urbach's empirical rule (Urbach, 1953) through the following equation:

$$\alpha_{eT} = A_0 \exp\left(\frac{B_0}{\lambda}\right) \tag{2.15}$$

where A_0 and B_0 are the material constants, determined from the absorbance spectrum. Those constants can be found in **Table 2.6** for the three most used polymers.

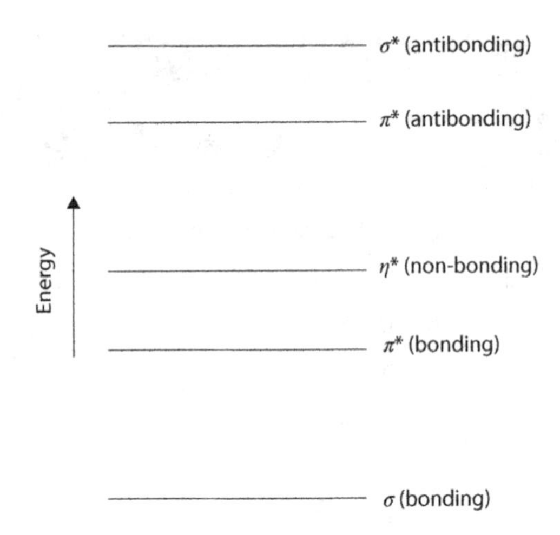

FIGURE 2.18 Energy levels for electrons according to molecular theory.

Table 2.6 A_0 and B_0 Material Constants for PMMA, PS, and PC

Material Constants	Polymer Material		
	PMMA (Kaino, Fujiki, & Jinguji, 1984)	PS (Kaino et al., 1981)	PC (Yamashita & Kamada, 1993)
A_0	1.58×10^{-12}	1.10×10^{-5}	1.87×10^{-2}
B_0	11.50×10^{3}	8.00×10^{3}	5.30×10^{3}

The calculated electronic transition loss based on the values expressed in **Table 2.6** can be seen in **Figure 2.19**.

Examining **Figure 2.19**, it can be observed that the ultraviolet absorption edge decreases as the wavelength increases, where for PMMA the loss is almost negligible after 450 nm. This is explained by the weak absorption at 220–230 nm of the $\eta \rightarrow \pi^*$ transition of the double bond of the ester group, and $\pi \rightarrow \pi^*$ transitions of low intensity near 200 nm due to the azo group in a polymerization initiator when azo compounds are used (Gupta et al., 1980; Kaino, 1985a).

Contrary to PMMA, the electronic transition loss curves of PS and PC extend further to the near infrared region. A closer look reveals losses of 100 dB/km at 500 nm and 7 dB/km at 600 nm for PS, and loss values as high as 750 dB/km at 500 nm and 130 dB/km at 600 nm for PC. The explanation is the result of the $\pi \rightarrow \pi^*$ electronic transitions of the phenyl groups in PS and PC (Kaino, 1985a) (see **Figures 2.10** and **2.12**). However, the total conjugate system in a molecule should also be considered. Additionally, PC presents two phenyl groups contrary to PS, which has only one, this being the reason for its higher losses (Tanaka et al., 1988).

The high losses presented by PS and PC prove that these two polymers are not the preferred choice for white light transmission, even considering the inexistence of other loss contributions. Another relevant aspect with these results is related to the design of an optical fiber. At first, one may think to use these two materials for the core of an optical fiber due to the higher refractive index that they offer when compared with other polymers (see **Table 2.2**). However, the theoretical electronic transition losses can compromise the application for which the fiber is intended.

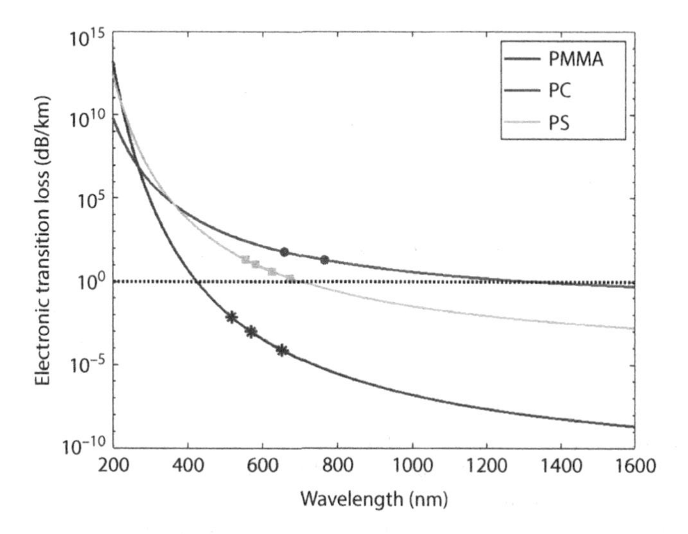

FIGURE 2.19 Calculated electronic transition absorption for the three most used polymers, PMMA, PS, and PC. The markers define the low loss regions of each polymer.

2.6.1.2 Molecular Vibration Absorption (IR Absorption)

The molecular vibration absorption is another type of intrinsic loss. This is the most predominant factor to contribute to the fiber attenuation at the IR and visible spectral regions.

The individual bonds and groups in a large molecule can be considered isolated anharmonic oscillators (Groh, 1988). The potential curves belonging to these oscillators can be approximated by the "Morse Potential," where the energy levels can be written as:

$$G(\upsilon)=\nu_0\left(\upsilon+\frac{1}{2}\right)-\nu_0\chi\left(\upsilon+\frac{1}{2}\right) \tag{2.16}$$

where υ is the quantum number, χ the anharmonic constant, and ν_0 the vibration frequency of an absorption band, which is dependent to a first approximation upon the masses of the atoms involved and the force constant of the interatomic bond, in accordance with the classical equation for the harmonic oscillator (Emslie, 1988; Kaye, 1954):

$$\nu_0=\sqrt{\frac{K}{4\pi^2\mu}} \tag{2.17}$$

where K is the force constant of the interatomic bond and μ the reduced mass of atoms involved in the vibration, defined as (Kaye, 1954):

$$\mu=\frac{m_1m_2}{m_1+m_2} \tag{2.18}$$

m_1 and m_2 being the masses of the atoms. The position of the fundamental vibration ν_1 or an overtone ν_υ ($\upsilon = 2,3,4,\ldots$) is now given by (Groh, 1988):

$$\nu_\upsilon=G(\upsilon)-G(0)=\nu_0\upsilon-\chi\nu_0\upsilon(\upsilon+1) \tag{2.19}$$

From the absorption spectra it is only possible to access $\nu_1, \nu_2, \ldots$, thus, ν_0 needs to be rewritten as:

$$\nu_0 = \frac{\nu_1}{1-2\chi} \tag{2.20}$$

and replacing ν_0 into (2.19), to produce:

$$\nu_\upsilon = \frac{\nu_1\upsilon - \nu_1\chi\upsilon(\upsilon+1)}{1-2\chi}, \qquad (\upsilon = 2,3,4...) \tag{2.21}$$

If ν_1 and χ are known, the spectral positions of the overtones can be calculated using (2.21). On the other hand, χ can be calculated from the fundamental vibration, ν_1 and the first overtone ν_2 (Groh, 1988).

The fundamental vibration absorption for different bonds, obtained from the absorption spectra collected by different authors, is shown in **Table 2.7**.

The estimated attenuation overtone due to molecular vibration absorption for a bond B (B=carbon-hydrogen (C–H), carbon-deuterium (C–D), carbon-fluorine (C–F), carbon-chlorine (C–Cl), carbon=oxygen (C=O), oxygen-hydrogen (O–H)) is given by Groh (1988) as:

$$\alpha_{mv}(\nu_\upsilon^B) = 3.2\times10^8 \frac{\rho N_B}{M_G}\left(\frac{E_\upsilon}{E_1^{C-H}}_B\right) \tag{2.22}$$

where ρ is the polymer density, N_B the number of B bonds, M_G the molecular weight of the monomer unit, E_υ/E_1^{C-H} the vibration energy ratio of each bond to the fundamental frequency of the C–H bond. **Figure 2.20** shows the spectral overtone positions and normalized band strengths for different bond vibrations, obtained from the data presented in Groh (1988). If one considers $\rho = 1.19$ g/cm^3, $M_G = 114$ g/mol, and $N_{C-H} = 8$ for PMMA, then, from (2.22), the value $E_\upsilon/E_1^{C-H} = 3.3 \times 10^{-8}$ refers to an attenuation of 1 dB/km.

Regarding **Figure 2.20**, it can be seen that the overtones decrease in intensity when the order of the harmonic increases [the intensity decreases one order of magnitude when the overtone (υ) increases one unit (Tanaka et al., 1988)]. Since the fundamental vibration absorption occurs for longer wavelengths (see **Table 2.7**), the occurrence of high attenuation at the infrared region is obvious when compared to the visible region.

The overtones for carbon deuterium and carbon-halogen bonds are several orders of magnitude lower than the overtone of carbon-hydrogen in the visible region. Thus, they should

Table 2.7 Fundamental Vibration Absorption of Different Bonds

Bond	Wavelength (nm)
C–H	3300–3500
C–D	4400
C–F	8000
C–Cl	11700–18200
C=O	5300–6500
O–H	2800

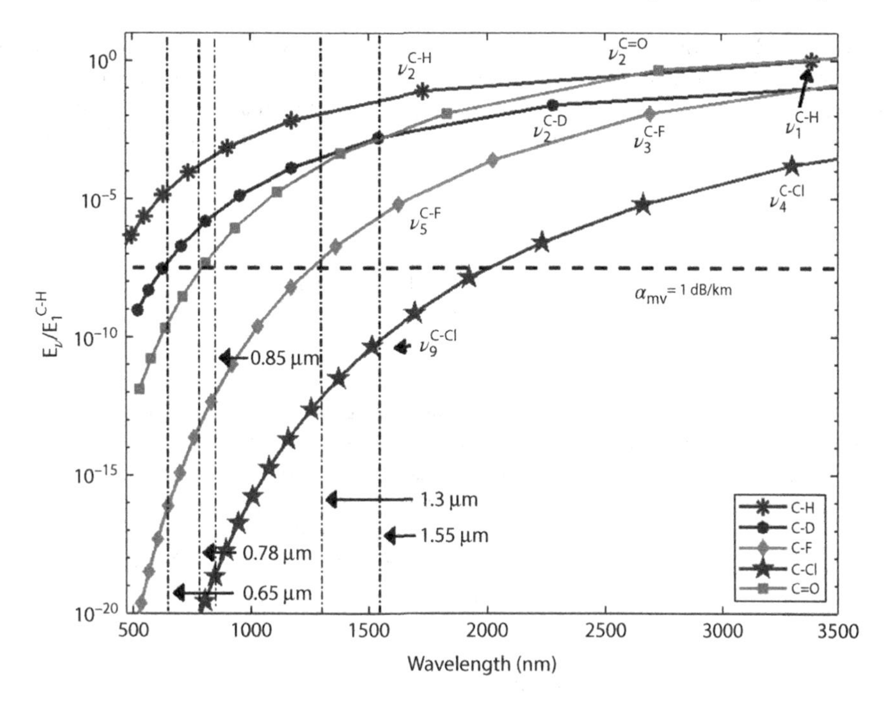

FIGURE 2.20 Absorption loss due to C-X vibrations versus spectral overtone position (based on the table values found in Groh (1988) and [2.22]).

contribute negligibly to the fiber loss in the visible and near IR. The same applies to the C=O vibration absorption (i.e. in PMMA and PC), especially because a monomeric unit usually contains only one carbonyl group (Groh, 1988).

The behavior presented by C–D, C–F, and C–Cl overtones suggests a way to improve the PMMA-POFs transparency by replacing the hydrogen atoms with these heavier elements, shifting the fundamental absorption frequency to longer wavelengths (see [2.17] and [2.18]). Taking this into account, some PMMA-POFs have been deuterated (PMMA-d8), achieving experimental and theoretical loss values of 20 and 10 dB/km respectively (Kaino, 1987; Kaino, Jinguji, & Nara, 1983). Deuterated polymers, besides being costly and difficult to synthesize, also absorb water. This will impose strong vibrational absorption of the O–H groups, especially in the near infrared region (Kaino, 1985b). In order to suppress the water absorption, the solution was fluorine substitution for the hydrogen core polymer, preventing the water penetration into the core polymer (Zubia & Arrue, 2001). Another great advantage is that carbon-halogen bonds present low vibration absorption in the wavelength between 850 and 1300 nm (see **Figure 2.20**), allowing their use with emitters and receivers already developed for glass optical fibers.

2.6.1.3 Rayleigh Scattering

Rayleigh scattering loss (α_{Rs}) is caused by inhomogeneities of random nature that occur in a small scale compared to light wavelengths. The order of size of these irregularities is about 1/10 of a wavelength. These inhomogeneities are fundamental and cannot be avoided, imposing an intrinsic loss limit on the material. They manifest as density fluctuations, orientation, and composition of the material. For the density fluctuations, they are related with βT (the isothermal compressibility) and with $\partial\varepsilon / \partial\rho|T$, where ε and ρ are respectively the dielectric constant and density (Zubia & Arrue, 2001). The orientation fluctuations are caused by the anisotropy of the monomers and by the crystallinity of the polymer links. Finally the composition fluctuations arise from the addition of substances to achieve the desired refractive index profiles (in case of GI-POFs) (Zubia & Arrue, 2001). These effects will modify the refractive index in distances on the order of the wavelength. The subsequent scattering of light (in almost all directions) due to these inhomogeneities produces an attenuation that follows the characteristic λ^{-4} Rayleigh scattering law. The optical power attenuation due to Rayleigh scattering can be written as (Koike & Gaudino, 2013):

$$\alpha_{Rs} = 10\log(e)\left(\frac{8\pi^3}{3\lambda^4}\right)n^2 p_e^2 K_B T_c \beta_{T_c} \tag{2.23}$$

where n is the material refractive index of the medium, λ is the wavelength of light, p_e the photoelastic coefficient, β_{Tc} the isothermal compressibility at the fictive temperature T_c, and K_B is the Boltzmann's constant (Koike, 1996). The fictive temperature is defined as the temperature at which the material can reach a state of thermal equilibrium and is closely related to the anneal temperature. **Equation (2.23)** indicates that the power attenuation can be reduced if low-temperature, low-index, and low-compressibility materials are used. Furthermore, as the scattering intensity is proportional to λ^{-4}, the Rayleigh scattering loss can be roughly estimated at an arbitrary wavelength by (Koike & Gaudino, 2013):

$$\alpha_{Rs}(\lambda) = \alpha_{Rs}(\lambda_0)\left(\frac{\lambda_0}{\lambda}\right)^4 \tag{2.24}$$

where $\alpha_{Rs}(\lambda_0)$ is the Rayleigh scattering loss at the specific wavelength λ_0. From this equation, it can be seen that the scattering loss rapidly diminishes with increasing wavelength. In order to compare the Rayleigh scattering loss evolution of different polymer materials in a wide wavelength range, (2.24) and the experimental values found in **Table 2.8** were used. The results are shown in **Figure 2.21**.

As can be seen from **Figure 2.21**, α_{Rs} for PS and PC are much higher than those of PMMA, deuterated PMMA, and PF materials. This finding may be explained with reference to their molecules represented in **Figures 2.10** and **2.12**, for PC and PS, respectively. First, the presence of phenyl groups gives rise to the refractive index (~1.59 for PC and PS, which is higher than 1.49 for PMMA and 1.34 for CYTOP®, @ 589 nm). Thus, α_{Rs} increases as may be deduced from (2.23). Secondly, the flat physical geometry of the ring increases the molecular anisotropy and hence scattering, when compared to the tetrahedral methyl group (CH_3), in PMMA, shown in **Figure 2.9**, which creates a more three-dimensional structure (Emslie, 1988).

Table 2.8 Isotropic Scattering Loss (dB/km)

Polymer Material	λ_0 (nm)	$\alpha_{Rs}(\lambda_0)$ (dB/km)
PMMA (Kaino, 1987)	633	8.8
PMMA-d8 (Kaino, 1987)	633	10
PC (Yamashita & Kamada, 1993)	488	277
PS (Kaino et al., 1981)	633	55
CYTOP® (Koike, 1996)	650	4.2

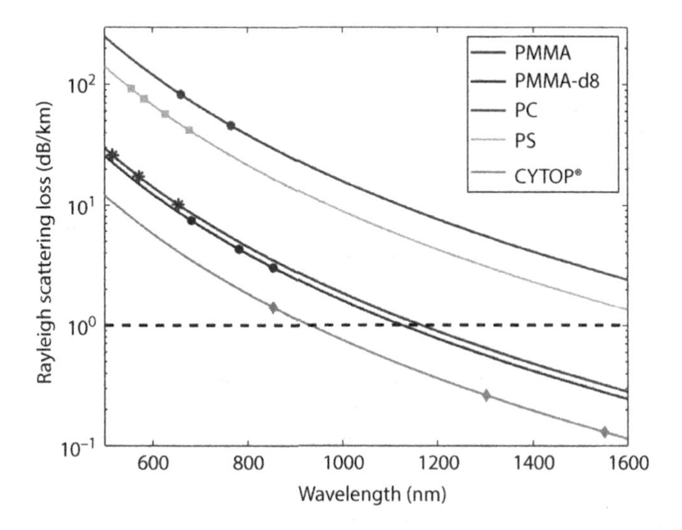

FIGURE 2.21 Calculated Rayleigh scattering loss for PMMA, PMMA-d8, PC, PS, and CYTOP® materials. The markers define the wavelength regions where the polymers offer lower losses.

2.6.2 Extrinsic Losses

Extrinsic loss factors arise from imperfections and impurities in the core of optical fibers. They can be, for instance, a variation in the fiber core diameter, roughness, inhomogeneities introduced during the fabrication, presence of pollutants, dust particles, O–H groups, bubbles, micro fractures, etc. These loss factors are related to the manufacturing process and thus, in principle, they can be reduced and/or even avoided. Additional external contributions may also appear under the form of bending losses, temperature, pressure, strain, humidity, and ageing, among others.

2.6.2.1 Absorption by Impurities

During the manufacturing process of an optical fiber, some contaminants can be unintentionally introduced into the polymer material, either during the polymerization or during the extrusion process. These impurities are essentially transition metal ions such as nickel (Ni), cobalt (Co), chromium (Cr), manganese (Mg), and iron (Fe) (Zubia & Arrue, 2001). In general, these impurities present the fundamental vibration absorption frequency whose overtones are centered in the visible and infrared regions of the spectrum (see **Table 2.9**). Co ions, for instance, have absorption bands centered at 530, 590, and 650 nm, resulting in a single large peak absorption. For this

Table 2.9 Loss Induced by Contaminants

Metal Ion	Concentration (ppb)	Loss (dB/km)
Co	2	10 (visible)
Cr	1	1 (650 nm)
Fe	1	0.7 (1100 nm)
Cu	1	0.4 (850 nm)

Source: Zubia, J. and Arrue, J., *Opt. Fiber Technol.*, 7, 101–140, 2001.

particular case, it is reported that only 2 ppb* of these ions increase the loss to about 10 dB/km (Kaino, 1985a). **Table 2.9** shows the contribution of different impurities.

However, another major extrinsic loss mechanism is caused by absorption due to water. The molecular vibration absorption of the O–H bond appears near the C–H vibrational absorptions of PMMA. One of the most affected transmission regions is at 766 nm (one of the optical windows of PMMA), where losses as high as 250 dB are expected for 90% relative humidity (Kaino, 1985b). The case becomes worse if one considers PMMA-d8 materials. For that, losses as high as 600 dB/km and 800 dB/km are observed for 746 nm and 840 nm (Kaino, 1985b). For this reason, PMMA-d8 is not the perfect candidate from the point of view of transmission property stability. Compared with fluoride polymers, there exists a huge drawback since for those, the water absorption is almost negligible.

Material impurities can also be introduced on the fabrication of mPOFs. This is especially relevant considering the coolant used on the preform drilling process. It has been found that switching from an oil-based emulsion to a semi-synthetic cutting fluid provided a 28% loss reduction (van Eijkelenborg et al., 2004).

2.6.2.2 Dispersion due to Physical Imperfections

The presence of large inclusions or high frequency variations in the diameter of the core fiber will result in a scattering loss that is independent of the incident wavelength. Large inclusion have imperfections with dimensions higher than 1 μm, and the inclusions cover defects such as bubbles, cracks, or dust. Those physical imperfections are shown in **Figure 2.22**.

When light traveling in an optical fiber encounters one of these imperfections, it is deviated mainly by surface reflection. However, refraction, diffraction, and absorption are also present, depending on the nature of the physical defect (Emslie, 1988).

2.6.2.3 Radiation Losses

Radiation losses may occur when fibers are bent. The type of bend can be classified into two groups: microbends and macrobends. The latter are referred to as macro scale bends, similar to the ones presented when rolling a fiber on a reel. Microbends are referred to as small scale fluctuations in the fiber axis, which can be considered smaller than the diameter of the fiber. The interface between core and cladding is probably the best candidate for this kind of imperfections (see example on **Figure 2.22**). Indeed, during the manufacturing process, the

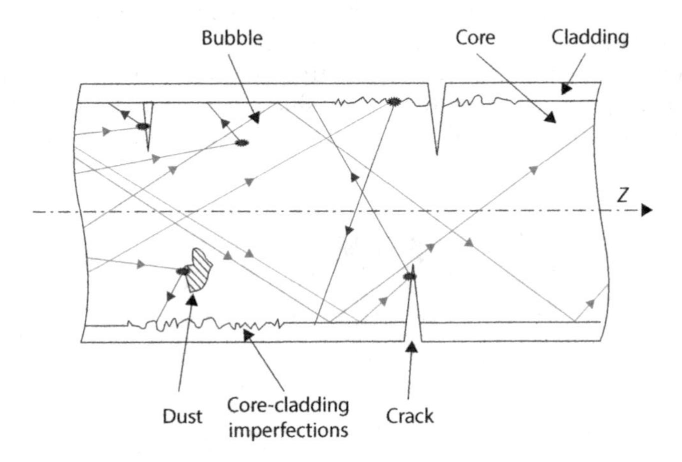

FIGURE 2.22 Scattering loss due to different physical imperfections (dust, cracks, bubbles, core-cladding imperfections).

* ppb means 1 impurity atom per 10^9 atoms.

different expansion coefficients of the core and cladding materials lead to the formation of tension on the interface, promoting the occurrence of small imperfections. On the other hand, the drawing of the fiber can cause the cladding to acquire a rough surface (Ziemann et al., 2008). Non-uniform lateral pressures during the cabling process is another type of defect associated with radiation loss.

For mPOFs, the imperfections created on the mechanical drilling process also are an important source of scattering loss at the holes boundary region.

All the contributions presented for the extrinsic loss factors can be considered typically as about 4 dB/km at 1300 nm for the best quality CYTOP®-POFs, 20 dB/km for a PMMA POF at 680 nm (Zubia & Arrue, 2001) and 45 dB/km for PS-based POFs at the visible region (Kaino et al., 1981).

2.6.2.4 Losses due to Ageing

Ageing is another external characteristic that gives rise to fiber losses. This process can be sped up at higher temperatures. Indeed the formation of monomeric and dimeric compounds leads to high absorption in the UV-visible region of the spectrum (Appajaiah & Jankowski, 2001; Takezawa, Tanno, Taketani, Ohara, & Asano, 1991), resulting in the yellowing of the polymer over time (especially under the exposure of UV light). Ageing can also be due to the mechanical degradation of POFs. This can be observed in POFs that are strained for a long time, leading to chain scission. Additionally, micro cracks formed during the POF manufacturing process may also increase their size due to stress relaxation for a long period of time (Appajaiah & Jankowski, 2001).

2.6.3 Total Loss Limit

The total loss limit accounts for the net contribution of both intrinsic and extrinsic losses. Concerning the intrinsic loss limits of POFs, for the most used polymer (PMMA), α_e is negligible in the visible region (see **Figure 2.19**), and the intrinsic loss limit is only due to α_{mv} and α_{Rs}. This imposes a loss of 35 dB/km in the best low loss region of PMMA (Kaino, 1987). The majority of the intrinsic loss is due to the C–H molecular vibration absorption, where high overtones fall into the visible region. The solution is the substitution of hydrogen by heavier atoms such as deuterium, fluoride, or chloride, where the higher overtones are far away from the visible region (see **Table 2.7** and **Figure 2.1**). For the first, it is possible to achieve an intrinsic low loss of 9 dB/km at 680 nm. However, deuterated PMMA is expensive and presents a huge problem related with water absorption that leads to high losses due to the overtones of the O–H absorption, which can be as high as 100 and 300 dB/km at regions of 780 and 850 nm for a humidity change of 20% RH to 90% RH, respectively (Kaino, 1985b). The solution can pass through the use of fluorine or chloride instead of deuterium. CYTOP® is an amorphous fully fluorinated polymer in which hydrogen no longer exists (see **Figure 2.13**). The theoretical loss limit at 850, 1300 and 1550 nm is only due to the scattering loss, since these wavelengths are found in minima of the C–C, C–F, and C–O overtones, specifically for the 5th through 7th peak absorptions. Therefore, loss minima of 0.26 and 0.15 dB/km may be found for 1300 and 1550 nm, respectively, which is comparable with that of silica fibers (Koike & Gaudino, 2013; Tanio & Koike, 2000).

POFs based on PC or PS both show higher losses than PMMA, and the main reason is due to the C–H electronic transitions and molecular vibrations in the phenyl group that increase both α_{eT} and α_{mv}. The phenyl group increases the refractive index and thus the scattering (see [2.23]). Additionally, the geometry of the ring increases the molecular anisotropy and hence scattering α_{Rs}. The total intrinsic losses in the low loss region of PC and PS are known to be 166 dB/km (Yamashita & Kamada, 1993) and 84 dB/km (Kaino et al., 1981) for 656 and 672 nm respectively. Even with such constrains, these materials have been considered for the integration in POFs due to their special physical properties (see **Table 2.2**). Considering the above discussion, **Table 2.10** presents the relevant loss contribution for the low loss region of different materials.

Table 2.10 Intrinsic Loss Factors and Limits, for Different Polymer Optical Fiber Materials

Polymer	PMMA (Kaino, 1987)			PMMA-d8 (Kaino, 1987)			PC (Yamashita & Kamada, 1993)			PS (Kaino et al., 1981)			PF (Koike, 1996; Tanio & Koike, 2000)		
λ (nm)	516	568	650	680	780	850	656	764	552	580	624	672	850	1300	1550
α_{eT} (dB/km)	0	0	0	0	0	0	62	19	22	11	4	2	0	0	0
α_{mv} (dB/km)	11.3	17.7	95.2	1.6	9.7	36	28	165	0	4	22	24	0	0	0
α_{Rs} (dB/km)	26	17.2	10.3	7.5	4.3	3.1	76	40	95	78	58	43	1.45	0.26	0.15
Intrinsic loss	37	35	106	9	14	39	166	224	117	93	84	69	1.45	0.26	0.15

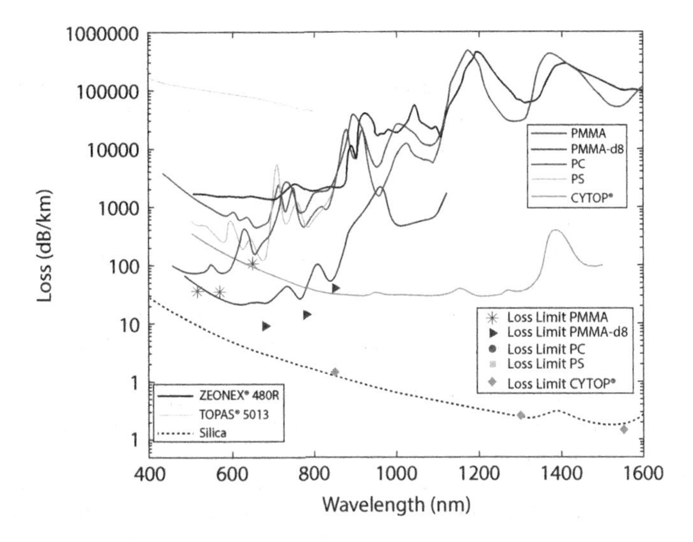

FIGURE 2.23 Transmission loss spectra of different fiber materials: PMMA (Argyros, 2009), PMMA-d8 (Naritomi, 1996), PC (Yamashita & Kamada, 1993), PS (Paradigm, 2016), CYTOP® (Lethien, Loyez, Vilcot, Rolland, & Rolland, 2011), ZEONEX®480R (Woyessa et al., 2017), TOPAS®5013 (Khanarian & Celanese, 2001), and silica (Naritomi, 1996). The marker points refer to the low loss limit of different materials, obtained from Table 2.10.

In addition to the intrinsic losses, POFs have also to deal with extrinsic losses. These are essentially related with the manufacturing process. This type of loss can in theory be suppressed; however, in the "physical world" there are always limits to deal with. **Figure 2.23** shows the loss spectra of different POF materials and silica-based fiber, collected from different authors. In the same figure, it is plotted the intrinsic loss limit (marker points), presented in **Table 2.10** for the interesting region of each material.

From the spectra shown in **Figure 2.23**, it is obvious that CYTOP® and PMMA-d8 based POF materials are the best candidates for low loss transmission waveguides. However, CYTOP® presents an intrinsic loss limit that is clearly distinguished from the others, achieving a values at the near infrared region, similar to those presented by conventional silica fibers (represented as black dashed line). Regarding that, the possibility to use the already developed silica based systems is possible for this kind of POF.

2.7 Conclusions

This chapter provided a brief summary of the principle of operation of optical fibers, including step index, graded index, and microstructured. The quality requirements needed for the development of an optical fiber were also analyzed, with special focus for polymer materials. Therefore, optical, mechanical, thermal, and chemical properties were introduced, and whenever possible, a comparison between different polymer materials was given. The most used POF-based materials were also presented. The most used techniques for the fabrication of different POFs (i.e. SI-POF, GI-POF, and mPOF) were also described. At the end of this chapter, the most significant factors involved on the attenuation in the most used polymer materials currently employed for the production of POFs were analyzed. The analysis was made for the visible and near infrared regions, and a comparison between the different polymer materials was also made.

Acknowledgments

This work was funded by FCT – Fundação para a Ciência e Tecnologia through Portuguese national funds through hiPOF project (PTDC/EEI-TEL/7134/2014), AQUATICsens project (POCI-01-0145-FEDER-032057), investigator grant IF/01664/2014 and project INITIATE.

References

AmericasStyrenics LLC. (2008). STYRON 666D General Purpose Polystyrene Resin. Retrieved from http://www.amstyrenics.com.

Anthony, J., Leonhardt, R., Argyros, A., & Large, M. C. J. (2011). Characterization of a microstructured Zeonex terahertz fiber. *Journal of the Optical Society of America B, 28*(5), 1013–1018. https://doi.org/10.1364/JOSAB.28.001013

Antunes, P., Lima, H., Monteiro, J., & Andre, P. S. (2008). Elastic constant measurement for standard and photosensitive single mode optical fibres. *Microwave and Optical Technology Letters, 50*(9), 2467–2469. https://doi.org/10.1002/mop.23660

Appajaiah, A., & Jankowski, L. (2001). A review on aging or degradation of polymer optical fibers (POFs): Polymer chemistry and mathematical approach. In C. Bastiaansen (Ed.), *10th international conference on plastic optical fibres* (pp. 317–324). Amsterdam, the Netherlands.

Argyros, A. (2009). Microstructured polymer optical fibers. *Journal of Lightwave Technology, 27*(11), 1571–1579. https://doi.org/10.1109/JLT.2009.2020609

Argyros, A. (2013). Microstructures in polymer fibres for optical fibres, THz waveguides, and fibre-based metamaterials. *ISRN Optics*, 1–22. https://doi.org/10.1155/2013/785162

Arimondi, M., Macchetta, A., Asnaghi, D., & Castaldo, A. (2005). United States patent No. *US 20050286847A1*. Retrieved from https://patents.google.com/patent/US20050286847.

Arrue, J., Jiménez, F., Ayesta, I., Illarramendi, M. A., & Zubia, J. (2011). Polymer-optical-fiber lasers and amplifiers doped with organic dyes. *Polymers, 3*(4), 1162–1180. https://doi.org/10.3390/polym3031162

Asahi Glass Co. Ltd. (2009). Amorphous Fluoropolymer (CYTOP). Tokyo, Japan.

Barton, G., van Eijkelenborg, M. A., Henry, G., Large, M. C. J., & Zagari, J. (2004). Fabrication of microstructured polymer optical fibres. *Optical Fiber Technology, 10*(4), 325–335. https://doi.org/10.1016/j.yofte.2004.05.003

Beckers, M., Schlüter, T., Vad, T., Gries, T., & Bunge, C. A. (2015). An overview on fabrication methods for polymer optical fibers. *Polymer International, 64*(1), 25–36. https://doi.org/10.1002/pi.4805

Bhowmik, K., Peng, G.-D., Ambikairajah, E., Lovric, V., Walsh, W. R., Prusty, B. G., & Rajan, G. (2015). Intrinsic high-sensitivity sensors based on etched single-mode polymer optical fibers. *IEEE Photonics Technology Letters, 27*(6), 604–607. https://doi.org/10.1109/LPT.2014.2385875

Bhowmik, K., Peng, G.-D., Luo, Y., Ambikairajah, E., Lovric, V., Walsh, W. R., & Rajan, G. (2015). Experimental study and analysis of hydrostatic pressure sensitivity of polymer fibre Bragg gratings. *Journal of Lightwave Technology, 33*(12), 2456–2462. https://doi.org/10.1109/JLT.2014.2386346

Bhowmik, K., Peng, G.-D., Luo, Y., Ambikairajah, E., Lovric, V., Walsh, W. R., & Rajan, G. (2016a). Etching process related changes and effects on solid-core single-mode polymer optical fiber grating. *IEEE Photonics Journal, 8*(1), 1–9. https://doi.org/10.1109/JPHOT.2016.2524210

Bhowmik, K., Peng, G.-D., Luo, Y., Ambikairajah, E., Lovric, V., Walsh, W. R., & Rajan, G. (2016b). High intrinsic sensitivity etched polymer fiber Bragg grating pair for simultaneous strain and temperature measurements. *IEEE Sensors Journal, 16*(8), 2453–2459. https://doi.org/10.1109/JSEN.2016.2519531

Bilro, L., Oliveira, J. G., Pinto, J. L., & Nogueira, R. N. (2011). A reliable low-cost wireless and wearable gait monitoring system based on a plastic optical fibre sensor. *Measurement Science and Technology, 22*(4), 045801. https://doi.org/10.1088/0957-0233/22/4/045801

Birks, T. A., Knight, J. C., & Russell, P. S. (1997). Endlessly single-mode photonic crystal fiber. *Optics Letters, 22*(13), 961–963. https://doi.org/10.1364/OL.22.000961

Bosc, D., & Toinen, C. (1992). Full polymer single-mode optical fiber. *IEEE Photonics Technology Letters, 4*(7), 749–750. https://doi.org/10.1109/68.145260

Brekner, M.-J., Deckers, H., & Osan, F. (1997). United States patent No. *US 5637400A*. Retrieved from https://patents.google.com/patent/US5637400.

Carroll, K. E., Zhang, C., Webb, D. J., Kalli, K., Argyros, A., & Large, M. C. (2007). Thermal response of Bragg gratings in PMMA microstructured optical fibers. *Optics Express, 15*(14), 8844–8850. https://doi.org/10.1364/OE.15.008844

Chen, X., Zhang, W., Liu, C., Hong, Y., & Webb, D. J. (2015). Enhancing the humidity response time of polymer optical fiber Bragg grating by using laser micromachining. *Optics Express, 23*(20), 25942–25949. https://doi.org/10.1364/OE.23.025942

Cook, K., Leon-Saval, S., Canning, J., Reid, Z., Hossain, M. A., & Peng, G.-D. (2015a). Air-structured optical fibre drawn from a 3D-printed preform. *Optics Letters, 40*(17), 3966–3969. https://doi.org/10.1117/12.2195466

Cook, K., Leon-Saval, S., Canning, J., Reid, Z., Hossain, M. A., & Peng, G.-D. (2015b). Air-structured optical fibre drawn from a 3D-printed preform. *OFS'24, 9634*(17), 96343E. https://doi.org/10.1117/12.2195466

Cui, J., Yang, J. X., Li, Y. G., & Li, Y. S. (2015). Synthesis of high performance cyclic olefin polymers (COPs) with ester group via ring-opening metathesis polymerization. *Polymers, 7*(8), 1389–1409. https://doi.org/10.3390/polym7081389

Cusano, A., Cutolo, A., & Albert, J. (2011). *Fiber Bragg grating sensors: Research advancements, industrial applications and market exploitation.* (A. Cusano, A. Cutolo, & J. Albert, Eds.) (1st ed.). Sharjah: Bentham Science Publishers Ltd. https://doi.org/10.2174/97816080508401110101

Dudley, J. M., Genty, G., & Coen, S. (2006). Supercontinuum generation in photonic crystal fiber. *Reviews of Modern Physics, 78*(4), 1135–1184. https://doi.org/10.1103/RevModPhys.78.1135

Dupont. (1979). *United Kingdom patent No. GB 2006790.* Retrieved from http://patents.google.com/patent/GB2006790B/en20.

Ebendorff-Heidepriem, H., Monro, T. M., van Eijkelenborg, M. A., & Large, M. C. J. (2007). Extruded high-NA microstructured polymer optical fibre. *Optics Communications, 273*(1), 133–137. https://doi.org/10.1016/j.optcom.2007.01.004

Ebendorff-Heidepriem, H., Monro, T., van Eijkelenborg, M. A., & Large, M. J. C. (2006). Extruded polymer preforms for high-NA polymer microstructured fiber. In *Optical Fiber Communication Conference* (p. OTh4). Optical Society of America. https://doi.org/10.1109/OFC.2006.215736

Emiliyanov, G., Jensen, J. B., Bang, O., Hoiby, P. E., Pedersen, L. H., Kjær, E. M., & Lindvold, L. (2007). Localized biosensing with Topas microstructured polymer optical fiber. *Optics Letters, 32*(5), 460–462. https://doi.org/10.1364/OL.32.000460

Emslie, C. (1988). Polymer optical fibres. *Journal of Materials Science, 23*(7), 2281–2293. https://doi.org/10.1007/BF01111879

Fasano, A., Woyessa, G., Stajanca, P., Markos, C., Nielsen, K., Rasmussen, H. K., ... Bang, O. (2016). Fabrication and characterization of polycarbonate microstructured polymer optical fibers for high-temperature-resistant fiber Bragg grating strain sensors. *Optical Materials Express, 6*(2), 649–659. https://doi.org/10.1364/OME.6.000649

Garvey, D. W., Zimmerman, K., Young, P., Tostenrude, J., Townsend, J. S., Zhou, Z., ... Dirk, C. W. (1996). Single-mode nonlinear-optical polymer fibers. *Journal of the Optical Society of America B, 13*(9), 2017. https://doi.org/10.1364/JOSAB.13.002017

Goodfellow Corporation. (2016). *Standard Price List for all Polymers.* Retrieved from www.goodfellowusa.com.

Groh, W. (1988). Overtone absorption in macromolecules for polymer optical fibers. *Macromolecular Chemistry and Physics, 189*(12), 2861–2874. https://doi.org/10.1002/macp.1988.021891213

Guerrero, H., Guinea, G. V., & Zoido, J. (1998). Mechanical properties of polycarbonate optical fibers. *Fiber and Integrated Optics, 17*(3), 231–242. https://doi.org/10.1080/014680398244966

Guerrero, H., Zoido, J., Escudero, J. L., & Bernabeu, E. (1993). Characterization and sensor applications of polycarbonate optical fibers. *Fiber and Integrated Optics, 12*(3), 257-268. https://doi.org/10.1080/01468039308204228

Gupta, A., Liang, R., Tsay, F., & Moacanin, J. (1980). Characterization of a dissociative excited state in the solid state: Photochemistry of poly (methyl methacrylate). Photochemical processes in polymeric systems. 5. *Macromolecules, 13*, 1696–1700. https://doi.org/10.1021/ma60078a060

Harmon Julie, P. (2001). Polymers for optical fibers and waveguides: An overview (ACS Symposium Series). In Harmon J. P. Harmon & G. K. Noren (Eds.), *Optical Polymers* (pp. 1–23). Washington, DC: American Chemical Society. https://doi.org/10.1021/bk-2001-0795.ch001

Hirose, R., Asai, M., Kondo, A., & Koike, Y. (2008). Graded-index plastic optical fiber prepared by the coextrusion process. *Applied Optics, 47*(22), 4177–4185. https://doi.org/10.1364/AO.47.004177

Im, S. H., Suh, D. J., Park, O. O., Cho, H., Choi, J. S., Park, J. K., & Hwang, J. T. (2002). Fabrication of a graded-index polymer optical fiber preform without a cavity by inclusion of an additional monomer under a centrifugal force field. *Applied Optics, 41*(10), 1858–1863. https://doi.org/10.1007/BF02697164

Ioannides, N., Chunga, E. B., Bachmatiuk, A., Gonzalez-Martinez, I. G., Trzebicka, B., Adebimpe, D. B., … Rümmeli, M. H. (2014). Approaches to mitigate polymer-core loss in plastic optical fibers: A review. *Materials Research Express, 1*(3), 032002. https://doi.org/10.1088/2053-1591/1/3/032002

Jiang, C., Kuzyk, M. G., Ding, J.-L., Johns, W. E., & Welker, D. J. (2002). Fabrication and mechanical behavior of dye-doped polymer optical fiber. *Journal of Applied Physics, 92*(1), 4–12. https://doi.org/10.1063/1.1481774

Kaino, T. (1985a). Absorption losses of low loss plastic optical fibers. *Japanese Journal of Applied Physics, 24*(12), 1661–1665. https://doi.org/10.1143/JJAP.24.1661

Kaino, T. (1985b). Influence of water absorption on plastic optical fibers. *Applied Optics, 24*(23), 4192–4195. https://doi.org/10.1364/AO.24.004192

Kaino, T. (1987). Preparation of plastic optical fibers for near-IR region transmission. *Journal of Polymer Science: Part A: Polymer Chemistry, 25*(1), 37–46. https://doi.org/10.1002/pola.1987.080250105

Kaino, T., Fujiki, M., & Jinguji, K. (1984). Preparation of plastic optical fibers. *Review of the Electrical Communication Laboratories, 32*(3), 478–488.

Kaino, T., Fujiki, M., & Nara, S. (1981). Low-loss polystyrene core-optical fibers. *Journal of Applied Physics, 52*(12), 7061–7063. https://doi.org/10.1063/1.328702

Kaino, T., Jinguji, K., & Nara, S. (1983). Low loss poly(methylmethacrylate-d8) core optical fibers. *Applied Physics Letters, 42*(7), 567–569. https://doi.org/10.1063/1.94030

Kaiser, P., Marcatili, E. A. J., & Miller, S. E. (1973). A new optical fiber. *Bell System Technical Journal, 52*(2), 265–269. https://doi.org/10.1002/j.1538-7305.1973.tb01963.x

Kang, L., Wang, L., Yang, X., & Wang, J. (2008). Mass-fabrication of air-core microstructured polymer optical fiber. *2008 Conference on Quantum Electronics and Laser Science Conference on Lasers and Electro-Optics, CLEO/QELS*, 8–9. https://doi.org/10.1109/CLEO.2008.4551724

Kaye, W. (1954). Near-infrared spectroscopy. *Spectrochimica Acta, 6*(4), 257–287. https://doi.org/10.1016/0371-1951(54)80011-7

Khanarian, G., & Celanese, H. (2001). Optical properties of cyclic olefin copolymers. *Optical Engineering, 40*(6), 1024–1029. https://doi.org/10.1117/1.1369411

Kiesel, S., Peters, K., Hassan, T., & Kowalsky, M. (2007). Behaviour of intrinsic polymer optical fibre sensor for large-strain applications. *Measurement Science & Technology, 18*(10), 3144–3154. https://doi.org/10.1088/0957-0233/18/10/s16

Knight, J. C., Birks, T. A., Russell, P. S. J., & Atkin, D. M. (1996). All-silica single-mode optical fiber with photonic crystal cladding. *Optics Letters, 21*(19), 1547–1549. https://doi.org/10.1364/OL.21.001547

Koike, Y. (1992). High bandwidth and low loss polymer optical fiber. In *First international conference on plastic optical fibres and applications* (pp. 15–19). Paris, France.

Koike, Y. (1996). Progress of plastic optical fiber technology. In *22nd European Conference on Optical Communications – ECOC'96* (p. MoB.3.1, 41–46). Oslo, Norway.

Koike, Y., & Gaudino, R. (2013). Plastic optical fibers and Gb/s data links. In I. Kaminow, L. Tingye, & A. E. Willner (Eds.), *Optical Fiber Telecommunications VIB: Components and Subsystems* (6th ed., Vol. VIA). Cambridge, MA: Academic Press.

Koike, Y., & Ishigure, T. (2006). High-bandwidth plastic optical fiber for fiber to the display. *Journal of Lightwave Technology, 24*(12), 4541–4553. https://doi.org/10.1109/JLT.2006.885775

Koike, Y., Ishigure, T., & Nihei, E. (1995). High-bandwidth graded-index polymer optical fiber. *Journal of Lightwave Technology, 13*(7), 1475–1489. https://doi.org/10.1109/50.400716

Koike, Y., Kimoto, Y., & Ohtsuka, Y. (1982). Studies on the Light-Focusing Plastic Rod. 12: The GRIN Fiber Lens of Methyl Methacrylate-Vinyl Phenylacetate Copolymer. *Applied Optics, 21*(6), 1057–1062. https://doi.org/10.1364/AO.21.001057

Koike, Y., & Koike, K. (2011). Progress in low-loss and high-bandwidth plastic optical fibers. *Journal of Polymer Science Part B: Polymer Physics, 49*(1), 2–17. https://doi.org/10.1002/polb.22170

Koike, Y., & Narutomi, M. (1998). United States patent No. *5783636*.

Kuang, K. S. C., Quek, S. T., Koh, C. G., Cantwell, W. J., & Scully, P. J. (2009). Plastic optical fibre sensors for structural health monitoring: A review of recent progress. *Journal of Sensors, 2009*, 1–13. https://doi.org/10.1155/2009/312053

Kuhlmey, B. T., McPhedran, R. C., & de Sterke, C. M. (2002). Modal cutoff in microstructured optical fibers. *Optics Letters, 27*(19), 1684–1686. https://doi.org/10.1364/OL.27.001684

Kuriki, K., Kobayashi, T., Imai, N., Tamura, T., Koike, Y., & Okamoto, Y. (2000). Organic dye-doped polymer optical fiber lasers. *Polymers for Advanced Technologies, 11*(8–12), 612–616. https://doi.org/10.1002/1099-1581(200008/12)11:8/12<612::AID-PAT11>3.0.CO;2-T

Kuzyk, M. G., Paek, U. C., & Dirk, C. W. (1991). Guest-host polymer fibers for nonlinear optics. *Applied Physics Letters, 59*(8), 902–904. https://doi.org/10.1063/1.105271

Lacraz, A., Polis, M., Theodosiou, A., Koutsides, C., & Kalli, K. (2015). Bragg grating inscription in CYTOP polymer optical fibre using a femtosecond laser. In K. Kalli (Ed.), *Micro-structured and specialty optical fibres IV* (Vol. 9507, p. 95070K). https://doi.org/10.1117/12.2185161

Large, M. C. J., Poladian, L., Barton, G. W., & van Eijkelenborg, M. A. (2008). *Microstructured polymer optical fibres:Microstructured polymer optical fibres*. Boston, MA: Springer US. https://doi.org/10.1007/978-0-387-68617-2

Large, M. C. J., Ponrathnam, S., Argyros, A., Pujari, N. S., & Cox, F. (2004). Solution doping of microstructured polymer optical fibres. *Optics Express, 12*(9), 1966–1971. https://doi.org/10.1364/OPEX.12.001966

Leon-Saval, S. G., Lwin, R., & Argyros, A. (2012). Multicore composite single-mode polymer fiber. *Optics Express, 20*(1), 141–148. https://doi.org/10.1364/OE.20.000141

Lethien, C., Loyez, C., Vilcot, J. P., Rolland, N., & Rolland, P. A. (2011). Exploit the bandwidth capacities of the perfluorinated graded index polymer optical fiber for multi-services distribution. *Polymers, 3*(4), 1006–1028. https://doi.org/10.3390/polym3031006

Liehr, S., Lenke, P., & Wendt, M. (2009). Polymer optical fiber sensors for distributed strain measurement and application in structural health monitoring. *IEEE Sensors Journal, 9*(11), 1330–1338. https://doi.org/10.1109/JSEN.2009.2018352

Makino, K., Akimoto, Y., Koike, K., Kondo, A., Inoue, A., & Koike, Y. (2013). Low loss and high bandwidth polystyrene-based graded index polymer optical fiber. *Journal of Lightwave Technology, 31*(14), 2407–2412. https://doi.org/10.1109/JLT.2013.2266671

Marcatili, E. (1973). United States patent No. *3712705*. Retrieved from https://patents.google.com/patent/US3712705.

Marcuse, D. (1974). *Theory of dielectric waveguides* (1st ed.). Academic Press.

Marcuse, D. (1978). Gaussian approximation of the fundamental modes of graded-index fibers. *Journal of the Optical Society of America, 68*(2), 103–109. https://doi.org/10.1364/JOSA.68.000103

Markos, C., Stefani, A., Nielsen, K., Rasmussen, H. K., Yuan, W., & Bang, O. (2013). High-Tg TOPAS microstructured polymer optical fiber for fiber Bragg grating strain sensing at 110 degrees. *Optics Express, 21*(4), 4758–4765. https://doi.org/10.1364/OE.21.004758

Markos, C., Yuan, W., Vlachos, K., Town, G. E., & Bang, O. (2011). Label-free biosensing with high sensitivity in dual-core microstructured polymer optical fibers. *Optics Express, 19*(8), 7790–7798. https://doi.org/10.1364/OE.19.007790

Minami, K. (2010). *Handbook of plastic optics*. (S. Bäumer, Ed.) (2nd ed.). Weinheim, Germany: Wiley-VCH Verlag GmbH & Co. KGaA. https://doi.org/10.1002/9783527635443

Minami, S. (1994). The development and applications of POF: Review and forecast. In *POF'94* (pp. 27–31). Yokowama, Japan.

Mitsubishi Rayon Co. Ltd. (2015). General Properties of Acrypet™. Retrieved from http://www.acrypet.com

Mortensen, N. A. (2002). Effective area of photonic crystal fibers. *Optics Express, 10*(7), 341–348. https://doi.org/10.1364/OE.10.000341

Mortensen, N. A. (2005). Semianalytical approach to short-wavelength dispersion and modal properties of photonic crystal fibers. *Optics Letters, 30*(12), 1455–1457. https://doi.org/10.1364/OL.30.001455

Mortensen, N. A., Folkenberg, J. R., Nielsen, M. D., & Hansen, K. P. (2003). Modal cutoff and the V parameter in photonic crystal fibers. *Optics Letters, 28*(20), 1879–1881. https://doi.org/10.1364/OL.28.001879

Nagasawa, M., Kondo, S., Ishigure, T., & Koike, Y. (2005). Perfluorinated polymer photonic crystal fiber. Paper presented at the *54th SPSJ Symposium on Macromolecules*, Yamagata, Japan, 20-22 September (Vol. 54, p. 4047). Japan: Polymer Preprints.

Nakamura, K., Mizuno, Y., & Hayashi, N. (2013). Improved technique for etching overcladding layer of perfluorinated polymer optical fibre by chloroform and water. *Electronics Letters, 49*(25), 1630–1632. https://doi.org/10.1049/el.2013.3267

Naritomi, M. (1996). "CYTOP®" Amorphous Fluoropolymers for Low Loss POF. In *POF Asia-Pacific Forum'96* (pp. 23–27). Boston, MA: IGI Consulting Inc.

Nielsen, K., Rasmussen, H. K., Adam, A. J., Planken, P. C., Bang, O., & Jepsen, P. U. (2009). Bendable, low-loss Topas fibers for the terahertz frequency range. *Optics Express, 17*(10), 8592–8601. https://doi.org/10.1364/OE.17.008592

Obuchi, K., Komatsu, M., & Minami, K. (2015). High performance optical materials cyclo olefin polymer zeonex. In *Proceedings of SPIE* (Vol. 6671, p. 66711I–1–9). https://doi.org/10.1117/12.749910.

Oikawa, S., Fujiki, M., & Katayama, Y. (1979). Plastic optical fibre with improved transmittance. *Electronics Letters, 15*(25), 829–830. https://doi.org/10.1049/el:19790589

Ortigosa-Blanch, A., Knight, J. C., Wadsworth, W. J., Arriaga, J., Mangan, B. J., Birks, T. A., & Russell, P. S. J. (2000). Highly birefringent photonic crystal fibers. *Optics Letters, 25*(18), 1325–1327. https://doi.org/10.1364/OL.25.001325

Paradigm. (2016). *The polymer division of Incom*, Inc. Retrieved from http://www.paradigmoptics.com/

Peng, G. D., & Chu, P. L. (2000). Polymer optical fiber photosensitivities and highly tunable fiber gratings. *Fiber and Integrated Optics, 19*(4), 277–293. https://doi.org/10.1080/014680300300001662

Polyplastics Co. Ltd. (2011). TOPAS®: Thermoplastic olefin polymer of amorphous structure (COC). Retrieved from htttp://www.polyplastics.com

Shao, Y., Cao, R., Huang, Y.-K., Ji, P. N., & Zhang, S. (2012). 112-Gb/s Transmission over 100 m of graded-index plastic optical fiber for optical data center applications. In *Optical Fiber Communication Conference* (p. OW3J.5). Los Angeles, CA: IEEE. https://doi.org/10.1364/OFC.2012.OW3J.5

Shin, B.-G., & Park, J.-H. (2004). Plastic photonic crystal fibers drawn from stacked capillaries. *Journal of Nonlinear Optical Physics & Materials, 13*(3 & 4), 519–523.

Snyder, A. W., & Love, J. D. (1984). *Optical waveguide theory* (2nd ed.). Boston, MA: Springer US. https://doi.org/10.1007/978-1-4613-2813-1

Spigulis, J. (2005). Side-emitting fibers brighten our world. *Optics & Photonics News*. Optical Society of America. https://doi.org/10.1364/OPN.16.10.000034

Sultanova, N., Kasarova, S., & Nikolov, I. (2009). Dispersion properties of optical polymers. In *PHOTONICA09* (Vol. 116, pp. 585–587). https://doi.org/10.12693/APhysPolA.116.585

Takezawa, Y., Tanno, S., Taketani, N., Ohara, S., & Asano, H. (1991). Analysis of thermal degradation for plastic optical fibers. *Journal of Applied Polymer Science, 42*, 2811–2817. https://doi.org/10.1002/app.1991.070421019

Tanaka, A., Sawada, H., Takoshima, T., & Wakatsuki, N. (1988). New plastic optical fiber using polycarbonate core and fluorescence-doped fiber for high temperature use. *Fiber and Integrated Optics, 7*(2), 139–158. https://doi.org/10.1080/01468038808222981

Tanio, N., & Koike, Y. (2000). What is the most transparent polymer? *Polymer Journal, 32*, 43–50. https://doi.org/10.1295/polymj.32.43

Teijin Limited. (2016). Panlite® AD-5503 – Polycarbonate. Retrieved from https://www.teijin.com

Topas Advanced Polymers GmbH. (2013). Cycloolefin Copolymer (COC). https://doi.org/10.1021/bi100733q

Urbach, F. (1953). The long-wavelength edge of photographic sensitivity and of the electronic absorption of solids. *Physical Review, 92*(5), 1324–1324. https://doi.org/10.1103/PhysRev.92.1324

van Eijkelenborg, M. A., Argyros, A., Bachmann, A., Barton, G., Large, M. C. J., Henry, G., … Zagari, J. (2004). Bandwidth and loss measurements of graded-index microstructured polymer optical fibre. *Electronics Letters, 40*(10), 592. https://doi.org/10.1049/el:20040371

van Eijkelenborg, M. A., Argyros, A., Barton, G., Bassett, I. M., Fellew, M., Henry, G., … Zagari, J. (2003). Recent progress in microstructured polymer optical fibre fabrication and characterisation. *Optical Fiber Technology, 9*(4), 199–209. https://doi.org/10.1016/S1068-5200(03)00045-2

van Eijkelenborg, M. A., Argyros, A., & Leon-saval, S. G. (2008). Polycarbonate hollow-core microstructured optical fiber. *Optics Letters, 33*(21), 2446–2448. https://doi.org/10.1364/OL.33.002446

van Eijkelenborg, M. A., Large, M., Argyros, A., Zagari, J., Manos, S., Issa, N., … Nicorovici, N. A. (2001). Microstructured polymer optical fibre. *Optics Express, 9*(7), 319–327. https://doi.org/10.1364/OE.9.000319

Villegas, A. G., Ocampo, M. A., Luna-Bárcenas, G., & Saldívar-Guerra, E. (2009). Obtainment of graded index preforms by combined frontal co-polymerization of MMA and BzMA. *Macromolecular Symposia, 283–284*(1), 336–341. https://doi.org/10.1002/masy.200950939

Watanabe, T., Ooba, N., Hida, Y., & Hikita, M. (1998). Influence of humidity on refractive index of polymers for optical waveguide and its temperature dependence. *Applied Physics Letters, 72*(13), 1533. https://doi.org/10.1063/1.120574

Webb, D., Aressy, M., Argyros, A., Barton, J. S., Dobb, H., van Eijkelenborg, M. A., … Silva-López, M. (2005). Grating and interferometric devices in POF. In *The 14th international conference on polymer optical fibers*, Hong Kong, China, 19-22 September (pp. 325–328).

Woyessa, G., Fasano, A., Markos, C., Stefani, A., Rasmussen, H. K., & Bang, O. (2017). Zeonex microstructured polymer optical fiber: Fabrication friendly fibers for high temperature and humidity insensitive Bragg grating sensing. *Optical Materials Express*, *7*(1), 286–295. https://doi.org/10.1364/OME.7.000286

Woyessa, G., Fasano, A., Stefani, A., Markos, C., Nielsen, K., Rasmussen, H. K., & Bang, O. (2015). Humidity insensitive step-index polymer optical fibre Bragg grating sensors. In *OFS'24* (Vol. 9634, p. 96342L). https://doi.org/10.1117/12.2194963

Woyessa, G., Fasano, A., Stefani, A., Markos, C., Rasmussen, H. K., & Bang, O. (2016). Single mode step-index polymer optical fiber for humidity insensitive high temperature fiber Bragg grating sensors. *Optics Express*, *24*(2), 1253–1260. https://doi.org/10.1364/OE.24.001253

Yamamoto, K., & Ogawa, G. (2005). Structure determination of the amorphous perfluorinated homopolymer: Poly[perfluoro(4-vinyloxyl-1-butene)]. *Journal of Fluorine Chemistry*, *126*(9–10), 1403–1408. https://doi.org/10.1016/j.jfluchem.2005.07.015

Yamashita, T., & Kamada, K. (1993). Intrinsic transmission loss of polycarbonate core optical fiber. *Japanese Journal of Applied Physics*, *32*(6A), 2681–2686. https://doi.org/10.1143/JJAP.32.2681

Yang, D. X., Yu, J., Tao, X., & Tam, H. (2004). Structural and mechanical properties of polymeric optical fiber. *Materials Science and Engineering A*, *364*(1–2), 256–259. https://doi.org/10.1016/j.msea.2003.08.025

Ye, C. C., Dulieu-Barton, J. M., Webb, D. J., Zhang, C., Peng, G.-D., Chambers, A. R., … Eastop, D. D. (2009). Applications of polymer optical fibre grating sensors to condition monitoring of textiles. In J. D. C. Jones (Ed.), *20th international conference on optical fibre sensors* (Vol. 178, p. 75030M). Bellingham: SPIE. https://doi.org/10.1117/12.833472

Yuan, W., Khan, L., Webb, D. J., Kalli, K., Rasmussen, H. K., Stefani, A., & Bang, O. (2011). Humidity insensitive TOPAS polymer fiber Bragg grating sensor. *Optics Express*, *19*(20), 19731–19739. https://doi.org/10.1364/OE.19.019731

Yuan, W., Stefani, A., Bache, M., Jacobsen, T., Rose, B., Herholdt-Rasmussen, N., … Bang, O. (2011). Improved thermal and strain performance of annealed polymer optical fiber Bragg gratings. *Optics Communications*, *284*(1), 176–182. https://doi.org/10.1016/j.optcom.2010.08.069

Zagari, J., Argyros, A., Issa, N. A., Henry, G., Large, M. C. J., Poladian, L., & van Eijkelenborg, M. A. (2004). Small-core single-mode microstructured polymer optical fiber with large external diameter. *Optics Letters*, *29*(8), 818–820. https://doi.org/10.1364/OL.29.000818

Zeon Chemicals. (2016). ZEONEX® Cyclo Olefin Polymer (COP). Retrieved from http://www.zeonex.com

Zhang, W., & Webb, D. J. (2014). Humidity responsivity of poly(methyl methacrylate)-based optical fiber Bragg grating sensors. *Optics Letters*, *39*(10), 3026–3029. https://doi.org/10.1364/OL.39.003026

Zhang, Y., Li, K., Wang, L., Ren, L., Zhao, W., Miao, R., … van Eijkelenborg, M. A. (2006). Casting preforms for microstructured polymer optical fibre fabrication. *Optics Express*, *14*(12), 5541–5547. https://doi.org/10.1364/OE.14.005541

Zhou, G., Pun, C. F. J., Tam, H. Y., Wong, A. C. L., Lu, C., & Wai, P. K. A. (2010). Single-mode perfluorinated polymer optical fibers with refractive index of 1.34 for biomedical applications. *IEEE Photonics Technology Letters*, *22*(2), 106–108. https://doi.org/10.1109/LPT.2009.2036377

Ziemann, O., Krauser, J., Zamzow, P. E., & Daum, W. (2008). *Optical short range transmission systems:POF handbook*. Berlin, Germany: Springer Berlin Heidelberg. https://doi.org/10.1007/978-3-540-76629-2

Zubia, J., & Arrue, J. (2001). Plastic optical fibers: An introduction to their technological processes and applications. *Optical Fiber Technology*, *7*(2), 101–140. https://doi.org/10.1006/ofte.2000.0355

Optical Fiber Sensing Principles

Daniel André Pires Duarte, Rogério N. Nogueira, and Lúcia Bilro

Contents

3.1 Introduction

Sensors based on fiber optic present several different advantages in all-in-one "packaging" that other sensing technologies cannot offer, which include immunity to electromagnetic interference, compactness, lightweight, multiplexing capabilities, higher sensitivity, etc. Much of the research on this field was established with the initially developed and wide spread silica optical fibers due to their usage in communications. However, polymer optical fibers (POFs) gradually have become an interesting alternative as a sensing device in the last decade. The main reason is their superior physical properties including higher elastic strain limits, higher fracture toughness, higher flexibility to bending, higher sensitivity to strain, and higher thermo-optic coefficients. Another reason is their compatibility with organic materials, giving them great potential for biomedical applications by not producing sharp edges when broken. Depending on the polymer used for the manufacturing of the fiber, different properties can also be explored such as the ability to have humidity sensitivity or resistance to higher temperatures.

Several methodologies can be used as fiber sensing that exploit the properties of light and its interaction with the surrounding medium. Variations of intensity, evanescent field absorbance, wavelength shifts, interferometry pattern changes, plasmon resonance, and polarimetry interaction are some of the physical phenomena that are possible with POF as a sensor and that were researched and developed by the scientific community working in this field. The advancements led to the possibility to measure, directly and indirectly, a wide range of fundamental properties such as refractive index, temperature, humidity, strain, pressure, bending, color, turbidity, etc. To accomplish this, several arrangements with or without morphology and chemical modification to the POF were tested as transducers in a variety of setups. These are the cases of bending, polishing, etching, drilling, tapering, photopolymerization, thin film deposition, etc.

In this chapter, a broad view of some of the published and most used techniques and schemes for sensing with POF will be presented. It is important to note that we only present a small part of the sensing devices using POF since the same techniques can, in many cases, be used to measure several different parameters with slight modifications. A deep physical and mathematical description of each device will also not be presented here, but can be further explored in the following chapters of this book.

3.2 Intensity-Based Fiber Optic Sensors

In the quest for low-cost ways of parameter measurement using POF, different approaches can be found in academia. The most common method of detecting is based on intensity variation which requires simple operation principles and instrumentation. In general, only a light source, a fiber, and a photodetector is needed, which can be low-cost elements like LEDs and photodiodes. This enables the manufacture of easy implementations devices with high portability. Optical fiber based sensors can be divided in two broad sensing categories which are characterized by light leaving the fiber to an external medium (extrinsic sensing) or by staying within the fiber (intrinsic sensing) with most of the interaction being done with the evanescent field with the external medium [1] (**Figure 3.1**).

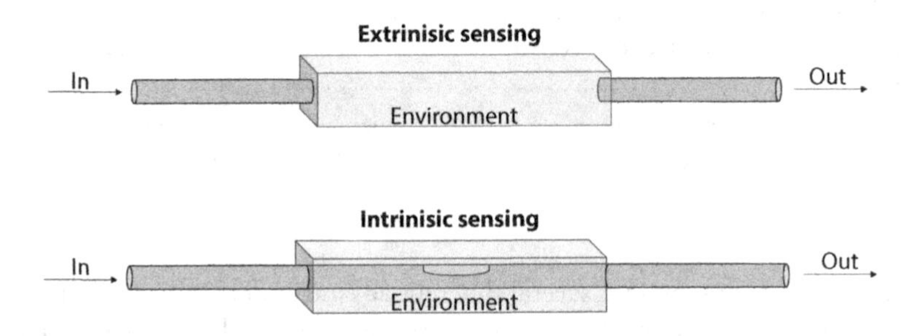

FIGURE 3.1 Extrinsic and intrinsic sensing scheme where in the former, light needs to leave the fiber to interact with the environment before it is captured again and in the latter, the light interacts with the environment without leaving the fiber.

3.2.1 Extrinsic Sensing

One of the extrinsic sensing techniques employed and commonly used is the light transmission configuration. Here the intensity variation is related to the optical signal coupling between two polymer optical fibers. This technique was used by Bilro et al. [2] for turbidity measurement in liquids and is a technique usually named Air-Gap design. It takes advantage of the light opening cone generated by the emission fiber that will depend on the medium to measure. Refractive index, color, and suspended particles are some parameters that can be detected using this configuration (**Figure 3.2a**). Another type of design observed is the displacement based sensors that use the reflection of light as the main phenomena (**Figure 3.2b**). This design proposed by Binu et al. [3] and Govindan et al. [4] consists of a two-fiber element embedded in the environment to measure the refractive index of a liquid. One of the fibers is used to emit light that will reflect in a reflective surface placed at some distance to the fiber tip. Then the second fiber, which can be at a certain distance to the first fiber, will receive the reflected light, the intensity of which will depend on the refractive index of the liquid medium. A 13.5% light intensity variation was obtained between a range of 1.3322 and 1.3627 refractive index units (RIU). A quasi-intrinsic sensing technique was also reported by Shin and Park [5] with its cascaded in-line holes POF for liquid refractive index measurement. The holes are created by micro drilling. While the sensing mechanism is done in-line, light propagating in the fiber will technically leave the fiber to interact with the solution flowing in the holes. Because the light leaves the fiber, it is an extrinsic sensing technique. Shin and Park obtained a sensitivity of 62.9 dB/RIU between the 1.33 and 1.42 ranges in a 3-hole structure.

3.2.2 Intrinsic Sensing

In the intrinsic sensing techniques, light does not physically leave the fiber but tries to promote the interaction of its evanescent field with the external medium. One of these techniques driving the evanescent field interaction uses a macrobending fiber system that can, or not, be assisted by side polishing of the fiber. Early experimental testing of this concept to measure refractive index can be found in the work of Zubia et al. [6] where a step-index POF was laterally polished with a slight curvature until the core were exposed (**Figure 3.3a**). When submersed in a solution, a resolution of 5×10^{-3} refractive index units was reported for indices of refraction from 1.30 to 1.59. This concept was further studied by Ana Cao-Paz et al. [7,8] with a high variation in the curvature of the fiber with radius from 0.1 to 1 cm, but without polishing the fiber (**Figure 3.3b**). Their work was implemented in the measurement of wine fermentation and electrolyte density in lead-acid batteries. Different geometries of side polishing were also used to try to improve

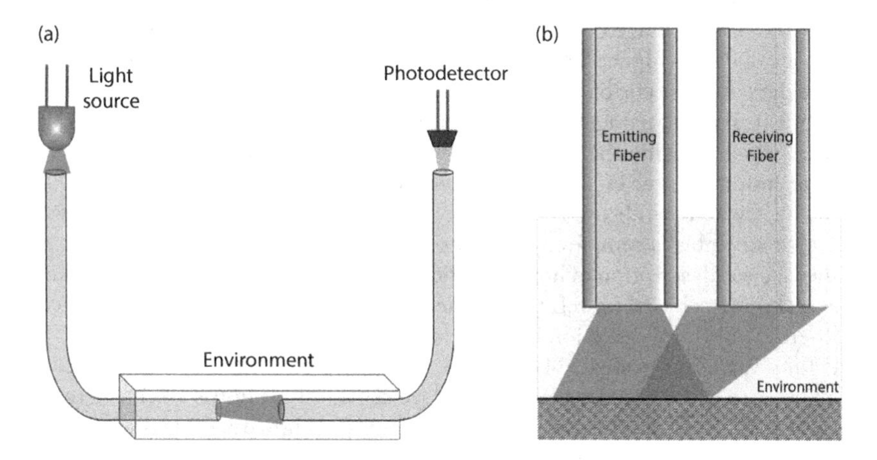

FIGURE 3.2 Schematic setup of the fiber-based sensors with (a) air-gap design and (b) displacement fiber.

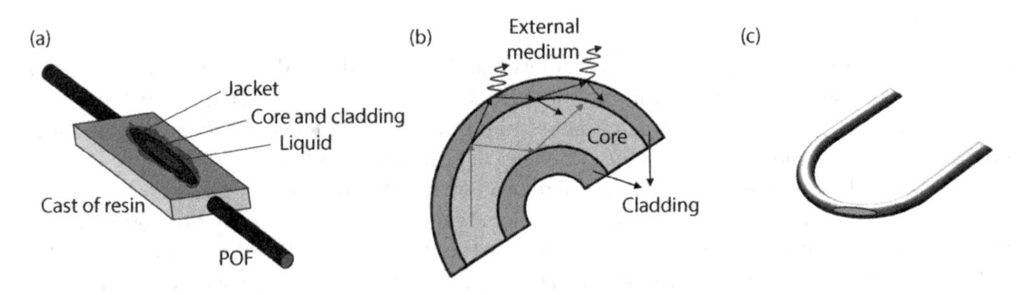

FIGURE 3.3 (a) Setup of the macrobending polished fiber intrinsic sensing in a cast of resin. (b) Bending fiber-based sensor without the need of cladding removal or etching. (c) D-shape fiber-based structure for sensing that can be designed with different groove depth and radii.

results. Bilro et al. [9] has developed an analytical model, verified by experimental evaluation—the concave shape side polishing in stretched fiber, similarly obtained by Zubia et al. The same side polishing configuration was used by Jing et al. [10] for refractive index measurement but maintaining the curvature of the fiber. Sensitivity of 154 dB/RIU power loss was obtained in a range from 1.33 to 1.44 RIU. Another technique was reported by De-Jun et al. [11] for its study using a D-shape fiber format with different groove depth and fiber radius with its optimized values in relation to sensitivity (**Figure 3.3c**). Sequeira et al. [12] also studied different lengths of the sensitive D-shaped fiber region obtained by mechanical polishing, obtaining a resolution of 6.48×10^{-3} RIU for the higher length of 6 cm. They also concluded, in a parallel work, that the sensitivity of the D-shape morphology was also strongly dependent on the roughness of the sensing surface which could be adjusted by polishing the surface with sandpaper having diverse grit sizes. Higher sizes produce higher losses for lower refractive index, which produces higher sensitivities when the measured refractive index approaches the refractive index of the fiber cladding [13]. A similar concept and results were developed and obtained by Teng et al. [14] with its multi-notched structure imprinted like a LPG with die-press-print mechanism. To increase sensitivity to refractive index measurements, the structure was used in a U-shaped configuration. Sensitivities of 1130%/RIU were obtained in changes of transmittance (resolution of 8.44×10^{-4}) in a range of 1.33–1.41.

Another fiber structural modification, also commonly used as intrinsic sensing, is based on tapering the fiber to promote the interaction of the evanescent light wave with the external medium. This technique can be used on glass optical fiber from the early years of fiber sensing [15], but with the disadvantage of being quite fragile. To overcome this issue, polymer optical fiber can be used due to its higher flexibility, but it has low ductility, which can be a difficulty factor to create tapers on this type of fibers with the standard methods used in glass fibers. A theoretical, numerical, and experimental analysis of optical fiber tapering in POF was reported by Xue et al. [16], which became the foundation for other authors to develop further research in this sensing technique. Arrue et al. [17] reported the analysis of tapered graded-index polymer optical fibers for refractive index sensing and proved this concept for different narrowing ratios highlighting the advantages compared to other sensing mechanisms and materials as glass fiber (**Figure 3.4a**). A clear description of how to fabricate tapered POF was reported by Gravina et al. [18] and several sensors based on refractive index change were created with this procedure. As examples, there is a salinity detection based sensor [19], with a resolution of 2×10^{-3} refractive index units for a variation from 1.334 to 1.3576; a biosensor testing [20], using taper fiber and bending the taper in a U-shaped format to improve sensitivity (**Figure 3.4c**), where resolution in the magnitude 10^{-3} was obtained; a double tapper sensor (**Figure 3.4b**) [21] with sensitivity of 950 μW/RIU up to 1.42 RIU; and a cladded removed taper in a U-shaped format (**Figure 3.4d**) [22] that has a range measurement to 1.45 RIU and a resolution of 5×10^{-4} with a waist diameter of 250 μm. An unconventional way of tapering a fiber was also developed by Shimada et al. [23]

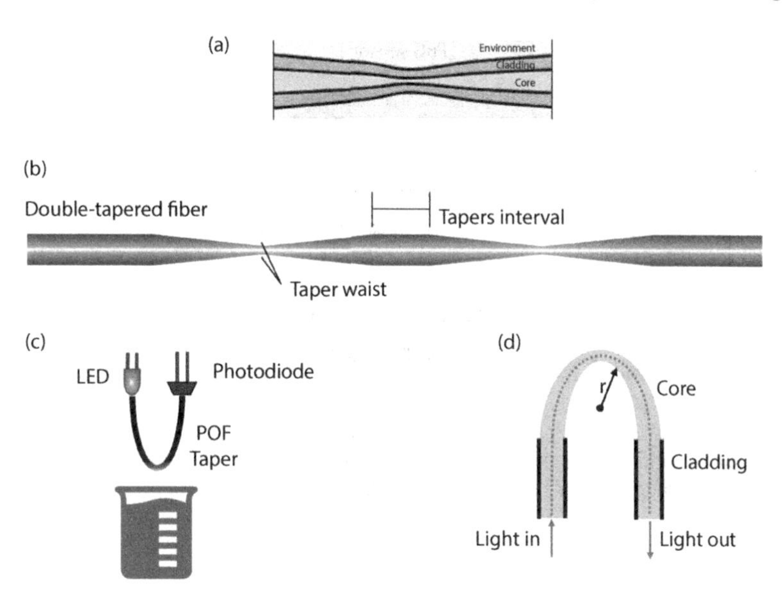

FIGURE 3.4 (a) Straight polymer optical fiber taper used for RI sensing with its cladding present. (b) Double taper fiber approach for increased sensitivity. (c) U-shaped bended POF fiber with a taper for improved sensitivity. (d) Cladding removed POF in a U-shaped format with taper.

by ultrasonically crushing it in a cost-effective method pressing a horn connected to an ultrasonic transducer. When submerged in a solution for refractive index measurements up to 1.36 RIU, a sensitivity of −62 dB/RIU was obtained. By fixing the refractive index and changing the temperature, a sensitivity of −0.094 dB/°C was also observed.

3.3 Wavelength Based Fiber Optic Sensors

Intrinsic sensors can be produced not only by changing the structure of the fiber but also by altering the refractive index of the fiber waveguide in the sensing area by UV or infrared-light exposure. This is usually attained using phase mask or interferometric patterning methods for small grating pitch and using direct point-by-point writing. Practical challenges related to stability of the writing system are found for higher wavelengths due to the required long inscription times in the order of tens of minutes, but it was already demonstrated for the first time by Oliveira et al. [24] that this required time can be lowered to less than 30 seconds with the use of a 248 nm UV light source. Fiber Bragg grating (FBG) and long period grating (LPG) are technologies that use this principle by creating a periodic refractive index change in the core of the fiber. External perturbations will promote changes in the interaction of light with these structures that can be detected with high resolution spectrum analyzers, fundamental equipment for wavelength-based fiber optic sensors.

3.3.1 Fiber Bragg Gratings

The principle of operation of the FBG relies on the dependence of the Bragg resonance with the effective refractive index (η_{eff}) and the grating pitch (Λ) of the modified area. This will work as a dielectric-mirror for some very specific wavelengths, mainly the Bragg wavelength (λ_{Bragg}) where the Bragg condition needs to be satisfied:

$$\lambda_{Bragg} = 2\eta_{eff}\Lambda \tag{3.1}$$

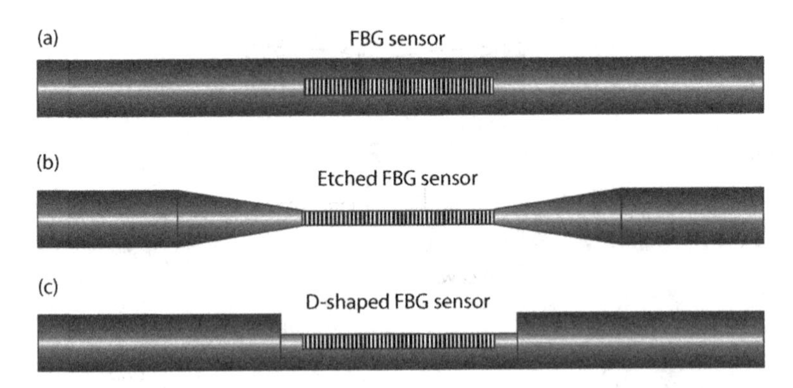

FIGURE 3.5 (a) Schematic design of a fiber Bragg grating in a POF for strain measurement sensing. (b) POF Bragg grating in an etched section of the fiber for refractive index sensing. (c) POF D-shaped fiber with Bragg grating inscribed for bend sensing.

The dependence of the fiber grating with temperature and humidity by changing its effective refractive index, and strain by changing its grating pitch, makes this a very common technique studied and applied as optic sensing elements, usually using refractometry wavelength analysis. The simpler sensor fiber structure only uses an inscribed FBG without any other fiber modifications (**Figure 3.5a**). Yuan et al. [25] reported a polymer optical fiber Bragg grating (POFBG) based sensor for strain measurement, with fiber annealing for increased sensitivity, obtaining a strain sensitivity of about 1.3 pm/με. One of the common structural changes applied to the fiber is cladding removal mainly by chemical etching (**Figure 3.5b**). This increases sensitivity and allows the interaction with the external medium. Thus, the measurement of refractive index changes is possible due to its now effective refractive index dependence producing shifts in the Bragg wavelength. Sensor applications of POFBG with etching can be found by Chen et al. [26] for strain, bend, and temperature sensing and by Oliveira et al. [27] for strain, temperature, and refractive index. Chen obtained Bragg shifting sensitivities of 1.13 pm/με for strain, –50.1 pm/°C for temperature, and 63.3 pm/m^{-1} for bending. Oliveira obtained sensitivities of 1.51 pm/με for strain, –64.6 pm/°C for temperature, and –4.49 nm/RIU. Other structural change found in literature uses a D-shaped section of a POF (**Figure 3.5c**), created by laser ablation, and an FBG was inscribed. Hu et al. [28] used this methodology for the development of a bend sensor achieving a sensitivity of –28.2 pm/m^{-1}.

3.3.2 Long Period Gratings

Long period gratings (LPG) are structures similar to the FBG but presenting a grating pitch period much larger than the wavelength used for the device. This structure will allow the coupling of the core propagation mode to the co-propagating cladding modes in specific wavelengths, which overall can be explained by the following equation:

$$m\lambda = \Lambda_{LPG}(n_{co} - n_{cl}) \tag{3.2}$$

with m being the order of the coupling, λ the wavelength, Λ_{LPG} the grating period, and n_{co} and n_{cl} the core and cladding effective mode indices, respectively. Few sensing applications can be found using LPG in POF. Direct strain and temperature measurements were performed by Min et al. [29] with its LPG also inscribed in a Polymethyl methacrylate (PMMA) microstructured POF (mPOF) with doped core for photosensitivity enhancement to the 248 nm UV laser (**Figure 3.6a**). The point-by-point technique was employed to inscribe with a pitch of 1 mm a LPG comprising a total of 25 mm. With strain applied, the pitch distance will change and

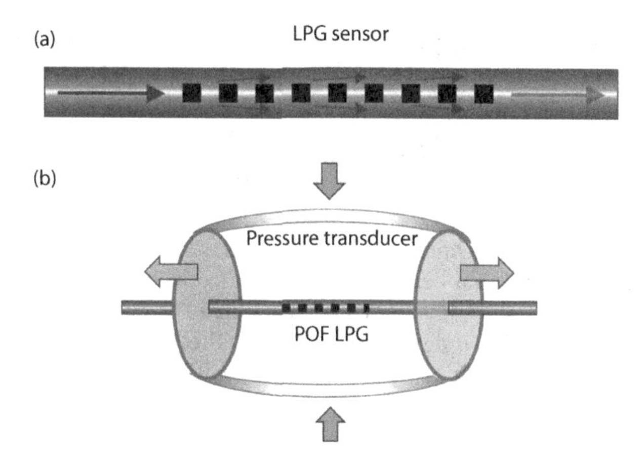

FIGURE 3.6 (a) LPG schematic structure of the core propagation mode coupling to the cladding modes. When strain is applied to the fiber, the wavelength of the coupling modes changes. (b) Application of a pressure transducer to a POF LPG. When under pressure, the transducer will apply a strain to the POF that can be measured.

consequentially the wavelength of the coupling modes will also suffer a variation. As with the FBG, with changes of temperature, the effective refractive index will also be affected and therefore a different wavelength will be coupled. Strain and temperature sensitivity of –2.3 nm/mε and 0.276 nm/°C, respectively, were obtained. Bundalo et al. [30] developed an LPG-based pressure sensor in a single mode mPOF using a transducer to convert strain to pressure (**Figure 3.6b**). The inscription was made using a CO_2 laser in point-by-point mode with a pitch of 1 mm. A sensitivity of 10.5 pm/mbar was obtained.

3.3.3 Tilted Fiber Bragg Gratings

A different spectral signature can be obtained if the grating is created with some tilt in relation to the axis of the fiber. Because of this tilt, the light that is reflected in the grating structure can access the cladding layer, creating new modes of propagation and the possibility of interaction with the external medium (**Figure 3.7**). The wavelength of the core (λ_{TFBG}) will now be dependent of the tilt angle (θ) of the grating as:

$$\lambda_{TFBG} = 2n_{co,eff}\frac{\Lambda}{\cos(\theta)} \tag{3.3}$$

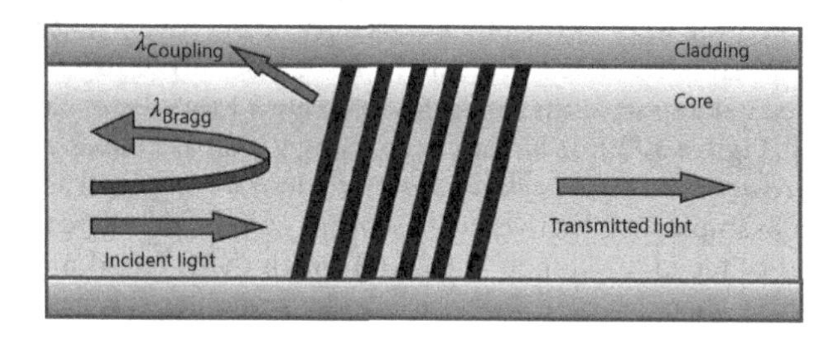

FIGURE 3.7 Schematic operation of the tilted fiber Bragg grating for RI measuring. The light coupled in the cladding will interact with the external medium.

The i cladding mode resonances $\left(\lambda_{cl}^{i}\right)$ are determined by phase matching conditions that have dependency on the effective i fiber core $\left(n_{co,eff}^{i}\right)$ and cladding $\left(n_{cl,eff}^{i}\right)$ refractive indices:

$$\lambda_{cl}^{i} = \left(n_{co,eff}^{i} + n_{cl,eff}^{i}\right)\frac{\Lambda}{\cos(\theta)} \tag{3.4}$$

Because of the insensibility of the core mode to the external medium, the presence of the Bragg wavelength provides the additional benefit of yielding a convenient temperature reference in refractometric sensing. Hu et al. [31] used this principle in a POF with tilts of 1.5, 3.0, and 4.5°, testing this device for refractive index measurements. The change of the external medium showed a sensor sensitivity of about 13 nm/RIU (7×10^{-5} RIU resolution) in the range of 1.42 to 1.49 RIU.

3.4 Interferometric Fiber Optic Sensors

Most common interferometric fiber optic sensors use the interference principle between two beams that have propagated through different optical paths. This can be accomplished by using a single fiber or two different fibers, requiring beam splitting and beam combining components in any configuration. The sensing mechanism will only work if at least one of the beams interact with a perturbation of the external medium. The different paths will create different interferometric spectral signatures that can be analyzed by changes in the wavelength, phase, intensity, frequency, etc. In POF, however, interferometric setups present some disadvantages in comparison to the ones developed with silica fibers, mainly due to their bulk and difficulty of incorporation into detection systems. Four types of traditional fiber optic interferometers can be found in academia as sensing mechanisms: Fabry-Perot, Mach-Zehnder, Michelson, and Sagnac. Yet, with POF, only the Fabry-Perot and Mach-Zehnder configurations have published results. In recent years, a new trend of interferometric fiber-based sensors have been developed using the fiber modal interferometry, commonly known as multimode interference. This device comprises a single-mode-multimode-single-mode fiber structure where the POF plays its sensing role as the multimode section.

3.4.1 Fabry-Perot Interferometer

A Fabry-Perot interferometer (FPI) is composed by two parallel reflecting surfaces in the same light path. The reflecting surfaces have some distance from each other forming the interferometric cavity due to the multiple superpositions of both reflected and transmitted beams at the surfaces. Grating elements on fiber are perfect to be used for interferometric sensing schemes like a fiber Fabry-Perot cavity interferometer, since it works as a reflective surface. This type of interferometer was reported using POF by Ferreira et al. [32] with an FBG in the first side of the cavity and the end tip face of a the fiber as the other extremity of the cavity (**Figure 3.8a**). This system, tested to measure refractive index, presented a resolution of 1×10^{-3} RIU up to 1.47 RIU. Webb et al. and Dobb et al. reported the first steps to create a Fabry-Perot cavity using a double FBGs in POF [33,34] (**Figure 3.8b**), as already seen using glass based fibers. This type of FPI is expected to have narrower resonance peaks and is more desirable for high accuracy wavelength measurement. As an example, a refractive index sensor could be developed using this scheme if the cavity is etched or taped to promote interaction with the external medium. This higher resolution can be proven when this type of device is under strain measurements where measurement noise was reduced by 4 times in comparison to the usage of a single FBG which increases its resolution by the same value [35]. A more recent work published by Theodosiou et al. [36] shows a strain sensitivity in a double FBG FPI of 1.35 pm/$\mu\varepsilon$ and temperature sensitivity of

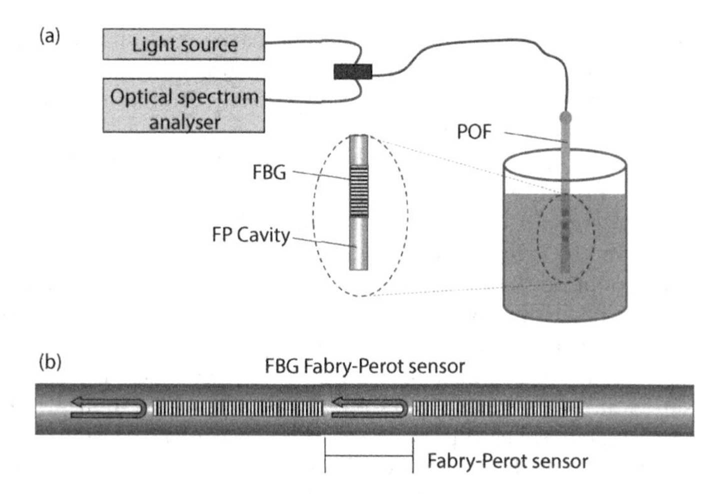

FIGURE 3.8 (a) Scheme of a POF Fabry-Perot cavity interferometer using an FBG and the tip of the fiber for refractive index measurement. (b) Fabry-Pérot interferometer made with two FBGs having increased resolution than the single FBG strain sensor.

26.4 pm/°C. FPI can also be constructed with photopolymerizable resins which are hardened with UV exposure creating the Fabry-Perot cavity as a polymeric waveguide. Oliveira et al. [37] used this technique with different resins and characterized the devices for humidity, temperature, pressure, and refractive index. Depending on the resin used, the best sensitivities obtained were 573.8 pm/%RH, 458.9 pm/°C, –15.7 pm/bar, and –1320 nm/RIU.

3.4.2 Mach Zehnder Interferometer

Mach-Zehnder interferometers (MZIs) have been used for sensing applications with very diverse configurations in both non-fiber and fiber based fields. The basic principle of this interferometer is the splitting of light into two separated optical paths, where one of the paths has a perturbation by a parameter to measure. These paths are then recombined to create the interferometric pattern. If one of the paths is essentially composed of a POF, then any perturbation that the POF suffers can be detected (**Figure 3.9a**). This is the most common MZI configuration found in

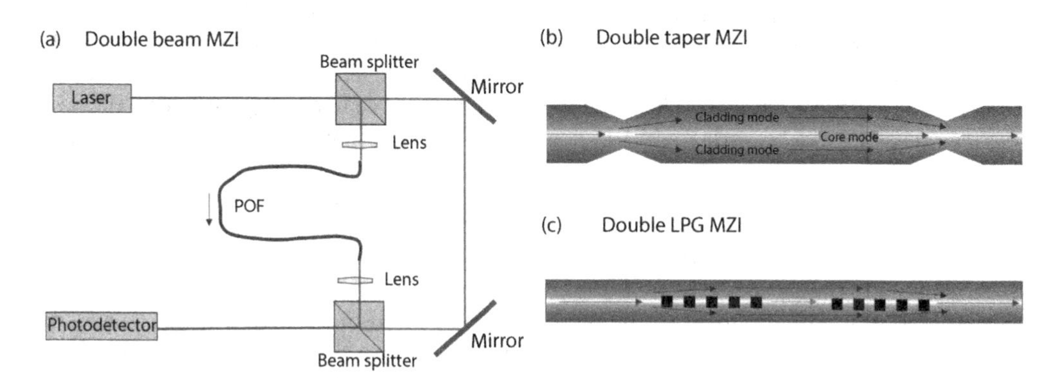

FIGURE 3.9 (a) Scheme of a double beam Mach-Zehnder interferometer setup using a POF as a waveguide path of one of the beams for measurements. (b) In-line Mach-Zehnder interferometer made with a double taper structure for strain and refractive index measurements. (c) In-line Mach-Zehnder interferometer made with a double LPG structure for strain measurements.

academia using POF. Silva-López et al. [38] used this approach for strain and temperature measurements based on the phase changes of the beam. Sensitivities of 131×10^5 rad/m and −212 rad m^{-1} K^{-1} for strain and temperature were obtained respectively. Similar approaches and results were also obtained by other groups [39–42].

Although the two-arm scheme is the most common found using POF, the same is not true for silica fibers. This approach was rapidly replaced with a configuration of in-line waveguide interferometers. Here, it is taking advantage of the fiber morphology to couple part of the energy of the core modes into the cladding modes, which can interact with the external medium, and then recoupling the modes to the core for interferometric patterns. This was achieved in glass fibers in distinct ways: by core offset [43], using different core mode segments [44], using double or more tapers [45], and using double LPGs [46]. Few publications of in-line MZI can be found based on POF. One of the exceptions is the double taper MZI developed by Jasim et al. [47] for strain and refractive index measurements (**Figure 3.9b**). The double taper was created by heating the fiber with the help of a soldering iron. Light propagating in the fiber will find the first taper, coupling core modes to the cladding and creating the opportunity of interaction with the external medium. The second taper will then recouple the cladding modes to the core creating the interferometric pattern. A refractive index sensitivity of 3.44 nm/RIU was obtained. Strain-sensing capability with a sensitivity of 0.2 pm/με was also confirmed. The other case of in-line MZI in POF was developed by Ferreira et al. [48] with two mechanically imprinted LPGs (**Figure 3.9c**). The fiber was placed on a 70°C hot plate with periodic grooves of 1 mm for LPG patterning. The space between the LPGs was 40 mm. A sensitivity of -3.9×10^{-3} nm/με was obtained with strain measurements. These interferometric elements could be adapted to be used as Michelson interferometers by placing mirrors in the fiber end tip, as can be seen in the same way for many applications using silica fibers [49,50] where gold layers are coated. In this case, the interferometric pattern would need to be measured in a reflectometry mode.

3.4.3 Multimode Interference

Fiber multimode interference (MMI) can be mostly seen in structures comprising single-mode-multimode-single-mode (SMS) configurations. The MMI structure and theory was deeply studied by Wang et al. [51]. The MMI theory can be understood by the phenomenon of self-imaging where an input single mode field profile is periodically reproduced along the multimode fiber (MMF) by exciting multiple modes with different propagation constants, resulting in constructive interference patterns with maxima or minima in its transmission spectrum. This pattern will be dependent on the diameter, length, and effective refractive index of the MMF [52]. Therefore, it can be used for strain, temperature, and also for refractive index measurements if the multimode sections have its cladding removed. Because of its advantageous properties, POF is used as the multimode section of the MMI between the single mode silica fibers (**Figure 3.10a**). The first application using an MMI with PMMA POF was published by Huang et al. [52] for large strain measurement, obtaining a sensitivity of −1.72 pm/με in 2% elongation. Numata et al. [53] used a perfluorinated graded-index POF, with a core diameter of 62.5 μm, for strain and temperature sensing obtaining sensitivities of −112 pm/με and 49.8 nm/°C respectively. The same approach was used with a partially chlorinated graded-index POF [54] (core diameter of 120 μm and 0.7 m length) obtaining fiber sensitivities of −4.47 pm/με for strain and 9.66 nm/°C for temperature. Later, an enhancement of the perfluorinated graded-index POF MMI sensor (core diameter 120 μm and a length of 0.3 m) was performed by Kawa et al. [55] by a fiber annealing process. It was proven an improvement of 2.9 times relative to the same fiber without annealing, with sensitivities of 2.17 nm/°C achieved. Oliveira et al. [27] conducted measurements with a PMMA POF in an MMI for strain, temperature, and refractive index, with the latter requiring the chemical etching of the POF for interaction with the external medium (**Figure 3.10b**). The POF section presented a length of 10 mm and core diameter of 73.5 μm. Sensitivity obtained

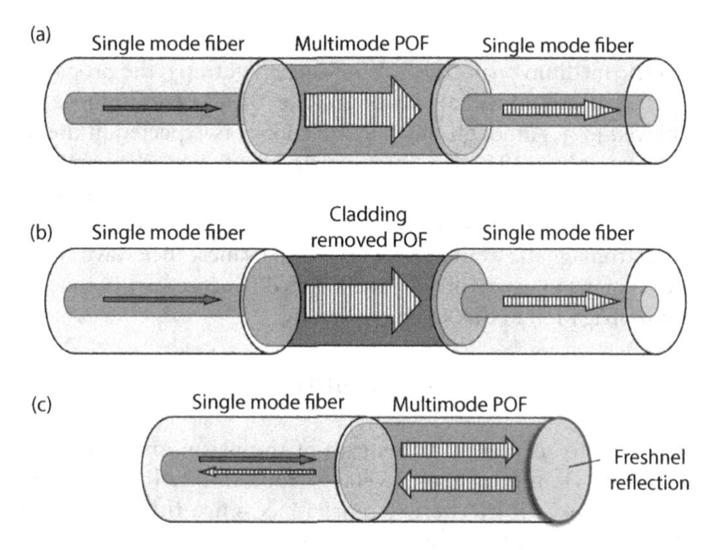

FIGURE 3.10 (a) Scheme of the SMS structure with the POF being used as the multimode section for interference pattern. (b) SMS structure with a cladded removed POF for refractive index measurements. (c) Reflectance mode detection of an SMS structure based sensor using Fresnel reflection in the tip end of the POF.

from strain was –3.03 pm/με for a total elongation of 1.5%, 103.6 pm/°C for temperature in a 45°–80° range, and 193.9 nm/RIU in a range from 1.333 to 1.354 and 515.3 nm/RIU from 1.354 to 1.368 RIU. A new MMI setup using a Fresnel reflection element at the distal open end of the POF was proposed by Kawa et al. [56] (**Figure 3.10c**). This MMI setup resembles a Michelson interferometer where the light pattern formed in the MMF will reflect on the end fiber without any single mode connector, which therefore can be detected in reflectance mode due to a circulator connected to the first single mode fiber. Kawa used this setup to measure strain and temperature, obtaining sensitivities of –122.2 pm/με and 10.1 nm/°C respectively with a perfluorinated graded-index POF, having a core diameter of 62.5 μm and a length of 0.7 m. The obtained results are comparable to the ones obtained by two-end-access configurations.

3.5 Surface Plasmon Resonance Fiber Optic Sensors

Another technique that became a hot topic in the science community for some time is the development of sensing mechanisms based on plasmon resonance phenomena. Surface plasmon resonance (SPR) occurs when collective electron oscillations propagate in metal-dielectric interface by interacting with light. This propagation will depend on the incident light frequency and on the natural resonant frequency of the electrons. These waves are also transverse magnetically (TM) polarized and get excited by only TM polarized light. By analyzing the Maxwell's equations for electromagnetism in this application, the coupling of light to form SPR cannot happen by direct incidence of light in the metal-dielectric interface. For this to be true, a condition must be met which requires equality between the propagation constant of the incident light and that of the surface plasmons at the interface $\left(k_{sp}\right)$ given by:

$$k_{sp} = k_0 \left(\frac{\varepsilon_m \varepsilon_s}{\varepsilon_m + \varepsilon_s} \right)^{1/2} \tag{3.5}$$

where, k_0 is the wave vector of light in free space and ε_m and ε_s are the dielectric constants of the metal and the dielectric medium respectively. With direct lightning, the propagation constant of the light in the dielectric medium will always be smaller, therefore techniques that increase this value must be employed [57]. Although this phenomenon was expected in the beginning of the twentieth century, it was only in 1968 that Otto reported the first mechanism to excite SPR [58]. Otto used a prism to create an electromagnetic evanescent wave based on the idea that the properties of both type of waves are similar, hence a strong possibility that SPR could be excited by coupling with the electromagnetic evanescence wave. An evanescence wave is created when TM light is totally reflected in the interface of two media with distinct refractive index. The evanescent wave (k_{ev}) constant propagation in this setup is given by:

$$k_{ev} = k_p \sin(\theta) \tag{3.6}$$

where k_p is the wave vector of light, in this case at the prism and θ the resonance angle. A metallic surface in the vicinity of this interface (≈200 nm) will interact with the created evanescent wave and excite SPR by energy transference since the condition $k_{ev} = k_{sp}$ is met (**Figure 3.11a**). The reflected light will have a loss in its intensity due to the energy transference, a technique that was named the attenuated total reflection (ATR). Since total reflection is angle dependent, SPR excitation is also angle dependent. However, this configuration presents the big disadvantage of having a very short distance between the metal surface and prism, and thus is difficult to apply in practice. This practical problem was solved by Kretschmann e Raether [59] in 1968 by using the same principle as Otto, but instead of having a metal surface at some distance, the metal is coated at the prism surface and, by guaranteeing that this layer is thin enough (about 50 nm), the evanescent light can propagate thought the metal layer until it reaches the metal-dielectric interface (**Figure 3.11b**). Several metals can be used as a coating for SPR coupling, but the most commonly used are gold and silver due to their better coupling properties [60].

Thin metal films are not the only way to promote plasmonic resonance. This can be achieved also by metal nanoparticles with lower sizes than the wavelength of the incident light. When incident light interacts with these metal nanoparticles, the electromagnetic field of light induces a collective coherent oscillation of the surface electrons with the frequency of light. Free electrons in the nanoparticle are attracted due to the electromagnetic field, leading to a charge separation between the free electrons and the ionic metal core. In turn, the Coulomb repulsion forces among the free electrons push them to the opposite direction acting as a restoring force, which

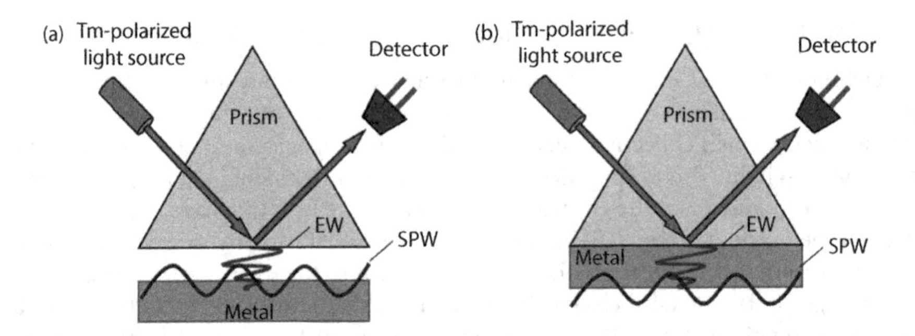

FIGURE 3.11 (a) Otto's prism configuration to create a evanescent wave (EW) that will promote a surface plasmon wave (SPW) in the vicinity of a metal surface. (b) Kretschmann's prism configuration where the metal layer is in contact with the prism and the evanescent wave crosses it, creating an SPW in the other side.

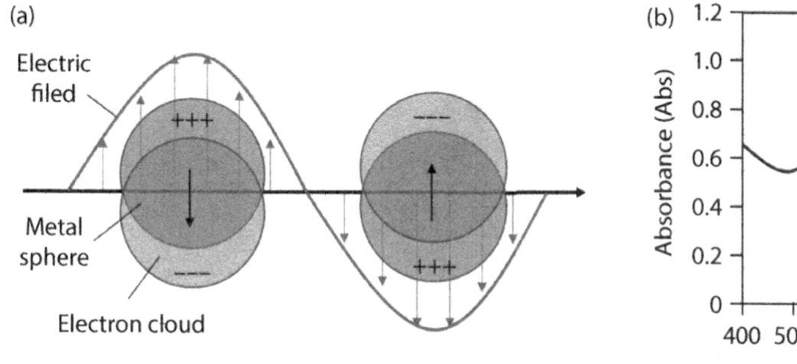

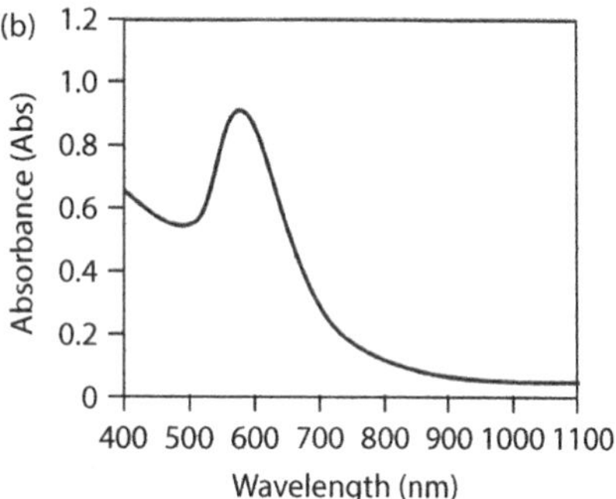

FIGURE 3.12 (a) Metal particle interaction with light creating localized surface plasmon resonance (LSPR). (b) Typical absorbance spectrum of a solution containing metal nanoparticles with the peak obtained due to the LSPR energy transference.

results in the collective oscillation of electrons (**Figure 3.12**). Because of these highly localized oscillations, it is called localized surface plasmon resonance (LSPR) [61,62].

The resonance wavelength is dependent on the size, shape, and type material of metal nanoparticles and, like in SPR, on the external refractive medium. It is, however, possible for it to be excited directly by incident light, contrary to thin film SPR. This gives it high flexibility in sensor conceptions since it can be used on simple glass surfaces or even in liquid suspension. Spectral analyzers to detect the shifts of the resonance wavelength are usually the detection methodology used for both sensing mechanisms, but absorbance measurement of a specific spectral region is also possible and an employed detection method.

Both SPR and LSPR have become very common with their integration in optic fiber in the last years, because of their advantages of miniaturization, simpler designs, fast response, and remote sensing. Optical fiber acts just like a prism in the Kreshman's configuration, and a very specific deep is detected in the transmitted spectrum, which is related to the attenuated total reflection. Most of the fiber components discussed before can be used as a base for SPR and LSPR sensors with the U-bent, etching, lateral polishing, tapering, and tilted grating the most common procedures applied in POF SPR based sensors. Examples of application of these techniques will be presented in this chapter.

With their high sensitivity to refractive index changes, these types of sensors are ideal candidates for the detection of ultra-low concentrations of bio-analytes in aqueous mediums when materials with high specificity to the analytes are used as functionalized intermediary layers, which changes their refractive index with the concentration of analytes.

3.5.1 U-Bent Fiber

SPR can be excited using the POF as its medium of light propagation. Without removing the cladding or tapering the fiber, only by fiber bending will the input light achieve the necessary angles to propagate through the cladding layer. Here the cladding can be used as the propagating medium where the metal layer can be deposited for an SPR excitation. This was the approach done by Arcas et al. [63] to fabricate their bending loss-based U-shaped POF biosensor for *Escherichia coli* detection (**Figure 3.13**). An 8-mm diameter bent was made with a hot-air blower gun, and layers of 3.5, 7.0, 10, 18, 30, 50, 70, and 100 nm were deposited in several probes by a radio-frequency magnetron sputtering system. The probes were then immobilized with rabbit anti-*Escherichia coli* polyclonal antibody for functionalization. Detection limit of 1.5×10^3 colony-forming units was achieved with the 70-nm probe by light intensity variation detection.

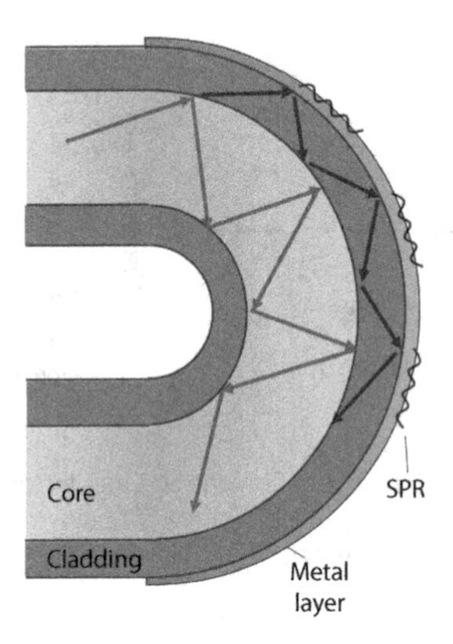

FIGURE 3.13 Schematic designs of U-bent SPR based sensor where SPR is excited using the cladding of the fiber as the dielectric medium for the metal thin film deposition.

3.5.2 Etched Fiber

In a straight fiber the SPR excitation can be easily achieved by creating a boundary with the core of the fiber and the metallic thin film. To deposit the metal layer over the core, etching the fiber is one of the mechanisms possible to be performed. Al-Qazwini et al. [64] used this approach by depositing a gold thin film to a symmetrically chemical etched POF, creating a sensor for refractive index measurements (**Figure 3.14a**). The gold layer was deposited by sputtering in four different fibers with thickness of 45, 55, 65, and 75 nm in 10-mm sections. As the thickness increases, sensitivities from 1147 to 2185 nm/RIU were obtained with refractive index in the 1.3353–1.3653 RIU range. The sensitivity of the sensor was also tested with the increase of the sensing SPR length section to 20 mm revealing a decrease in its value from 1600 to 1400 nm/RIU due to the higher curve's full width at half maximum (FWHM). Non-cladded PMMA straight fibers were also used by Jin et al. [65] as the dielectric waveguide for LSPR using deposited 19 nm diameter gold nanoparticles by chemical modification of fibers and buffer immobilization of the nanoparticles (**Figure 3.14b**). Refractive index measurements were carried on obtaining a sensitivity of 42.9 nm/RIU from 1.33 to 1.44 RIU.

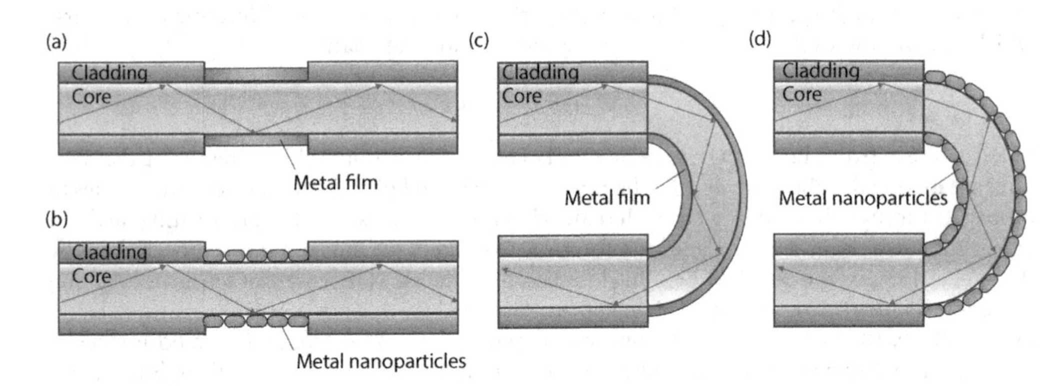

FIGURE 3.14 Schematic designs of etched SPR based sensor using (a) metal layer in a straight fiber; (b) metal nanoparticles in a straight fiber; (c) metal film in a U-bent fiber; and (d) metal nanoparticles in a U-bent fiber.

To increase sensitivity, U-bent etched POFs were also reported as SPR and LSPR sensors. The U-bent morphology increases the interaction angles with the deposited metal layers, increasing the possible SPR excitation and therefore its sensitivity (**Figure 3.14c** and **d**). Christopher et al. [66] used both sensing mechanisms by controlling the quantity of gold deposited in a POF, forming first nanoparticles and, with enough time, creating a thin film. The chemical etched fiber gained the bent form by placing it inside a glass capillary and keeping it in a hot air oven at 80°C for 10 min to obtain a bend diameter of 2.25 mm. Gold deposition was carried on by sputtering the fiber for 30, 60, 90, and 120 seconds. Spectral analysis clearly shows an LSPR predominant activity to the 30 second fiber while SPR activity was observed for the 120 second fiber. The latter presented a film thickness of 60 nm and, under refractive index characterization, a sensitivity of 1040 nm/RIU in a range 1.33–1.361 RIU. Absorbance measurement was used for the 30 second fiber obtaining a sensitivity of 15.5 ΔAbs_{457nm}/RIU. Other published works using decladed U-bent POF include LSPR sensors with gold nanoflowers (~37 nm size) with variation of both fiber core diameter (250, 500, 750, and 1000 μm) and bending diameter with the highest absorbance sensitivity obtained for the 500 μm core and 1 mm bend diameter with 5.57 ΔAbs_{560nm}/RIU with RI changes from 1.33 to 1.47 RIU [67] and the successful demonstration of an SPR base sensor with 35 nm gold nanospheres used as labels for the detection of different nucleic acid concentrations, showing the potentiality of these type of sensors for bio-analytes detection [68].

Silver is also an alternative metal used to promote SPR or LSPR excitation. Using POF in an etched U-bent configuration, Jiang et al. [69] used graphene and silver nanoparticle hybrid structure to fabricate a refractive index sensor. At first a thin film of silver is deposited into the POFs by thermal evaporation and then it is dip-coated in a solution containing the hybrid nanoparticles. The thickness of the silver films used were 3, 5, ~7, and 10 nm. Measurements of RI from 1.330 to 1.3657 RIU were conducted, and sensitivity of 700.3 nm/RIU was obtained for the 5-nm film thickness fiber.

Etched fiber tips with metal thin film coating or nanoparticles for back reflection spectral analysis [70,71] are also structures observed in silica fibers but not in POF, having the potentiality to be also developed and used with POF.

3.5.3 Lateral Polished Fiber

Lateral polishing of fiber can also be used for posterior metal coating, obtaining similar results as fiber etching. This technique can be applied for both SPR and LSPR. Cennamo et al. used exclusively D-shaped structures that can be combined with other structures or using molecularly imprinted polymer (MIP) film for selective sensing [72]. To create a D-shaped structure, the fiber is embedded in a resin block where the cladding and part of the core are removed by polishing it with polishing paper in an "8" pattern. A 1.5 μm buffer photoresist is then deposited by spin-coating, which will promote high adherence to the gold layer that will be sputtered by using a sputtering machine, achieving a thickness of 60 nm (**Figure 3.15a**). A fiber with a 1000 μm diameter proved to have a sensitivity of 1.325×10^3 nm/RIU (0.001 RIU resolution) under refractive index characterization tests from 1.333 to 1.371 RIU [72]. A comparison of different photoresist buffers was also performed [73]. Cennamo et al. also used a metal bilayer composed by palladium and gold in its D-shaped configuration, with and without the photoresist buffer layer [74] (**Figure 3.15b**). A higher sensitivity of 3×10^3 nm/RIU in a 1.38–1.42 RIU range was obtained without the photoresist and for a 11.5 nm thickness for both palladium and gold. Other variations of the initial configuration include the usage of POFs with fluorescent optical fibers as light sources with both gold and silver deposition [75]. Sensitivities of 1044 nm/RIU and 523 nm/RIU were obtained for gold and silver respectively. Other research groups have also published similar works using this base structure. Gasiour et al. [76] published its development using a gold layer deposited on a single-mode birefringent polymer being able to control the polarization of light guided in the fiber and excite only the TM mode strongly coupled to the plasmon. Sensitivities

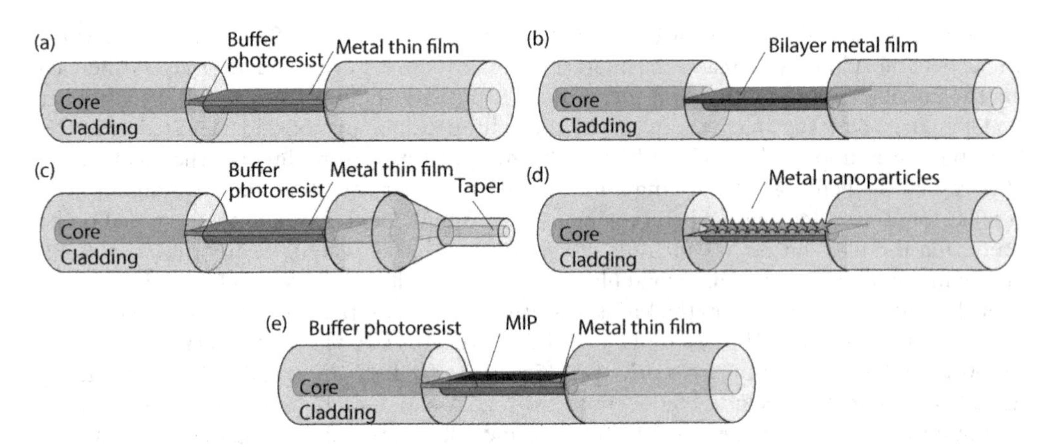

FIGURE 3.15 Designs of lateral polished D-shaped fibers for SPR-based sensing using (a) metal layer thin film; (b) bilayer metal thin film; (c) metal layer thin film with taper for mode filtering; (d) metal nanoparticles; and (e) MIP deposited over a metal thin film layer for high specification.

of 1300 and 2800 nm/RIU for the refractive index of external medium with 1.362 and 1.410 RIU respectively were obtained. Cennamo et al. also proposed the usage of a taper at the input or at the output of the gold coated SPR system for higher-order modes filtering to improve the operation range of SPR-based sensors in multimode POF fibers [77] (**Figure 3.15c**). Wavelength shifts of 86.5, 94, and 98 nm were obtained for no taper, taper-before-SPR, and taper-after-SPR configuration, respectively, for a refractive index range from 1.330 to 1.390 RIU.

Using the same initial D-shaped POF embedded approach but without the buffer photoresist, a direct deposition and immobilization of five-branched gold nanostars into the polished core was also performed for LSPR based sensing [78] (**Figure 3.15d**). Measurements of refractive index solutions in a range between 1.333 and 1.371 RIU revealed a sensitivity of 84 nm/RIU. Other SPR/LSPR structures on POF were also proposed with numerical simulations only, as the cases of the nano-antenna arrays [79] and the bimetallic nanowire gratings [80].

Selective sensing using this base structure is now the researching goal being pursued with the help of molecularly imprinted polymer (MIP) films deposited over the metal layer (**Figure 3.15e**). MIPs are synthetical porous solids containing specific sites interacting with the molecule of interest according to a "key and lock" model and obtained by the molecular imprinting methods, having good stability, reproducibility, and low cost. Cennamo et al. used an MIP for the detection of trinitrotoluene in its D-shaped SPR sensing structure [81]. The sputtered gold layer presented a thickness of 60 nm and the spin-coated prepolymeric mixture stayed under polymerization for 16 hours at 70°C. Highly selective detection of trinitrotoluene, down to about 50 μM was achieved. Other MIPs for the detection of different contaminant elements in power transformers were also reported with successful selective detection: dibenzyl disulfide [82], furfural [83], and the two elements simultaneously with a cascaded SPR POF configuration [84].

3.5.4 Taper Fiber

To increase the interaction of the electromagnetic evanescent wave without the need of cladding removal, the fabrication of fiber tapers can be a good alternative. Thin metal films or nanoparticles can also be deposited in this type of structure for SPR measurements. Cennamo et al. have also used this approach by imbedding a taper POF in resin block [85]. The taper was fabricated by exposing it at 150°C and stretching it with a motorized linear positioning stage. Polishing was performed before the deposition of a 60-nm gold thin film by sputtering. An MIP, with specificity to capture of L-nicotine, was deposited over the gold layer as the sensing goal (**Figure 3.16a**).

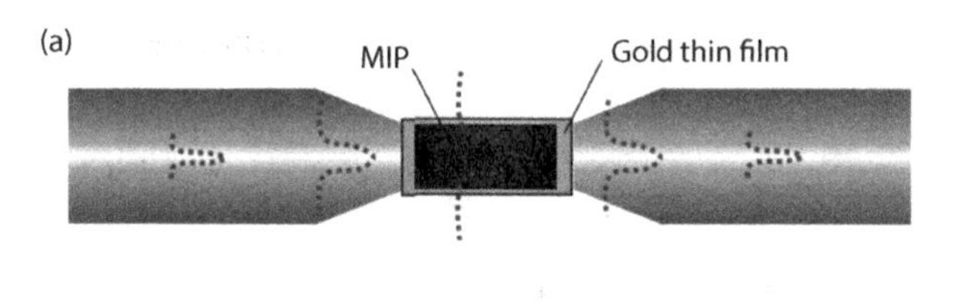

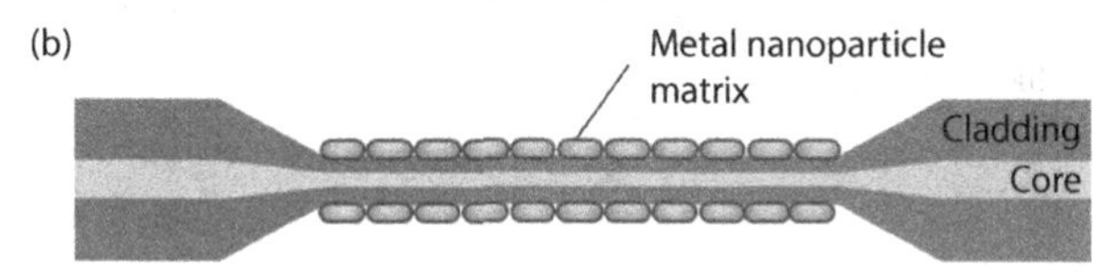

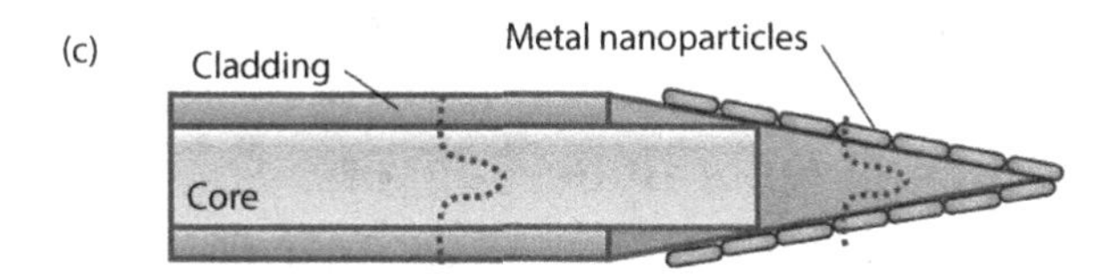

FIGURE 3.16 POF taper designs for SPR based sensing using (a) MIP deposited over a metal thin film layer for high specification; (b) metal nanoparticles matrix; and (c) fiber tip with metal nanoparticles for surface-enhanced Raman scattering sensing.

They have proven an increased sensitivity using the taper structure in opposition to the non-taper version by reaching a sensitivity value of 1.3×10^4 nm/M with a taper ratio 1.8 from a value of 1.7×10^3 nm/M for the non-taper configuration.

An ammonia sensing application was proposed by Rithesh Raj et al. by using silver nanoparticles imbedded in polyvinyl alcohol matrix and deposited by dip-coating over a POF tapered fiber for LSPR sensing with a 3-cm length [86] (**Figure 3.16b**). The thickness of the coating was about 80 μm with an average particle size of 21 nm. Higher sensitivities with higher silver nanoparticles concentration in the matrix were obtained.

A surface-enhanced Raman scattering sensor (SERS) taper fiber tip was also proposed by Xie et al. having gold nanorods, with a width and length of about 15 and 45 nm, immobilized in the fiber tip [87] (**Figure 3.16c**). When excited with 514.5 nm argon ion laser, a wavelength near the LSPR of that nanorod, light coming from the tip is detected with a Raman spectrometer due to molecular vibrations. Each molecule will have its own "fingerprint" spectrum due to the different vibration mechanisms. The nanorods have the role of amplifying the electromagnetic fields by means of the localized surface plasmons [88].

3.5.5 Tilted FBG

Tiled fiber Bragg grating are also good fiber elements to couple energy to SPR when a thin metal layer is coating this element. As previously seen, TFBG will reflect light to the cladding modes of propagation, and these modes, when in the presence of a metal layer, will excite SPR by attenuated total reflection. It will create a specific spectral signature with the extinction of some modes and being sensible to the external refractive index medium. Hu et al. [89] reported the only TFBG for SPR sensing in POF (**Figure 3.17**). This type of structure allows the excitation of surface plasmon waves at near-infrared telecommunication wavelengths. Over the angled 6° grating inscribed with UV light and a phase mask, a 30-nm gold layer was deposited by sputtering. Refractive index measurements between 1.408 and 1.428 revealed a sensitivity of around 550 nm/RIU.

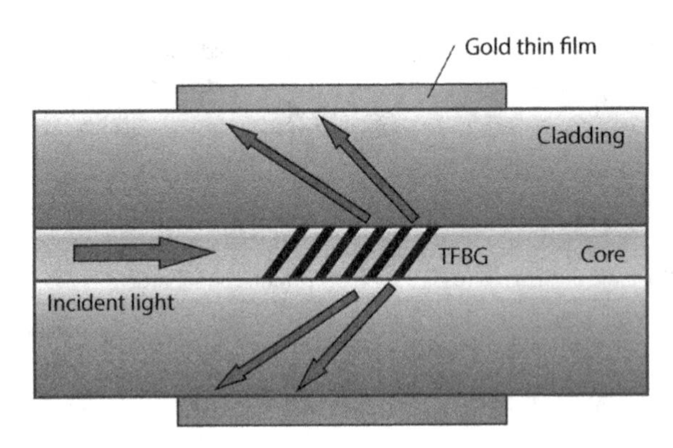

FIGURE 3.17 Usage of a tilted fiber Bragg grating to promote the formation of plasmonic surface resonant wave in the metal-external medium boundary of a deposited thin film metal layer.

3.6 Polarimetric-Based Fiber Optic Sensors

Polarimetric-based fiber sensing uses highly birefringent (HiBi) fibers as its fundamental element sensing. HiBi fibers have proper geometry arising from microstructural asymmetries that imposes two effective refractive index values, one for each of the orthogonal axes with high birefringence. This can be achieved by braking the symmetry of the hexagonal structure in the microstructured cladding which can be easier to implement than with silica fiber by techniques such as drilling and casting. This optical device has the possibility to measure several physical quantities due to its properties, which include hydrostatic pressure, strain, vibration, temperature, acoustic wave, etc. The common methodology of measurement uses an interferometric system composed of a broad band light source, polarizer, and an analyzer placed at the end of the tested fiber, with a 45° alignment to its polarization axes (**Figure 3.18**). Such alignment causes interference at its output by the pairs of orthogonally polarized modes to be excited in the fiber. The obtained interference pattern is then analyzed for sensing due to perturbations occurring to the fiber where phase difference between polarization modes will be detected.

The polarimetric sensitivity of the fiber to an external parameter (X), represents the phase difference between polarization modes $(\varphi_x - \varphi_y)$ induced by unit change of the parameter over unit length (L) of the fiber and can be expressed by:

$$K_x = \frac{1}{L}\frac{d(\varphi_x - \varphi_y)}{dX} = \frac{2\pi}{\lambda}\left(\frac{d\Delta n}{dX} + \frac{\Delta n}{L}\frac{dL}{dX}\right) \tag{3.7}$$

Szczurowski et al. [90] used this approach to measure the polarimetric sensitivity to hydrostatic pressure, strain, and temperature. For pressure changes in the range from 0 to 2 MPa, a sensitivity 72 rad/(MPa × m) was obtained. To strain a maximum value up to 8 mε was achieved

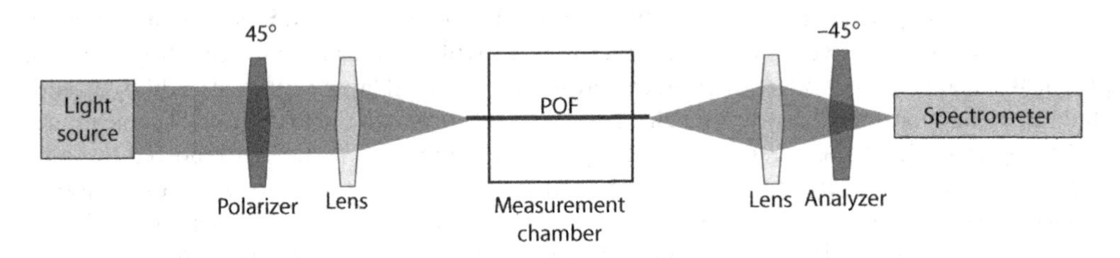

FIGURE 3.18 Typical polarimetric-based fiber optic sensing setup where parameters such as strain, temperature, and pressure are applied to the POF in the measurement chamber.

obtaining a sensitivity of 3.1 rad/(mε × m). Temperature measurements revealed very different polarimetric responses between fibers, typically with high nonlinearity and hysteresis, in the 3°C–60°C range, depending on the thermal history of the specific piece of the fiber. Nevertheless, sensitivities of −0.6 rad/K × m to 0.7 rad/K × m were reported for certain measurement ranges. Martynkien et al. [91] also presented his study with a HiBi POF for the measurement of hydrostatic pressure. Sensitivities of 48 rad/MPa × m were obtained with a range up to 8.5 MPa.

3.7 Summary Table

A summary table with all the POF based structures used for sensing and described in this chapter is presented in **Table 3.1**.

Table 3.1 Summary Table With Most of the Sensing Structures and Devices Used Based in POF

Fiber Extrinsic Based Sensing		
	Measured Parameters	**References**
Air-gap design	Refractive index	[2]
Displacement based	Refractive index	[3]
Cascaded in-line holes	Refractive index	[5]
Intrinsic Intensity Based Sensors		
	Measured Parameters	**References**
Macrobending	Refractive index	[6]
Side polishing	Refractive index	[10]
D-shape	Refractive index	[12]
Multi-notched LPG	Refractive index	[14]
Taper	Refractive index	[19]
U-shaped taper	Refractive index	[20]
Double taper	Refractive index	[21]
U-shaped clad removed taper	Refractive index	[22]
Ultrasonic crushed fiber	Refractive index	[23]
Grating Based Sensors		
	Measured Parameters	**References**
FBG	Strain	[25]
Etched FBG	Strain; Bend; Temperature	[26]
D-shaped FBG	Bend	[28]
LPG	Strain; Temperature	[29]
Transducer based LPG	Pressure	[30]
TFBG	Refractive index	[31]

(Continued)

Table 3.1 (*Continued*) Summary Table With Most of the Sensing Structures and Devices Used Based in POF

Interferometric Based Sensors		
	Measured Parameters	**References**
Semi-FBG Fabry-Perot	Refractive index	[32]
FBG Fabry-Perot	Strain	[35]
Photopolymerizable resin Fabry-Perot	Humidity; Temperature; Pressure; Ref. index	[37]
Double beam Mach-Zehnder	Strain; Temperature	[38]
Double taper Mach-Zehnder	Strain; Refractive index	[47]
Double LPG Mach-Zehnder	Strain	[48]
MMI	Strain	[52]
Etched MMI	Strain; Temperature; Refractive index	[27]
Fresnel reflection MMI	Strain; Temperature	[56]
Fiber Based SPR Sensing		
	Measured Parameters	**References**
U-bent		
Gold film	Refractive index (*Escherichia coli*)	[63]
Etched fiber		
Gold film	Refractive index	[64]
Gold nanoparticles	Refractive index	[65]
U-bent gold film	Refractive index	[66]
U-bent gold nanoparticles	Refractive index	[67]
U-bent silver film and nanoparticles	Refractive index	[69]
Lateral polished		
D-shaped gold film	Refractive index	[72]
D-shaped palladium-gold bilayer	Refractive index	[74]
D-shaped silver film	Refractive index	[75]
D-shaped gold film—Taper filtering	Refractive index	[77]
D-shaped gold nanostars	Refractive index	[78]
D-shaped gold film—MIP	Refractive index (trinitrotoluene)	[81]

(*Continued*)

Table 3.1 (*Continued*) Summary Table With Most of the Sensing Structures and Devices Used Based in POF		
Taper		
Gold film—MIP	Refractive index (L-nicotine)	[85]
Silver nanoparticles matrix	Refractive index (ammonia)	[86]
Fiber tip gold nanorods	SERS	[87]
Tiled fiber Bragg grating		
Gold film	Refractive index	[89]
Polarimetric Based Sensors		
	Measured Parameters	**References**
HiBi mPOF	Hydrostatic pressure; Strain; Temperature	[90]

Acknowledgments

This work is funded by FCT/MEC through national funds and, when applicable, co-funded by the FEDER–PT2020 partnership agreement under the projects AQUATICsens - POCI-01-0145-FEDER 032057 and INITIATE IF/FCT IF/01664/2014/CP1257/CT0002, investigator grant IF/01664/2014 and PhD fellowship (Daniel Duarte: SFRH/BD/130966/2017).

References

1. L. Bilro, N. Alberto, J. L. Pinto, and R. Nogueira, "Optical sensors based on plastic fibers," *Sensors*, vol. 12, no. 12, pp. 12184–12207, 2012.
2. L. Bilro, S. A. Prats, J. L. Pinto, J. J. Keizer, and R. N. Nogueira, "Design and performance assessment of a plastic optical fibre-based sensor for measuring water turbidity," *Meas. Sci. Technol.*, vol. 21, no. 10, p. 107001, 2010.
3. S. Binu, V. P. Mahadevan Pillai, V. Pradeepkumar, B. B. Padhy, C. S. Joseph, and N. Chandrasekaran, "Fibre optic glucose sensor," *Mater. Sci. Eng. C*, vol. 29, no. 1, pp. 183–186, 2009.
4. G. Govindan, S. G. Raj, and D. Sastikumar, "Measurement of refractive index of liquids using fiber optic displacement sensors," *J. Am. Sci.*, vol. 5, no. 2, pp. 13–17, 2009.
5. J.-D. Shin and J. Park, "High-sensitivity refractive index sensors based on in-line holes in plastic optical fiber," *Microw. Opt. Technol. Lett.*, vol. 57, no. 4, pp. 918–921, 2015.
6. J. Zubia, G. Garitaonaindía, and J. Arrúe, "Passive device based on plastic optical fibers to determine the indices of refraction of liquids," *Appl. Opt.*, vol. 39, no. 6, p. 941, 2000.
7. A. M. C. y Paz, J. M. Acevedo, J. D. Gandoy, A. del Rio Vazquez, C. M.-P. Freire, and M. Soria, "Plastic optical fiber sensor for real time density measurements in wine fermentation," *2007 IEEE Instrumentaion and Measurement Technology Conference IMTC 2007*, pp. 1–5, 2007.
8. A. M. C. Paz et al., "A multi-point sensor based on optical fiber for the measurement of electrolyte density in lead-acid batteries," *Sensors*, vol. 10, no. 4, pp. 2587–2608, 2010.
9. L. Bilro, N. Alberto, L. M. Sa, J. de L. Pinto, and R. Nogueira, "Analytical analysis of side-polished plastic optical fiber as curvature and refractive index sensor," *J. Light. Technol.*, vol. 29, no. 6, pp. 864–870, 2011.
10. N. Jing, J. Zheng, X. Zhao, and C. Teng, "Refractive index sensing based on a side-polished macro-bending plastic optical fiber," *IEEE Sens. J.*, vol. 15, no. 5, pp. 2898–2901, 2015.

11. F. De-Jun, Z. Mao-Sen, G. Liu, L. Xi-Lu, and J. Dong-Fang, "D-shaped plastic optical fiber sensor for testing refractive index," *IEEE Sens. J.*, vol. 14, no. 5, pp. 1673–1676, 2014.
12. F. Sequeira et al., "Refractive index sensing with D-shaped plastic optical fibers for chemical and biochemical applications," *Sensors*, vol. 16, no. 12, p. 2119, 2016.
13. F. Sequeira et al., "Analysis of the roughness in a sensing region on D-shaped POFs," *Proc. International Conference on Plastic Optical Fibers – POF*, p. OP36, 2016.
14. C. Teng, F. Yu, Y. Ding, and J. Zheng, "Refractive index sensor based on multi-mode plastic optical fiber with long period grating," *Opt. Sens.*, 2017, vol. 10231, p. 102311M.
15. A. Leung, P. M. Shankar, and R. Mutharasan, "A review of fiber-optic biosensors," *Sensors Actuators B Chem.*, vol. 125, no. 2, pp. 688–703, 2007.
16. S. Xue, M. A. van Eijkelenborg, G. W. Barton, and P. Hambley, "Theoretical, numerical, and experimental analysis of optical fiber tapering," *J. Light. Technol.*, vol. 25, no. 5, pp. 1169–1176, 2007.
17. J. Arrue et al., "Analysis of the use of tapered graded-index polymer optical fibers for refractive-index sensors," *Opt. Express*, vol. 16, no. 21, pp. 16616–16631, 2008.
18. R. Gravina, G. Testa, and R. Bernini, "Perfluorinated plastic optical fiber tapers for evanescent wave sensing," *Sensors*, vol. 9, no. 12, pp. 10423–10433, 2009.
19. H. A. Rahman et al., "Tapered plastic multimode fiber sensor for salinity detection," *Sensors Actuators A Phys.*, vol. 171, no. 2, pp. 219–222, 2011.
20. C. Beres, F. V. B. de Nazaré, N. C. C. de Souza, M. A. L. Miguel, and M. M. Werneck, "Tapered plastic optical fiber-based biosensor—Tests and application," *Biosens. Bioelectron.*, vol. 30, no. 1, pp. 328–332, 2011.
21. F. De-Jun, L. Guan-Xiu, L. Xi-Lu, J. Ming-Shun, and S. Qing-Mei, "Refractive index sensor based on plastic optical fiber with tapered structure," *Appl. Opt.*, vol. 53, no. 10, p. 2007, 2014.
22. C. Teng, N. Jing, F. Yu, and J. Zheng, "Investigation of a macro-bending tapered plastic optical fiber for refractive index sensing," *IEEE Sens. J.*, vol. 16, no. 20, pp. 7521–7525, 2016.
23. S. Shimada et al., "Refractive index sensing using ultrasonically crushed plastic optical fibers," *POF 2016—25th International Conference on Plastic Optics Fibres*, vol. 012201, pp. 3–5, 2017.
24. R. Oliveira, L. Bilro, and R. Nogueira, "Bragg gratings in a few mode microstructured polymer optical fiber in less than 30 seconds," *Opt. Express*, vol. 23, no. 8, p. 10181, 2015.
25. W. Yuan et al., "Improved thermal and strain performance of annealed polymer optical fiber Bragg gratings," *Opt. Commun.*, vol. 284, no. 1, pp. 176–182, 2011.
26. X. Chen, C. Zhang, D. J. Webb, G. D. Peng, and K. Kalli, "Bragg grating in a polymer optical fibre for strain, bend and temperature sensing," *Meas. Sci. Technol.*, vol. 21, no. 9, p. 094005, 2010.
27. R. Oliveira, T. H. R. Marques, L. Bilro, R. Nogueira, and C. M. B. Cordeiro, "Multiparameter POF sensing based on multimode interference and fiber Bragg Grating," *J. Light. Technol.*, vol. 35, no. 1, pp. 3–9, 2017.
28. X. Hu, X. Chen, C. Liu, P. Mégret, and C. Caucheteur, "D-shaped polymer optical fiber Bragg grating for bend sensing," in *Advanced Photonics 2015*, OSA Technical Digest (online) (Optical Society of America, 2015), p. SeS2B.5.
29. R. Min, C. Marques, K. Nielsen, O. Bang, and B. Ortega, "Fast inscription of long period gratings in microstructured polymer optical fibers," *IEEE Sens. J.*, vol. 18, no. 5, pp. 1919–1923, 2018.
30. I.-L. Bundalo, R. Lwin, S. Leon-Saval, and A. Argyros, "All-plastic fiber-based pressure sensor," *Appl. Opt.*, vol. 55, no. 4, p. 811, 2016.
31. X. Hu, C.-F. J. Pun, H.-Y. Tam, P. Mégret, and C. Caucheteur, "Tilted Bragg gratings in step-index polymer optical fiber," *Opt. Lett.*, vol. 39, no. 24, p. 6835, 2014.
32. M. F. S. Ferreira, G. Statkiewicz-Barabach, D. Kowal, P. Mergo, W. Urbanczyk, and O. Frazão, "Fabry-Perot cavity based on polymer FBG as refractive index sensor," *Opt. Commun.*, vol. 394, pp. 37–40, 2017.
33. D. Webb, M. Aressy, and A. Argyros, "Grating and interferometric devices in POF," *14th International Conference on Polymer Optical Fiber*, Tsimshatsui, Hong Kong, pp. 325–8, 2005.
34. H. Dobb et al., "Grating based devices in polymer optical fibre," *Opt. Sens. II*, vol. 6189, p. 618901, 2006.

35. G. Statkiewicz-Barabach, P. Mergo, and W. Urbanczyk, "Bragg grating-based Fabry–Perot interferometer fabricated in a polymer fiber for sensing with improved resolution," *J. Opt.*, vol. 19, no. 1, p. 015609, 2017.
36. A. Theodosiou, X. Hu, C. Caucheteur, and K. Kalli, "Bragg gratings and Fabry-Perot cavities in low-loss multimode CYTOP polymer fibre," *IEEE Photonics Technol. Lett.*, vol. 1135, no. c, pp. 1–1, 2018.
37. R. Oliveira, L. Bilro, and R. Nogueira, "Fabry-Pérot cavities based on photopolymerizable resins for sensing applications," *Opt. Mater. Express*, vol. 8, no. 8, p. 2208, 2018.
38. M. Silva-López et al., "Strain and temperature sensitivity of a single-mode polymer optical fiber," *Opt. Lett.*, vol. 30, no. 23, p. 3129, 2005.
39. S. Kiesel, K. Peters, T. Hassan, and M. Kowalsky, "Large deformation in-fiber polymer optical fiber sensor," *IEEE Photonics Technol. Lett.*, vol. 20, no. 6, pp. 416–418, 2008.
40. D. Gallego and H. Lamela, "High-sensitivity ultrasound interferometric single-mode polymer optical fiber sensors for biomedical applications," *Opt. Lett.*, vol. 34, no. 12, p. 1807, 2009.
41. A. Minardo, R. Bernini, and L. Zeni, "Distributed temperature sensing in polymer optical fiber by BOFDA," *IEEE Photonics Technol. Lett.*, vol. 26, no. 4, pp. 387–390, 2014.
42. D. Gallego, D. Sáez-Rodríguez, D. Webb, O. Bang, and H. Lamela, "Interferometric microstructured polymer optical fiber ultrasound sensor for optoacoustic endoscopic imaging in biomedical applications," *23rd Int. Conf. Opt. Fibre Sens.*, vol. 9157, p. 91574X, 2014.
43. Z. Tian, S. S.-H. Yam, and H.-P. Loock, "Single-mode fiber refractive index sensor based on core-offset attenuators," *IEEE Photonics Technol. Lett.*, vol. 20, no. 16, pp. 1387–1389, 2008.
44. M. Shao, X. Qiao, H. Fu, H. Li, Z. Jia, and H. Zhou, "Refractive index sensing of SMS fiber structure based Mach-Zehnder interferometer," *IEEE Photonics Technol. Lett.*, vol. 26, no. 5, pp. 437–439, 2014.
45. T. Zhu, D. Wu, M. Deng, D.-W. Duan, Y.-J. Rao, and X. Bao, "Refractive index sensing based on Mach-Zehnder interferometer formed by three cascaded single-mode fiber tapers," *Appl. Opt.*, vol. 50, no. 11, p. 77532P, 2011.
46. J.-F. Ding, A. P. Zhang, L.-Y. Shao, J.-H. Yan, and S. He, "Fiber-taper seeded long-period grating pair as a highly sensitive refractive-index sensor," *IEEE Photonics Technol. Lett.*, vol. 17, no. 6, pp. 1247–1249, 2005.
47. A. A. Jasim et al., "Refractive index and strain sensing using inline Mach–Zehnder interferometer comprising perfluorinated graded-index plastic optical fiber," *Sens. Actuators A Phys.*, vol. 219, pp. 94–99, 2014.
48. M. Ferreira, O. Frazão, and M. Marques, *Polymer Fiber based Sensors*, University of Porto, Porto, Portugal, 2016.
49. Z. Tian, S. S. Yam, and H. Loock, "Refractive index sensor based on an abrupt taper Michelson interferometer in a single-mode fiber," *Opt. Lett.*, vol. 33, no. 10, p. 1105, 2008.
50. W. B. Ji, H. H. Liu, S. C. Tjin, K. K. Chow, and A. Lim, "Ultrahigh sensitivity refractive index sensor based on optical microfiber," *IEEE Photonics Technol. Lett.*, vol. 24, no. 20, pp. 1872–1874, 2012.
51. Q. Wang, G. Farrell, and W. Yan, "Investigation on single-mode–multimode–single-mode fiber structure," *J. Light. Technol.*, vol. 26, no. 5, pp. 512–519, 2008.
52. J. Huang et al., "Polymer optical fiber for large strain measurement based on multimode interference," *Opt. Lett.*, vol. 37, no. 20, p. 4308, 2012.
53. G. Numata, N. Hayashi, M. Tabaru, Y. Mizuno, and K. Nakamura, "Ultra-sensitive strain and temperature sensing based on modal interference in perfluorinated polymer optical fibers," *IEEE Photonics J.*, vol. 6, no. 5, pp. 1–7, 2014.
54. G. Numata, N. Hayashi, M. Tabaru, Y. Mizuno, and K. Nakamura, "Strain and temperature sensing based on multimode interference in partially chlorinated polymer optical fibers," *IEICE Electron. Express*, vol. 12, no. 2, pp. 20141173–20141173, 2015.
55. T. Kawa, G. Numata, H. Lee, N. Hayashi, Y. Mizuno, and K. Nakamura, "Temperature sensing based on multimodal interference in polymer optical fibers: Room-temperature sensitivity enhancement by annealing," *Jpn. J. Appl. Phys.*, vol. 56, no. 7, p. 078002, 2017.
56. T. Kawa, G. Numata, H. Lee, N. Hayashi, Y. Mizuno, and K. Nakamura, "Single-end-access strain and temperature sensing based on multimodal interference in polymer optical fibers," *IEICE Electron. Express*, vol. 14, no. 3, pp. 20161239–20161239, 2017.
57. S. K. Srivastava, "Fiber optic plasmonic sensors: Past, present and future," *Open Opt. J.*, vol. 7, no. 1, pp. 58–83, 2013.
58. A. Otto, "Excitation of nonradiative surface plasma waves in silver by the method of frustrated total reflection," *Zeitschrift für Phys. A Hadron. Nucl.*, vol. 216, no. 4, pp. 398–410, 1968.

59. E. Kretschmann and H. Raether, "Notizen: Radiative decay of non radiative surface plasmons excited by light," *Zeitschrift für Naturforsch. A*, vol. 23, no. 12, pp. 2135–2136, 1968.
60. J. Homola, S. S. Yee, and G. Gauglitz, "Surface plasmon resonance sensors: Review," *Sens. Actuators B Chem.*, vol. 54, no. 1–2, pp. 3–15, 1999.
61. K. A. Willets and R. P. Van Duyne, "Localized surface plasmon resonance spectroscopy and sensing," *Annu. Rev. Phys. Chem.*, vol. 58, no. 1, pp. 267–297, 2007.
62. J. Cao, T. Sun, and K. T. V. Grattan, "Gold nanorod-based localized surface plasmon resonance biosensors: A review," *Sensors Actuators B Chem.*, vol. 195, pp. 332–351, May 2014.
63. A. Arcas, F. Dutra, R. Allil, and M. Werneck, "Surface plasmon resonance and bending loss-based U-shaped plastic optical fiber biosensors," *Sensors*, vol. 18, no. 2, p. 648, 2018.
64. Y. Al-Qazwini, A. S. M. Noor, Z. Al-Qazwini, M. H. Yaacob, S. W. Harun, and M. A. Mahdi, "Refractive index sensor based on SPR in symmetrically etched plastic optical fibers," *Sensors Actuators A Phys.*, vol. 246, pp. 163–169, 2016.
65. Y. Jin, K. H. Wong, and A. M. Granville, "Developing localized surface plasmon resonance biosensor chips and fiber optics via direct surface modification of PMMA optical waveguides," *Colloids Surfaces A Physicochem. Eng. Asp.*, vol. 492, pp. 100–109, 2016.
66. C. Christopher, A. Subrahmanyam, and V. V. R. Sai, "Gold sputtered U-bent plastic optical fiber probes as SPR- and LSPR-based compact plasmonic sensors," *Plasmonics*, vol. 13, no. 2, pp. 493–502, 2018.
67. A. Gowri and V. V. R. Sai, "Development of LSPR based U-bent plastic optical fiber sensors," *Sens. Actuators, B Chem.*, vol. 230, pp. 536–543, 2016.
68. A. Gowri and V. V. R. Sai, "U-bent plastic optical fiber based plasmonic biosensor for nucleic acid detection," *Opt. Sens.*, vol. 10231, p. 1023113, 2017.
69. S. Jiang et al., "A novel U-bent plastic optical fibre local surface plasmon resonance sensor based on a graphene and silver nanoparticle hybrid structure," *J. Phys. D. Appl. Phys.*, vol. 50, no. 16, p. 165105, 2017.
70. P. Hlubina, M. Kadulova, D. Ciprian, and J. Sobota, "Reflection-based fibre-optic refractive index sensor using surface plasmon resonance," *J. Eur. Opt. Soc. Rapid Publ.*, vol. 9, p. 14033, 2014.
71. J. Chen et al., "Optimization and application of reflective LSPR optical fiber biosensors based on silver nanoparticles," *Sensors*, vol. 15, no. 6, pp. 12205–12217, 2015.
72. N. Cennamo, D. Massarotti, R. Galatus, L. Conte, and L. Zeni, "Performance comparison of two sensors based on surface plasmon resonance in a plastic optical fiber," *Sensors*, vol. 13, no. 1, pp. 721–735, 2013.
73. N. Cennamo, M. Pesavento, L. De Maria, R. Galatus, F. Mattiello, and L. Zeni, "Comparison of different photoresist buffer layers in SPR sensors based on D-shaped POF and gold film," *2017 25th Opt. Fiber Sens. Conf. (OFS)*, p. 103234F, 2017, IEEE.
74. N. Cennamo, P. Zuppella, D. Bacco, A. J. Corso, M. G. Pelizzo, and L. Zeni, "SPR sensor platform based on a novel metal bilayer applied on D-shaped plastic optical fibers for refractive index measurements in the range 1.38–1.42," *IEEE Sens. J.*, vol. 16, no. 12, pp. 4822–4827, 2016.
75. N. Cennamo, F. Mattiello, R. V. Galatus, E. Voiculescu, and L. Zeni, "Plasmonic sensing in D-shaped POFs with fluorescent optical fibers as light sources," *IEEE Trans. Instrum. Meas.*, vol. 67, no. 4, pp. 754–759, 2018.
76. K. Gasior, T. Martynkien, M. Napiorkowski, K. Zolnacz, P. Mergo, and W. Urbanczyk, "A surface plasmon resonance sensor based on a single mode D-shape polymer optical fiber," *J. Opt.*, vol. 19, no. 2, p. 025001, 2017.
77. N. Cennamo et al., "Modal filtering for optimized surface plasmon resonance sensing in multimode plastic optical fibers," *IEEE Sens. J.*, vol. 15, no. 11, pp. 6306–6312, 2015.
78. N. Cennamo et al., "Localized surface plasmon resonance with five-branched gold nanostars in a plastic optical fiber for bio-chemical sensor implementation," *Sensors*, vol. 13, no. 11, pp. 14676–14686, 2013.
79. N. Cennamo, R. Galatus, F. Mattiello, R. Sweid, and L. Zeni, "Design of surface plasmon resonance sensor in plastic optical fibers based on nano-antenna arrays," *Procedia Eng.*, vol. 168, pp. 880–883, 2016.
80. D. Feng, G. Liu, Q. Li, J. Cui, J. Zheng, and Z. Ye, "Design of infrared SPR sensor based on bimetallic nanowire gratings on plastic optical fiber surface," *IEEE Sens. J.*, vol. 15, no. 1, pp. 255–259, 2015.

81. N. Cennamo, G. D'Agostino, R. Galatus, L. Bibbò, M. Pesavento, and L. Zeni, "Sensors based on surface plasmon resonance in a plastic optical fiber for the detection of trinitrotoluene," *Sensors Actuators B Chem.*, vol. 188, pp. 221–226, 2013.
82. N. Cennamo, L. Zeni, L. De Maria, C. Chemelli, M. Pesavento, and A. Profumo, "Surface plasmon resonance in a D-shaped plastic optical fibre: Influence of gold layer thickness in monitoring molecularly imprinted polymers," *2016 IEEE Sensors Applications Symposium (SAS)*, 2016, pp. 1–5.
83. N. Cennamo, L. De Maria, C. Chemelli, A. Profumo, L. Zeni, and M. Pesavento, "Markers detection in transformer oil by plasmonic chemical sensor system based on POF and MIPs," *IEEE Sens. J.*, vol. 16, no. 21, pp. 7663–7670, 2016.
84. M. Pesavento, L. De Maria, D. Merli, S. Marchetti, L. Zeni, and N. Cennamo, "Towards the development of cascaded surface plasmon resonance POF sensors exploiting gold films and synthetic recognition elements for detection of contaminants in transformer oil," *Sens. Bio-Sensing Res.*, vol. 13, pp. 128–135, 2017.
85. N. Cennamo, G. D'Agostino, M. Pesavento, and L. Zeni, "High selectivity and sensitivity sensor based on MIP and SPR in tapered plastic optical fibers for the detection of l-nicotine," *Sens. Actuators B Chem.*, vol. 191, pp. 529–536, 2014.
86. D. R. Raj, S. Prasanth, T. V. Vineeshkumar, and C. Sudarsanakumar, "Ammonia sensing properties of tapered plastic optical fiber coated with silver nanoparticles/PVP/PVA hybrid," *Opt. Commun.*, vol. 340, pp. 86–92, 2015.
87. Z. Xie et al., "Polymer optical fiber SERS sensor with gold nanorods," *Opt. Commun.*, vol. 282, no. 3, pp. 439–442, 2009.
88. W. Li, X. Zhao, Z. Yi, A. M. Glushenkov, and L. Kong, "Plasmonic substrates for surface enhanced Raman scattering," *Anal. Chim. Acta*, vol. 984, pp. 19–41, 2017.
89. X. Hu, P. Mégret, and C. Caucheteur, "Surface plasmon excitation at near-infrared wavelengths in polymer optical fibers," *Opt. Lett.*, vol. 40, no. 17, p. 3998, 2015.
90. M. K. Szczurowski, T. Martynkien, G. Statkiewicz-Barabach, W. Urbanczyk, and D. J. Webb, "Polarimetric sensitivity to hydrostatic pressure and temperature in birefringent dual-core microstructured polymer fiber," *Fourth European Workshop on Optical Fibre Sensors*, vol. 18, no. 12, p. 76530D, 2010.
91. T. Martynkien, P. Mergo, and W. Urbanczyk, "Sensitivity of birefringent microstructured polymer optical fiber to hydrostatic pressure," *IEEE Photonics Technol. Lett.*, vol. 25, no. 16, pp. 1562–1565, 2013.

LED-POF-Photodiode as Sensing Elements in High Voltage Environment

Marcelo Martins Werneck

Contents

4.1 POF as Transmission Media

As opposed to electrical sensors, fiber optic sensors are very convenient for the electric power industry as they are dielectric and therefore can be directly applied to high voltage without electrical hazards problems.

In the oil and gas industry, electrical sensors face another problem when deployed in places where there are chances of gas and fuel leaks, also known as intrinsic security locations. In such cases passive fiberoptic sensors also play an important role for many parameters to be sensed.

In general, POF media presents the following advantages as compared to silica fibers:

- Good transmission for VIS
- High numerical aperture: Goes well with LED
- Highly multimodal: Low bend loss
- Easy maintenance (D.I.Y. in the field)

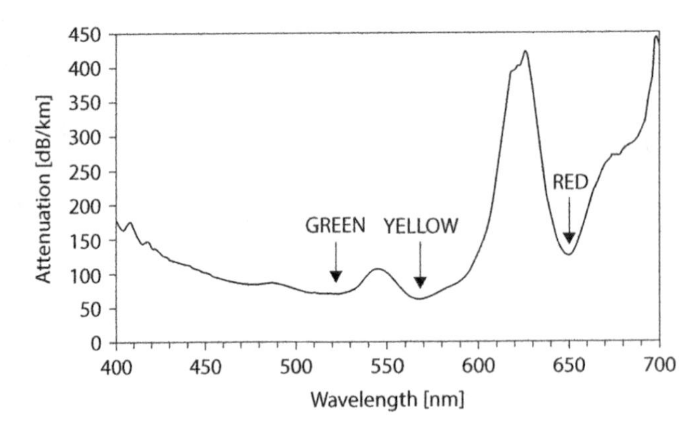

FIGURE 4.1 Attenuation spectra of the POF ESKA (Mitsubishi, Japan) for different grades. (Adapted from Joncic, M. et al., Investigation on spectral grids for VIS WDM applications over SIPOF, *2013 ITG Symposium Proceedings-Photonic Networks*, Leipzig, Germany, pp. 1–6, 2013.)

The main disadvantage of POF is the relatively high attenuation that POF presents. **Figure 4.1** shows a typical attenuation spectrum of a POF in the visible spectral range.

Notice that the attenuation increases with wavelength with more than 10 dB/m at near infrared. We have, however, three transmission windows, in green, in yellow, and in red. Figure 4.1 shows the spectrum with arrows indicating the three preferable bands. These bands should be chosen when a long-distance data relay is necessary.

In **Table 4.1** we selected these three wavelengths and calculated the attenuation a POF would present for several distances.

For a 100 m link, for instance, if we choose a wavelength of 570 nm we will get at the end of the fiber only 20% of the total power injected at the beginning.

For a few centimeters' length sensor, such as a biosensor, it does not matter the wavelength of the source as the attenuation is negligible. For a 10 cm sensor, for instance, the power lost is about 2% for any one of the three wavelengths. For this case, we can even use an infrared LED, at say, 850 nm, with an attenuation of 3.2 dB/m. This would provide a 92% attenuation with

Table 4.1 Total Attenuation for the GH 4001 Datacom Grade POF at Three Transmission Windows for Different Lengths

Wavelength	520 nm	570 nm	650 nm
Color[a]	Ultra-Pure Green	Ultra Green	Super Red
Attenuation (dB/m)	0.08 dB/m	0.07 dB/m	0.12 dB/m
Attenuation (%)/m	98%	98.4%	97.3%
Attenuation @ 10 cm	99.82	99.84	99.72
Attenuation @ 5 m	91%	92.3%	87.1%
Attenuation @ 10 m	83%	85%	75.9%
Attenuation @ 20 m	69.2%	72.4%	57.5%
Attenuation @ 40 m	47.9%	52.5%	33.1%
Attenuation @ 100 m	15%	20%	6.31%

[a] As per Lumex Color Guide (www.lumex.com).

a loss of only 8% of power. However, if we use a silicon photodetector, its responsivity at this wavelength is 0.5 A/W, that is, about the double of that at 550 nm. This means that we end up in advantage, as we gain more in light conversion through the photodetector than we lose in attenuation due to the higher fiber attenuation at this wavelength.

4.2 POF in a High Voltage Environment

Insulator leakage current is the current flowing from high voltage conductor to ground over the outside surface of the insulator. Leakage current occurs in any high voltage insulator, either in transmission lines or in distribution lines installed outdoors, due to the progressive coating of conductive deposits from environment pollution.

There are many types of pollution that affect the insulation property of transmission lines. These are classified on basis of contaminant substance like salt, cement, fertilizers, coal, dust, volcanic ash, chemicals, smoke etc. All these contaminants are conductive and will provide some leakage currents over the insulator surface.

Near coastal areas most insulator pollution is due to airborne salt particles from a nearby ocean. Small water bubbles are launched in the air from the waves and storms and are transported by the wind to the coast.

In the presence of mist, fog, or water condensation, particles on the insulator surface will dissolve with the water droplets and will provide an alternative path from high-voltage to ground potential.

What happens on an insulator also happens with any insulating surface that connects high voltage potential to ground potential, such as an optical fiber. However, contrary to the supportability of the insulator material to temperature, e.g. glass, polymer, or porcelain, the plastic optical fiber will not withstand the temperature produced by the currents and will melt. For this reason, we have to deal with the POF the same way we deal with insulators to avoid or to control the leakage currents.

There are some standards that help technicians to choose the best insulator for a specific region, such as the IEC 60815 (2008), which defines parameters for insulator selection, for instance, pollution severity level, creepage, or arcing distance.

Arcing distance means the shortest path which the voltage applied across the two extremities of the insulator can puncture the air outside the insulator in order to short-circuit to ground potential. Creepage distance means the shortest path between two conductive parts measured along the surface of the insulation. Therefore, creepage distance is the shortest distance, or the sum of the shortest distances, along the contours of the external surfaces of the insulator between which the operating voltage is applied.

Increasing the creepage distance by increasing the number of sheds of an insulator or by augmenting its length gives the insulator more protection when it is applied on polluted areas. The amount of pollution is defined by the IEC as site pollution severity (SPS) into five classes of pollution, characterizing qualitatively the site severity from very light pollution to very heavy pollution as follows:

- A – Very light
- B – Light
- C – Medium
- D – Heavy
- E – Very heavy

The pollution classification, A to E, is defined by the salt deposit density (SDD), which is the amount of sodium chloride (NaCI) in an artificial deposit on a given surface of the insulator divided by the area of this surface, generally expressed in mg/cm^2. Then, according to the IEC 60815 (2008), the equivalent salt deposit density (ESDD) is the amount of NaCl that, when dissolved in demineralized water, gives the same volume conductivity as that of the natural deposit removed from a given surface of the insulator divided by the area of this surface.

The ESDD allows calculation of the creepage distance or the unified specific creepage distance (USCD) in mm/kV an insulator must attend in order to be installed in such location. **Figure 4.2** shows a graph of USCD versus SPS that helps define the creepage distance.

The same goes for optical fiber—the length between the point where it touches the high voltage to the point at ground potential must attend the creepage distance of the insulators installed in that particular location.

Let's, as an example, calculate the fiber optic cable length that should be used for an application to be installed at 230 kV for a Class A pollution area. According to the graph in Figure 4.2, the USCD for Class A is 23 mm/kV. Then, for 230 kV the creepage distance of such insulator should be about 5.3 m. This creepage distance is, of course, much longer than the insulator length, which is about the same length as the arcing distance, also standardized by the IEC, and is about 2 meters.

This incompatibility poses a problem to the designers of fiber optic systems installed in high voltage environments. One cannot let the fiber hanging sort of 5 meters inside a substation area not only because the wind will swing it around but also because the space allocated for equipment inside a substation is limited.

There are only two options to solve this problem. The first one is to make the fiber cross inside a specially designed insulator with the adequate creepage distance, that is, an insulator just the same type as the one used in the application. This poses another problem, since polymeric insulators are made of a solid fiberglass core surrounded by polymer sheds. It is very difficult to open a hole of 2 mm in diameter inside a 2-meter long fiberglass rod as there are no drills that long. Insulators would have to be manufactured just for this application, which would became a very expensive process as the production quantity would be only a few samples. Another solution was to contract a company to fabricate such a drill and make the hole by ourselves. It took one week to drive a 2-meter long hole in our workshop due to the need to keep a very low speed, otherwise the drill curves and misses the core's center. **Figure 4.3** shows the result, a 2-m long insulator with a POF inside it.

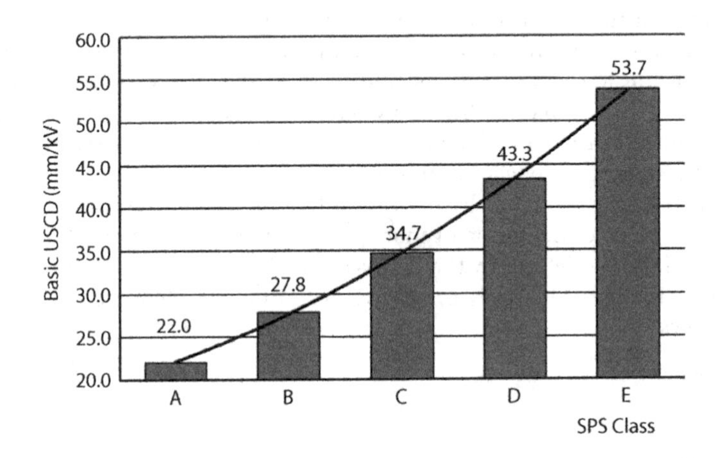

FIGURE 4.2 Selection and dimension of high voltage insulators intended for use in polluted conditions according to IEC 60815 (2008). The graph shows the unified specific creepage distance (USCD) in mm/kV an insulator must attend in order to be installed in classified locations (site pollution severity—SPS).

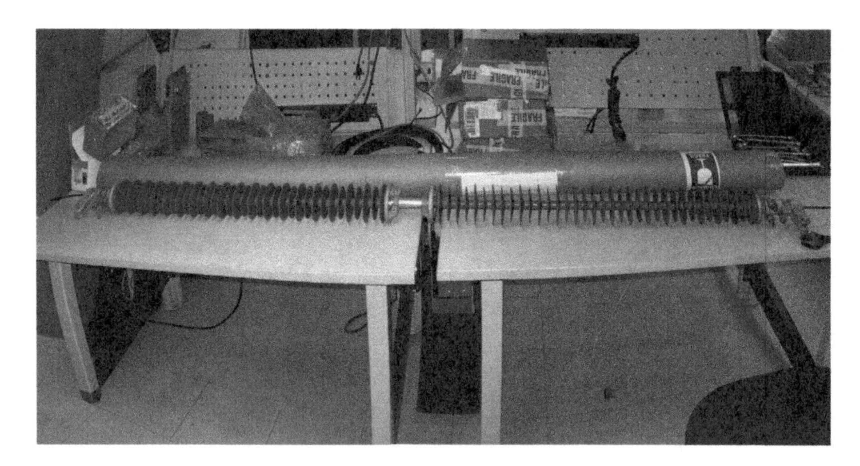

FIGURE 4.3 Two-meter long insulator with a POF inside it, adequate to work at 230 kV.

FIGURE 4.4 Two steps in the fabrication of a polymeric insulator.

As the solution we found was not practical for series manufacturing, we developed another method together with a local polymeric insulator factory just to attend a few prototypes made in our laboratory.

Figure 4.4 shows the two steps in the fabrication of a regular 13.8 kV polymeric insulator. The first step (top picture) is to clamp the galvanized cast iron terminals at each extremity of the fiberglass core. The second step is to place the core with terminals inside a mold and inject the monomer that polymerizes to form the polymer sheds of the insulator, as shown in the lower picture.

The idea of this solution is to take advantage of the established fabrication steps and insert an optical fiber inside the insulator. We carved a helical groove around the surface of the fiberglass core as shown in **Figure 4.5a**. Then, the fiber was wound inside the groove around the fiberglass core. **Figure 4.5b** shows the extremity where the fiber goes into the core and exits at its top.

Then, two cast iron extremities as the one shown in **Figure 4.5c** were clamped to each side of the core with the fiber passing inside them and leaving the assembly by each extremity. Finally, the assembly is inserted into the mold and filled with the monomer. The final fiber optic insulator is shown in **Figure 4.5d. Figure 4.6** shows a picture of the insulator without its ends where it is possible to see two fibers coming out of the insulator.

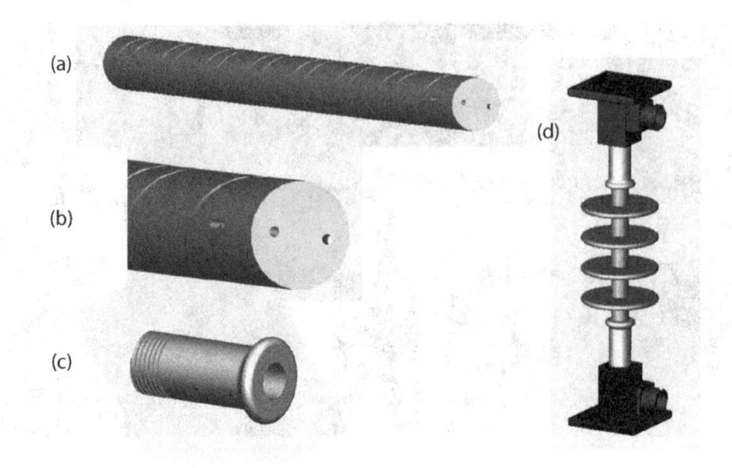

FIGURE 4.5 (a–d) Steps in the fabrication of a POF insulator.

FIGURE 4.6 Insulator for an optical fiber. Notice the two plastic optical fibers coming out of the insulator in one of its extremities.

As said above, there are two methods to protect the fiber against the leakage current. One was to make it pass inside an insulator as show in the last example. The other method is to expose the whole length of the creepage distance to the air, but in a way that it would occupy a small space.

For this we used a spiral plastic tube used for compressed air. The advantage of this spiral tubing is that it is very rigid and therefore does not need to be fixed by each loop, but rather only by each 10 loops and by the extremities. The properties for the one we used are presented in **Table 4.2**.

In order to achieve the necessary five meters of creepage distance, we used 16 loops. The fiber is then inserted into the tubing and the tubing is filled with liquid silicone for eliminating all air inside the tubing. Each extremity is then sealed inside a fiber optic splice box. **Figure 4.7** shows one of the applications of this system installed on one section of a 230 kV insulator. Notice one fixing point at the middle of the insulator.

High voltage equipment or any equipment to be installed in contact with high voltage must comply with several standards. One of the standards that insulators must comply with is the IEC 383 (1993). A fiber optic cable or fiber optic system to be installed in a high voltage environment must also comply with this standard.

Table 4.2 Proprieties of the Spiral Tubing	
Inside diameter	8 mm
Outside diameter	12 mm
Ambient temperature	−35°C to 60°C
Coil diameter	95 mm
Material	TPE-U(PU)—Thermoplastic Polyurethane
Number of windings	As many as necessary, up to 20
Shore hardness	D 52±3
Maximum pressure	10 bar
Maximum working length	2 m
High resistance to UV radiation and stress cracks	

FIGURE 4.7 Achieving 5 m of creepage distance for an optical fiber by the use of spiral tubing.

For instance, the following tests are among those specified by the IEC 383.

- Dry lightning impulse withstand voltage, meaning the lightning impulse voltage that the insulator withstands dry, under the prescribed conditions of test.
- 50% dry lightning impulse flashover voltage, meaning the value of the lightning impulse voltage that, under the prescribed conditions of test, has a 50% probability of producing flashover on the insulator, dry.

- Wet power-frequency withstand voltage, meaning the power-frequency voltage that the insulator withstands wet, under the prescribed conditions of test.
- Puncture voltage, meaning the voltage that causes puncture of a string insulator unit or a rigid insulator under the prescribed conditions of test.

Another recommendation for those who intend to install polymeric materials, such as optical fibers, in high voltage environment it that no raw polymer will give satisfactory performance in an outdoor environment without some additives that modify their behavior. Typically, such additives include anti-tracking agents, UV screens and stabilizers, antioxidants, ionic scavengers, etc. The ideal additive will help the polymer material to exhibit hydrophobicity and the capability to transfer hydrophobicity to the layer of pollution. For more information on these topics the reader should refer to the IEC/TS 62073 (2016).

4.3 LED Properties

In this section we will see how to use an LED as an amplitude modulated (AM) current sensor. As an AM sensor, the measuring parameter is the optical power amplitude which varies according with the measurand. For this reason any variation on the light power will be interpreted as a measurand variation and therefore we must study all parameters that might interfere with the LED light power. As one of these parameters is, of course, the temperature, let's start studying the drifts of the LED with temperature.

In an LED, the internal quantum efficiency depends on the junction temperature, and the temperature modifies the junction forward voltage. In reality, all LEDs exhibit forward voltage variation under temperature changes, and the temperature coefficient (TC) depends on the junction type. InGaAlP LEDs (yellow) have a TC between −3.0 and −5.2 mV/K, and the InGaN LEDs (blue, green, and white) present the TC between −3.6 and −5.2 mV/K. **Figure 4.8** shows the junction forward voltage as a function of temperature for different LEDs.

Notice in the graphs of Figure 4.8 that in any LED the direct voltage decreases with temperature and the smaller the LED's current, the smaller will be the TC.

The performance of the LED is also affected by the junction temperature, i.e. the temperature on the LED die. An increase in the junction temperature of the LED has an adverse effect on the power output as much as in its forward voltage. The rising in junction temperature results in a decrease of light output from the LED at a specified current.

From a simple experiment of installing a LED inside an oven and by injecting a known current and measuring the total light output, one can determine its power drift. In order to avoid an increase of the LED temperature due to the input current, we decided to use a pulsed current with a very low duty cycle (<1%) in order to allow the LED to cool down between pulses. The graphs in **Figure 4.9** show the power drift of several LEDs against the temperature.

Figure 4.10 shows the power output versus ambient temperature from Nichia (Japan) LEDs where we notice the same behavior—the red LED is the one with larger drift and the blue is the one with the smallest drift.

The spectrum of the emitted light is concentrated in and around a specific wavelength defined by the energy gap Eg such that $Eg = hc/\lambda$, where h is the Planck constant, c is the speed of light, and λ is the center wavelength of the spectrum.

By increasing the junction temperature, the band gap energy decreases and the emitted wavelength increases. It follows that the peak wavelength shifts to longer wavelength. In the case of a white LED, where the white color is due to a blue LED and a fluorescence of a dopant inside the

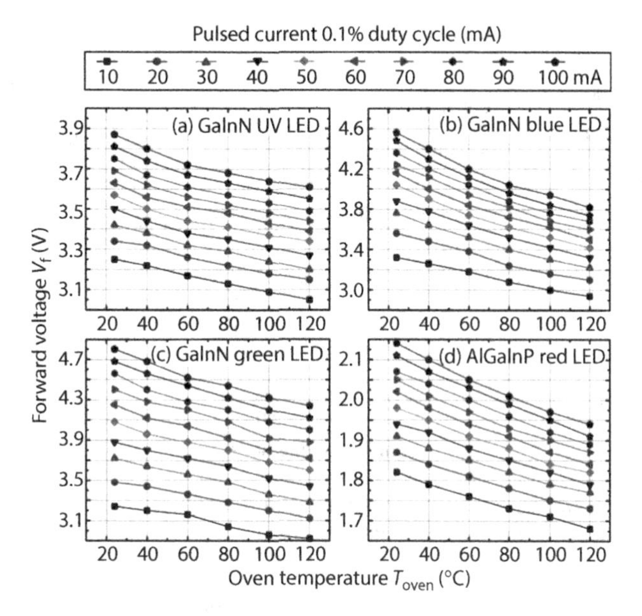

FIGURE 4.8 Junction forward voltage versus temperature for different LEDs at different currents. (Adapted from Chhajed, S. et al., Junction temperature in light-emitting diodes assessed by different methods, *Proceedings of the SPIE 5739, Light-Emitting Diodes: Research, Manufacturing, and Applications IX*, March 25, 2005.)

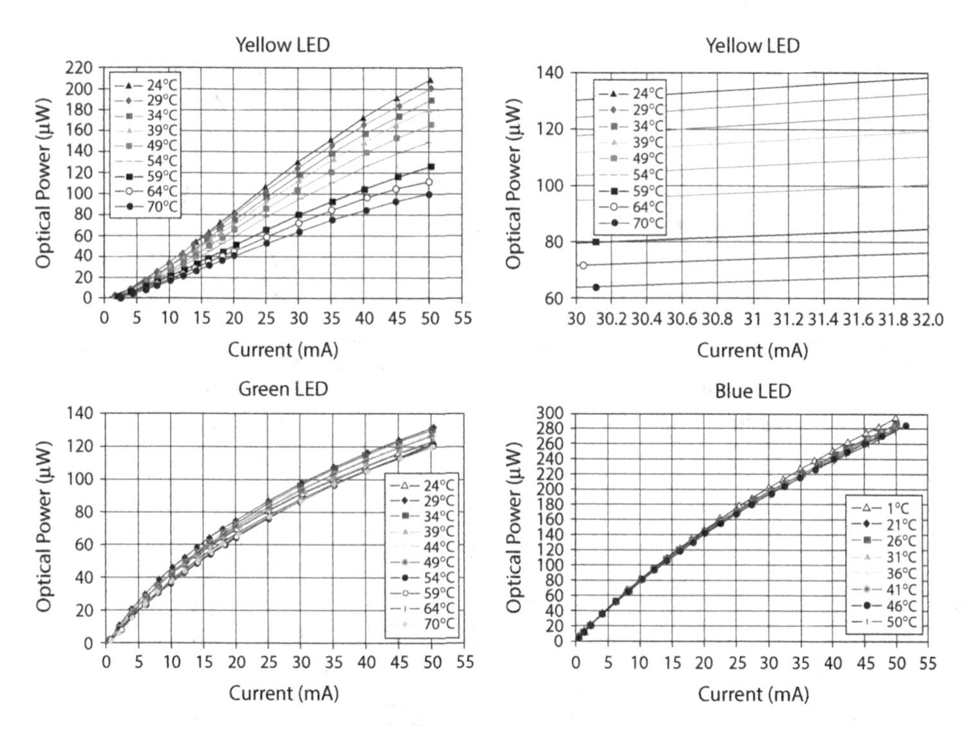

FIGURE 4.9 Power drift over temperature of blue, yellow, and green LEDs. Top Left: Yellow LED; Top Right: Same as at left but expanded between 30°C and 32°C; Bottom Left: Green LED; Bottom Right: Blue LED. Notice the smallest drift of the blue LED.

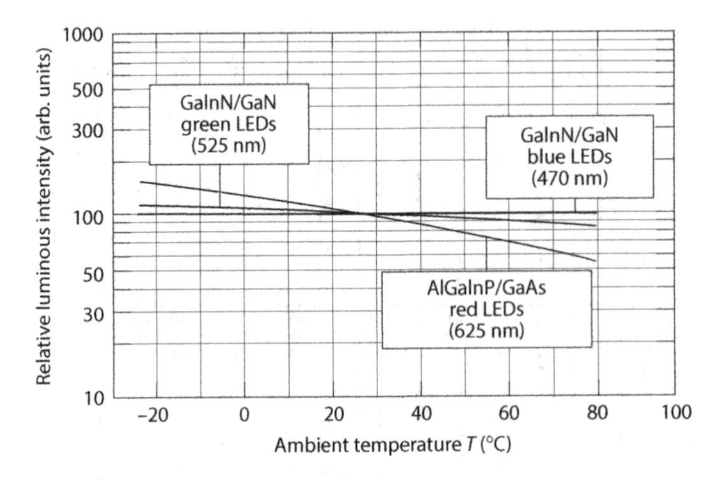

FIGURE 4.10 Output power versus ambient temperature for Nichia LEDs. Notice the smallest drift of the blue LED (with permission from Nichia Co., Japan).

LED's dome, the chromaticity coordinates would also change with rising junction temperature. From this point of view, one can use the temperature to shift the center wavelength of an LED.

Figure 4.11 shows the center wavelength drift for the three LEDs shown in Figure 4.9.

In conclusion, the junction temperature of a LEDs directly or indirectly affects its internal efficiency, maximum output power, reliability, peak wavelength, and spectral width. From the graphs shown above, we notice that the blue LED is the one with smallest drift, either in amplitude or in peak wavelength.

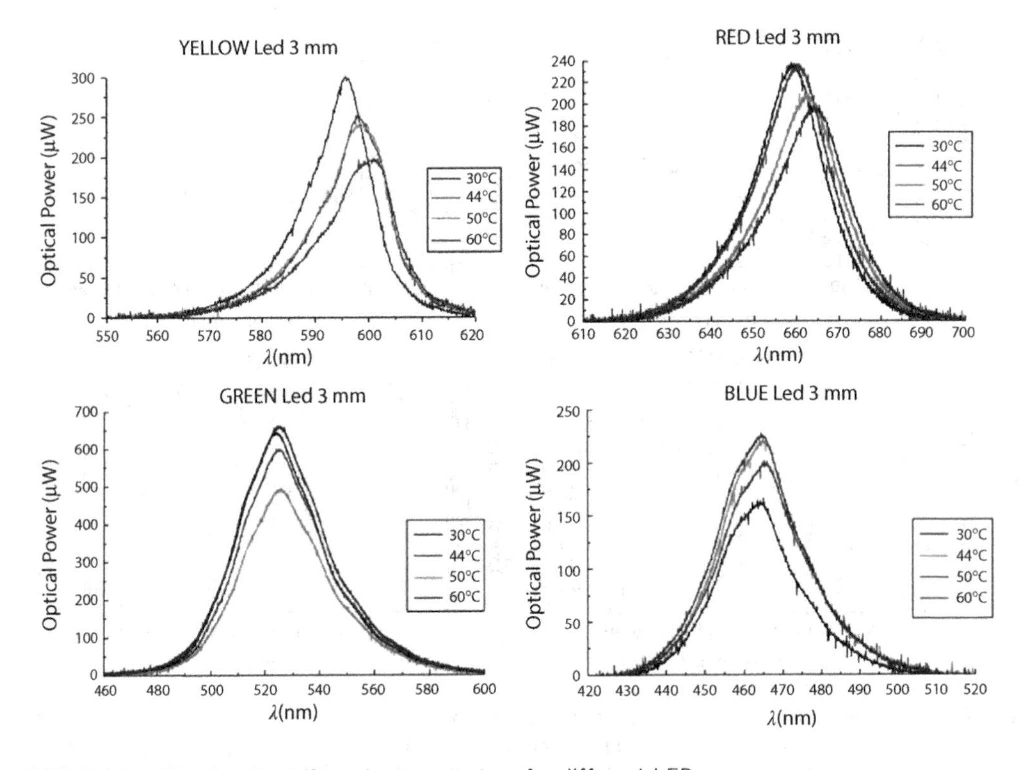

FIGURE 4.11 Wavelength drift under temperature for different LEDs.

4.4 Silicon Photodetector Properties

The spectral responsivity of a silicon photodiode is a measure of its sensitivity to light, and it is defined as the ratio of the photocurrent to the incident light power at a given wavelength. In other words, it is a measure of the effectiveness of the conversion of the light power into electrical current. It varies with the wavelength of the incident light (**Figure 4.12**) as well as the applied reverse bias and temperature.

Notice that the peak sensitivity does not match with any transmission window shown in Figure 4.1. Therefore, for a long link there will be always a tradeoff between the used wavelength, the transmission windows, and the peak sensitivity of the photodetector.

For instance, for the 570 nm transmission window, the fiber attenuation is minimal; (0.07 dB/m) however, at this wavelength the photodiode response is only about 0.32 A/W, whereas for 670 nm wavelength the photodiode response is about 0.45 A/W with a fiber attenuation 0.12 dB/m. For small distances this discussion is irrelevant, but for long distances one must study carefully the pros and cons.

The responsivity variations due to change in temperature are shown in **Figure 4.13**. This happens because the increase or decrease in the temperature causes a decrease or increase of the band gap, respectively.

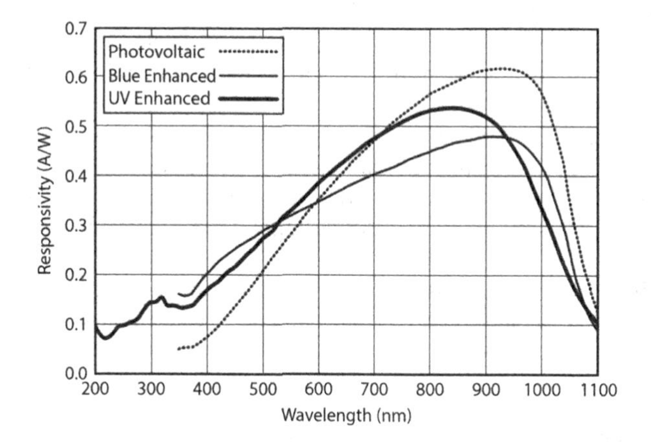

FIGURE 4.12 Responsivity of a photodetector. (From OSI Optoelectronics—www.osioptoelectronics.com. With permission.)

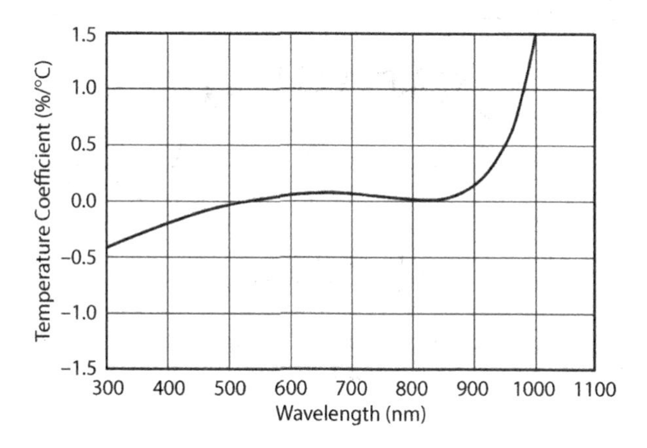

FIGURE 4.13 Temperature coefficient of a photodetector responsivity. (From OSI Optoelectronics—www.osioptoelectronics.com. With permission.)

This drift is important to be taken into account because if the center wavelength of the transmission light reaches the photodetector anywhere in the left slope of its sensitivity, as it happens with all transmission windows, as the photodetector drifts its peak, according to the graph in Figure 4.13, this effect will be the same as an edge filter, producing variable outputs when temperature varies.

4.5 LED-POF-PD Compound

Now that we studied the LED, the POF, and the photodetector, we can now put all together and check for its performance under temperature variation.

Figure 4.14 shows a setup used for measuring current in a high voltage environment, which we refer as a LED-POF-PD assembly. This configuration will appear in many projects and applications in this book, being useful to measure current, voltage, or simply as a transmission media. At the left side of the optical fiber is the measurand, in this case, the current. At the right side is the interrogation end with a photodetector that is driven by a transimpedance amplifier. This is a typical situation when one needs to measure a parameter located in a high voltage environment. Notice that the LED side of the system does not need to be energized.

In case of using the LED-POF-PD assembly for an amplitude modulated sensor, both the LED, as a light source, and the photodetector need to be as stable as possible. The main factor to change these characteristics, as we have seen above, is the temperature. Therefore, one has to choose the right LED for such application.

Now, by employing a blue LED as light source and a transimpedance amplifier to drive the photodetector, we can monitor the voltage output against the LED input current for different temperatures. **Figure 4.15** shows the obtained family of curves. All curves drift; however, for 5 mA input current, the output voltage drifts negligibly in the room temperature range.

By applying the LED-POF-PD assembly, several setups can be developed for measuring voltage, current, displacement, etc. in high voltage environments as will be seen in later chapters.

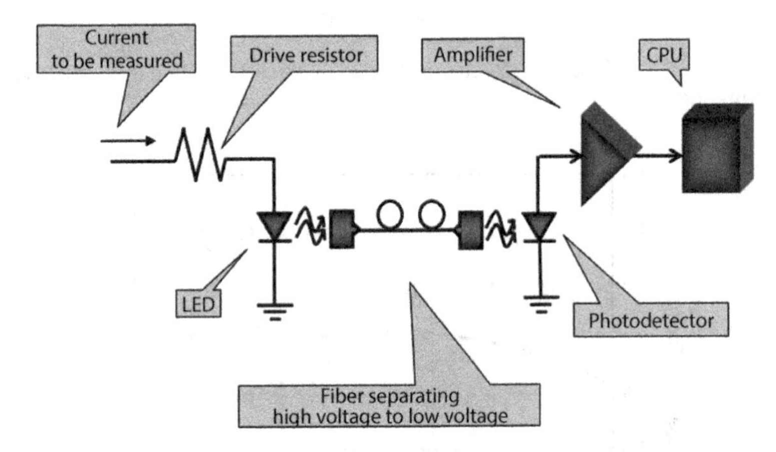

FIGURE 4.14 LED-POF-PD assembly in a typical installation where the left side of the POF is the measurand and at the right side is the interrogation end, where the photodetector is driven by a transimpedance amplifier.

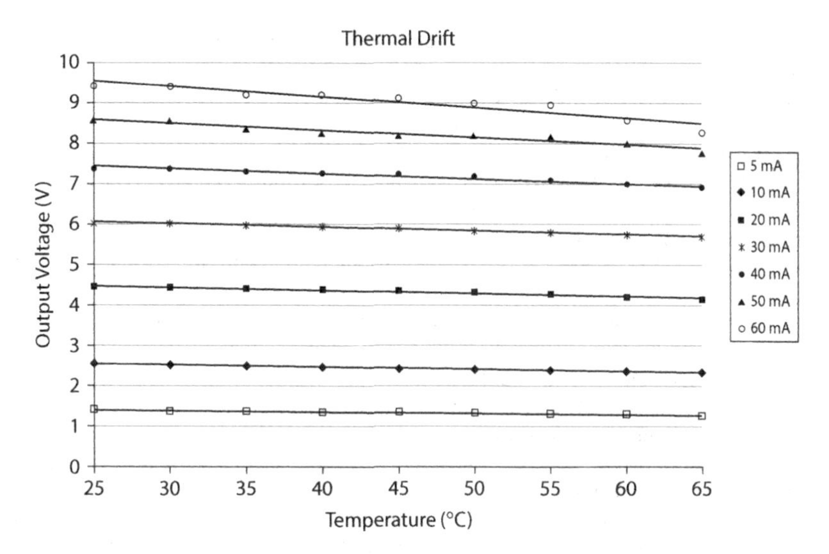

FIGURE 4.15 Family of curves showing the temperature drift of the photodetector output voltage versus the LED input current.

References

Chhajed, S., Xi, Y., Gessmann, T., Xi, J.Q., Shah, J.M., Kim, J.K., and Schubert, E.F., "Junction temperature in light-emitting diodes assessed by different methods", *Proceedings of the SPIE 5739, Light-Emitting Diodes: Research, Manufacturing, and Applications IX*, March 25, 2005. doi:10.1117/12.593696.

IEC 60815. Guide for the selection and dimensioning of high-voltage insulator for polluted conditions, Edition 1.0, International Electrotechnical Commission, 2008.

IEC 383. Insulators for overhead lines with a nominal voltage above 1000 V, International Electrotechnical Commission, 1993.

IEC/TS 62073. Guidance on the measurement of hydrophobicity of insulator surfaces, International Electrotechnical Commission, 2016.

Joncic, M., Haupt, M. and Fischer, U.H.P., "Investigation on spectral grids for VIS WDM applications over SIPOF," *2013 ITG Symposium Proceedings-Photonic Networks*, Leipzig, Germany, pp. 1–6, 2013.

Further Reading

Werneck, M., Santos, D., Carvalho, C., de Nazaré, F., and Allil, R., "Detection and monitoring of leakage currents in power transmission insulators", *IEEE Sensors Journal*, 2014. doi: 10.1109/JSEN.2014.2361788.

Werneck, M.M., "POF sensors and systems at the photonics and instrumentation laboratory", *IV International Symposium on Non-Crystalline Solids, VII Brazilian Symposium on Glass and Related Materials 4th International School on Glasses and Related Materials*, Aracaju - Sergipe, Brasil, 21 a 28 de Outubro de 2007.

Werneck, M.M., "Application of POF sensors in energy, oil and biotechnology", *3rd International POF Modelling Workshop 2015*, September 21, 2015, Nuremberg, Germany, joint with the 24th International Conference on Plastic Optical Fibers, held at Technische Hochschule Nürnberg Georg Simon Ohm, Nuremberg, Germany, 22–24 September, 2015.

Plastic Optical Fiber Sensors

Werneck, M.M., "POF sensors and systems at photonics and instrumentation laboratory", *Proceedings of the 15th International Conference on Polymer Optical Fibre - ICPOF2006*, "The Joint International Conference on Plastic Optical Fiber & Microoptics 2006", held in the Grand Hilton Seoul, Seoul, South Korea, pp. 265–271, 11–14 September, 2006.

Werneck, M.M., Carvalho, C.C., Santos, D.M., da Silva-Neto, J.L., Maciel, J.L., and Allil, R.C., "Development and field tests of na LED/POF-based leakage current sensor industrial prototype for 13.8 kV distribution lines", *Proceedings of the 21st International Conference on Plastic Optical Fibers - POF 2012*, Atlanta, GA, 10–12 September, 2012.

Werneck, M.M., Santos, D.M., de Nazarél, F.V.B., da Silva Neto, J.L., Allil, R.C., Ribeiro, B.A., Carvalho, C.C., and Lancelotti, F., "Detection and Monitoring of Leakage Currents in Distribution Line Insulators", *Proceedings of the IEEE International Instrumentation and Measurement Technology Conference (I2MTC 2014)*, pp. 468–472, Radisson Montevideo Victoria Plaza Hotel & Conference Center, Montevideo, Uruguay, May 12–15, 2014.

Werneck, M.M., Silva, A.V., Carvalho, C.C., Souza, N.C.C., Miguel, M.A.L., Beres, C., Yugue, E.S. et al., "Fiberoptic applications in sensors and telemetry for the electric power industry". *1st Workshop on Specialty Optical Fiber and their Applications - 1st WSOF 2008*, Volume 1055, pp. 43–45, São Pedro-SP, August 20–22, 2008.

Werneck, M.M., Zubia, J., Poisel, H., Kalymnios, D., Krebber, K., Sully, P., "POF Sensors applications in every day's life", *33rd European Conference and Exhibition on Optical Communication, (ECOC 2007) - International Congress Center (ICC)*, Berlin, Germany, September 16–20, 2007.

Current and Voltage Sensing

Marcelo Martins Werneck

Contents

5.1 Fiber Optic–Based Current and Voltage Measuring System for High Voltage Distribution Lines

5.1.1 Introduction

The monitoring of current, voltage, and power in distribution lines of medium-voltage (13.8 kV) is a routine procedure for planning and quality evaluation of electric delivery systems. In densely populated areas, electric networks can grow without limits, and the load is constantly changing. This leads to the constant need to monitor the line with the appropriate periodicity.

All monitoring systems to date still use conventional, magnetic current transformer (CT) and voltage transformer (VT) with iron core and copper winding. Both are instrument transformers and are used for metering as well as protection purposes. Conventional CTs and VTs, although low-cost, still suffer from an enormous drawback, which is the difficulty of installation.

The installation of such equipment demands hours of work requiring the line to be switched off, which is obviously expensive and not desirable for social and economic reasons.

From these observations came the motivation for developing a practical system with the following characteristics: simultaneous measurement of voltage, current and power factor, low cost, low weight, easy installation using a hot stick without the need to de-energize the line, and especially, electric safety.

This section describes the development and testing of a fiber optic system that simultaneously measures and records current, voltage, and power factor in distribution lines that can be installed and de-installed in a few minutes by live-line techniques.

5.1.2 System Description

The system is comprised of a voltage transducer, a current transducer, a transmitter, an optical link, a receiver, and a data logger, one set for each phase. Each module simultaneously measures the current in one phase and the voltage between that phase and another one. The voltage and the current measured in each phase are modulated and sent down via a POF cable.

The voltage transducer measures the voltage between two phases by a high value resistive divider (47 MΩ) located inside a 25 kV-rated hollow polymeric insulator. At the top of the insulator, a hot-line clamp is fixed in such a way that it can be tightened onto the conductor by a hot stick. The resistors inside the insulator are mounted on a fiberglass printed circuit board with their ends at a suitable distance to avoid sparks. **Figure 5.1** shows the measuring scheme.

The voltage signal from the resistive divider modulates in amplitude a 10 kHz sinusoidal carrier that is added to the current signal. The resulting signal modulates in frequency a 70 kHz train of 20 ns pulses, thus generating a pulse-frequency modulated (PFM) signal. Finally, this signal is fed to a current amplifier that drives a green LED, which emits the light pulses through a plastic optical fiber. The reason for using 20 ns light pulses is for energy saving, since all power comes from a 3200 mA-h lead battery. The PFM scheme also guarantees a stable transmission, independent of fiber attenuation of bending losses.

With this scheme, all circuits in the transmitter require about 8 mA, and the unit is capable of monitoring the high voltage line for a period of up to 10 days without the need to recharge the battery. The transmitter PCB is housed in a robust weatherproof box under IP65 protection, capable of withstanding adverse environmental conditions. This box is fixed at the bottom end of the hollow insulator. The transmitter is also equipped with an IP-65 protected on/off push button and a "normally open" test button. When this button is pressed, prior to deployment, the transmitter generates a test signal proportional to the battery voltage, therefore testing not only the battery charge, but also the transmitter circuits, the fiber optic link, and the receiver. There are three interconnected modules like the one shown in **Figure 5.1**, one for each phase, as shown in **Figure 5.2**.

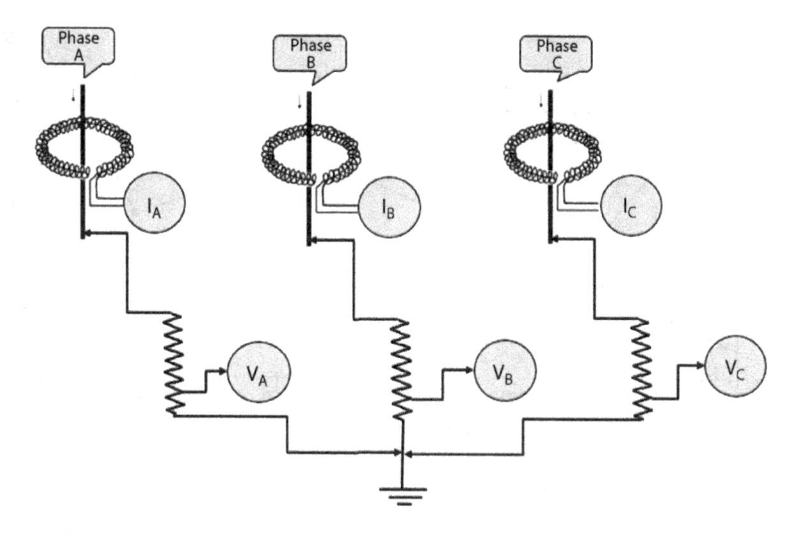

FIGURE 5.1 The measuring setup.

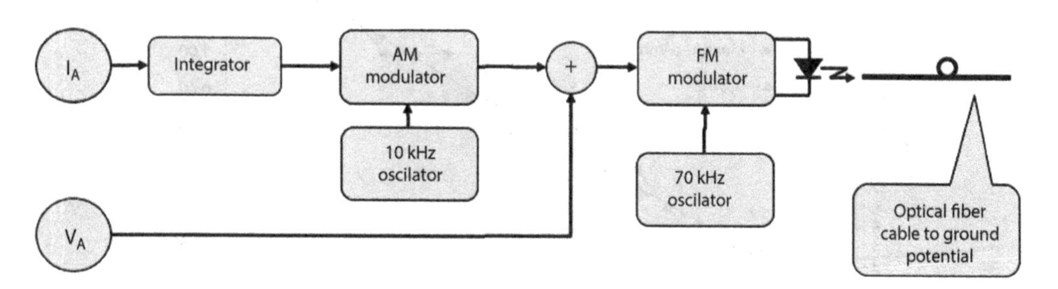

FIGURE 5.2 Block diagram of the sensing module.

Each module contains one current probe and one resistive divider inside the hollow insulator. The connection between the modules and the main board is made through copper wires carrying a low voltage representative of the line voltage and the current signal, representative of the current line.

5.1.3 The Receiver

At the receiver, a digital phase locked loop (PLL) restores the original composite signal from the train of pulses received by a PIN photodiode. A band pass filter centered at 10 kHz and an envelope detector restores the voltage signal. A low pass filter restores the current signal directly from the output of the PLL. Both signals are made available to the microprocessor PCB. After being filtered by a second order 60 Hz low-pass filter, both signals are injected in an RMS converter and displayed by an LCD at the instrument's front panel.

5.1.4 Calibration

The system is able to measure up to 800 A and 15 kV, corresponding to the +5 V full scale of the A/D converter. To calibrate the system in current mode, a current rig was constructed using two serially connected CTs and a 200-mm^2 conductor in a loop. The current was varied from zero to 800 A, and the measured current was compared with the direct measurement from a calibrated current meter. The output read and the respective error against the input current is shown in **Figure 5.3**. The linearity in this range was better than 1.5%.

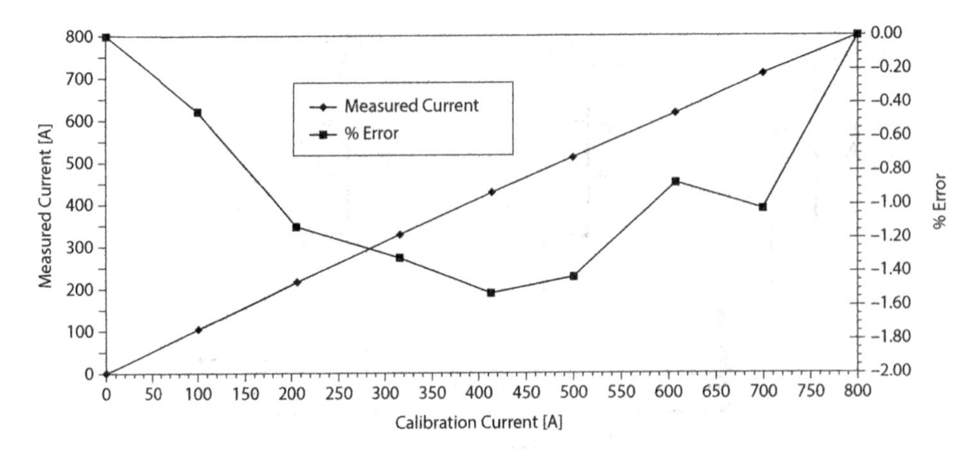

FIGURE 5.3 Current calibration and respective error vs. calibration current.

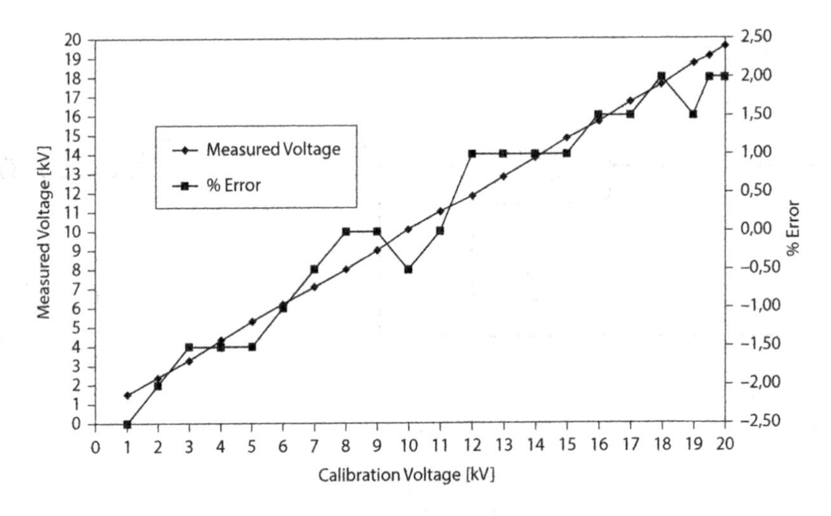

FIGURE 5.4 Voltage calibration and respective error vs. calibration voltage.

For the calibration in voltage mode, a high voltage rig was constructed and isolated in a restricted area. The transmitter was connected to the high voltage feeder of the transformer, and the voltage was varied from zero to 15 kV. The output read and the respective error against the input voltage is shown in **Figure 5.4**. The errors range between +2.5% and −2.5%.

Crosstalk tests were made using full scale in one channel and measuring the output signal at all the remaining channels of the receiver. The errors due to crosstalk were less than 1.5%.

5.1.5 Installation in the Field

After calibration procedures, a first prototype was taken to the field. This prototype was composed of one current sensor and one voltage sensor. In this way we deployed three prototypes for monitoring the three phases at the same time.

The installation was carried out by the maintenance staff of the local electric utility in the city of Niterói, state of Rio de Janeiro.

The measuring scheme of this prototype is shown in **Figure 5.5** (top) and the prototype ready for deployment in **Figure 5.5** (bottom).

The prototype is composed of the voltage sensor to be installed in one phase and a current sensor to be installed on another phase. The control box receives the POF cable and contains a data logger.

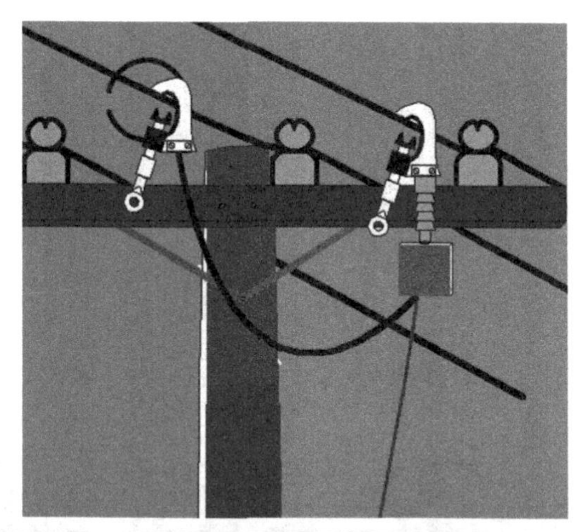

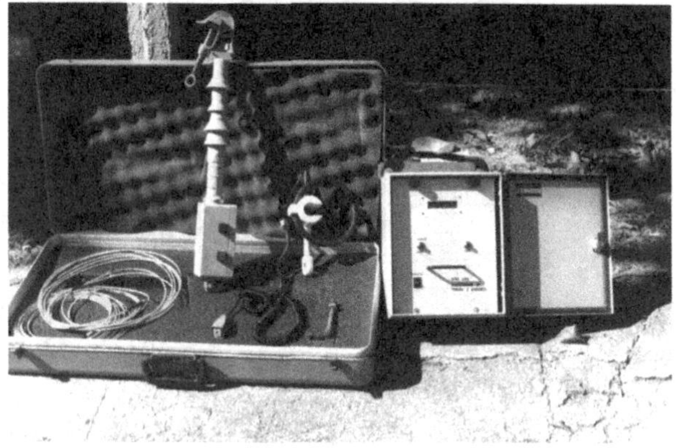

FIGURE 5.5 Top: The measuring scheme of the first prototype. Bottom: Prototype ready for deployment.

Figure 5.6 shows the installation procedure in which three independent systems were installed in order to monitor the three phases. Notice that for each unit, the voltage sensor is installed in one phase and its associated current sensor in another phase. In this way all phases are monitored both for current and voltage.

The deployed system has shown to be adequate and reached the proper sensitivity needed for the application. In view of the successful operation, we constructed a new prototype composed of three voltage sensors and three current sensors and one single datalogger that receives one optical cable from each sensor. **Figure 5.7** (left) shows the complete system, composed of the three modules, ready to be deployed. **Figure 5.7** (right) shows the installation procedure.

We decided to test the unit at the main feeder of the substation because each feeder is provided with a current and a voltage transformer for measurement purposes, and therefore these measurements could be used as calibration for our system. **Figure 5.8** shows the unit being deployed at a 13.8 kV distribution line using live-line techniques. The installation procedure is comprised of the following steps: First the whole system is clamped to the central phase (Phase B). Then, each module is detached from the main assembly and clamped to the other phases, one after another. **Figure 5.8** (left) shows the first module being clamped to Phase C and **Figure 5.8** (right) shows the system completely installed.

FIGURE 5.6 Three independent systems were installed as a first performance test.

FIGURE 5.7 Left: The complete system ready to be deployed. Right: Installation procedure.

The system was left on the line for a period of 10 days, programmed to record minimum and maximum values of current and voltage. After this period, the measurements stored in memory were retrieved and compared with the hourly records made from that particular feeder by the substation staff. Unfortunately, the resolution of the CT and PT meters present in the substation have poor resolutions (50 A and 100 V respectively) and therefore it

FIGURE 5.8 Installation of the unit at a high voltage line. Left: The central module already installed on Phase B and the first module being installed on Phase C. Right: The complete system installed.

was very difficult to make comparisons. Nevertheless, the data stored in the system memory showed good correspondence with the records made at the substation.

5.1.6 Discussion

Plastic optical fiber was used in this project not as sensor but as a transmitting medium, necessary to electrically isolate high potential to ground potential.

The system described here is designed to substitute conventional techniques of measuring current and voltage in high voltage lines. In all tests, either in the laboratory or in the field, the system performed well. One can mention here advantages and drawbacks of the system.

CTs are not only employed to measure current and voltage (and consequently, energy), but also for the protection of transformers and other equipment against surges and short-circuits. The price of a conventional CT grows exponentially with rated voltage. For a medium sized substation, one can spend tens of thousands of dollars in protection, the major part of that in CTs. Therefore, the use of the system described in this work in new substations could substantially reduce construction and maintenance prices.

The system power consumption is about 15 mW, too high to allow a relatively light battery last more than 10 days. This would be acceptable for a temporary system but quite inadequate for the present day "deploy and forget" tendency for permanent monitoring systems. In reality, in order to use this equipment in substitution of conventional CTs, at least one year of autonomy would be necessary.

This problem can be solved through two approaches. One is by minimizing power consumption of the transmitter. Since the transmitter has only a few components, with a microchip in use, it is expected the power consumption to drop from the present 15 mW to a value as low as 200 μW.

The other approach is the complete elimination of the battery by optically powering the transmitter. This is carried out by transmitting light through the same fiber optic that connects the transmitter to the receiver, with conversion into electrical power by a photodiode in the transmitter box.

5.2 Techniques to Measure Current and Voltage with POF

5.2.1 Current Sensing

By applying the LED-POF-PD assembly discussed in the chapter "LED-POF as Sensing Elements," it is possible to measure current or voltage. Contrary to the last application, with this scheme the measurement is possible without the need for electric energy at the measuring point. At the same time, because the POF cable is an insulator, we guarantee a complete isolation between the high voltage potential and ground potential.

Figure 5.9 shows the setup for measuring low current in high voltage. The current to be measured is made to cross the transmitting LED after being rectified by four diodes in bridge. The two transorbs protect the system against transients, always present in high voltage transmission lines. We used a fiber optic insulator to increase the creepage distance between high voltage potential and ground potential to avoid leakage currents over the fiber cable surface.

5.2.2 High Current Sensing

Figure 5.10 shows a modified setup for measuring high currents. In this case, the high current is divided to an appropriate value by the current transformer. The current transformer is an open magnetic circuit capable to be installed around the conductor without opening the electric circuit. **Figure 5.11** shows an example of transducer for measuring high current. The screw on the bottom of the transducer is used for closing or opening the magnetic core around the high voltage conductor by the use of a hot-stick.

5.2.3 Voltage Sensing

Until now, in order to measure current, it is only necessary to capture the magnetic field created by the flowing current. When we need to measure voltage, it is necessary to link the two points between which we need to measure the difference of potential. In other words, we need to convert voltage to current in order to measure voltage. For this, we use the same setup as the one shown in **Figure 5.9**, but with an added resistor in order to create a current proportional to

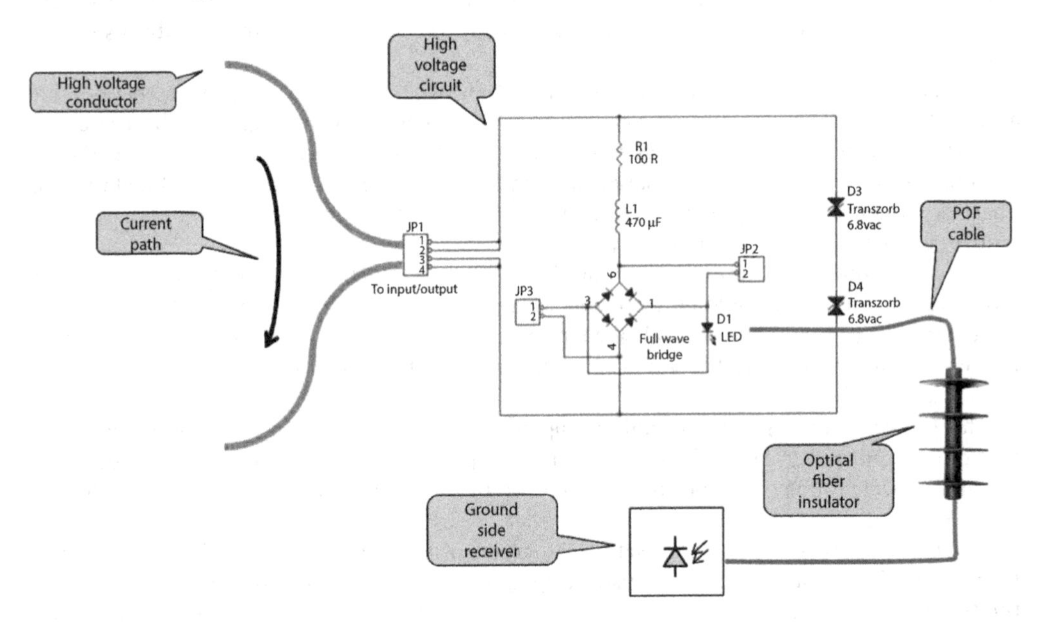

FIGURE 5.9 Technique to measure low current in a high voltage potential.

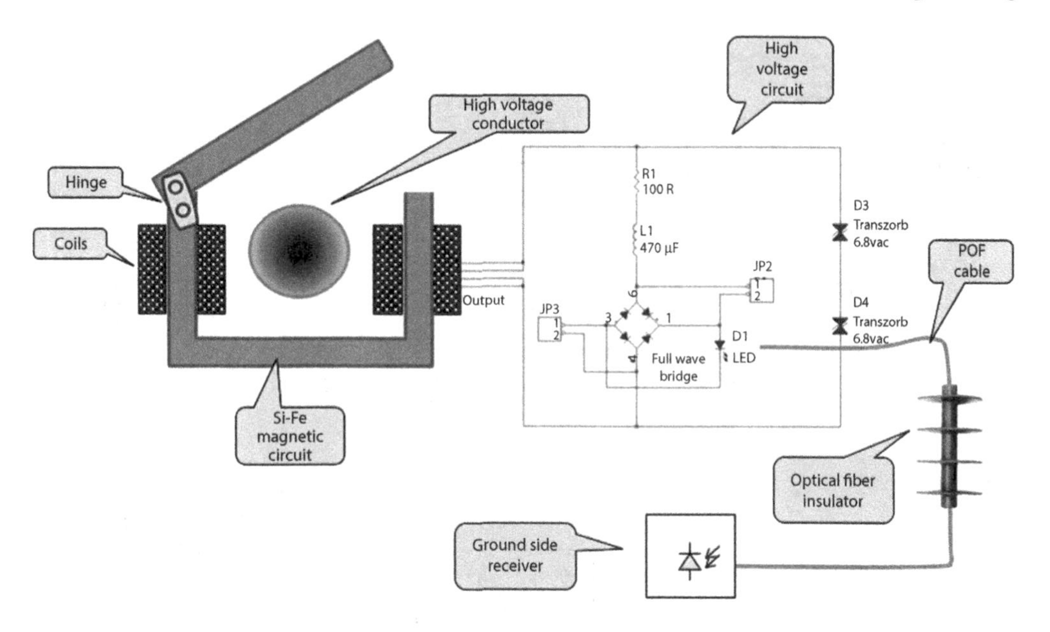

FIGURE 5.10 Technique to measure high current in a high voltage potential.

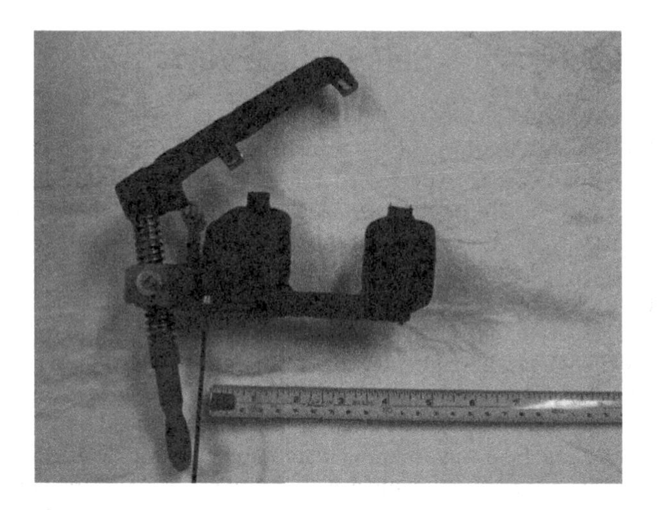

FIGURE 5.11 Prototype to measure high current in high voltage potential.

the voltage to be measured. The series resistor is used to control the current for the transmitting LED. **Figure 5.12** shows the idea. In this case, the output optical power will be proportional to the voltage.

There is a drawback, though, with this configuration. Note that the voltage applied to LEDs is very small due to the voltage drop (V) caused by the current (I) crossing the resistor ($V = RI$). However, if the current is zero, there will be no voltage drop and the high voltage will appear on the lower side of the circuit, which is obviously a real danger either to the circuitry or to the operators. Zero current can be due to an open or bad earthing or open circuit such as a burned LED.

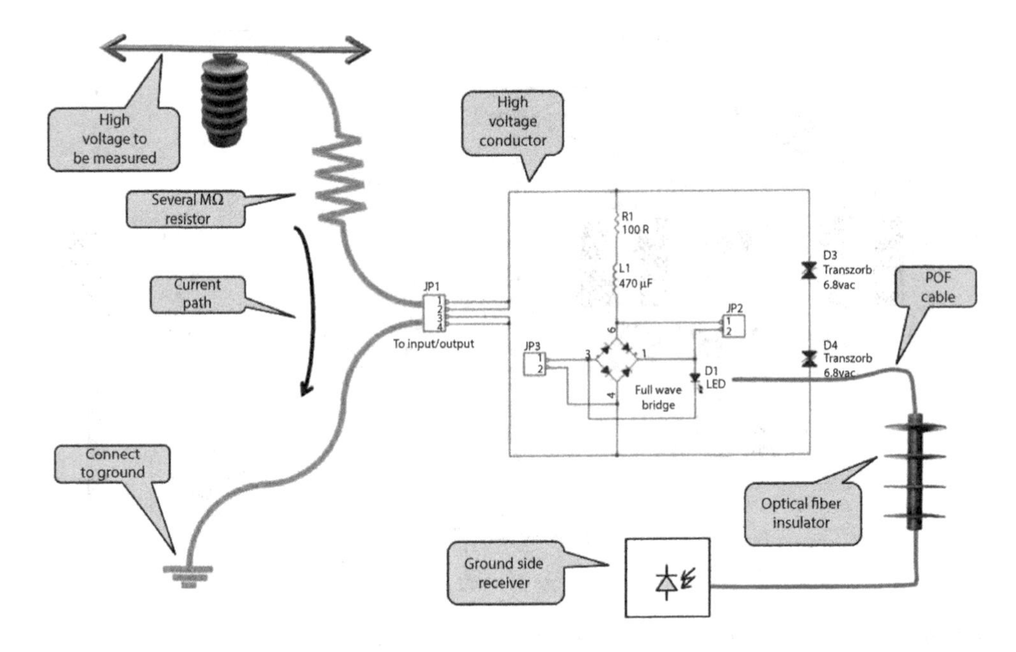

FIGURE 5.12 Measuring high voltage. The resistor is used to control the current through the LED.

5.3 Current Monitoring in High Voltage Lines

5.3.1 Introduction

The conventional technology used in voltage transformer (VT) and current transformer (CT) by the electricity companies has experienced an enormous evolution and is employed by them with reliability and effectiveness. However, new technologies of electric current sensing emerged which are better for solving inherent problems found on the 100-year-old iron/copper technology.

One of those technologies is the optoelectronics associated with plastic optical fiber (POF) that present advantages particularly to the power industry, which is the immunity to electromagnet interference (EMI), small weight and volume, permit passive sensing (no electric energy required), and particularly, does not allow electric path from the power line to ground, which means high personnel security. The optic technology has demonstrated simplicity and reliability.

Actually, there is no practical way to know instantaneously the flowing current in a power line if it is not equipped with a CT. The only way to do so is to turn off the power, install a temporary CT, turn on the line, measure the current, turn it off again, uninstall the CT, and power on the line again.

The objective of the system described in this section is to demonstrate a POF-based current sensor which has been used to measure currents on a 13.8-kV distribution line. The sensor can be installed and uninstalled in a few minutes without the need do de-energize the line, or it can be left at the power line for a few days monitoring and storing current data.

5.3.2 System Description

Two versions have been developed: one prototype for quick deployment and short-term measurements, which can be taken to the field and check consumption or power outages, and a permanent fit version appropriated for substations. This version can be used for monitoring power lines and protection of power transformers against short circuit and over-currents.

The project has been divided into several parts: the sensor itself (current transformer), the electronic hardware for transmission, reception and signal processing, and the software.

The CT uses the sensing principle shown in Section 5.2.2 and is made of a magnetic circuit to be closed around the power line, producing a sample of the line current which modulates an LED that injects the light on a POF. The POF (multimode, PMMA) performs insulation between the high potential and ground potential while transmitting the current signal amplitude-modulated by the light. This CT does not need local energy as all the energy it needs comes from the power line through the magnetic field.

Two types of sensors were developed. One is the temporary version, shown in **Figure 5.11** that can be quickly deployed. The other version is the permanent CT, to be installed for long-term monitoring. **Figure 5.13** shows the prototype, which also contains the silicon-steel loop that can be closed around the conductor to be monitored.

The monitoring end of both sensors contains the optoelectronic conversion composed of a PIN photodiode driven by a transconductance amplifier and a microcontroller with enough memory to monitor the current for several days. It also contains an infrared transmitter in order to be possible to download the stored data to a notebook or smartphone. **Figure 5.14** shows a picture of the control box which is powered by 127 Vac.

FIGURE 5.13 Prototype of the permanent current sensor. Notice the four small rectangles, two at each side of the clamp that are the ends of the magnetic circuit.

FIGURE 5.14 Picture of the control box where it is possible to see the POF cable and the AC power cord. The BNC connector provides an output of the current signal.

5.3.3 Calibration

A current rig has been constructed for testing and calibrating the units. The current rig was built around an open-core current transformer driven by a high-power rheostat and a high ampacity conductor crossing the open core in a closed loop. The current sensor clamp was tightened around the conductor, and a calibrated ammeter was used in series with the current sensor. Then, the current was varied in small increments from zero to 800 A and a calibration curve was built. **Figure 5.15** shows the test rig and **Figure 5.16** the calibration curve showing a very good linearity over the entire range.

5.3.4 Field Tests

Both systems were taken to the field for tests. The short-term version was installed on the live line by a hot-stick as shown in **Figure 5.17**. The measured current was compared with the current measured by the company's monitoring equipment showing excellent agreement.

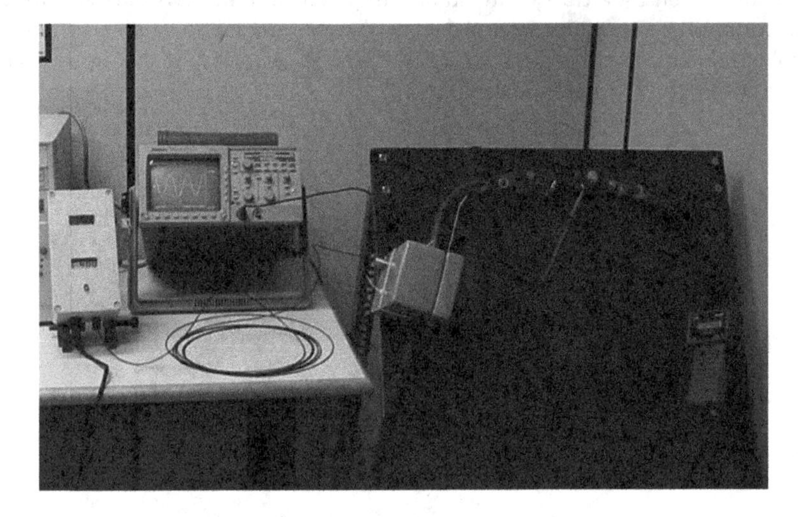

FIGURE 5.15 The test rig is comprised of a current loop capable to generate a current up to 900 A. The unit under test was connected around the current loop and a commercial ammeter was used for calibration.

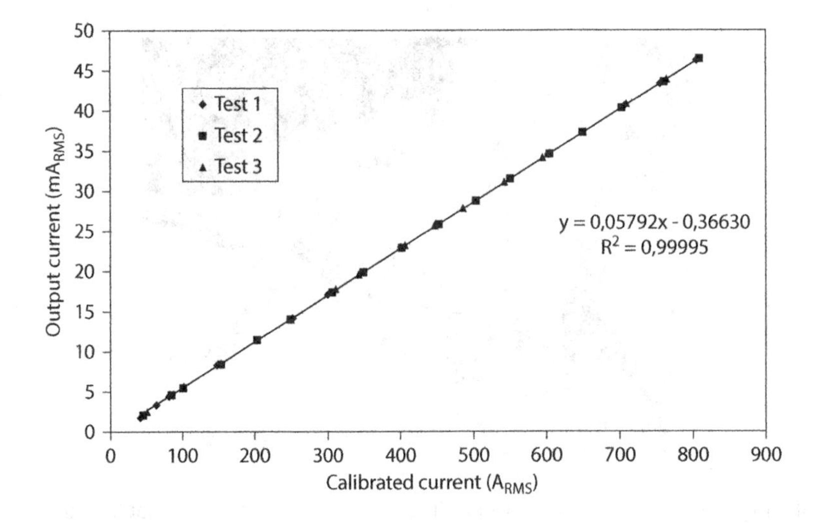

FIGURE 5.16 The calibration curve showing a very good linearity in all testes performed.

FIGURE 5.17 The current sensor being installed on a 13.8-kV live line by a hot-stick. Left: Open magnetic loop being installed. Right: The hot-stick is rotated to close the magnetic loop and disconnected from the unit.

FIGURE 5.18 The system being installed on the live line by the company's personnel.

The receiver of the permanent version is comprised of an optical interface that processes the optical signal, an AC-to-RMS converter, a datalogger that stores the signal, and a battery that guarantees autonomy for up to 10 days. The box is protected against harsh environments (see **Figure 5.14**). After the acquisition period, the data is transferred to a notebook through an infrared port located on the bottom of the box.

The system was deployed in the live 13.8-kV line by the company's personnel as shown in **Figure 5.18**. **Figure 5.19** (left) shows the complete system after deployment and (right) shows the control box/datalogger installed inside the pole.

FIGURE 5.19 Left: The complete system after deployment. Right: The control box/datalogger installed inside the pole.

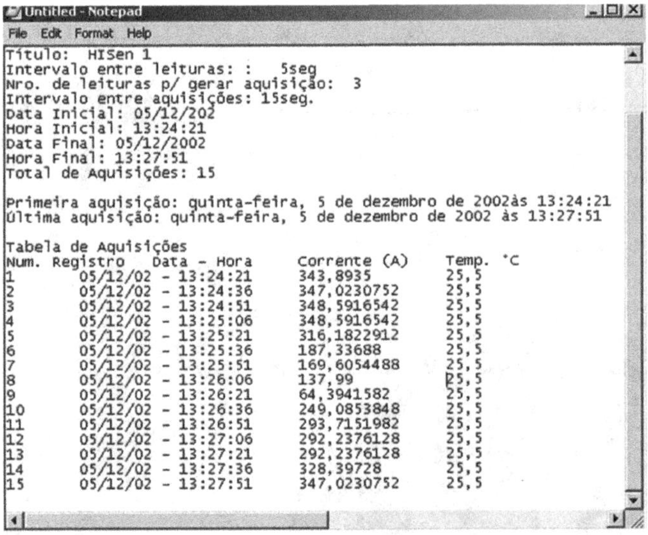

```
Untitled - Notepad
File  Edit  Format  Help
Título:  HISen 1
Intervalo entre leituras: :   5seg
Nro. de leituras p/ gerar aquisição:  3
Intervalo entre aquisições: 15seg.
Data Inicial: 05/12/202
Hora Inicial: 13:24:21
Data Final: 05/12/2002
Hora Final: 13:27:51
Total de Aquisições: 15

Primeira aquisição: quinta-feira, 5 de dezembro de 2002às 13:24:21
Última aquisição: quinta-feira, 5 de dezembro de 2002 às 13:27:51

Tabela de Aquisições
Num. Registro   Data - Hora       Corrente (A)     Temp. °C
1        05/12/02 - 13:24:21      343,8935         25,5
2        05/12/02 - 13:24:36      347,0230752      25,5
3        05/12/02 - 13:24:51      348,5916542      25,5
4        05/12/02 - 13:25:06      348,5916542      25,5
5        05/12/02 - 13:25:21      316,1822912      25,5
6        05/12/02 - 13:25:36      187,33688        25,5
7        05/12/02 - 13:25:51      169,6054488      25,5
8        05/12/02 - 13:26:06      137,99           25,5
9        05/12/02 - 13:26:21      64,3941582       25,5
10       05/12/02 - 13:26:36      249,0853848      25,5
11       05/12/02 - 13:26:51      293,7151982      25,5
12       05/12/02 - 13:27:06      292,2376128      25,5
13       05/12/02 - 13:27:21      292,2376128      25,5
14       05/12/02 - 13:27:36      328,39728        25,5
15       05/12/02 - 13:27:51      347,0230752      25,5
```

FIGURE 5.20 Printout of data collected in the field tests.

The system was left monitoring the line for 12 days. After this period a notebook was used to download the data from the datalogger. A printout of the acquired data is shown in **Figure 5.20**. This information is used for detecting trends of the line current and possible unacceptable values such as currents above the limits of the local transformer.

Figure 5.21 shows a plot of the current acquired by the system after 12 days in the field. Notice the alternating values of the RMS currents showing that maximum power consumption occurs mainly during the day, particularly in the evening. Lowest currents are observed during the night.

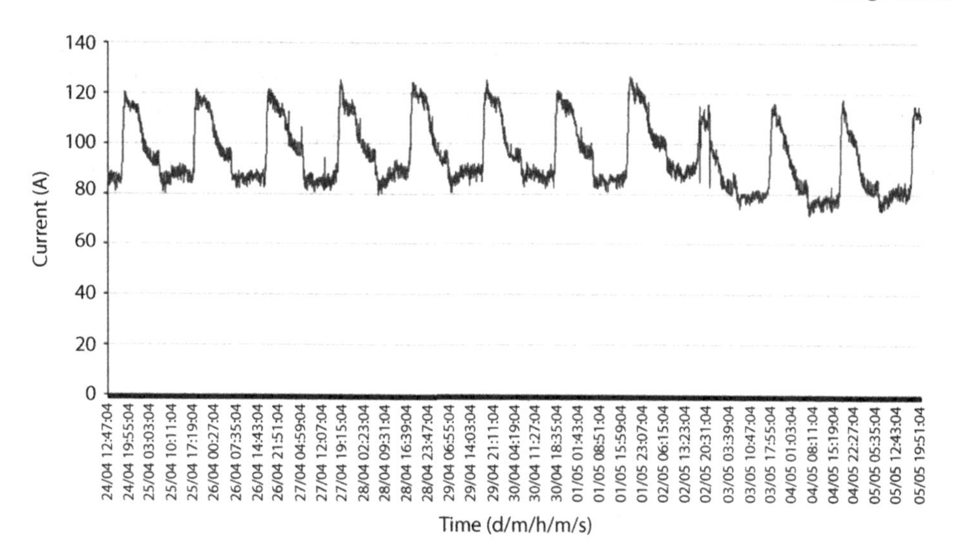

FIGURE 5.21 A plot of the signal collected at the substation.

5.4 Leakage Current Monitoring in Insulators

Generally, high-voltage transmission line insulators are subjected to different conductive pollution sources, which often cause an insulator's flashover and subsequent supply breakdown. Flashover of insulator strings is a very serious disaster affecting transmission lines with heavy fines applied to the system operator. It is also dangerous to the power grid because of the potential for short-circuits.

Contamination originates from diverse substances present in the environment surrounding high power transmission lines, for example, sea salt, dust from nearby dirt roads, agricultural fires, and by-products from local industries. These substances are deposited on the insulating surface. In dry conditions, these layers do not cause great problems; however, in wet situations, such as in the presence of rain, humidity, fog, or dew, the substances start to conduct electricity from the high voltage potential to ground potential; these currents are called leakage current. The pollution layer keeps increasing until it causes a cut-off from the switch gear of the substation that drives the transmission line. The probability of this failure occurrence depends on the nature of insulating materials, the climatic conditions of the area, the types of pollutants, the degree of contamination, and the voltage level of the transmission line.

In order to mitigate this problem, transmission line operators often contract a third-party service to wash the insulator with a jet of treated water. As this service has to be continually performed and it is very expensive, the operators have to balance the washing time based on previous experience in order to not wash insulators uselessly and yet, in good time to avoid flashovers. The Photonic and Instrumentation Laboratory was contacted to find a solution in order to inform the company when the leakage currents reach a dangerous level.

5.4.1 13.8 kV Distribution Line

The leakage current occurred in a 13.8 kV distribution line in a locale close to the sea shore. During the day, the wind brings the sea salt diluted in the air humidity that deposits on the insulators' surface. During the night fog, the dew point is reached, air humidity condenses over the insulator surface, and the salt becomes conductive starting the leakage current. We decided to install a current monitoring system on one of the insulators. Considering that all insulators are installed in similar meteorological conditions, it is sufficed one sensor in each micro-climate location.

The transducer shown in **Figure 5.22** contains the electronics, already described in previous sections, enclosed by a specially designed ceramic cup. The polymeric insulator provides an increase of the creepage distance avoiding leakage currents over the surface of the POF cable.

The POF cable is directed to the remote unit (RU) which contains the optical receiver, a humidity sensor, a temperature sensor, and a GPRS transmitter that uses a SIM card from the local telephone operator. The RU is shown in **Figure 5.23**.

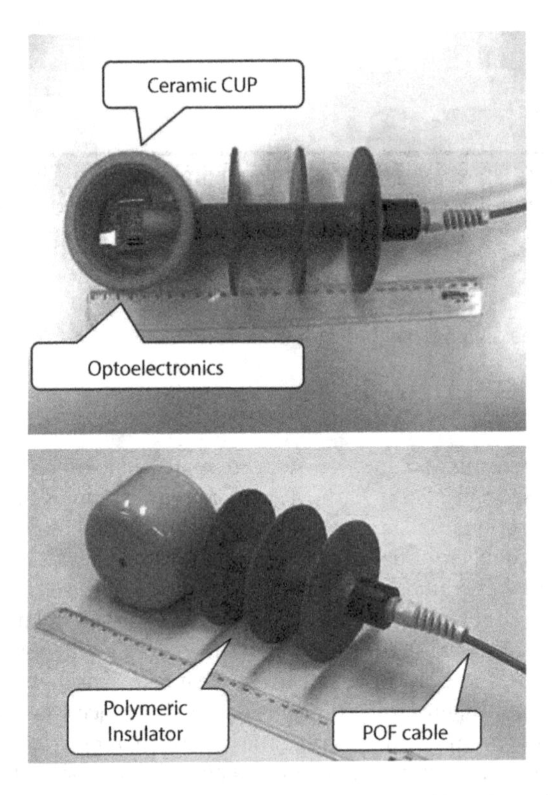

FIGURE 5.22 Current transducer for 13.8 kV distribution lines. Top: Transducer seen from below, with the cup without the filling resin. It is possible to see the optoelectronic circuit inside. Bottom: Transducer seen from the top.

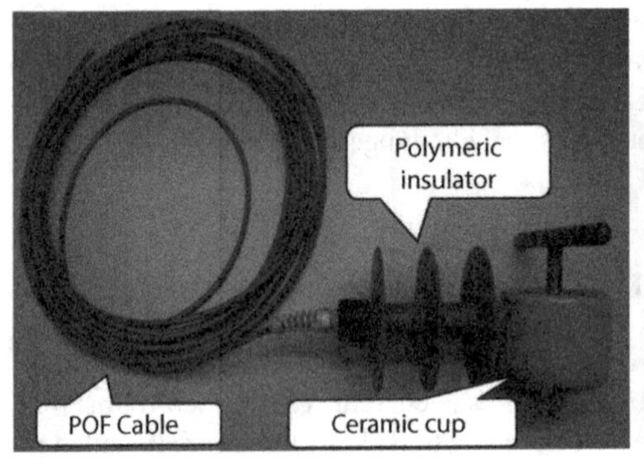

FIGURE 5.23 Left: The sensor and the POF cable. Right: The remote unit containing a humidity sensor, a temperature sensor, and a GPRS transmitter.

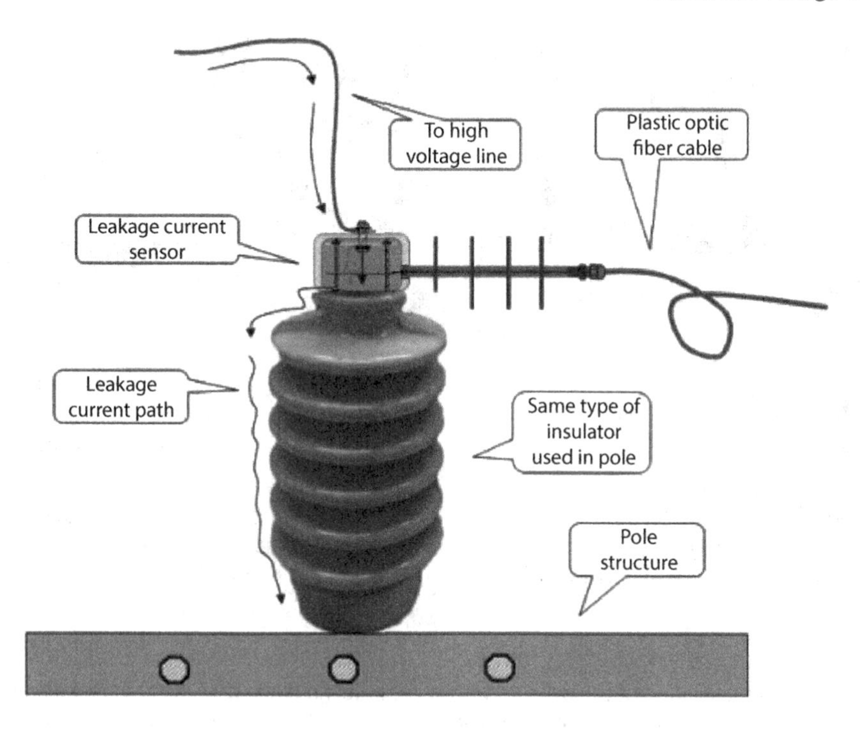

FIGURE 5.24 Installation setup showing the transducer over the insulator so as to measure all currents flowing from the high voltage to ground potential.

The installation setup is shown in **Figure 5.24**. Notice that the leakage current is made to cross the transducer before flowing through the insulator surface to ground potential. In **Figure 5.25** we can see the installation process by the use of an insulated crane that allows the technicians touch the line for maintenance or installation without the need to turn off the power line.

Figure 5.26 shows the graph obtained with such setup. In the graph it is possible to see the leakage current, the local air humidity, the local temperature, and the dew point. When the temperature drops during the night and reaches the dew point, condensation occurs on the insulator surface. This makes the insulator deposited salt become conductive. Notice that the leakage current increases just when the dew point is reached. As the average leakage current increases, in can eventually reach a previously set current limit which triggers the warning signal for washing the insulators.

5.4.2 500 kV Transmission Line Leakage Current Monitoring

As much as leakage currents occur in medium voltages, they also occur in high voltage lines. In these cases, the consequences can be worse as currents are higher and, as there are many more insulators, the total current can be so much higher that eventually they can trigger the transmission line protection and a shutdown may occur.

The same transducer used in medium voltages can be used for monitoring the leakage current on high voltage transmission lines, such as 500 kV. **Figure 5.27** (top) shows a picture of one of these towers in which a leakage current sensor was installed. At the right it is shown the installation setup on the tower steel structure. One 500 kV insulator string is made of up 25 sheds like those depicted in **Figure 5.27** (bottom). A copper conductor is used to bypass

FIGURE 5.25 (1) Installation of the sensors with hot-line techniques. (2) Sensor is installed over the insulators in a 13.8 kV distribution line. (3) Remote unit is installed on the pole in ground potential. (4) Complete system installed on the pole ready to transmit the local parameters to the remote website.

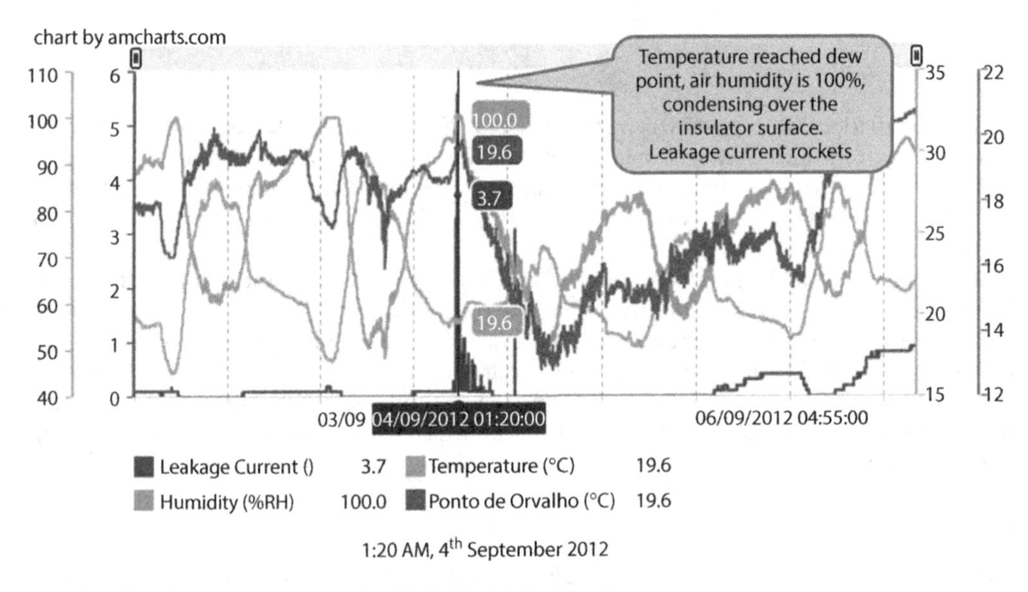

FIGURE 5.26 Current transducer for 13.8 kV distribution lines.

the leakage current from the last shed of the string to cross the sensor and then, to ground potential.

Figure 5.28 shows the control box with the optoelectronic receiver (small circuit board on the top of the batteries). The amount of energy used by the optoelectronic decoder and the GPRS transmitter is so small that two automobile batteries were enough to power the system for two years without the need of a photovoltaic panel.

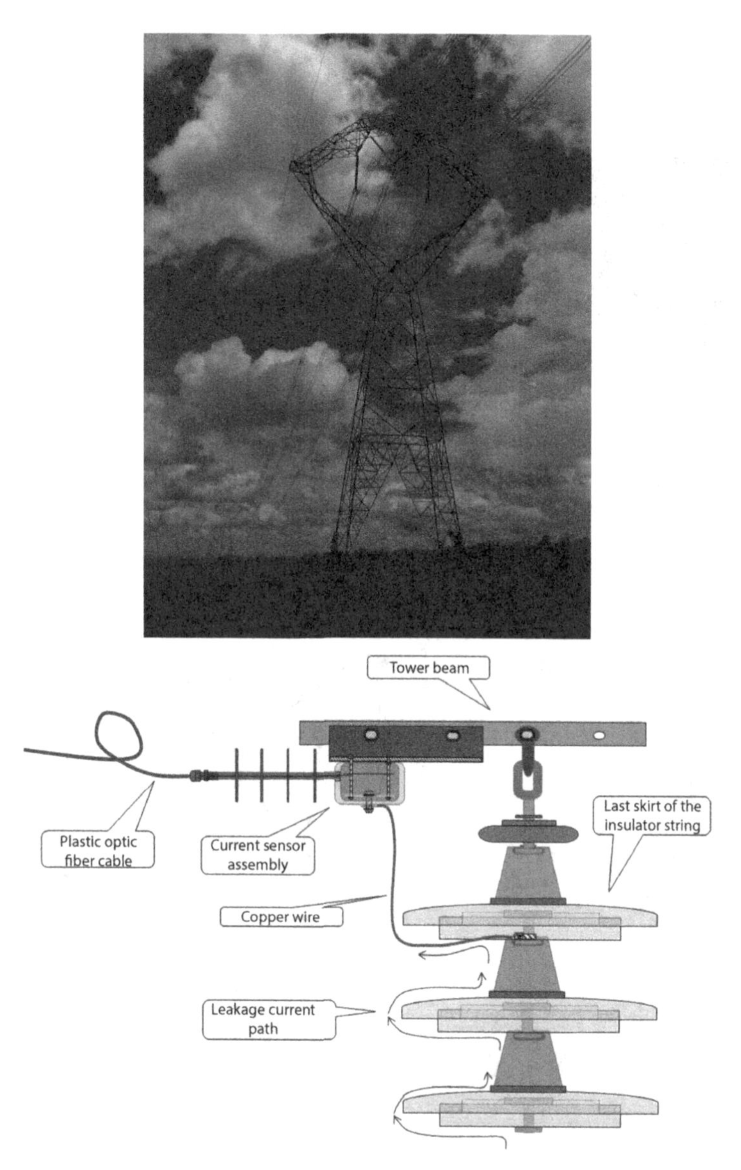

FIGURE 5.27 Top: A 500 kV transmission line tower in which leakage sensors are installed. Bottom: Setup used to bypass the leakage current from the insulator string to cross the sensor end then directed to ground potential.

Figure 5.29 shows a complete view of the installed system. The control box at bottom center is dwarfed by the large dimensions of the tower. Between the sensor and the control box we used about 25 m of optical fiber cable.

Figure 5.30 shows the obtained results for a 500-kV transmission line. The upper trace is the temperature that goes from 25°C during the night to up to 40°C at noon. The leakage current, in turn, is nearly zero during the day as the salt deposit is dry. However, during the night, as temperature drops eventually reaching the dew point, the leakage current increases because the humidity dissolves the salt into ions, therefore becoming conductive. Notice that at the center of the graph the system captured a current peak of 32.2 mA. Looks like a relatively small current,

FIGURE 5.28 The control box with the optoelectronic receiver and the batteries. The two batteries were enough to power the system for two years.

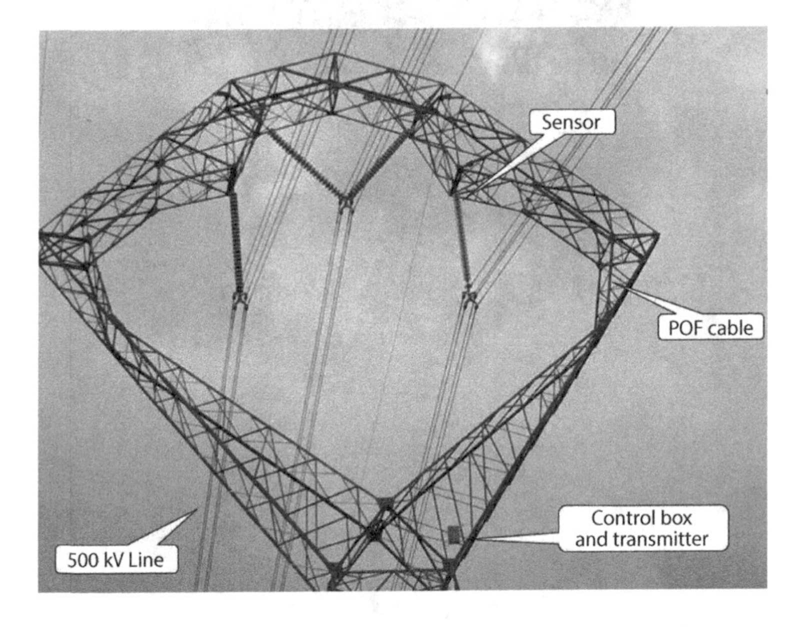

FIGURE 5.29 Installation setup for a steel tower structure.

but when converted to dissipated power at 500 kV, we get 16.1 kW. These power peaks occur stochastically, but considering that in a several hundred kilometers long transmission line there are thousands of insulators, we noticed that the power losses due to leakage currents can be really huge.

Figure 5.31 shows that, after a heavy rain, the insulator gets free of deposited salts, causing a reduced leakage current. As pollution keeps depositing on the insulator surface, there is a slow increase of leakage currents on a day-to-day basis.

The effect described above can be clearly seen in the graph shown in **Figure 5.32** with a two-month history of leakage current recording. Rainy days, identified by the temperature drop during the day, washes the insulator reducing the leakage currents. Then, as pollution keeps depositing on insulator surface, there is a daily increase of leakage currents. The leakage

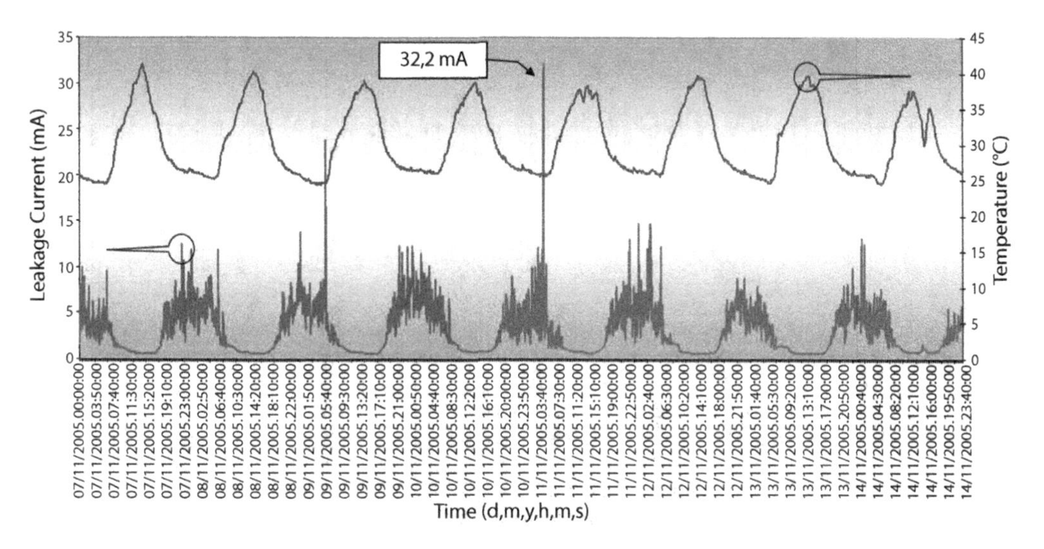

FIGURE 5.30 Local temperature (upper trace) and leakage current (lower trace) graphs of a 500-kV transmission line. The temperature goes from 25°C during the night to up to 40°C at noon. The leakage current is nearly zero during the day as salt deposit is dry and goes high during the night as the dew point is reached and the humidity dissolves the salt into ions becoming conductive.

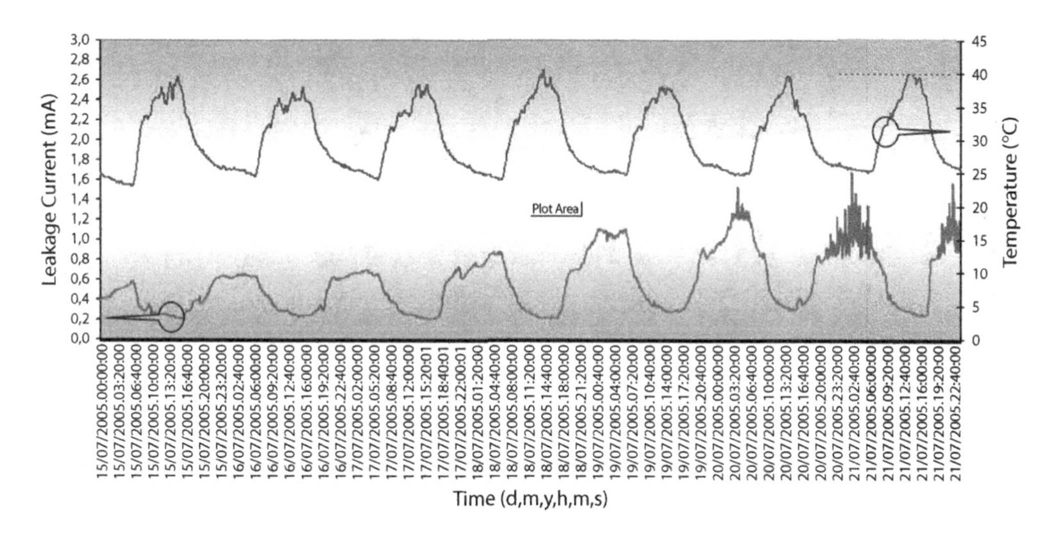

FIGURE 5.31 After a wash-out caused by the rain, the insulator gets free of deposited salt leading to a reduced leakage current. As pollution keeps depositing on insulator surface, there is a slow increase of leakage currents in a day-to-day basis.

currents keep increasing until the next torrential rain, after which a leakage current crescendo occurs again and again.

The likelihood of rain decreases in some seasons and in consequence, the pollution builds up. **Figure 5.33** shows the increase of leakage currents in an insulator, announcing an imminent flashover. This is exactly the situation that the line operators need to avoid by an alarm signal indicating the urgent need to wash the insulators.

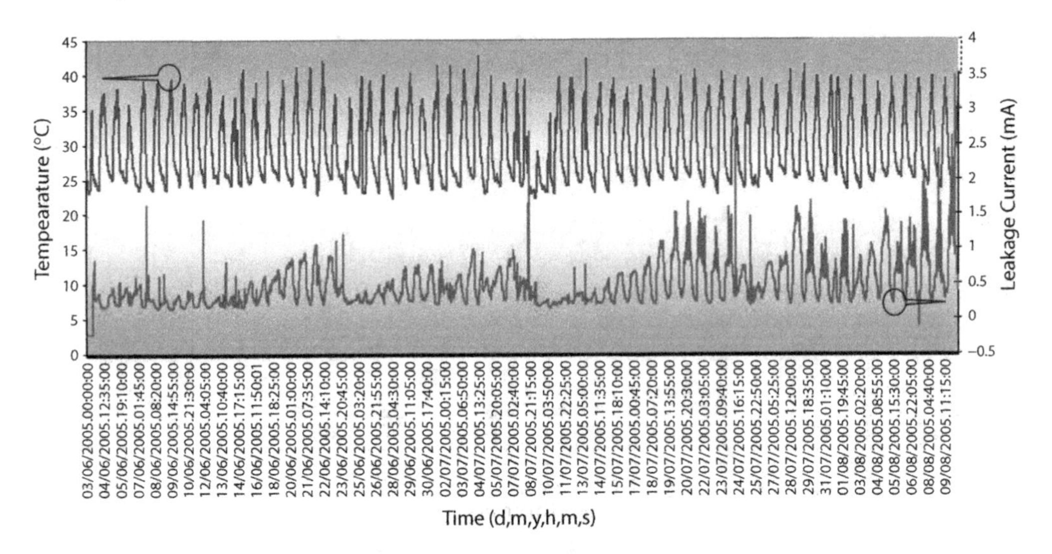

FIGURE 5.32 Two-months temperature and leakage current history. Rainy days, identified by the temperature drop during the day, washes the insulator reducing the leakage currents. Then, as pollution keeps depositing on insulator surface, there is slow increase of leakage currents on a day-to-day basis.

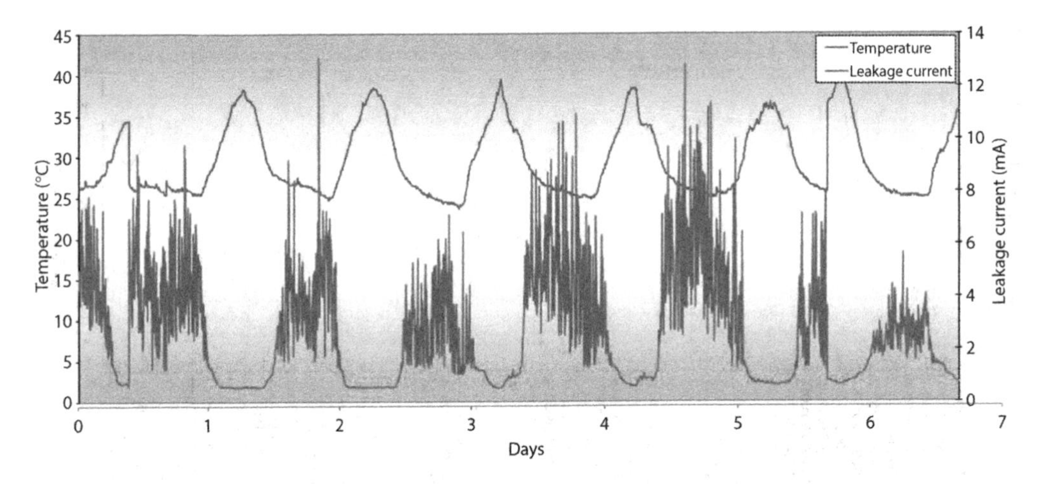

FIGURE 5.33 Heavy leakage currents captured in an insulator, announcing an imminent flashover.

5.5 Conclusions

The results demonstrated the viability of using a POF-based sensor to measure high voltage distribution and transmission line parameters. The system was patented since it has proven to be of high value to the power industry. The POF cable components have proven to be an excellent insulator and capable to withstand the 13.8 kV of the line and also ultraviolet radiation.

The advantage of this technique over the conventional technology is the low-cost and small footprint, very important for today's need for increasing efficiency of the electric network.

References

Ribeiro, R.M., Werneck, M.M., "Plastic optical fibre technology for high voltage current measurements", *11th International Plastic Optical Fibres Conference 2002 (POF 2002)*, September 18–20, Tokyo, Japan.

Werneck, M., Santos, D., Carvalho, C., de Nazaré, F., Allil, R., "Detection and monitoring of leakage currents in power transmission insulators", *IEEE Sensors Journal*, 15(3), 1338–1346, 2014. doi:10.1109/JSEN.2014.2361788.

Werneck, M.M., Abrantes, A.C.S., "Fiber-optic-based current and voltage measuring system for high-voltage distribution lines", *IEEE Transactions on Power Delivery*, 19(3), 947–951, 2004. doi:10.1109/TPWRD.2004.829916.

Werneck, M.M., Carvalho, C.C., Ribeiro, R.M., and Maciel, F.L., "High-voltage current sensing based hybrid technology", *Proceedings of the 12th International POF Conference 2003*, pp. 50–53, University of Washington, Seattle, WA, September 14–17, 2003.

Werneck, M.M., Carvalho, C.C., Ribeiro, R.M., Maciel, F.L., "Application of a POF-based current sensor for measuring leakage current in 500 kV transmission line", *Proceedings of the 13th International Conference on Polymer Optical Fibre - ICPOF2004*, pp. 345–350, Nürnberg, Germany, September 27–30, 2004.

Werneck, M.M., Germano, S.B., Ribeiro, R.M., Maciel, F.L., Porciúncula, P., Almeida, A., and Martins, L., "Plastic optical fiber technology for high voltage leakage current monitoring", *Proceedings of the 11th International POF Conference 2002*, pp. 271–273, Hotel New Otani, Tokyo, Japan, September 18–20, 2002.

Werneck, M.M., Maciel, F.L., Carvalho, C.C., Ribeiro, R.M., "Development and field tests of a 13.8 kV leakage current LED/POF based sensor", *Proceedings of the 12th International POF Conference 2003*, pp. 54–57, University of Washington, Seattle, WA, September 14–17, 2003.

POF Bragg Gratings

David Webb

Contents

6.1 Introduction

Optical fiber Bragg gratings (FBGs) usually take the form of a periodic modulation of the effective index of the fiber core along a short (mm to cm) length of fiber. This structure has the effect of reflecting back down the fiber light which is close to a resonant wavelength (known as the Bragg wavelength) given by [1]

$$\lambda_B = 2n\Lambda, \tag{6.1}$$

where n is the effective index and Λ the period of the spatial modulation. FBGs find numerous applications, such as mirrors for fiber lasers [2], dispersion compensation [3], and add-drop filters in optical communications systems [4]. From a sensing perspective, they are of interest because environmentally induced changes to the fiber can affect both the effective index and the period; strain and temperature are the main factors, though the fiber is also slightly sensitive to magnetic field via the Faraday effect, and other parameters can often be transduced into changes in strain, allowing FBGs to be used, for example to detect pressure, acceleration, and voltage.

FBGs in silica fiber have existed since the late 1970s, when they were discovered fortuitously by Ken Hill [5], thanks to an innate photosensitivity of the doped core of the fiber to UV light. Originally produced by counter-propagating light beams in the fiber, they remained a curiosity until the side-writing technique was invented by Meltz and Morey [6]. This involved interfering two beams of UV light incident from the side of the fiber at different angles and allowed the fabrication of gratings with controllable length and essentially arbitrary periods (determined by the intersection angle of the two beams). Control of the grating period allowed FBGs to be constructed with any Bragg wavelength, allowing them to be matched to suitable light sources in low loss regions of the fiber's transmission spectrum.

In the last quarter century, FBGs in silica fiber have become a mature and commercialized technology, with a multitude of companies springing up that manufacture sensors, interrogation systems, or provide complete sensing solutions. They find application where their unique features give them an advantage over conventional electronic sensors. The main features are:

- Low fiber loss allows multi-kilometer distances between the interrogation unit containing the light source and detector and the sensing elements
- Sensing elements are located in a dielectric material and hence are electrically passive
- Various multiplexing techniques have been developed, allowing up to several hundred sensors to be placed at arbitrary points along a sensing fiber
- Optical fiber is small (typically 125-micron diameter) and lightweight

The motivation for the development of polymer optical fiber Bragg gratings (POFBGs) comes from the very different material properties of polymers compared to silica. As we will see later, compared to FBGs in silica fiber, the use of polymer fiber can enable

- The sensing of higher strains
- Greater sensitivity to fiber stress
- Higher temperature sensitivity
- Intrinsic sensitivity to water

By using polymer materials, which are drawn to fiber at a much lower temperature than is the case with silica, it is also potentially feasible to use organic chemistry techniques to modify the base polymer structure to provide added functionality, perhaps by enhancing the non-linear properties of the fiber, providing optical gain or sensitizing the fiber to specific chemical species.

These potential advantages over silica FBGs come at a cost: the relative immaturity of the technology, whether that be in the lack of fundamental knowledge, for example about the grating recording mechanisms, or the lack of availability of single mode components. Addressing this technological immaturity has been the focus of much of the research on POFBGs over the last decade, and the progress made will form the central theme of this chapter.

Before going any further, it is appropriate to acknowledge the existing review articles relevant to this topic. There have been both book chapters [7,8] and journal papers [9,10] directly addressing POFBGs, but the interested reader may also want to study some of the reviews of silica FBGs [1,11–15] and polymer fiber sensing technology [16].

6.2 Overview of Major Developments

What follows is a rather subjective quasi-historical perspective of the major landmarks in the development of POFBGs. Unless otherwise stated, the work described here was carried out on fiber based on polymethyl methacrylate (PMMA), sometimes with dopants added to increase the refractive index, enabling the realization of step index fiber and/or providing enhanced photosensitivity.

The first POFBG was reported by Gang-Ding Peng's team at the University of New South Wales in Sydney, Australia, in 1999 [17] using step index PMMA-based POF they had manufactured themselves. Using a tunable optical parametric oscillator source, grating inscription was studied at 248 and 325 nm, with an interference pattern created by overlapping two beams of light in a modified Sagnac interferometer. The 325 nm radiation produced a volume grating whereas the 248 nm light formed a surface relief structure; this seems to be a consequence of the very large absorption of PMMA below 300 nm [18], which prevents the light penetrating a significant distance. For CW grating inscription in POF, 325 nm has become a commonly used inscription wavelength as it can be obtained from a relatively inexpensive and reliable HeCd laser. Peng's team was able to demonstrate a wide (20 nm) strain tuning range for POFBGs [19] along with an 18 nm thermal tuning range for a temperature variation of 50°C [20]. Later studies have reported much smaller thermal sensitivities than this, and Peng's result may have been affected by humidity cross-sensitivity, which was first discussed in the PhD thesis of Harbach [21].

In 2001, Ejkelenborg reported the manufacture of photonic crystal or microstructured POF (mPOF) [22] at the Optical Fibre Technology Centre of the University of Sydney, and the team were soon producing a wide range of types of mPOF to commercial standards. Dobb et al. (2005) at Aston University in the UK were the first to report POFBG fabrication in mPOF, using a 30 mW CW HeCd laser to produce gratings in the 1550 nm spectral region (convenient for compatibility with test equipment designed for telecommunications, but not ideal due to the 1 dB/cm attenuation of POF in this region). By gluing the mPOF to a silica fiber down-lead, the Aston team were able to take their POFBGs off the optical bench to begin exploring applications research in such areas as textile strain monitoring [23], smart elastic skins [24], and combined temperature and humidity sensors [25]. The same group were also the first to show that POF could support gratings in the lower loss 800 nm wavelength range [26] and produce the first multiplexed POFBGs, exploiting fiber annealing to produce POFBGs at different wavelengths using a single phase mask [27] (a recent paper shows how annealing can also be used to shift the Bragg reflection to wavelengths that are longer than that of the original inscription [28]). Gratings in the visible spectrum were first demonstrated by Marques et al. [29]. The water sensitivity of PMMA can be a disadvantage in many strain sensing applications, but the Aston team, working with colleagues at the Technical University of Denmark (DTU) who were developing a polymer fiber fabrication facility, demonstrated the first gratings in TOPAS cyclic olefin copolymer mPOF and showed these to have a much reduced humidity sensitivity compared to PMMA POFBGs [30]. The DTU team went on to perform a detailed study of the strain behavior and hysteresis effects in POF [31], concluding that with proper choice of parameters, it was possible to fabricate POF-based accelerometers with negligible viscoelastic effects [32]. Abang and Webb showed that where the strain applied to a POFBG was in the range where hysteresis would occur, if the grating was embedded in the structure being monitored, the hysteresis was greatly reduced compared to the situation where the strain was applied to the grating in free space [33].

Researchers had also been addressing a number of the technical challenges associated with using single mode POF as well as the practicalities of grating inscription. Unlike silica fiber, POF does not cleave but needs to be cut or sawn, and several research groups have investigated how to carry this out repeatably and with high quality [34–37]. A technique enabling POF to be connectorized in FC/PC connectors has also been described [38]. Saez-Rodriguez et al. [39], Bundalo et al., and Bonefacino et al. [40] explored techniques for reducing the inscription time of POFBGs, through doping the fiber and improving the experimental arrangement. Pospori et al. were able to inscribe an 18dB grating using a single 15 ns laser pulse [41] and extended this to realize fast, point-by-point writing of long period gratings [42] and two-pulse writing of Moiré gratings [43].

The team at DTU had the capability to fabricate preforms as well as draw fiber, and they exploited this in investigating a range of new materials with which to fabricate fiber for POFBG inscription, including high-Tg TOPAS [44], humidity insensitive Zeonex [45], and polycarbonate [46], enabling strain sensing at temperatures of up to 125°C [47]. Addressing the problems around the transmission loss that are intrinsic to most POFs, Lacraz et al. used a femtosecond laser inscription technique to write gratings in perfluorinated fiber [48] and were then able to demonstrate a 7-sensor array along a 20 m length of fiber [49].

In the remainder of this chapter, we shall look in more depth at the topics of photosensitivity, grating inscription, measurand sensitivity, the challenges posed by polymer fibers, and applications.

6.3 Photosensitivity

6.3.1 Mechanisms

The photosensitivity allowing the creation of FBGs in POF is a complex subject that is probably not yet fully understood. Indeed, as with silica fibers, there are many factors that impact photosensitivity, depending on the material composition, the presence of dopants, or the characteristics of the inscribing light (wavelength, CW, or pulsed etc…). The first studies on bulk PMMA using 325 nm light date back to the 1970s and led to the conclusion that the observed increase in refractive index and density were probably a result of the photopolymerization of unreacted monomers in the material [50,51] (an interesting point relating to this early work is that the gratings produced formed over *hundreds of hours* after illumination finished [52], whereas POFBG inscription usually takes place in *10s of minutes* during inscription). Later studies investigated photodegradation effects under illumination with wavelengths from 260 to 320 nm, with both main chain and side chain scission being observed [53,54]. It is interesting to note that in this study, no modification of the polymer was found at wavelengths longer than 300 nm, despite 325 nm being a commonly used inscription wavelength.

Following the first report of the inscription of POFBGs by Peng's group at the University of New South Wales [19], they described the dynamics of grating formation in PMMA fiber doped with benzyl methacrylate illuminated by pulsed 325 nm light with average intensity 60 mW/cm^2 [55]. There were two regimes: an initial linear decrease in index over a period of about an hour, providing an index change of about -2×10^{-4}, followed by a rapid increase in index modulation up to a maximum of around 2×10^{-3}, accompanied by losses at wavelengths below the Bragg wavelength, related to visible damage at the core-cladding boundary. Asecond paper [56] reports very different results obtained with a slightly lower mean intensity of 45 mW/cm^2, where this time they observed an increase in reflectivity for 28 minutes after which the reflectivity remained roughly constant up to 48 minutes, the reflectivity then *decreasing* until after 88 minutes there was very little reflected signal visible. The UV light was then turned off and the reflectivity once again *increased* over 8 hours to a constant value. The authors suggested that the recovery might be due to the relaxation of thermal stresses.

It is clear from this brief description that photoinscription probably generally involves multiple processes, and Saez-Rodriguez et al. [57] have proposed that with pure PMMA the relevant mechanisms could be photodegradation due to main chain scission leading to index reduction and photopolymerization providing an index increase. They reported experiments where inscribing gratings using a CW laser at 325 nm under higher levels of strain led to larger index modulations and linked this to the enhancement of photodegradation by higher stresses [58]. By scanning a UV beam over an existing grating, they were able to confirm a net positive wavelength shift due to UV irradiation.

The use of ultra-short pulse lasers for inscription complicates the picture still further. A comprehensive review of the subject as of 2012 is provided by Scully et al. [59]. They report on the use of different pulse regimes capable of inducing both positive and negative index changes, with typical magnitudes in the range 10^{-4} to almost 10^{-2}. Importantly, Scully et al. showed that using 800 nm, 40 fs pulses with 1 kHz repetition rate, permanent refractive index changes of up to 4×10^{-3} could be induced in clinical grade PMMA (material that is undoped, maximally polymerized, and containing no UV inhibitors and minimal traces of initiator). A potentially significant recent development for POFBGs enabled by short pulse lasers has been the inscription of FBGs in perfluorinated fiber. While the loss of single-mode, PMMA based fiber is typically of the order of 10^5 dB/km at 1500 nm, perfluorination—replacement of hydrogen atoms with fluorine—shifts the responsible molecular absorption bands further into the infrared, reducing the loss at 1500 nm to less than 30 dB/km, see **Figure 6.2**. While still a high figure for a communications system, such losses would permit sensing fiber lengths of tens of meters—more than sufficient for many applications. The perfluorination process removes the absorption feature of PMMA around 300 nm, which enables the use of the 325 nm CW HeCd laser for inscription, and so far it has not proven possible to inscribe gratings in perfluorinated fiber with a CW source. Grating inscription has, however, been reported using a pulsed KrF (248 nm) laser [60,61] and a 1 kHz repetition rate, 220 fs pulse, 517 nm laser [48].

The production of POFBGs shares much in common with that of silica FBGs, and the interested reader is directed to the review articles and books on FBGs listed in the introduction, which summarize the various approaches. A number of POFBG researchers use arrangements in which the POF is supported by a horizontal v-groove while the UV beam is incident from above [62]. This was originally done to try to prevent sagging of the POF, which is much more compliant than the relatively stiff silica fiber. However, this vertical inscription geometry is not necessary, and we have successfully recorded gratings using a scanned horizontal beam directed at more than 15 cm of POF supported between two clamps.

6.4 Sensitivity

6.4.1 Temperature and Strain

The sensitivity of the Bragg wavelength to temperature and strain is given by [63,64]

$$\begin{aligned}\Delta\lambda &= 2\left(\Lambda\frac{\partial n}{\partial \varepsilon_z} + n\frac{\partial \Lambda}{\partial \varepsilon_z}\right)\Delta\varepsilon_z + 2\left(\Lambda\frac{\partial n}{\partial T} + n\frac{\partial \Lambda}{\partial T}\right)\Delta T \\ &= \lambda\left(\frac{1}{n}\frac{\partial n}{\partial \varepsilon_z} + 1\right)\Delta\varepsilon_z + \lambda(\varsigma + \alpha)\Delta T,\end{aligned} \tag{6.2}$$

where

$$\alpha = \frac{1}{\Lambda}\frac{\partial \Lambda}{\partial T} \quad (6.3)$$

is the fiber's thermal expansion coefficient while

$$\varsigma = \frac{1}{n}\frac{\partial n}{\partial T} \quad (6.4)$$

is the thermo-optic coefficient.

The variation of refractive index with axial strain is given by

$$\frac{1}{n}\frac{\partial n}{\partial \varepsilon_z} = -\frac{n^2}{2}\left(p_{12} - \nu\left(p_{11} + p_{12}\right)\right) \quad (6.5)$$

where ν is Poisson's ratio and p_{11} and p_{12} are components of the strain-optic tensor.

For PMMA and silica, the right-hand side of **Eq. (6.5)** turns out to be negative, meaning that tension in the fiber leads to a reduction in the refractive index and hence the Bragg wavelength. However, for both types of fibers this is more than compensated for by the physical elongation of the grating, so tensile strain leads to a net increase in Bragg wavelength. **Table 6.1** lists the strain and temperature sensitivities of typical Bragg gratings operating in the third telecommunication window.

For POF the figures should be understood to be typical values for PMMA-based fiber as the precise values are subject to cross sensitivity effects (for example the thermal sensitivity of PMMA-based fiber depends on the humidity) and can depend on the fiber drawing process, as discussed in Section 6.5. Note that the thermal sensitivity of POFBGs is negative, reflecting the fact that the effect of the polymer's negative thermo-optic coefficient dominates over the fiber thermal expansion (though there is one report of a small but positive thermal sensitivity [67]).

6.4.2 Pressure

Uniform pressure acting on the fiber leads to an isotropic strain, causing a refractive index variation given by:

$$\frac{1}{n}\frac{\partial n}{\partial P} = -\frac{n^2}{2E}(1-2\nu)\left(p_{11} + 2p_{12}\right) \quad (6.6)$$

where E is the Young's modulus of the material. Although this expression only strictly applies to static pressure, because of the small size of FBG sensors the equation is a good approximation for the effect on the fiber of most acoustic frequencies of interest.

Table 6.1 Typical Sensitivities of PMMA-Based POF and Silica FBGs to Temperature and Strain in the 3rd Telecommunications Window

Wavelength Range		3rd Window	References
Thermal sensitivity (pm/°C) at constant humidity	POF	−55 @ 1542 (PMMA)	[25]
	Silica	11.9 @ 1561 nm	[65]
Strain sensitivity (pm/με)	POF	1.3 @ 1533 nm (PMMA)	[66]
	Silica	1.20 @ 1561 nm	[65]

Unfortunately, the pressure sensitivity is not large; for example, for silica fiber we have measured it to be around -3.9×10^{-9} nm/Pa in the 1550 nm window [68], which compares reasonably well with a theoretically calculated value of -4.3×10^{-9} nm/Pa. However, in the same experiment we measured the sensitivity for POF as 130×10^{-9} nm/Pa, which is completely at odds with the theoretical value of -2.8×10^{-9} nm/Pa, both in magnitude and sign. We speculate that the reason for this discrepancy lies in the transversely isotropic fiber structure introduced by the drawing process, which means that the simple expression in **Eq. (6.6)** is no longer valid. More recent measurements in a multi-layered graded index perfluorinated POF revealed a long time constant of 150 minutes to reach a stable Bragg wavelength when the fiber was subject to 0.5 MPa pressure and a subsequent negative dependence of Bragg wavelength on pressure [69].

6.4.3 Humidity

Silica fiber is effectively insensitive to water in the surrounding environment, but this is not the case for certain polymers, including the commonly used PMMA, which have an affinity for water. Where water is present around the fiber, it will be drawn in, leading to a swelling of the fiber and an increase in its refractive index, both leading to a positive Bragg wavelength shift, as first reported by Harbach [21]. This is a reversible equilibrium process, with the amount of water absorbed by the fiber reflecting the concentration of water in the region around the fiber. More precisely, the equilibrium amount of water in the fiber is determined by the water activity of the surrounding medium, defined as the fraction of the maximum (saturated) water content. So compared to dry conditions, the wavelength shift obtained in 100% humidity air is similar to that obtained in aviation fuel saturated with water, despite the latter corresponding to only a few tens of ppm water by volume, i.e. a much smaller concentration of water in absolute terms.

At constant strain, the variation of the Bragg wavelength with humidity and temperature can be written as

$$\Delta\lambda_B = \lambda_B(\eta + \beta)\Delta H + \lambda_B(\zeta + \alpha)\Delta T \tag{6.7}$$

where η accounts for the dependence of refractive index on humidity and β the swelling related to humidity induced volumetric change $(\% \text{ RH})^{-1}$ (the other coefficients were defined earlier). From a sensing perspective, there are problems because η and β are not constants; the former decreasing approximately linearly with increasing temperature and the latter, while being independent of temperature, is only constant in the region 40%–100% RH, decreasing significantly below 40% [70]. Zhang and Webb [71] conducted a study of the humidity response of PMMA-based POFBGs that investigated the relative contributions of refractive index change and expansion to the overall Bragg wavelength shift. Due to fiber anisotropy induced by the drawing process, the degree of axial expansion for a given humidity change depends on the fiber drawing conditions. It was shown that this variability could be removed by applying a pre-strain to the fiber, at the expense of a reduced humidity sensitivity. The use of a pre-strain brought the additional advantages of a more linear response to humidity and a significantly faster response time as illustrated in **Figure 6.1**.

6.4.4 Magnetic Field

For completeness, it should be noted that FBGs in silica or POF are also intrinsically sensitive to magnetic field, via the Faraday effect, whereby circular birefringence is induced in the fiber proportional to the magnitude of any axial magnetic field. The presence of circular birefringence means that the Bragg wavelength will be different when the FBG is illuminated by left or right circularly polarized light, and this effect has been used to measure magnetic field with silica FBGs with a resolution of 4 gauss in a 0.15 Hz bandwidth [72]. The magnetic field sensitivity of PMMA is likely to be similar as it has a comparable Verdet constant to silica [73].

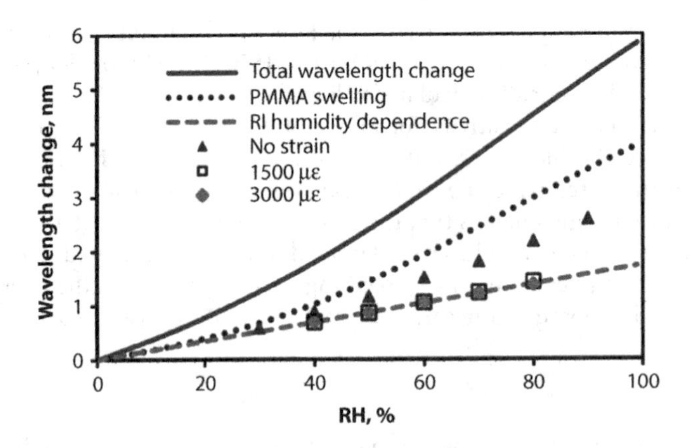

FIGURE 6.1 Dependence of Bragg wavelength on humidity for a PPMA-based POFBG subject to no strain and pre-strains of 1500 and 3000 με. (From Zhang, W. and Webb, D.J., *Opt. Lett.*, 39, 3026–3029, 2014.)

6.5 Challenges

The different material properties of polymers that provide motivation for the development of POFBG technology also pose a number of challenges that must be overcome if that technology is to be used outside the laboratory:

- Polymers have much higher loss than silica-based fiber
- The maximum usable temperature of POF is much less than that of silica
- Polymers can display inelastic behavior
- As we have seen, polymer behavior can be influenced by the presence of water
- The properties of polymers are hard to control precisely
- There has been no push to develop single mode POF components

Progress has been made in all these areas, though that is not say that further work isn't required.

6.5.1 Fiber Loss

The first POFBGs were fabricated in PMMA-based fiber with a Bragg wavelength in the 1550 nm region. This wavelength was likely chosen to be compatible with available components and equipment appropriate for work with silica fiber; however, it is a poor choice for PMMA-based fiber, as the loss in this region is of the order of 1 dB/cm; perfectly acceptable for laboratory studies, but of limited practical use. The problem for PMMA is absorption caused by the aliphatic hydrocarbons; a series of harmonics of the absorption band stretches down through the NIR spectral region and into the visible, as may be seen in **Figure 6.2**. PMMA has a minimum loss around 500–600 nm, below which attenuation rises due to Rayleigh scattering and electronic transitions.

PMMA has a recording resolution sufficient to permit recording of FBGs with a Bragg wavelength in the 600 nm spectral region [29]; however, single-mode mPOF at this wavelength exhibited a loss of 3.8 dB/km, significantly more than the 0.1 dB/km suggested by **Figure 6.2**. This discrepancy arises because of losses due to scattering at the boundary between different materials that become more relevant as the core size is reduced. As a consequence, the loss of

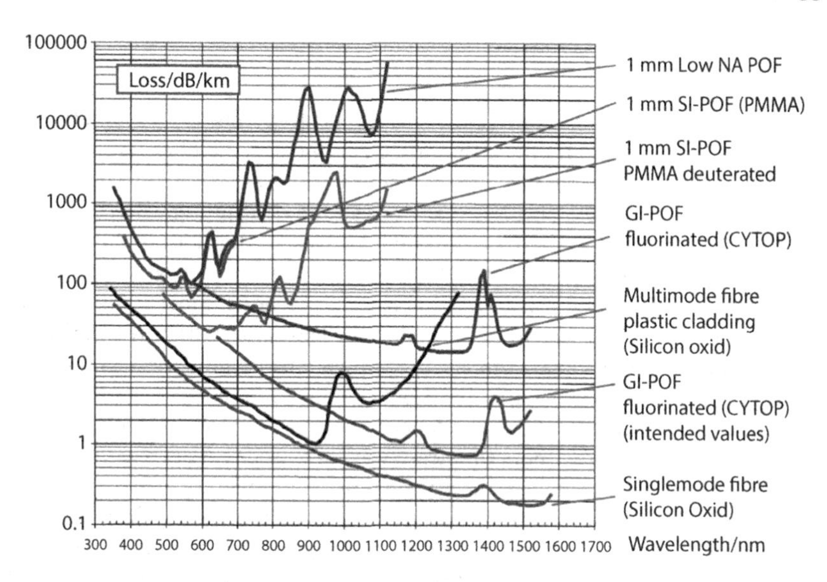

FIGURE 6.2 Transmission loss for different types of fibers. (From Webb, D.J. and Kalli, K., *Fiber Bragg Grating Sensors: Thirty Years from Research to Market*, Ed Cusano, A., Bentham eBooks, 2010.)

few- or single-mode POF has always been significantly worse than that measured for the bulk material or large core fibers, whether these are step index or microstructured in nature. Note though that mPOF with a loss of 0.16 dB/m at 650 nm has been reported [74].

There appears to be limited scope for improving the loss through the use of other polymer materials. The one exception is perfluorinated PMMA (one trade name for which is CYTOP); see **Figure 6.2**. This is a version of PMMA in which the hydrogen atoms have been replaced with more massive fluorine atoms, the consequence being that the molecular absorption bands and their harmonics are shifted to lower frequencies (higher wavelengths) leading to a minimum loss around 1300 nm of a few tens of dB, with the value at 1550 nm not being much greater.

Initial attempts at grating inscription in this material were made by Peng's group; using a pulsed Nd:YAG laser emitting at 355 nm for inscription, they succeeded in detecting gratings recorded in a CYTOP preform, but were unable to detect the presence of gratings in CYTOP fiber [75]. The first successful inscription of grating structures in this type of fiber was reported by Koerdt et al. [60] using a krypton fluoride excimer laser. The 50 micron core graded index fiber was heavily multimode, and the gratings had complex spectral structures and a spectral width of around 10 nm.

6.5.2 Temperature

Elevated temperatures pose two general problems for POFBGs: annealing of the fiber can occur (dealt with in the next section), but here we are referring to the survivability of the POFBG and indeed of the fiber sensor. The relevant temperature is the glass transition temperature at which the fiber is starting to become viscous. Above this point the polymer structure is not stable, and the sensor will not be practically useable. However, even at temperatures slightly below the glass transition, there is evidence for significant decay of gratings strength [76], though there does not appear to have been a definitive study of the mechanisms involved. PMMA has a glass transition temperature of 104°C [77], though in practice this can vary somewhat with the presence of any copolymers. In an effort to improve on the high temperature behavior, researchers at the Technical University of Denmark have created fibers from a range of other polymers, all of which demonstrate photosensitivity to CW 325 nm light. POFBGs have been demonstrated to survive up to 110°C in TOPAS [78] and 125°C in polycarbonate [46] fibers.

6.5.3 Inelastic Behavior

POF is a viscoelastic material and in general, if it is subject to a constant strain, the fiber stress will gradually reduce (stress relaxation). Equally, if subject to a constant tension, the fiber strain will gradually increase (creep), and once that force is removed, the fiber will not return to its original length. **Figure 6.3** illustrates this effect where a POF was repeatedly subject to the same strain increment at a rate of 0.5% per minute, with each cycle beginning when the applied stress was zero [79]. It may be seen that there is a gradual increase in the resting fiber length, though given sufficient time unloaded the fiber would relax back toward its initial length.

In many applications, the hysteresis shown in **Figure 6.3** is not as much a concern as might be imagined. When POFBG strain sensors are tested in the laboratory, they are usually held at two points on either side of the grating; strains are then applied by moving these points apart. If the fixed points are returned to their original position, the fiber can hang slack as a result of creep having taken place. An example of the dependence of the Bragg wavelength on strain for such a situation is shown in **Figure 6.4a**, where the strain was increased and decreased at a rate of 0.2% per minute. When the strain is reduced, there is no tension in the fiber below 0.5%.

This approach does not, however, necessarily reflect how POFBGs would be used in a strain sensing application. In many cases the sensor would be glued to (or embedded in) the structure under test, and in this case, the structure can force the POF back to its original length, even after creep has taken place. **Figure 6.4b** shows the strain response of a POFBG identical to that used for **Figure 6.4a**, with the strain varied at the same rate, but with the POFBG

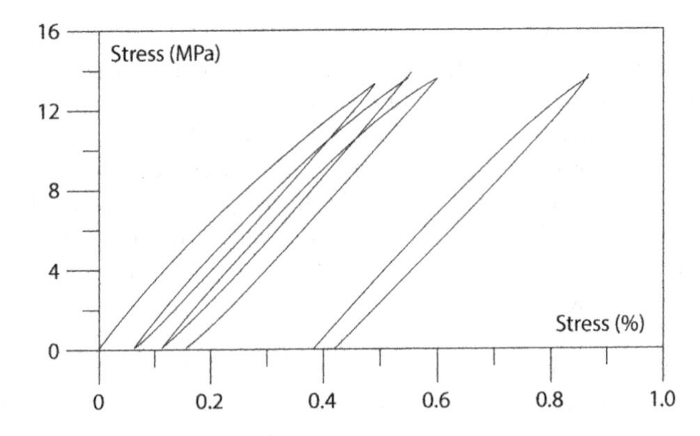

FIGURE 6.3 The first, second, third, and tenth cycles during a repetitive strain sensing experiment with PMMA fiber. (After Yang, D. et al., *Mater. Sci. Eng.*, A364, 256–259, 2004.)

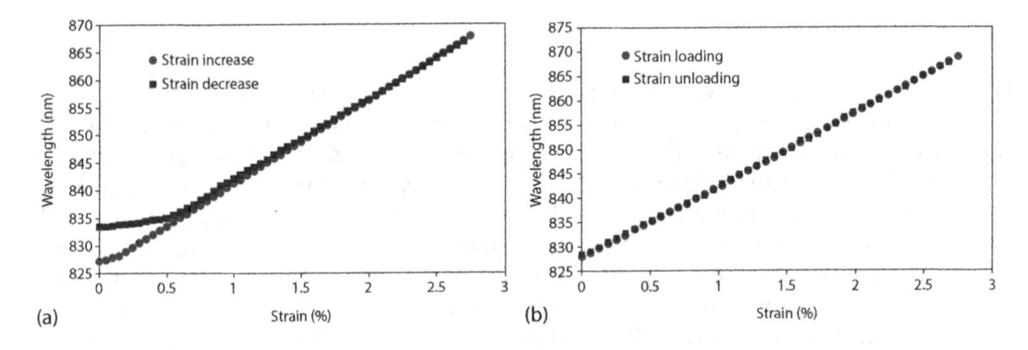

FIGURE 6.4 Strain response of two similar POFBGs. (a) Fiber suspended between two supports, the movement of which is used to impart strain. (b) Fiber glued to plastic cantilever, which is bent to apply strain. (From Abang, A. and Webb, D.J. *Opt. Lett.*, 38, 1376–1378, 2013.)

glued to a plastic cantilever beam that was bent to induce strain in the fiber. It may be seen that the hysteresis is considerably reduced (by a factor of more than 12 in this experiment).

6.5.4 Water Sensitivity

We have described earlier how the water affinity of PMMA can be utilized for humidity sensing. This feature becomes a problem if the intention is instead to develop a strain or temperature sensor. Fortunately, not all polymers share the water absorbing property of PPMA. Yuan et al. showed that gratings recorded in microstructured POF made from TOPAS [80], a cyclic olefin copolymer, were at least 50 times less sensitive to humidity than PMMA. POFBGs in humidity insensitive, single-mode, step-index POF were demonstrated by Woyessa et al., with the fiber being fabricated from TOPAS (core) and ZEONEX (cladding) [81].

6.5.5 Polymer Variability

PMMA, or indeed any other polymer, is composed of molecules with a range of different molecular weights. Depending on the polymerization process, the mean molecular weight and the distribution of weights can change radically. The situation is complicated still further since the polymer can contain various initiators, chain transfer agents, and plasticizers, added to control the polymer properties. As a consequence, certain material properties can vary quite widely; for example, the Young's modulus of PMMA is quoted in the literature at between 1.6 and 3.4 GPa [82,83]. The mean molecular weight of PMMA can vary over a wide range and it has empirically been found that around 60,000 to 100,000 is most suitable for drawing [84].

As an indication of the difficulties that polymer variability can bring, we can report that in the early stages of our POF research, we studied the mechanical properties of fiber drawn from commercially obtained samples of PMMA rod. Having eventually exhausted our stock, we reordered further supplies of ostensibly the same product but found these rods much easier to work with, in that they did not have the same tendency to form bubbles when heated, as was sometimes the case for the original rods. Samples of the two batches were submitted to Rapra Technology for characterization using gel permeation chromatography, and the results provided are shown in **Figure 6.5**; the rather different average molecular weights were 87,000 and 133,000, indicating the need to have control over all aspects of the POF sensor production process.

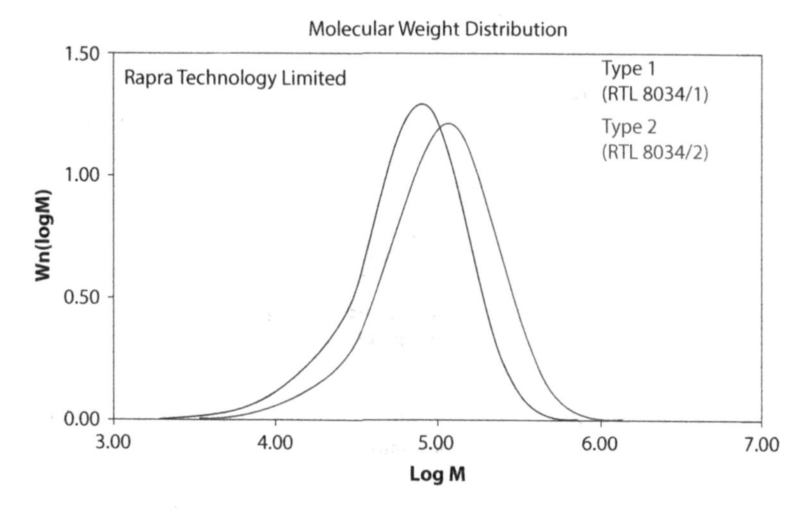

FIGURE 6.5 Molecular weight distributions of two batches of nominally identical PMMA rods. (From Webb, D.J. and Kalli, K., *Fiber Bragg Grating Sensors: Thirty Years from Research to Market*, Ed Cusano, A., Bentham eBooks, 2010.)

While bulk PMMA can usually be regarded as being essentially isotropic, that is not generally the case when drawn into a fiber. Tension during the drawing process tends to align the polymer molecules along the fiber axes, leading to an axially isotropic material. Higher drawing tensions lead to a greater degree of anisotropy, which is revealed both through variations in the mechanical properties of the fiber and via the presence of optical birefringence [86].

Figure 6.6 shows stress-strain curves for various fibers drawn from the same preform but at different tensions, controlled in practice by altering the oven temperature. Szczurowski et al. have measured the birefringence of otherwise similar fibers drawn under different tensions and thereby determined the differential stress-optic coefficient, ΔC, of the fibers. They showed that ΔC in PMMA fibers have a negative sign and ranged from -1.5 to -4.5×10^{-12} Pa^{-1}, depending on the drawing stress of the fibers examined. Higher drawing stresses result in larger initial fiber birefringence and a lower ΔC [86].

The anisotropy "frozen in" to the fiber during the drawing process can be reduced by heating the fiber above room temperature—essentially an annealing process. Consequently, the mechanical (and optical properties) of such a fiber will change with time when it is subjected to thermal treatment—a potentially serious issue if the fiber is being used as a sensor at the time. Carroll et al. showed that by raising the temperature of a POFBG to 95°C for a few minutes, relaxation of the polymer chain alignment caused the fiber to shrink, resulting in a permanent shift in the Bragg wavelength of nearly –20 nm [87]. Following this annealing, the POFBG was much more stable and furthermore its Bragg wavelength had a more linear response to temperature.

A more detailed study of fiber annealing has been carried out by Stajanca et al. [89]. They showed that fiber length shrinkage due to annealing at 95°C and 40% relative humidity could be modeled by a stretched exponential decay. The shrinkage after 300 hours was heavily dependent on the fiber drawing tension; with negligible tension, the shrinkage was less than 10% whereas with a tension of 0.8 N, the shrinkage was more than 70%. The conclusion was that for POFBG applications the drawing process should be carried out with minimal tension, and hence with a minimum of molecular alignment.

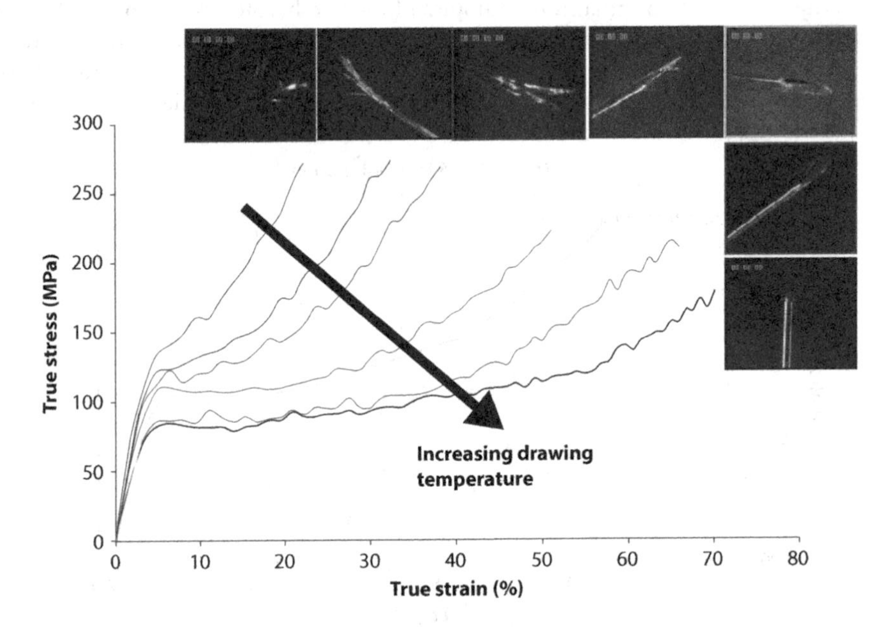

FIGURE 6.6 Stress versus strain for a range of fibers drawn from the same preform at varying tensions (increased drawing temperature equates to lower drawing tension here). (From Aressy, M., *Manufacturing optimisation and mechanical properties of polymer optical fibre*, Birmingham University, Birmingham, UK, 2006.)

Pospori et al. investigated the effects on POFBG properties of annealing the fiber (to reduce frozen in stress) and etching using acetone (to reduce the fiber diameter) [90]. They found that, with a high degree of confidence, annealing enhances both the strain and stress sensitivity of the POFBGs. The reduction in fiber diameter as a result of etching would be expected to render the devices more sensitive to fiber tension; however, the authors found that for the majority of sensors, the etching also increased the stress sensitivity. They proposed this latter observation was indicative of radial inhomogeneity of the fiber material. Bhowmik et al. suggested that etching affects the material properties of the POF and discovered that etched POF allowed the inscription of POFBGs with shorter inscription times and higher reflectivities than un-etched POFs [91] and furthermore the POFBGs produced in etched fiber had enhanced sensitivity to strain, temperature, and pressure [92]. Further information on the effects of annealing and etching has been provided by Leal-Junior et al [93].

To simplify the process of annealing, Fasano et al. noted that when saturated with water, the *Tg* of PMMA is reduced by about 20°C [94]. They therefore concluded that if the *Tg* could be reduced sufficiently, annealing could occur efficiently at room temperature. They found that with a high concentration of methanol in water (70% by volume), a grating recorded in PMMA-based mPOF at around 850 nm would experience a blue shift of 100 nm in a little over a day [95]. Annealing of the POF preform prior to drawing the fiber has also been shown to reduce the inscription time needed for POFBG fabrication [96].

6.5.6 Single Mode POF Technology

The commercial drivers for POF development have primarily come from communications and illumination applications. Both of these currently use large core, highly multimode fiber; in the former case to allow low cost termination and connectorization using injection molded plastic connectors, and in the latter to facilitate the collection of light from large area emitters. In contrast, Bragg grating sensors are almost always operated with single mode fiber, since the presence of other modes leads to additional peaks in the reflection spectrum, which significantly complicates the readout process (though Ganziy et al. have developed an algorithm that significantly improves the accuracy with which an FBG in a multimode fiber can be monitored [97]). As there are as yet no significant markets for POFBGs, there is little incentive for commercial development of single mode POF and hence most such fiber has been made in research laboratories. Two exceptions to this have been Kyriama, a former spin-out from the University of Sydney, and Shute Sensing Solutions, spun-out of the Technical University of Denmark.

Along with a lack of a reliable source of single mode fiber, other components have also been missing, such as single mode polymer couplers and connectors. While it is perfectly possible to use silica fiber couplers, single mode POF connection technology is essential, particularly where FBGs in the 1550 nm telecommunications window are used in order to take advantage of existing FBG interrogators. In this wavelength region, **Figure 6.2** shows that the loss is around 1 dB/cm, limiting the length of POF to perhaps 10 cm at most. This issue was first addressed by directly gluing the POF to a silica single mode fiber pigtail [98]; however, such joints are not very robust and not demountable. A technique for mounting POF in standard FC/PC connectors was developed by Abang and Webb [99]. There are two problems to be solved here: firstly, the diameter of the single and few-moded POF produced in research laboratories tends to vary quite widely, so finding a piece with a diameter that corresponds to one of the existing standards, e.g. 125 μm, is difficult. Secondly, even if you have a fiber with this diameter, feeding the POF into the ferrule of a connector, which typically has an internal diameter of 126 μm, turns out to be essentially impossible. This is because while silica fiber is rigid enough to maintain its shape as it is pushed through the connector, the much more elastic POF tends to jam in the ferrule.

The solution developed for both these problems was to etch a shallow taper over a few centimeters at the end of the POF by slowly inserting the fiber into acetone over a period of about 30 s. The thin end of the POF was then inserted into the ferrule and the fiber pulled through until it jammed and could be permanently fixed in the ferrule by gluing. Initially, the end face was prepared by polishing but later on it was discovered that the fiber could be simply cut with a razor blade at 77 C. Initially developed for use with large core, few-moded fibers [99], the approach was later extended to single-mode POF, where a connection loss of 8.5 dB was obtained [38] (our most recent work has demonstrated a best-case loss of less than 1 dB).

6.6 Applications

Applications of POFBGs have been suggested in a number of areas where their unique properties give them an advantage over silica-based FBGs, which may outweigh the challenges in using the more immature polymer technology.

6.6.1 Water-in-Fuel Sensing

There is some concern around the presence of dissolved water in aviation fuel. Although the solubility is low in absolute terms (around 70 ppm at room temperature [100]), when the aircraft is at altitude and the temperature drops, significant amounts of water can still condense out of solution and form a layer of pure water at the bottom of the fuel tank. Not only can this provide an environment supportive of the growth of anaerobic bacteria, but the water can freeze and the resulting solid ice can pose a risk to the fuel system; for example, this was the cause identified for the British Airways Flight 38 accident at Heathrow in 2008 (AAIB Bulletin S1/2008 Special).

Currently there are spot tests available providing a quasi-quantitative indication of water content, but an in-line, real-time, quantitative measurement system is not commercially available. Zhang et al. have exploited the water affinity of PMMA-based POF to quantify dissolved water in Jet-A1 aviation fuel [100,101] using the wavelength shift of a POFBG recorded at 1531 nm. They prepared aviation fuel samples with known water content by allowing the fuel to reach equilibrium with humidity-controlled air in an environmental chamber. The absolute water content was checked by assessing three fuel samples using a Karl-Fischer coulometer: one sample that had been exposed to air dried in a desiccator, one that was at equilibrium with thelaboratory environment, and one that had reached equilibrium with 100% RH air. The results are shown in **Figure 6.7** and the sensitivity reported was 59 ± 3 pm/ppm.

One of the limitations of the use of POFBGs for water activity sensing (visible in **Figure 6.7a**) is the relatively slow response time, governed by the diffusion process of water into the fiber core. The response time can be significantly improved by etching down the fiber diameter using acetone.

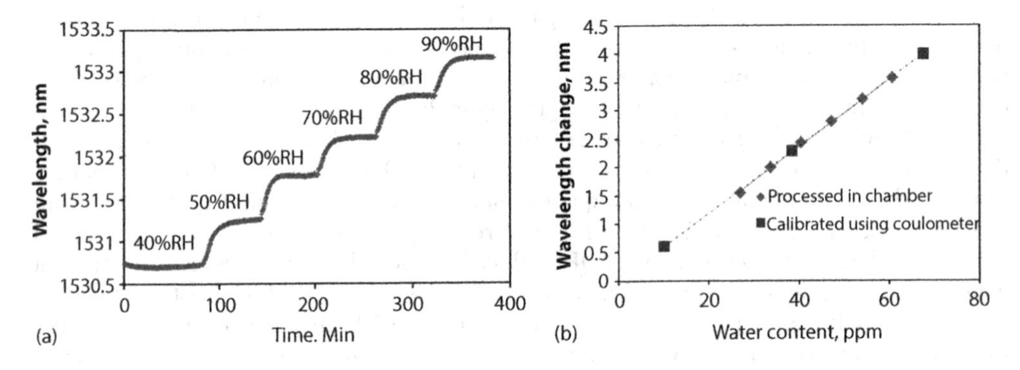

FIGURE 6.7 (a) Bragg wavelength of POFBG in PMMA based fiber in aviation fuel exposed to different relative humidities. (b) Wavelength shift as a function of water content showing the data from Figure 6.7a combined with samples calibrated with a Karl Fischer coulometer. (From Zhang, W. et al., Measuring water activity of aviation fuel using a polymer optical fiber Bragg grating, In Proceedings of SPIE 9157 pp 91574V-1, 2014.)

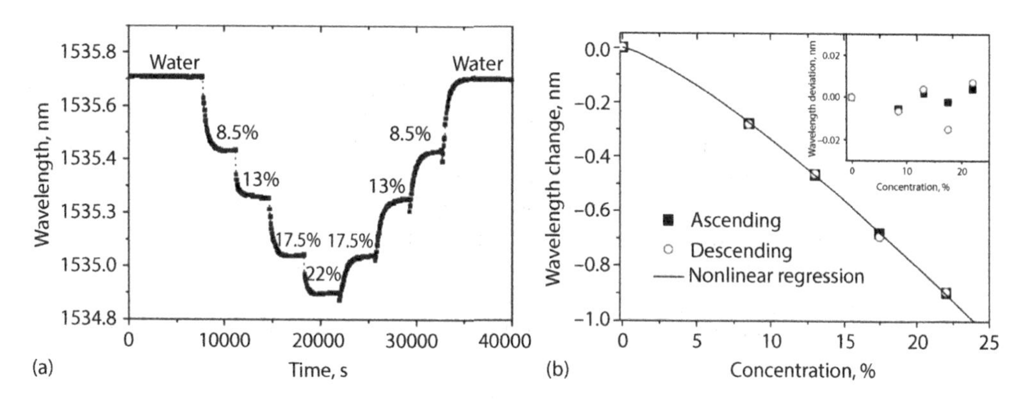

FIGURE 6.8 (a) Bragg wavelength response of PMMA-based POFBG in saline solutions of different concentrations. (b) Wavelength shift versus concentration (note the absence of hysteresis in the response when the concentration is being lowered). (From Zhang, W. et al., *Opt. Lett., 37,* 1370–1372, 2012.)

For example, in one experiment with PMMA-based fiber, the response time (defined as the time needed to achieve 90% of the wavelength shift resulting from a step-change in humidity) was reduced from 31 minutes to 12 minutes by reducing the fiber diameter from 191 to 135 μm [102]. Taking this approach further, Rajan et al. etched a PMMA-based fiber down to 25 μm and claimed a response time of under 10 s [103].

In the previous example of water-in-fuel sensing, the water is the solute. POFBGs can also be used to monitor concentrations of solutions where the water is the solvent [104]. For example, **Figure 6.8** shows the response of a PMMA-based POFBG sensor to changes in concentrations of salt and sugar solutions [105].

6.6.2 Dynamic Sensing

The much lower elastic modulus of POF compared to silica suggests that POF should have an advantage when used as the basis for an accelerometer. It must be recalled though that POF is a viscoelastic medium and this needs to be taken into account. Stefani et al. investigated sample fibers composed of PMMA and TOPAS and concluded that for relatively low strains of up to 0.28%, the Young's modulus of the fiber could be considered as constant, at least up to 100Hz, which was the limit of their dynamic strain testing system [31]. Little evidence of viscoelastic behavior was seen at excitation frequencies up to 10 Hz. However, for step changes in strain of 2.8%, significant stress relaxation was visible, with a time constant of about 5 seconds, corresponding to a viscosity of 20 GPa.s. Exploiting this knowledge, Stefani and his colleagues were able to design a high sensitivity accelerometer utilizing a POFBG, with a flat frequency response out to over 1kHz, the ability to measure up to 15 g and a sensitivity four times higher than an equivalent silica fiber based system [32]. In a later paper, the same group investigated long-term strain sensing. Having applied a strain of up to 0.9% for several minutes, they found that once the tension in the fiber was removed the strain relaxation occurred via two components, an initial elastic relaxation followed by a slower viscoelastic contraction [106]. They found that when the strain was applied for longer times (up to 50 minutes) the range of elastic behavior became smaller.

The more compliant nature of POF compared to silica also renders it more sensitive to pressure, whether quasi-static or in the form of acoustic waves. This was exploited by Marques et al. in a tunable filter, in which the reflection profile of a POFBG was controllably modified by launching ultrasonic acoustic waves into the fiber using an acoustic horn [107]. The ability of POFBGs to detect high frequency acoustic waves was confirmed by Broadway et al. [108] who studied the response of POFBGs to bursts of ultrasound at 1, 5, and 10 MHz, as part of work aimed at realizing a POF-based sensor for opto-acoustic imaging.

6.6.3 Fuel Gauging

With a desire to reduce weight and improve safety, there is a push to replace electrical sensors in aircraft fuel tanks with optical fiber-based ones. Several groups have developed different approaches for measuring the level of fuel in the tank. One approach involves locating a pressure monitor at the bottom of the tank, since the pressure experienced will be proportional to the density of the liquid multiplied by the depth. Pressure can be monitored via the deflection of a thin diaphragm exposed to the liquid on one side and ambient pressure (or vacuum) on the other. Conventionally, the deflection can be monitored using a capacitance-based sensor, but all-optical approaches are also possible and one way to do this is to attach an FBG to the surface of the diaphragm, such that as the diaphragm deflects, the FBG picks up an increasing strain.

To improve the sensitivity of such a structure the diaphragm can either be made thinner or of a more compliant material, but an issue with this approach is that as the diaphragm material is made more compliant, the stiffness of the attached optical fiber becomes a limiting factor, at least if the fiber is made from silica. Essentially the fiber restricts the movement of the diaphragm and returns a smaller strain reading than there would be if the fiber were absent [24]. Marques et al. investigated the advantage to be gained in such a situation from using a POFBG to monitor an elastic membrane [109]. In one experiment using a 15 mm diameter, 1.1 mm thick diaphragm made of silicone rubber with an embedded FBG, when the FBG was in silica fiber the sensitivity to water depth was 10.2 pm/cm, while for POF the figure was 57 pm/cm.

The gauging approach used by Marques et al. exploited multiple sensors at different depths as shown in **Figure 6.9**. The idea here is that the depth is calculated using a linear interpolation of the readings from sensors that are submerged, as shown in **Figure 6.9a**. At the expense of requiring several sensors, this approach brings with it some important advantages:

- Common mode temperature-induced wavelength shifts in the individual sensors are automatically compensated.
- Temperature-induced changes in the sensor pressure sensitivity are also compensated.
- The approach provides the possibility to detect and compensate for malfunctioning sensors.

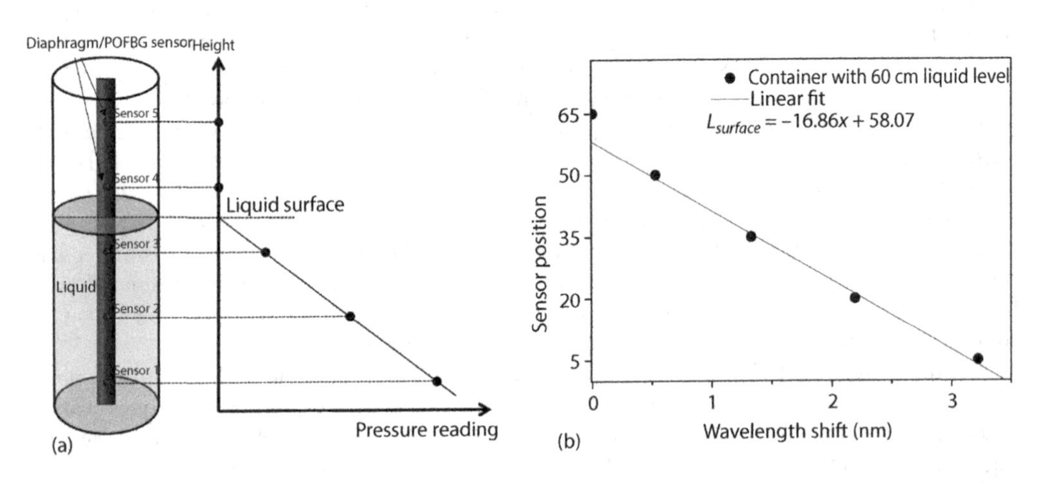

FIGURE 6.9 Fuel gauging system. (a) Multiple sensor approach to interpolating level. (b) Example of level detection where 4 out of 5 sensors are submerged. (From Marques, C.A.F. et al., *Opt. Expr.*, 23, 6058–6072, 2015.)

- The system is immune to changes in the density of the monitored fluid and even to changes in the effective force of gravity, as might be obtained in an aerospace application; changes in density or g-force affect the gradient of the interpolated line, but not its intercept.

Figure 6.9b shows experimental data from a five-sensor system responding to a depth of water of 60 cm. Level resolution in this system was a little under 1 cm, though this has been significantly improved in later research.

6.6.4 Other Applications

A number of other applications taking advantage of the unique properties of POFBGs have been demonstrated. Liu et al. exploited the wide strain tuning range of POF to create a tunable fiber laser where a POFBG formed one of the cavity mirrors [110]; a tuning rang of over 35 nm was demonstrated in the 1550 nm region. Cheng et al. have exploited the compliant nature of POF to realize a pH sensor in which a POFBG is strained by a coating of a UV-cured pH-sensitive hydrogel, poly(ethylene glycol) diacrylate (PEGDA) [111]. The device had a response time of 30 s and a sensitivity of up to −0.41 nm/pH over the pH range 3.5 to 6.5. Zubel et al. have demonstrated the feasibility of embedding POFBGs in 3D printed structures [112].

6.7 Related Work

In this chapter we have focused on polymer optical fiber Bragg gratings. There are a number of research areas that are closely related to this topic that deserve a mention. Firstly, Bragg gratings are not the only form of diffractive structure that can be realized. Long period gratings (LPGs) have been investigated by several groups [113] and have the advantage that their large-scale structure means that they can be thermo-mechanically imposed on the fiber [114]. LPGs have been used for strain sensing [115] and hydrostatic pressure monitoring [116] and have been photo-inscribed in less than a minute using a sequence of 15 ns pulses from a 248 nm KrF laser [42]. Other types of grating have also been investigated, such as chirped structures [117] and tilted gratings [118]. Polymer optical fibers can also be used within interferometric systems [119] and POFBGs have been used to fabricate the mirrors for POF-based Fabry-Perot systems [120–122] and an analysis carried out to determine where POF-based interferometers might outperform their silica-based cousins [123].

References

1. Othonos A and Kalli K 1996. *Fiber Bragg Gratings: Fundamentals and Applications in Telecommunications and Sensing* (Boston, MA: Artech House Publishers).
2. Zyskind J L, Mizrani V, DiGiovanni D J and Sulhoff J W 1992. Short single frequency erbium-doped fibre laser *Electronics Letters* **28** 1385–1387.
3. Hill K O, Bilodeau F, Malo B, Kitagawa T, Thériault S, Johnson D C, Albert J and Takiguchi K 1994. Chirped in-fiber Bragg gratings for compensation of optical-fiber dispersion *Optics Letters* **19** 1314–1316.
4. Bilodeau F, Johnson D C, Theriault S, Malo B, Albert J and Hill K O 1995. An all-fiber dense wavelength-division multiplexer/demultiplexer using photoimprinted Bragg gratings *IEEE Photonics Technology Letters* 7 388–390.
5. Hill K O, Fujii Y, Johnson D C and Kawasaki B S 1978. Photosensitivity in optical fiber waveguides: Application to reflection filter fabrication *Applied Physics Letters* **32** 647–649.
6. Meltz G, Morey W W and Glenn W H 1989. Formation of Bragg gratings in optical fibers by a transverse holographic method *Optics Letters* **14** 823–825.
7. Webb D J and Kalli K 2010. Polymer fibre Bragg gratings. In: *Fiber Bragg Grating Sensors: Thirty Years from Research to Market*, Ed A Cusano Oak Park, IL: Bentham eBooks.
8. Kalli K and Webb D J 2011. Polymer optical fibre based sensors. In: *Advanced Fiber Optics*, Ed L Thevenaz (EPFL Press).

9. Webb D J 2015. Fibre Bragg grating sensors in polymer optical fibres *Measurement Science and Technology* **26** 092004.
10. Nogueira R, Oliveira R, Bilro L and Heidarialamdarloo J 2016 [INVITED]. New advances in polymer fiber Bragg gratings *Optics and Laser Technology* **78** 104–109.
11. Kersey A D, Davis M A, Patrick H J, LeBlanc M, Koo K P, Askins C G, Putnam M A and Friebele E J 1997. Fiber grating sensors *Journal of Lightwave Technology* **15** 1442–1463.
12. Rao Y-J 1997. In-fibre Bragg grating sensors *Measurement Science and Technology* **8** 355–375.
13. Webb D J 2014. *Handbook of Optical Sensors*, Ed J L Santos JL and F Farahi (Boca Raton, FL: CRC Press, Taylor & Francis Group) pp. 504–533.
14. Campanella C, Cuccovillo A, Campanella C, Yurt A and Passaro V 2018. Fibre Bragg grating based strain sensors: Review of technology and applications *Sensors* **18** 3115.
15. Cusano A (Ed) 2010. *Fiber Bragg Grating Sensors: 30 Years from Research to Market* (Oak Park, IL: Bentham eBooks).
16. Peters K 2011. Polymer optical fiber sensors-a review *Smart Materials and Structures* **20** 013002.
17. Peng G D, Xiong Z and Chu P L 1999. Photosensitivity and gratings in dye-doped polymer optical fibers *Optical Fiber Technology* **5** 242–251.
18. Ahmed R M 2009. Optical study on poly(methyl methacrylate)/poly(vinyl acetate) blends *International Journal of Photoenergy* **2009** Article ID 150389, 7.
19. Xiong Z, Peng G, Wu B and Chu P 1999. Highly tunable Bragg gratings in single-mode polymer optical fibers *IEEE Photonics Technology Letters* **11** 352–354.
20. Liu H Y, Peng G D and Chu P L 2001. Thermal tuning of polymer optical fiber Bragg gratings *IEEE Photonics Technology Letters* **13** 824–826.
21. Harbach N G 2008. *Fiber Bragg Gratings in Polymer Optical Fibers* (Lausanne, Switzerland: EPFL).
22. Eijkelenborg M V, Large M, Argyros A, Zagari J, Manos S, Issa N A, Bassett I M et al 2001. Microstructured polymer optical fibre *Optics Express* **9** 319–927.
23. Ye C C, Dulieu-Barton J M, Webb D J, Zhang C, Peng G D, Chambers A R, Lennard F J and Eastop D D 2009. Applications of polymer optical fibre grating sensors to condition monitoring of textiles In: *20th International Conference on Optical Fiber Sensors*, Ed J D C Jones (Edinburgh: SPIE) p 75030M.
24. Chen X, Zhang C, Hoe B V, Webb D J, Kalli K, Steenberge G V and Peng G-D 2010. Photonic skin for pressure and strain sensing In: *Optical Sensors and Detection*, Ed F Berghmans and A G Mignani (Brussels, Belgium: SPIE) p. 3.
25. Zhang C, Zhang W, Webb D J and Peng G D 2010. Optical fibre temperature and humidity sensor *Electronics Letters* **46** 643–U63.
26. Johnson I P, Kalli K and Webb D J 2010. 827 nm Bragg grating sensor in multimode microstructured polymer optical fibre *Electronics Letters* **46** 1217–1218.
27. Johnson I P, Webb D J, Kalli K, Large M C and Argyros A 2010. Multiplexed FBG sensor recorded in multimode microstructured polymer optical fibre In: *Photonic Crystal Fibres*, Ed K Kalli and W Urbanczyk (Brussels, Belgium: SPIE) p. 10.
28. Pospori A, Marques C A F, Sagias G, Lamela-Rivera H and Webb D J 2018. Novel thermal annealing methodology for permanent tuning polymer optical fiber Bragg gratings to longer wavelengths *Optics Express* **26** 1411–1421.
29. Marques C A F, Bilro L B, Alberto N J, Webb D J and Nogueira R N 2013. Narrow bandwidth Bragg gratings imprinted in polymer optical fibers for different spectral windows *Optics Communications* **307** 57–61.
30. Yuan W, Khan L, Webb D J, Kalli K, Rasmussen H K, Stefani A and Bang O 2011. Humidity insensitive TOPAS polymer fiber Bragg grating sensor *Optics Express* **19** 19731–19739.
31. Stefani A, Andresen S, Yuan W and Bang O 2012. Dynamic characterization of polymer optical fibers *IEEE Sensors Journal* **12** 3047–3053.
32. Stefani A, Andresen S, Yuan W, Herholdt-Rasmussen N and Bang O 2012. High sensitivity polymer optical fiber-Bragg-grating-based accelerometer *IEEE Photonics Technology Letters* **24** 763–765.
33. Abang A and Webb D J 2013. Influence of mounting on the hysteresis of polymer fiber Bragg grating strain sensors *Optics Letters* **38** 1376–1378.
34. Law S H, Harvey J D, Kruhlak R J, Song M, Wu E, Barton G W, van Eijkelenborg M A and Large M C J 2006. Cleaving of microstructured polymer optical fibres *Optics Communications* **258** 193–202.
35. Stefani A, Nielsen K, Rasmussen H K and Bang O 2012. Cleaving of TOPAS and PMMA microstructured polymer optical fibers: Core-shift and statistical quality optimization *Optics Communications* **285** 1825–1833.

36. Saez-Rodriguez D, Nielsen K, Bang O and Webb D J 2015. Simple room temperature method for polymer optical fibre cleaving *Journal of Lightwave Technology* **33** 4712–4716.
37. Ghirghi M V P, Minkovich V P and Villegas A G 2014. Polymer optical fiber termination with use of liquid nitrogen *IEEE Photonics Technology Letters* **26** 516–519.
38. Abang A, Saez-Rodriguez D, Nielsen K, Bang O and Webb D J 2013. Connectorisation of fibre Bragg grating sensors recorded in microstructured polymer optical fibre In: *Proceedings of the Fifth European Workshop on Optical Fibre Sensors*, Ed L R Jaroszewicz (Bellingham, Washington: SPIE) **8794**.
39. Saez-Rodriguez D, Nielsen K, Rasmussen H K, Bang O and Webb D J 2013. Highly photosensitive polymethyl methacrylate microstructured polymer optical fiber with doped core *Optics Letters* **38** 3769–3772.
40. Bonefacino J, Tam H Y, Glen T S, Cheng X, Pun C F J, Wang J, Lee P H, Tse M L V and Boles S T 2018. Ultra-fast polymer optical fibre Bragg grating inscription for medical devices *Light: Science & Applications* **7** 17161.
41. Pospori A, Marques C, Bang O, Webb D and André P 2017. Polymer optical fiber Bragg grating inscription with a single UV laser pulse *Optics Express* **25** 9028–9038.
42. Min R, Marques C, Nielsen K, Bang O and Ortega B 2018. Fast inscription of long period gratings in microstructured polymer optical fibers *IEEE Sensors Journal* **18** 1919–1923.
43. Min R, Marques C, Bang O and Ortega B 2018. Moire phase-shifted fiber Bragg gratings in polymer optical fibers *Optical Fiber Technology* **41** 78–81.
44. Stefani A, Yuan W, Markos C, Rasmussen H K, Andresen S, Guastavino R, Nielsen F K et al. 2012. *22nd International Conference on Optical Fiber Sensors*, Pts 1-3, Ed Y Liao et al.
45. Woyessa G, Fasano A, Markos C, Stefani A, Rasmussen H K and Bang O 2017. Zeonex microstructured polymer optical fiber: Fabrication friendly fibers for high temperature and humidity insensitive Bragg grating sensing *Optical Materials Express* **7** 286–295.
46. Fasano A, Woyessa G, Stajanca P, Markos C, Stefani A, Nielsen K, Rasmussen H K, Krebber K and Bang O 2016. Fabrication and characterization of polycarbonate microstructured polymer optical fibers for high-temperature-resistant fiber Bragg grating strain sensors *Optical Materials Express* **6** 649–659.
47. Fasano A, Woyessa G, Stajanca P, Markos C, Stefani A, Nielsen K, Rasmussen H K, Krebber K and Bang O 2016. Creation of a microstructured polymer optical fiber with UV Bragg grating inscription for the detection of extensions at temperatures up to 125 degrees C. *Proc. SPIE*, (Bellingham, Washington: SPIE) **9996**.
48. Lacraz A, Polis M, Theodosiou A, Koutsides C and Kalli K 2015. Femtosecond laser inscribed Bragg gratings in low loss CYTOP polymer optical fiber *IEEE Photonics Technology Letters* **27** 693–696.
49. Lacraz A, Theodosiou A and Kalli K 2016. Femtosecond laser inscribed Bragg grating arrays in long lengths of polymer optical fibres; a route to practical sensing with POF *Electronics Letters* **52** 1626–1627.
50. Bowden M J, Chandross E A and Kaminow I P 1974. Mechanisms of the photoinduced refractive index increase in polymethyl methacrylate *Applied Optics* **13** 112.
51. Marotz J 1985. Holographic storage in sensitized polymethyl methacrylate blocks *Applied Physics B* **37** 181–187.
52. Moran J M and Kaminow I P 1973. Properties of holographic gratings photoinduced in polymethyl methacrylate *Applied Optics* **12** 1964–1970.
53. Mitsuoka T, Torikai A and Fueki K 1993. Wavelength sensitivity of the photodegradation of poly(methyl methacrylate) *Journal of Applied Polymer Science* **47** 1027–1032.
54. Wochnowski C, Metev S and Sepold G 2000. UV-laser-assisted modification of the optical properties of polymethylmethacrylate *Applied Surface Science* 706–711.
55. Liu H Y, Liu H B, Peng G D and Chu P L 2004. Observation of type I and type II gratings behaviour in polymer optical fiber *Optics Communications* **16** 159–161.
56. Liu H B, Liu H Y, Peng G D and Chu P L 2004. Novel growth behaviours of fiber Bragg gratings in polymer optical fiber under UV irradiation with low power *IEEE Photonics Technology Letters* **16** 159–1613.
57. Sáez-Rodríguez D, Nielsen K, Bang O and Webb D J 2014. Increase of the photosensitivity of undoped poly(methylmethacrylate) under UV radiation at 325 nm. In: *Micro-Structured and Speciality Optical Fibres, Photonics Europe* (Brussels, Belgium: SPIE) p. 91280P.
58. Tyler D R 2004. Mechanistic aspects of the effects of stress on the rates of photochemical degradation reactions in polymers *Polymer Reviews* **44** 351–388.

59. Scully P, Baum A, Liu D and Perrie W 2012. Refractive Index Structures in Polymers. In: *Femtosecond Laser Micromachining. Topics in Applied Physics*, Ed R Osellame, G Cerullo and R Ramponi (Springer, Berlin: Heidelberg) **123**.
60. Koerdt M, Kibben S, Hesselbach J, Brauner C, Herrmann A S, Vollertsen F and Kroll L 2014. Fabrication and characterization of Bragg gratings in a graded-index perfluorinated polymer optical fiber *Procedia Technology* **15** 138–146.
61. Marques C A F, Leal A G, Min R, Domingues M, Leitao C, Antunes P, Ortega B and Andre P 2018. Advances on polymer optical fiber gratings using a KrF pulsed laser system operating at 248 nm *Fibers* **6** 13.
62. Dobb H, Webb D J, Kalli K, Argyros A, Large M C J and van Eijkelenborg M A 2005. Continuous wave ultraviolet light-induced fiber Bragg gratings in few- and single-mode microstructured polymer optical fibers *Optics Letters* **30** 3296–3298.
63. Butter C D and Hocker G B 1978. Fiber optics strain gauge *Applied Optics* **17** 2867–2869.
64. Hocker G B 1979. Fiber-optic sensing of pressure and temperature *Applied Optics* **18** 1445–448.
65. Brady G, Kalli K, Webb D J, D.A. Jackson L R and Archambault J L 1997. Simultaneous measurement of strain and temperature using the first- and second-order diffraction wavelengths of Bragg gratings *IEE Proceedings-Optoelectronics* **144** 156–161.
66. Yuan S, Stefani A, Bache M, Jacobsen T, Rose B, Herholdt-Rasmussen N, Nielsen F et al. 2011. Improved thermal and strain performance of annealed polymer optical fiber Bragg gratings *Optics Communications* **284** 176–182.
67. Zhang Z F and Tao X M 2012. Synergetic effects of humidity and temperature on PMMA based fiber Bragg gratings *Journal of Lightwave Technology* **30** 841–845.
68. Johnson I P, Webb D and Kalli K 2012. Hydrostatic pressure sensing using a polymer optical fibre Bragg grating In: *3rd Asia Pacific Optical Sensors Conference* (Sydney, Australia: SPIE).
69. Ishikawa R, Lee H, Lacraz A, Theodosiou A, Kalli K, Mizuno Y and Nakamura K 2017. Pressure dependence of fiber Bragg grating inscribed in perfluorinated polymer fiber *IEEE Photonics Technology Letters* **29** 2167–2170.
70. Zhang Z F and Tao X M 2013. Intrinsic temperature sensitivity of fiber Bragg gratings in PMMA-based optical fibers *IEEE Photonics Technology Letters* **25** 310–312.
71. Zhang W and Webb D J 2014. Humidity responsivity of polymer optical fiber Bragg grating sensors *Optics Letters* **39** 3026–3029.
72. Kersey A D and Marrone M J 1994. Fiber Bragg grating high-magnetic-field probe. *Tenth International Conference on Optical Fibre Sensors* (Bellingham, Washington: SPIE) **2360** 53–56.
73. Eul Ha H and Byoung Yoon K 2006. Pulsed high magnetic field sensor using polymethyl methacrylate *Measurement Science and Technology* **17** 2015–2021.
74. Large M C, Lwin R, Argyros A and Leon-Saval S G 2011. Low loss microstructured polymer optical fibre (mPOF). In: *Proceedings of the Optical Fiber Communication Conference* (Optical Society of America) p paper OWS6.
75. HY L, Peng GD, Chu C, Koike Y and Watanabe Y 2001. Photosensitivity in low-loss perfluoropolymer (CYTOP) fibre material *Electronics Letters* **37** 347–348.
76. Allsop T, Carroll K, Lloyd G, Webb D J, Miller M and Bennion I 2007. Application of long-period-grating sensors to respiratory plethysmography *Journal of Biomedical Optics* **12** 064003.
77. Brandrup J 1999. *Polymer Handbook* vol 1&2 (New York: Wiley).
78. Markos C, Stefani A, Nielsen K, Rasmussen H K, Yuan W and Bang O 2013. High-T-g TOPAS microstructured polymer optical fiber for fiber Bragg grating strain sensing at 110 degrees *Optics Express* **21** 4758–4765.
79. Yang D, Yu J, Tao X and Tam H 2004. Structural and mechanical properties of polymeric optical fiber *Materials Science and Engineering* **A364** 256–259.
80. TOPAS cyclic olefin copolymer, www.topas.com.
81. Woyessa G, Fasano A, Stefani A, Markos C, Nielsen K, Rasmussen H K and Bang O 2016. Single mode step-index polymer optical fiber for humidity insensitive high temperature fiber Bragg grating sensors *Optics Express* **24** 1253–1260.
82. Kiesel S, Peters K, Hassan T and Kowalsky M 2007. Behaviour of intrinsic polymer optical fibre sensor for large-strain applications *Measurement Science & Technology* **18** 3144–3154.
83. Welker D J, Tostenrude J, Garvey D W, Canfield B K and Kuzyk M G 1998. Fabrication and characterization of single-mode electro-optic polymer optical fiber *Optics Letters* **23** 1826–1828.

84. Peng G D, Chu P K, Xiong Z, Whitbread T W and Chaplin R P 1996. Dye-doped step-index polymer optical fiber for broadband optical amplification *IEEE Journal of Lightwave Technology* **14** 2215–2223.
85. Dobb H, Carroll K, Webb D J, Kalli K, Komodromos M, Themistos C, Peng G D et al. 2006. Reliability of fibre Bragg gratings in polymer optical fibre - art. no. 61930Q. In: *Reliability of Optical Fiber Components, Devices, Systems, and Networks III* (Bellingham, Washington: SPIE) pp Q1930-Q.
86. Szczurowski M K, Martynkien T, Statkiewicz-Barabach G, Urbanczyk W, Khan L and Webb D J 2010. Measurements of stress-optic coefficient in polymer optical fibers *Optics Letters* **35** 2013–2015.
87. Carroll K E, Zhang C, Webb D J, Kalli K, Argyros A and Large M C J 2007. Thermal response of Bragg gratings in PMMA microstructured optical fibers *Optics Express* **15** 8844–8850.
88. Aressy M 2006. *Manufacturing optimisation and mechanical properties of polymer optical fibre* (Birmingham, UK: Birmingham University).
89. Stajanca P, Cetinkaya O, Schukar M, Mergo P, Webb D J and Krebber K 2016. Molecular alignment relaxation in polymer optical fibers for sensing applications *Optical Fiber Technology* **28** 11–17.
90. Pospori A, Marques C, Sáez-Rodríguez D, Nielsen K, Bang O and Webb D 2017. Thermal and chemical treatment of polymer optical fiber Bragg grating sensors for enhanced mechanical sensitivity *Optical Fiber Technology* **36** 68–74.
91. Bhowmik K, Peng G D, Luo Y H, Ambikairajah E, Lovric V, Walsh W R and Rajan G 2016. Etching process related changes and effects on solid-core single-mode polymer optical fiber grating *IEEE Photonics Journal* **8** 2500109.
92. Rajan G, Bhowmik K, Xi J T and Peng G D 2017. Etched polymer fibre Bragg gratings and their biomedical sensing applications *Sensors* **17** 2336.
93. Leal A, Frizera A, Marques C and Pontes M J 2018. Mechanical properties characterization of polymethyl methacrylate polymer optical fibers after thermal and chemical treatments *Optical Fiber Technology* **43** 106–111.
94. Smith L S A and Schmitz V 1988. The effect of water on the glass transition temperature of poly(methyl methacrylate) *Polymer* **29** 1871–1878.
95. Fasano A, Woyessa G, Janting J, Rasmussen H K and Bang O 2017. Solution-mediated annealing of polymer optical fiber Bragg gratings at room temperature *IEEE Photonics Technology Letters* **29** 687–690.
96. Marques C A F, Pospori A, Demirci G, Cetinkaya O, Gawdzik B, Antunes P, Bang O, Mergo P, Andre P and Webb D J 2017. Fast Bragg grating inscription in PMMA polymer optical fibres: Impact of thermal pre-treatment of preforms *Sensors* **17** 891.
97. Ganziy D, Jespersen O, Woyessa G, Rose B and Bang O 2015. Dynamic gate algorithm for multimode fiber Bragg grating sensor systems *Applied Optics* **54** 5657–5661.
98. Lwin R and Argyros A 2009. Connecting microstructured polymer optical fibres to the world. In: *International Plastic Optical Fibre Conference* (Sydney).
99. Abang A and Webb D J 2012. Demountable connection for polymer optical fiber grating sensors *Optical Engineering* **51** 080503.
100. Zhang W, Webb D J, Carpenter M and Williams C 2014. Measuring water activity of aviation fuel using a polymer optical fiber Bragg grating In: *Proceedings of SPIE* 9157 pp. 91574V-1.
101. Zhang W, Webb D J, Lao L, Hammond D, Carpenter M and Williams C 2019. Water content detection in aviation fuel by using PMMA based optical fiber grating *Sensors and Actuators B: Chemical* **282** 774–779.
102. Zhang W, Webb D J and Peng G D 2012. Investigation into time response of polymer fiber Bragg grating based humidity sensors *Journal of Lightwave Technology* **30** 1090–1096.
103. Rajan G, Noor Y M, Liu B, Ambikairaja E, Webb D J and Peng G-D 2013. A fast response intrinsic humidity sensor based on an etched single mode polymer fiber Bragg grating *Sensors and Actuators A: Physical* **203** 107–111.
104. Zhang W, Webb D and Peng G 2012. Polymer optical fiber Bragg grating acting as an intrinsic biochemical concentration sensor *Optics Letters* **37** 1370–1372.
105. Zhang W and Webb D 2014. Polymer optical fiber grating as water activity sensor In: *Micro-Structured and Specialty Optical Fibres* (Brussels, Belgium: SPIE).
106. Bundalo I L, Nielsen K, Woyessa G and Bang O 2017. Long-term strain response of polymer optical fiber FBG sensors *Optical Materials Express* **7** 967–976.

107. Marques C A F, Bilro L, Kahn L, Oliveira R A, Webb D J and Nogueira R N 2013. Acousto-optic effect in microstructured polymer fiber Bragg gratings: Simulation and experimental overview *Journal of Lightwave Technology* **31** 1551–1558.
108. Broadway C, Gallego D, Pospori A, Zubel M, Webb D J, Sugden K, Carpintero G and Lamela H 2016. A compact polymer optical fibre ultrasound detector. pp 970813.
109. Marques C A F, Peng G-D and Webb D J 2015. Highly sensitive liquid level monitoring system utilizing polymer fiber Bragg gratings *Optics Express* **23** 6058–6072.
110. Liu H Y, Liu H B, Peng G D and Chu P L 2006. Polymer optical fibre Bragg gratings based fibre laser *Optics Communications* 132–135.
111. Cheng X, Bonefacino J, Guan B O and Tam H Y 2018. All-polymer fiber-optic pH sensor *Optics Express* **26** 14610–14616.
112. Zubel M G, Sugden K, Webb D J, Saez-Rodriguez D, Nielsen K and Bang O 2016. Embedding silica and polymer fibre Bragg gratings (FBG) in plastic 3Dprinted sensing patches, *proceedings of the SPIE* (Bellingham, Washington: SPIE) **9886**.
113. Dobb H, Petrovic J S, Mezentsev V, Webb D J and Kalli K 2005. Long-period gratings fabricated in photonic crystal fibre In: *17th International Conference on Optical Fibre Sensors* Pts 1 and 2 pp 334–337.
114. Hiscocks M, Eijkelenborg M V, Argyros A and Large M 2006. Stable imprinting of long-period gratings in microstructured polymer optical fibre *Optics Express* **14** 4644–4648.
115. Lwin R, Argyros A, Leon-Saval S G and Large M C J 2011. *21st International Conference on Optical Fiber Sensors*, Ed W J Bock et al.
116. Statkiewicz-Barabach G, Kowal D, Szczurowski M K, Mergo P and Urbanczyk W 2013. Hydrostatic pressure and strain sensitivity of long period grating fabricated in polymer microstructured fiber *IEEE Photonics Technology Letters* **25** 496–499.
117. Marques C A F, Antunes P, Mergo P, Webb D J and Andre P 2017. Chirped Bragg gratings in PMMA step-index polymer optical fiber *IEEE Photonics Technology Letters* **29** 500–503.
118. Hu X, Pun C-F J, Tam H-Y, Megret P and Caucheteur C 2014. Tilted Bragg gratings in step-index polymer optical fiber *Optics Letters* **39** 6835–6838.
119. SIilva-Lopez M, Fender A, Macpherson W, Barton J and Jones J 2005. Strain and temperature sensitivity of a single-mode polymer optical fiber *Optics Letters* **30** 3129–3131.
120. Dobb H, Carroll K, Webb D J, Kalli K, Komodromos M, Themistos C, Peng G D, Argyros A et al. 2006. Grating based devices in polymer optical fibre - art. no. 618901. In: *Optical Sensing II* pp. 18901.
121. Statkiewicz-Barabach G, Mergo P and Urbanczyk W 2017. Bragg grating-based Fabry-Perot interferometer fabricated in a polymer fiber for sensing with improved resolution *Journal of Optics* **19** 015609.
122. Theodosiou A, Hu X H, Caucheteur C and Kalli K 2018. Bragg gratings and Fabry-Perot cavities in low-loss multimode CYTOP polymer fiber *IEEE Photonics Technology Letters* **30** 857–860.
123. Pospori A and Webb D J 2017. Stress sensitivity analysis of optical fiber Bragg grating-based Fabry–Pérot interferometric sensors *Journal of Lightwave Technology* **35** 2654–2659.

Temperature Sensing by Rubi Fluorescence

Marcelo Martins Werneck

Contents

7.1 Introduction

Temperature is a very important parameter for the electric power industry because insulators, copper conductors, iron core of transformers, insulator oil, and all equipment are very sensitive to the temperature, which has to be kept under strict control at all times. Thermal monitoring of conductors and equipment of substations is the best means of early detection of faults that are liable to lead to equipment faults or disruptive dielectric discharges.

At present, the most commonly used solution to protect against high temperature occurrences is monitoring by conventional sensors such as thermocouples and Pt100 or using infrared thermal cameras. Nevertheless, in some situations, either the infrared camera does not reach the local or the spot is under high voltage, which prevents the application of a conventional electronic sensor.

To circumvent this problem, optical fiber temperature sensors have been widely developed and many publications are available in the literature. However, most of these solutions are not commercially available due to their cost, which makes them a non-competitive solution.

In this chapter we demonstrate a simple and cheap POF temperature sensor that has been applied in real high-voltage energized substation equipment.

7.2 Motivation

Brazilian Itaipu power plant at Brazil's southwest border with Paraguay transmits from a total of 14 GW of power, 6.3 GW over two 800-km 600 kV HVDC lines. **Figure 7.1** shows the main Brazilian transmission lines and **Figure 7.2** shows the area inside the dotted box, in which the HVDC line is shown in detail.

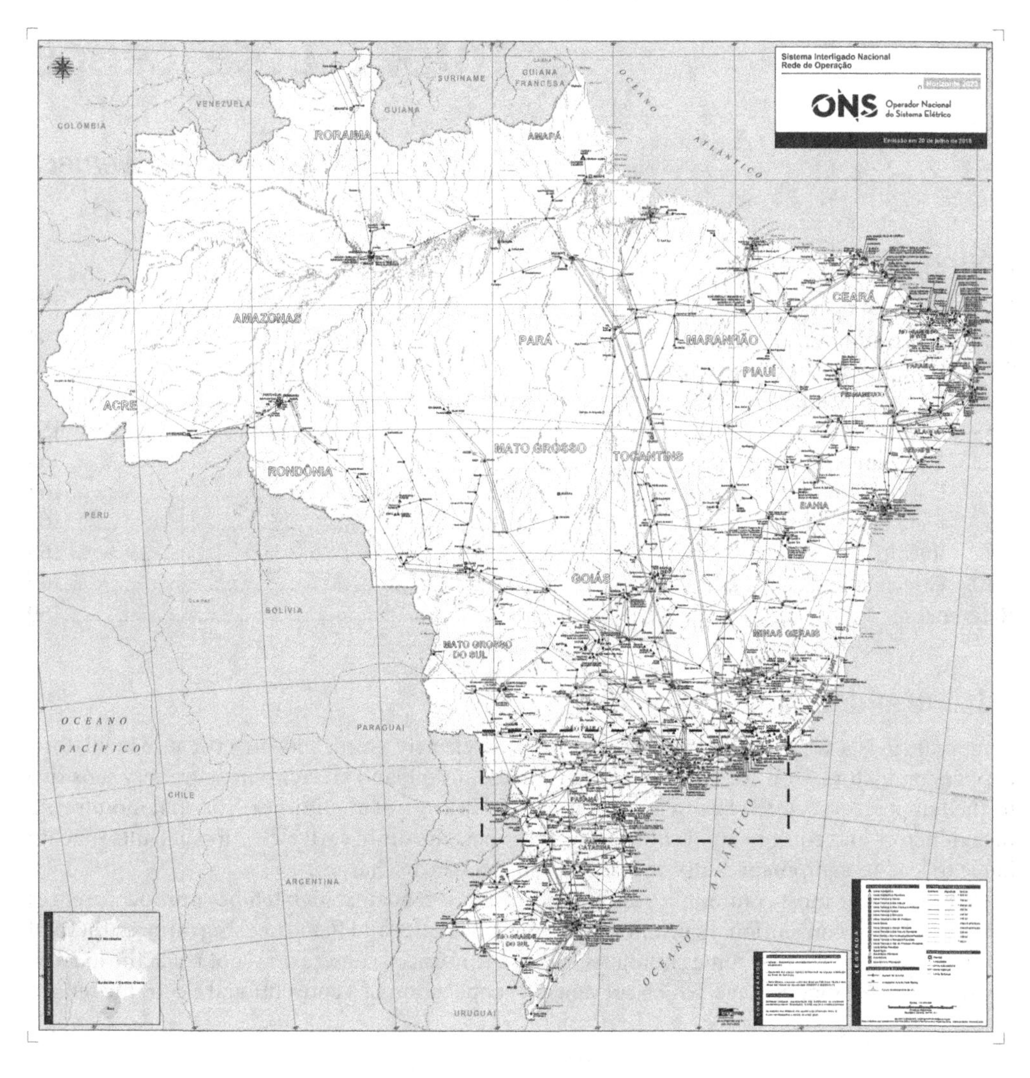

FIGURE 7.1 Main Brazilian transmission lines with Itaipu HVDC inside box. (From Operador Nacional do Sistema Elétrico, http://ons.org.br/paginas/sobre-o-sin/mapas.)

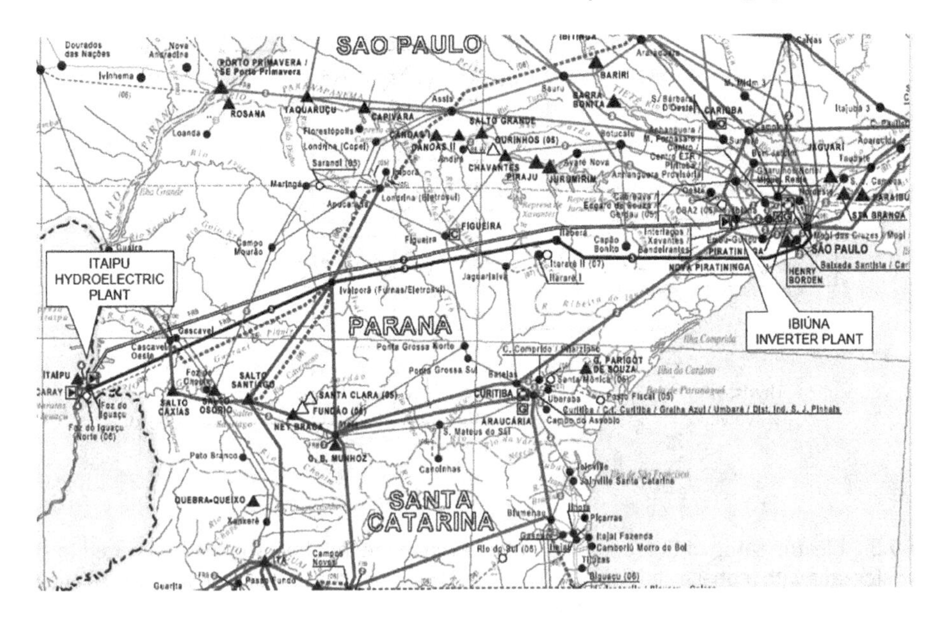

FIGURE 7.2 Enlarged section of Figure 7.1 showing the 800-km long 600 kV HVDC Itaipu transmission line.

The conversion from AC to DC occurs in a substation located close to the hydroelectric plant, at the State of Paraná.

At the load side of the DC line, 790 km away at the State of São Paulo, the DC is converted back to 345 kV AC. Since the converter used for HVDC is able to operate with power conversion in either direction, that is, AC to DC (rectification) and conversion from DC to AC (inversion), the two ends of the HVDC line are similar.

The converter is located inside a large building comprised of about 9,000 dish-like thyristors valves. This process is known as line commutated converter in which many thyristors are connected in series and parallel to form a thyristor valve, and each converter normally consists of twelve thyristor valves. The thyristor valves are joined in groups of four and stand on insulators on the floor inside the building. **Figure 7.3** shows the electric setup of the system.

The converter transformers step up the voltage using a star-to-delta connection of the transformer windings, producing 12 pulses for each cycle in the AC supply. On the other side of the DC line, the same process occurs in the opposite direction, producing the 60 Hz AC voltage. In this process the produced AC voltage is comprised of a stepwise sinusoidal wave shape that contains numerous harmonic components. For this reason, harmonic filters are necessary for the elimination of the harmonic components and for the production of the reactive power at line-commutated converter stations. At 12 pulse converter stations, only harmonic voltages or currents of the order $12n + 1$ result.

The resulting output sinusoidal wave shape is similar to the one shown in **Figure 7.4**. It is a staircase-like sinusoidal signal that contains odd harmonics, particularly the third and the fifth, as shown in the spectrum plotted at right.

The filters, consisting of series combinations of capacitors and inductors, lie in the substation close to the converter building and are tuned to the expected harmonic frequencies. The filter equivalent circuit is shown in **Figure 7.5**. Capacitors dielectric and coils core are made of air and open to the atmosphere.

The filter impedance spectrum is shown in **Figure 7.6**. Notice that the impedance modulus is zero with phase zero at 180 and 300 Hz, meaning that there is a virtual short-circuit do ground at these frequencies.

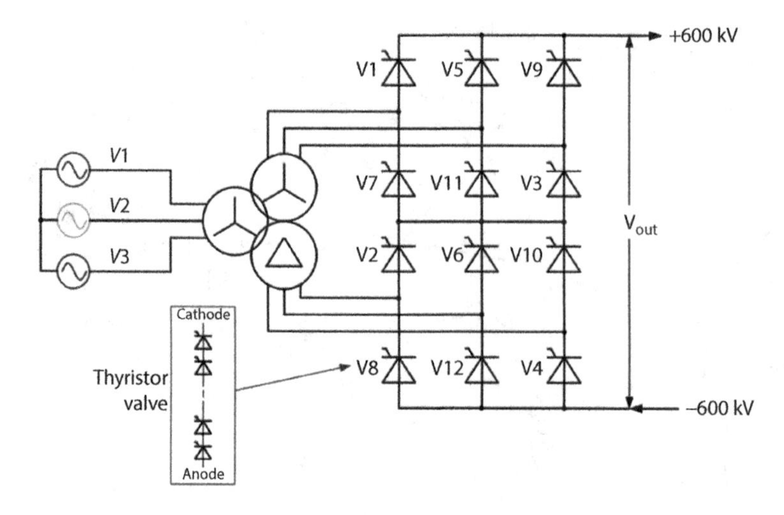

FIGURE 7.3 Electric setup of the rectifier system in which the input AC voltage is elevated to supply the thyristor rails with a phase shift of 30°. The inverter side of the DC line contains a similar system which is driven by a different set of synchronizing clocks.

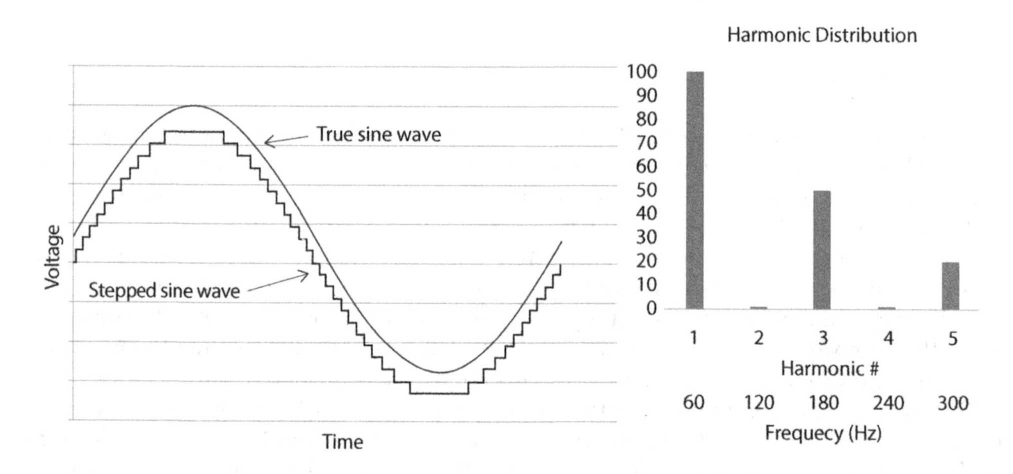

FIGURE 7.4 Left: Recovered 60 Hz sinusoidal waveform and its harmonic contents. Right: Notice that only odd harmonics are present.

Figure 7.7 shows a picture of the filter where the inductors, capacitors and resistors shown in the equivalent circuit of **Figure 7.5** are indicated by arrows.

Due to the power dissipation from high currents, the inductors of this filter tend to heat up and eventually burn the insulator in several spots. Since some of these hot-spots are located in inner layers of the inductor, infrared imagery was not efficient for detection and location of these hot-spots. Additionally, the inductors are located at a 345-kV line, which complicates even more the application of conventional temperature sensors.

The plant engineers decided that the best approach was to employ a specially designed fiberoptic sensor to measure the temperature and they looked for an experienced research group. Eventually, the Photonics and Instrumentation Laboratory was selected and hired to do the job.

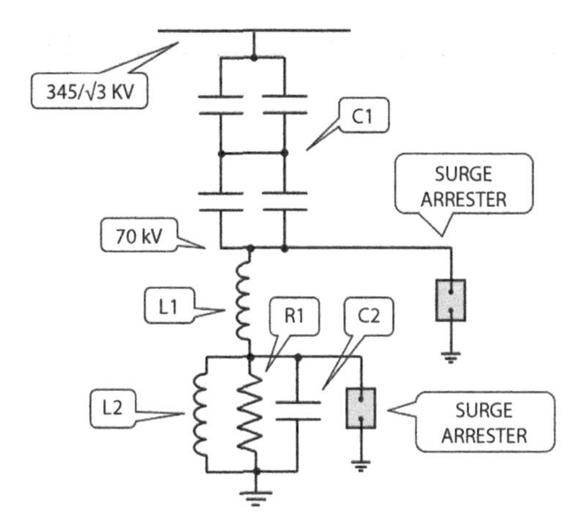

FIGURE 7.5 The harmonic filter equivalent circuit.

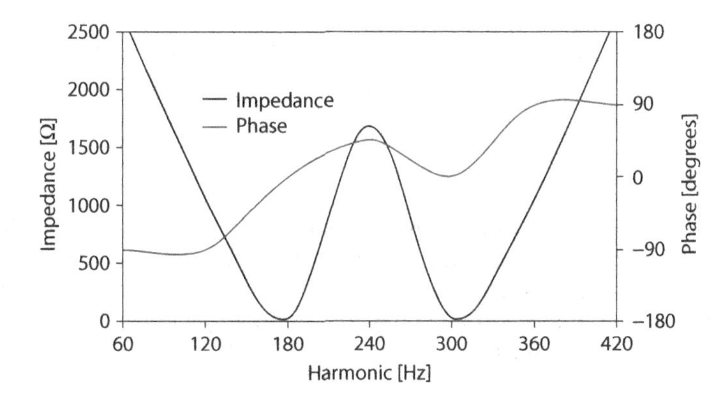

FIGURE 7.6 Modulus and phase of the harmonica filter. Notice that the impedance modulus is zero for the third (180 Hz) and for the fifth harmonic (300 Hz).

FIGURE 7.7 The harmonic filter of Ibiúna substation. The numbers in boxes refer to Figure 7.5.

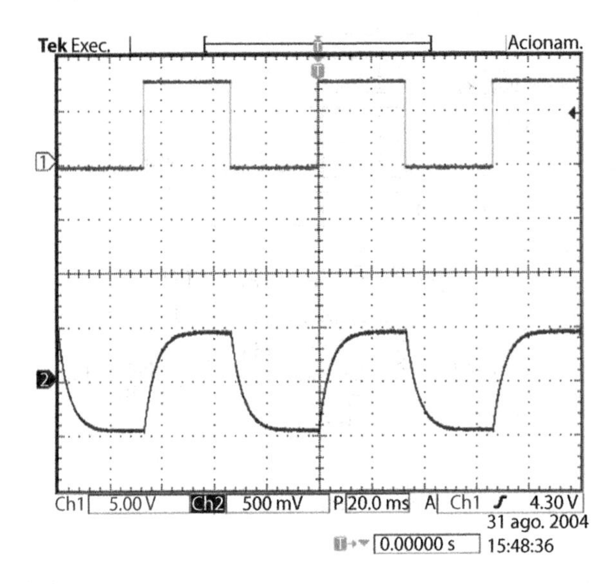

FIGURE 7.8 Ruby fluorescence when excited with blue light. Pump light (upper trace) and resulted fluorescence light (lower trace) of an artificial ruby.

7.3 Principle of Measurement

The solution to the temperature measurement should include several specific characteristics such as supportability of high voltage and leakage currents, possibility to be installed between two coil layers, robust enough to support all kinds of weather, and IP66 rating protection. Above all, the prototype should be made of components and technology allowing a mass production of a low-cost system so as to present a final price comparable with conventional sensors.

Therefore, we decided to test a temperature sensor based on the known fluorescence property of ruby. Rubies, either artificial or natural, are red in color which, when excited with blue light, produces red fluorescence, as shown in **Figure 7.8**.

7.4 Previous Studies in Fluorescence for Temperature Measurement

Studies have shown the dependence of the half-life time of fluorescent materials on temperature. Temperature measurements using the half-life of fluorescent materials have the advantage of being independent of possible fluctuations in the light intensity of the excitation source or losses in the fiber. In addition, using this technique, one can construct several sensors for which, in principle, the intrinsic characteristic of the fluorescence will be maintained from material to material. This has important implications for fiber optic sensors, since the sensor probe can be replaced promptly if it is damaged or defective without the need for calibration.

In this section we will show some studies and applications of fluorescence for temperature measurement.

Augousti et al. (1987 and 1988) developed and calibrated a fiber optic temperature sensor pumped by an LED in the visible range. In the experimental setup, a crystal of alexandrite was used as sensor element. The instrument was able to measure temperatures between 20°C to 150°C with accuracy of ±1°C and response time of 1 s. The sensing tip consists of a small, irregularly-shaped, 8 mm³ crystalline alexandrite crystal (using EPOTEK 514ND transparent epoxy resin) on two optical fibers (one for transmission and one for signal reception). The fluorescence of the crystal was around 675 nm.

Sun (1992) developed a sensor with magnesium fluorogermanate deposited on the end of an optical fiber of 300 μm in diameter. In this system a blue light was used to excite the material and a photodetector to capture the red emission intensity from the fluorescence after the blue pulse. The decay time of the emission was measured and correlated with the temperature in the sensor. For the temperature of 27.5°C and 40°C the decay was approximately 3.4 and 3.3 ms respectively, allowing an accuracy of ±0.2°C.

A large amount of temperature measurement techniques based on the half-life of the fluorescence is proposed by Zhang et al. (1992). A variety of fluorescent materials such as Nd: YAG, Nd: glass, ruby, and alexandrite were used. The best result was obtained with Cr^{3+}:$LiSrAlF_6$ (also known as Cr: LiSAF). The system consisted of a 670 nm–1 mW laser diode focused with lenses into the core of a 200/230 μm silica fiber which guided the light to the Cr:LiSAF crystal. The crystal, with dimensions 0.5 × 0.6 × 0.5 mm^3 was glued on the tip of the fiber. The returned light was deviated to a photodetector through a 3 dB fiber coupler.

The work of Alcala et al. (1996) was based on the excited state phosphorescence lifetime of ruby crystals. They used the system to monitor temperature in the physiological range from 15°C to 45°C with precision and accuracy less than 1°C. A 500 μm cubic ruby crystal bounded to the distal end of an optical fiber was excited with pulsed He-Ne laser of 9 μW average power. Due to the small amount of optical power at the fiber end, they used a photomultiplier as detector.

Fernicola et al. (1997) built a silica fiber optic thermometer based on the fluorescence of chromium in olivine ($CrMg_2SiO_4$) and forsterite (Cr:(Mg, Fe)SiO_4) crystals with a 3 mW–670 nm laser diode for excitation. As a detector they used a multichannel optical spectrum analyzer.

From the works described so far, one may notice the difficulty in dealing with the low light power guided by silica optical fibers. For circumventing this problem, the authors had to use laser diodes or He-Ne laser to excite the crystals and photomultipliers or an optical spectrum analyzer to read the return signal.

With the advent of large-scale plastic optical fibers manufacturing, prices went down which stimulated research in many different areas, including sensing. The main advantage of POF is, of course, the large diameter and the possibility to work with low-cost peripheral devices, such as LEDs and silicon photodiodes. A disadvantage is the high optical attenuation, which makes it complicated to work with long distances. Sensing, however, normally needs a few meters distance from sensor to monitoring equipment and therefore attenuation is not always a problem.

Based on these advantages we can mention the work of Persegol et al. (1999), which used a POF in their sensing system. A pulsed light beam from a green GaN LED is carried to the optical probe containing ruby powder by one arm of a duplex 1 mm plastic fiber. The second fiber brings back fluorescence from the probe to a silicon PIN photodiode operating in a photovoltaic mode. A long-pass filter attenuates the reflected excitation light in order to avoid saturation of the electronic analogue stage.

The sensor system described in this chapter is based on the work of Ribeiro et al. (2003) that applied a commercial polystyrene fiber with artificial ruby crystal for a temperature sensor in a high voltage environment.

7.5 Sensor Setup

High temperature polyethylene POF (Eska FH4001, Mitsubishi Rayon Co, Japan) was used in this prototype. Although polyethylene presents a higher attenuation as compared with PMMA, it has the advantage of a higher supportability to temperature; 105°C instead of 70°C.

In any case, since the temperature limit to most of electrical wires is less than 100°C, in normal operational conditions the sensor will work normally. However, since the real temperature of the reactor coil was unknown, in case of temperature elevation above the shutdown limit, the sensor will fail and eventually will melt down, a risk we decided to cope with.

Figure 7.9 shows a depiction of the sensor tip with the hemispherical ruby crystal on its tip. Epoxy resin was used to fix the crystal at the POF tip, and a heat shrink tubing protects the assembly.

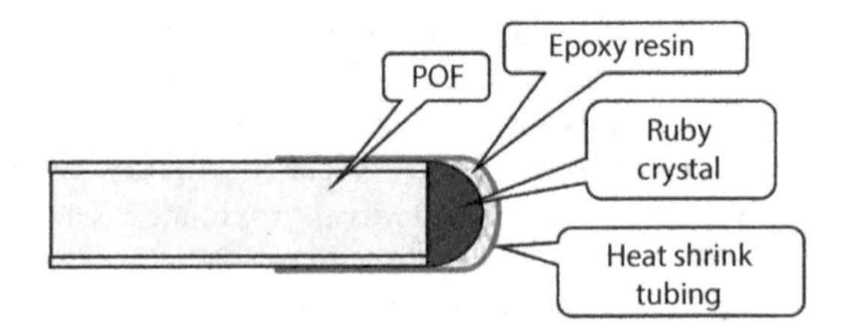

FIGURE 7.9 Illustration of the miniaturized POF probe with hemispherical ruby crystal.

The gemstone sapphire contains pure alumina (Al_2O_3) while the gemstone ruby contains the same aluminum oxide of sapphire but with chromium inside its molecule, known as Al_2O_3:Cr. Artificial sapphire is used for producing lens with a transmission window extending from 150 nm up to 4 µm, and artificial ruby is used in mechanical watches to reduce friction, increasing accuracy and bearings life.

Figure 7.10 shows in (a) a stone containing ruby incrustations. In (b) there is a picture of a raw ruby, as removed from the stone. In (c) is a ruby gemstone after lapidated. In (d) and (e) two methods of fixing an axle in a ruby bearing are shown. In (F) are pieces of synthetic ruby found in the watch-makers market.

Synthetic ruby keeps the same properties as natural ruby gemstone, particularly the hardness 9 in the Mohs scale and the low coefficient of friction.

Figure 7.11 shows several synthetic ruby pieces tested in our lab in the search for the best connection between the ruby and the POF.

Figure 7.12 shows a schematic of one of the four channels of the sensor. A pulsed light beam from a super-luminescent blue LED is guided to the optical probe containing the ruby crystal by one arm of a 1-mm datacom grade PMMA POF (Eska EH4001-Mitsubishi Rayon, Japan). This is the standard POF that can operate up to 70°C to 85°C.

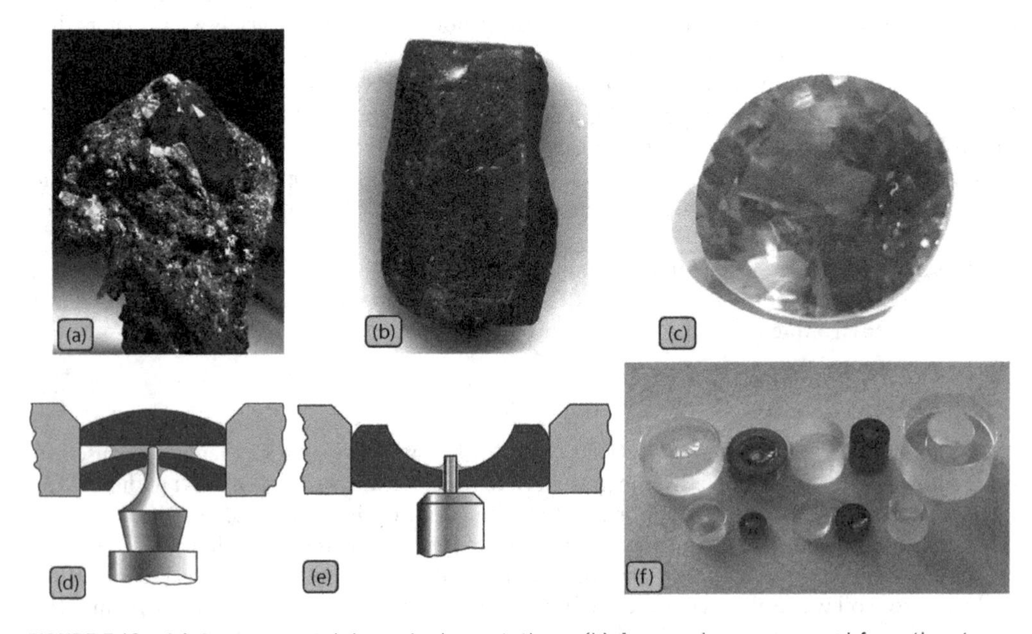

FIGURE 7.10 (a) A stone containing ruby incrustations. (b) A raw ruby, as removed from the stone. (c) Ruby gemstone after lapidated. (d and e) Two methods of fixing an axle in a ruby bearing. (f) Pieces of synthetic ruby available in the watch-makers' market.

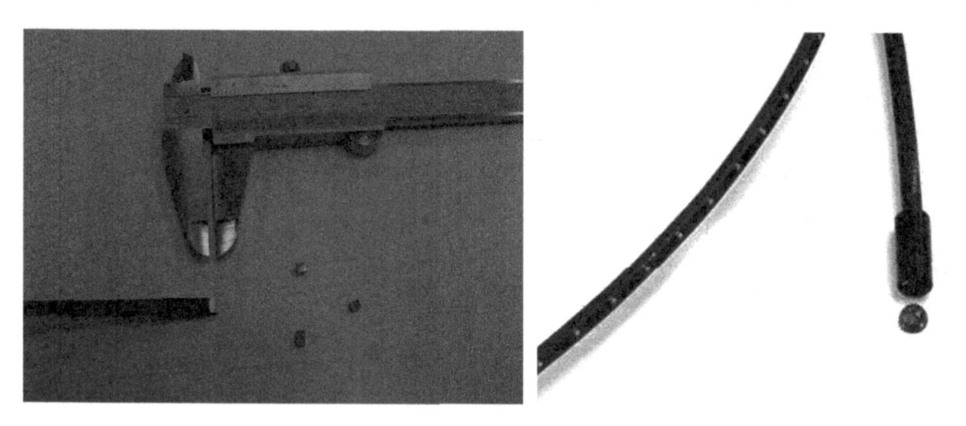

FIGURE 7.11 Several formats of ruby as available in the watch-makers' market.

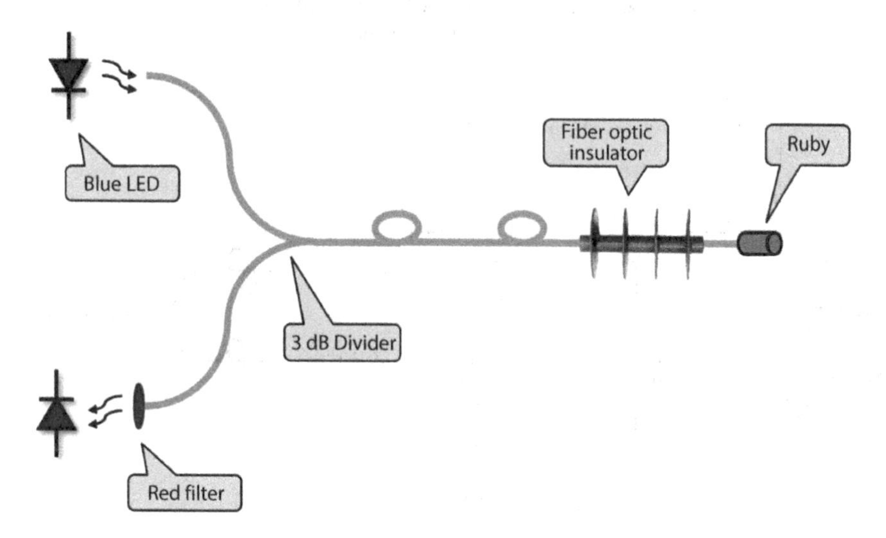

FIGURE 7.12 Setup of the system. The fiber optic insulator is used to increase the creepage distance of the fiber, from the high potential to ground potential, thus increasing surface resistance and avoiding large leakage currents.

The return light is deviated at the 3 dB coupler (DieMount, Germany) and brings back fluorescence from the probe to a silicon PIN photodiode operating in photoconductive mode. A long-pass red filter attenuates the reflected excitation light in order to avoid saturation of the electronic analogue stage.

Figure 7.13 shows the spectra of all components in the optical setup. The blue line is the LED spectrum, the red line is the fluorescent spectrum, the green line is the transmittance of the red filter, used to block some of the blue light that also returns to the photodetector. The black line is the ruby absorbance that presents low regions at blue and red. Notice the LED light centered at one of the low regions.

The optoelectronic setup is shown in **Figure 7.14**. Optical pulses of 32 ms from the LED were generated to pump the ruby crystal at 15.6 Hz. The decay time of ruby is calculated by the computer by interpolation of an exponential curve and then compared with a calibration curve.

Figure 7.15 shows a picture of the electronic board with four interfaces for four probes. A USB connector interfaces the control board with a computer.

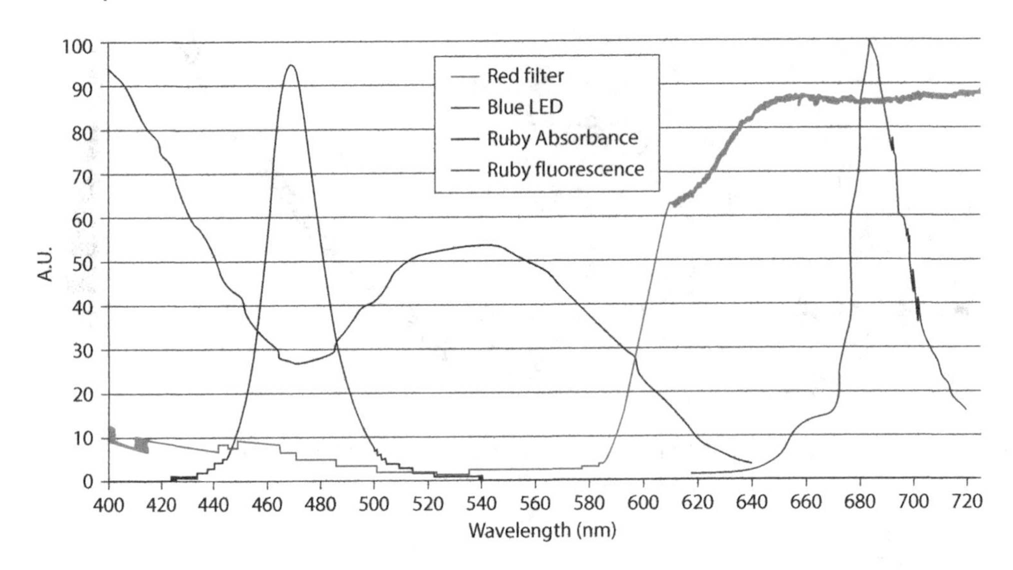

FIGURE 7.13 LED spectrum (blue line), fluorescence of ruby (red line), transmittance of the red filter (green line), and absorbance of ruby (black line).

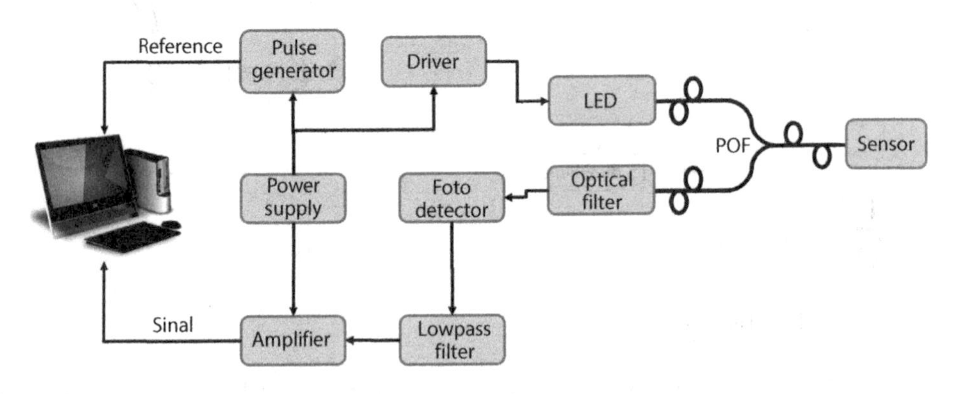

FIGURE 7.14 The optoelectronic setup.

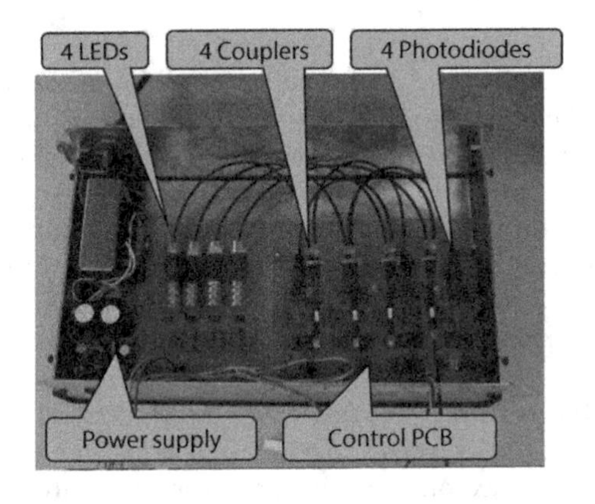

FIGURE 7.15 Picture of the electronic board, with four interfaces for four probes. A USB connector interfaces the control board with a computer.

7.6 The Temperature Probe

The construction of the probe should conform with several demands. The first one was to withstand a temperature up 100°C. Additionally, the distance from the sensor probe, installed on the coil, and the monitoring system, should be at least 10 m. The reason for this is that the monitoring system was to be installed outside de fence separating the filter, at high voltage, from the substation common area (see **Figure 7.8**). The other demand is that the sensor probe was to be installed between two adjacent coil layers, that is, within a space of about 2 cm.

As for the temperature, we decided to use the fiber type FH4001 (see **Table 7.1**), which would provide the highest available temperature supportability. On the other hand, as per **Table 7.1**, notice the high attenuation provided by this material as comparable to the EH4001, about 1.3 dB/m against 0.4 dB/m for the red wavelength. For a 10-m distance the total attenuation will be 13 dB for the red and 40 dB for the blue, which would make it inviable either for the blue LED to inject enough light to excite the ruby crystal and for the red fluorescent light to be sensible by the photodetector.

For this reason, we chose to use the FH4001 just at the tip of the sensor, which would be in contact with the copper coil. At a few centimeters away from the tip, the high temperature fiber will be connectorized to a FH4001 which will reach all the way to the control box 10 m away.

Due to the expected attenuation of 10 m of POF, we polished the LED case until almost reaching the semiconductor chip thus maximizing the light caption. Light launching was made through butt-coupling the polished LED and a carefully terminated POF. A miniature 3 dB coupler 1 × 2 POF-coupler (DyeMount, Germany) was used to collect the return fluorescence light toward the Si-photodetector.

Figure 7.16 shows details of the probe assembling. Red cellophane filter was carefully cut to fit inside the POF connector (left). At right we show the test setup. On one side the LED connected to the POF and on the other side, the return light passing inside the filter inside the commercial POF connector.

The final version of the probe is shown in **Figure 7.17**. Four probes were installed in the filter coil, two on the top and two on the bottom.

Table 7.1 Attenuation of Three POFs at Some Key Wavelengths

Fiber Type	470 nm	525 nm	694 nm
EH4001 (datacom-grade)	0.10 dB/m	0.10 dB/m	0.40 dB/m
DH4001 (heat-resistant, 115°C)	0.95 dB/m	0.48 dB/m	0.40 dB/m
FH4001 (heat-resistant, PC core, 125°C)	4.00 dB/m	2.70 dB/m	1.30 dB/m

Note: PC = polycarbonate.

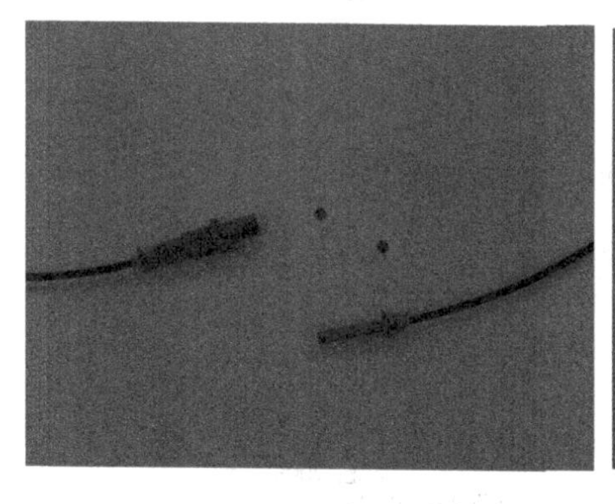

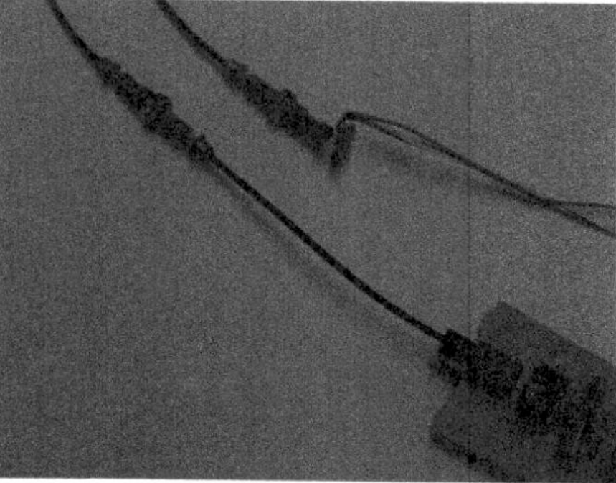

FIGURE 7.16 Details of the probe assembling. Red cellophane filter was carefully cut to fit inside the POF connector (left). At right we show the test setup with the LED connected to the POF and the return light passing through a red filter inside a commercial POF connector, directed to the PIN photodetector.

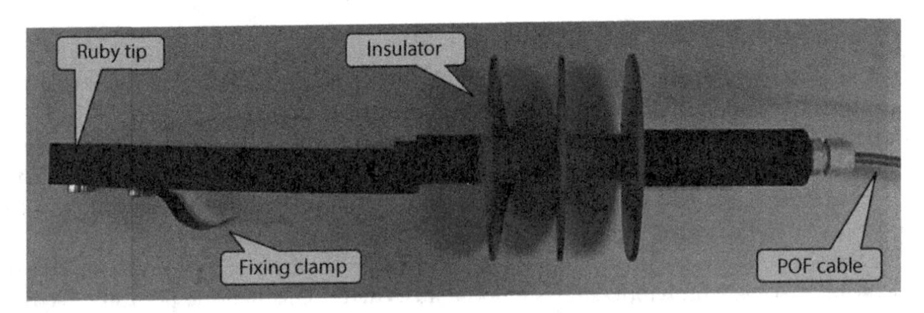

FIGURE 7.17 The probe.

7.7 Calibration

The calibration procedure consists of applying light pulses on the ruby and recording the corresponding fluorescent pulses that return to the photodetector. Then the sensor is exposed to several temperatures in an oven, the fluorescent pulses are recorded, and an exponential fitting is calculated by the algorithm.

Figure 7.18 shows the excitation train of pulses and the correspondent fluorescent pulses as seen in an oscilloscope.

Figure 7.19 shows superimposed pulses of different temperatures from about 30°C to 70°C.

The exponential decay seen in **Figure 7.19** can be expressed as:

$$P(t) = P_0 \exp[-t/\tau(T)]$$

where $P(t)$ is the output light power at a time t, P_0 is the light power at $t = 0$, and $\tau(T)$ is the time-decay constant at temperature T.

By applying a fitting on the decay curves of **Figure 7.19**, it is possible to calculate $\tau(T)$ for each temperature and plot then against temperature, as shown in **Figure 7.20**.

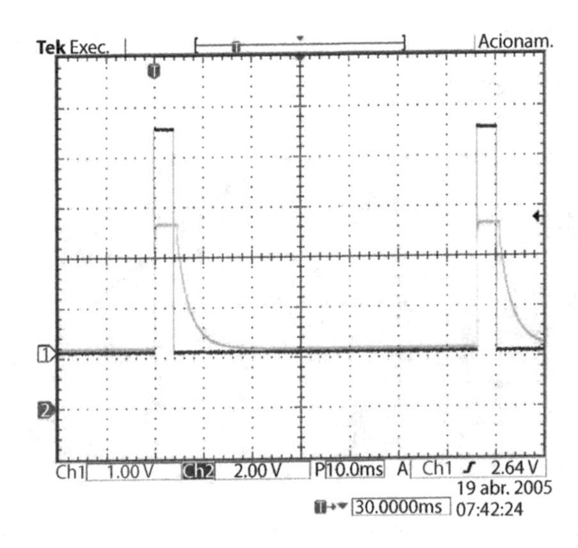

FIGURE 7.18 Excitation train of pulses and the correspondent fluorescent pulses as seen in the oscilloscope.

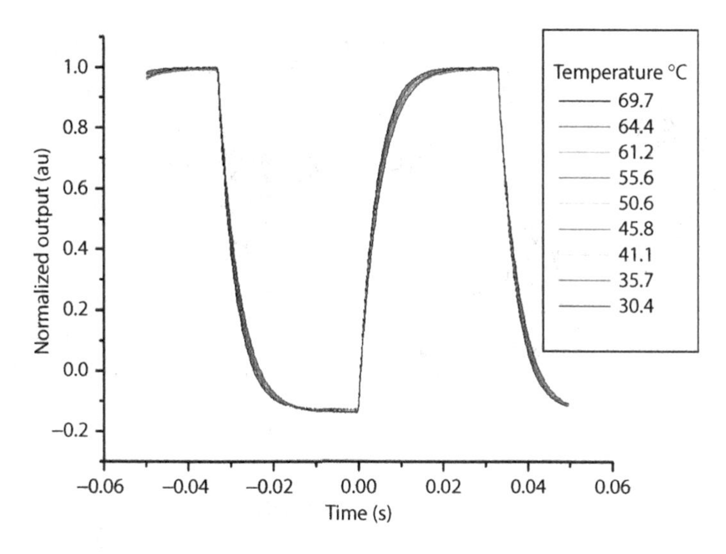

FIGURE 7.19 Superimposed pulses of different temperatures from about 30°C to 70°C.

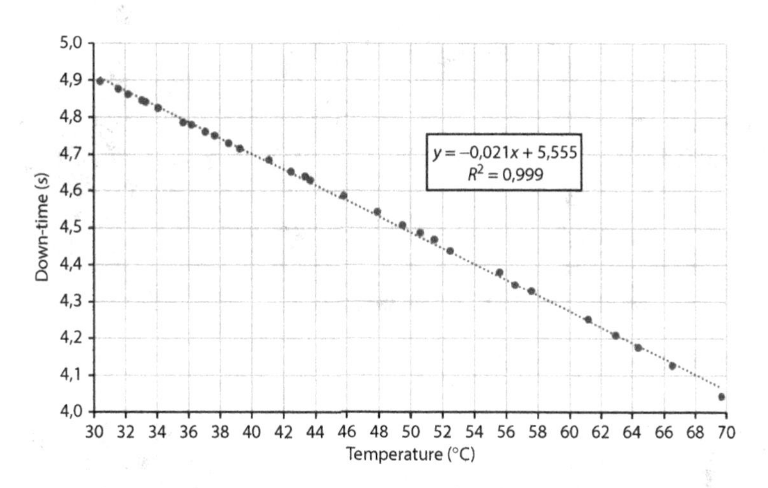

FIGURE 7.20 A plot of the decay time of ruby as a function of temperature. Notice a very good linear relationship with a regression coefficient $R^2 = 0.999$.

From the straight-line slope shown in **Figure 7.20** the sensitivity is calculated to be 22.5 μs/°C corresponding to an estimated temperature resolution of about 1°C.

We can now compare our sensitivity with 9 μs/°C, reported by Persegol et al. (1999). It is possible to speculate that, as they used ruby powder instead of a bulk crystal, this could have caused the lower sensitivity due to the high light dispersion inside the glue/powder composite.

7.8 Field Installation

The system was installed on the harmonic filter of Furnas Substation in the city of Ibiúna, State of São Paulo, Brazil, to allow the operators to monitor the temperature in four points of the coil of this reactor in real time. If the temperature reaches 80°C, an alarm is issued in order to shut down the transmission line. The following pictures show the sequence of work performed (**Figures 7.21** through **7.26**).

FIGURE 7.21 Installing an underground conduit for the optical fiber cable.

FIGURE 7.22 Installation for the sensor probe on the upper rim of the coil.

FIGURE 7.23 Sensors installed on top and on bottom of the coil.

FIGURE 7.24 Details of the sensor probes installed. Left: Sensor inserted inside the space between layers. Right: Spacer used to make the fiber cable hang from the top of the coil without touching its external surface.

FIGURE 7.25 Left: Control box installed outside the fencing. Right: Details of the control box. Connecting the control box with the control room there was a silica optical fiber operating a serial connection.

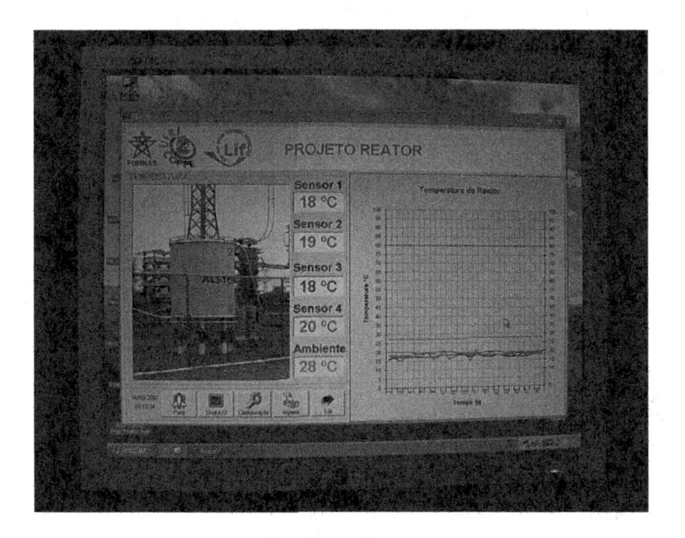

FIGURE 7.26 Screen shot of the control software showing the graph of the four temperatures as well as the ambient temperature.

7.9 Results

Temperature data from the four sensor probes together with ambient temperature was monitored 24 hours, seven days per week for six months. **Figure 7.27** shows some of the data collected. Notice the ambient temperature (lower trace) is normally the lowest temperature of the group. During the night it falls below 17°C, starts to rise by 8:00 AM, reaches a high by 2:00 PM, and decreases again. At the same time, three temperatures of different points of the coil follow the same pattern, just at a substantial elevated temperature. On the other hand, a case of study that puzzles the operators is the behavior of Transducer 2 (blue line). In reality it follows a completely inverse pattern, very hot during the night and cooler during the day.

It is also interesting to notice (lower graph) an electric power shortage by 2:00 AM in which all temperatures taken in the coil slowly reached the ambient temperature when the coil filter coolers. Then, by 10:00 AM the energy returns, and the patterns return to normal, with Transducer 2 always higher than the others.

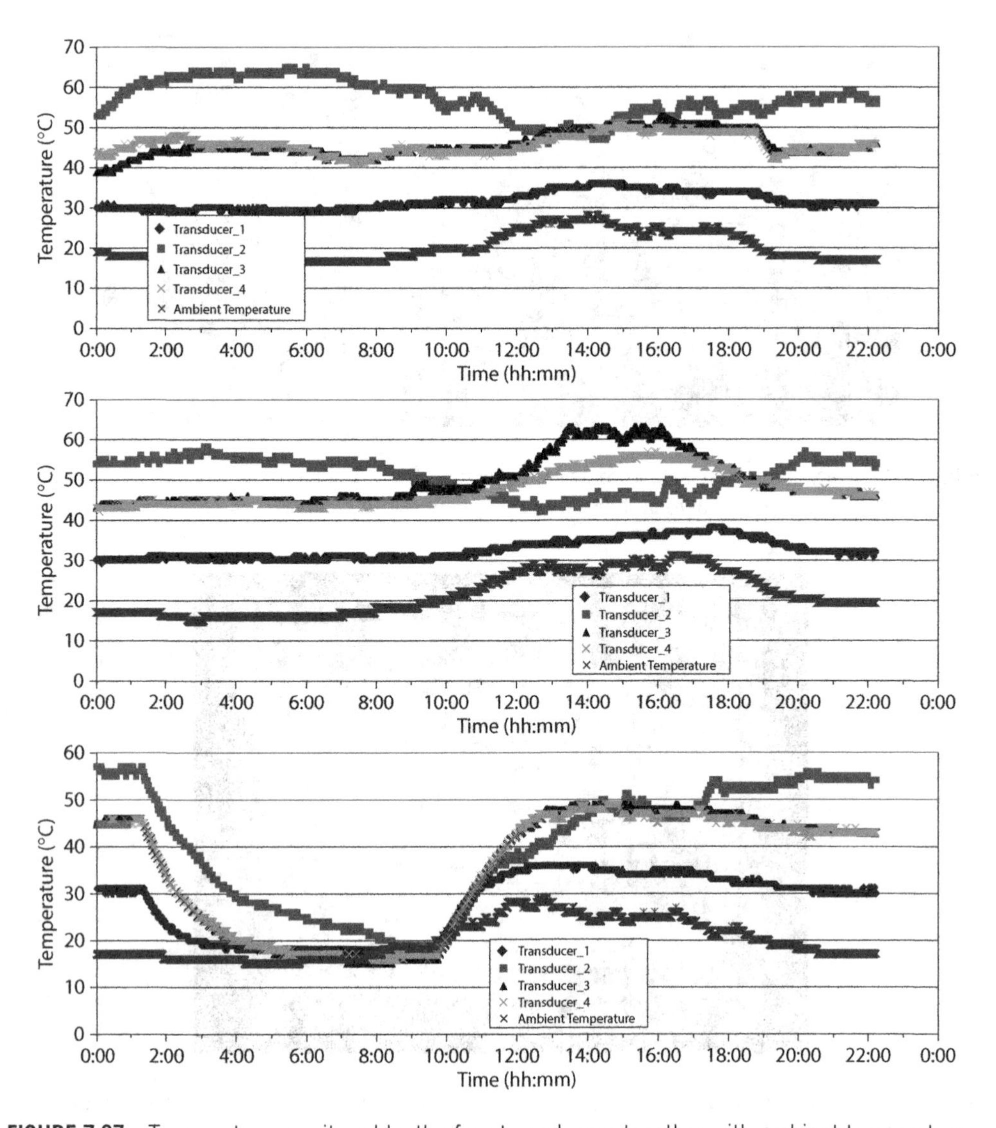

FIGURE 7.27 Temperature monitored by the four transducers together with ambient temperature.

7.10 Conclusion

Experimental results of a simple and low cost four-point temperature POF sensor prototype based on time-decay of the ruby fluorescence are presented. The best choice for temperatures up to 115°C was a small strand of polystyrene POF-probe to be in contact with the point to be measured and a PMMA POF to guide the pump and fluorescent light to and from the probe.

Ruby, as used for fluorescence thermometry, presents several advantages, such as its low cost, easy availability, POF compatible, requires low-cost light source and detector (LED and Si photodetector) and simple electronics. Additionally, it presents a strong intensity fluorescent signal and long decay time of 2–4 ms at a wavelength of 694 nm.

The system performed well during six months of continuous monitoring of temperature of a reactor coil and provided important data for the substation operation.

References

Alcala, J. R., Liao, S. C. and Zheng, J., "Real time frequency domain fibreoptic temperature sensor using ruby crystals," *Meb. Eng. Phys.* 18(1), 51–56, 1996.

Augousti, A. T., Grattan, K. T. V. and Palmer, A. W. "A laser-pumped temperature sensor using the fluorescence decay time of alexandrite," *J. Lightwave Technol.* 5, 759–762, 1987.

Augousti, A. T., Grattan, K. T. V. and Palmer, A. W. "Visible-LED pumped fiber-optical temperature sensor," *IEE Trans. Instrum. Meas.* 37(3), 470–472, 1988.

Fernicola, V. C., Zhang, Z. Y. and Grattan, K. T. V. "Fiber optic thermo metry on Cr-fluorescence in olivine crystals," *Ver. Sci. Instrum.* 68(6), 2418–2421, 1997.

Persegol, D., Lovato, J. L. and Minier, V., "Thermal diagnosis of medium voltage switchboards: A cost-effective multi-point POF sensor," *Proceedings of the 8th International Conference on Plastic Optical Fiber*, pp. 256–259, Chiba, Japan, July 14–16, 1999.

Ribeiro, R. M., Marques-Filho, L. A. and Werneck, M. M., "Fluorescent plastic optical fibers for temperature monitoring," *Proceedings of the 12th International Conference on Plastic Optical Fiber*, pp. 282–285, Seattle, WA, September 2003.

Sun, M., "Fiberoptic thermometry based on photoluminescent decay times," In: *Temperature: Its Measurement and Control in Science and Industry*, vol. 6, American Institute of Physics, New York, pp. 715–719, 1992.

Zhang, Z., Grattan, K. T. V. and Palmer, A. W., "Fiber-optic high temperature sensor based on the fluorescence lifetime of alexandrite," *Rev Sci. Instrum.* 63, 3869–3873, 1992.

Gas Sensing

Meysam M. Keley

Contents

8.1 Introduction

The exploration of optical fibers in chemical sensing area gained major interest during the last decades. In this type of sensor, the electromagnetic wave transmitted through the fiber interacts either directly with the detected substance or with an intermediate sensitive material. In the latter mechanism, the optical fiber sensor receives a thin film, which intentionally is selected to react (chemically or physically) with the measurand.

Plastic optical fibers (POFs) benefit from their higher mechanical strength which makes them more flexible than glass fibers to use in toughness-demanded cases. POFs also possess an extremely large core to cladding diameter ratio. This factor takes into account when the removal of cladding is required to fabricate a satisfactory sensor.

Gas sensors are a subcategory of chemical detectors, which are designated to sense and/or quantify the concentration of one or more gaseous species. Highly increasing demand to detect gases in low concentrations inspires scientists to develop more precise and fast gas sensors.

In this chapter, some examples of classic gas sensors are presented. A sort of gas-sensitive material is cited followed by the principles of optical gas sensors. In the final section, some successful POF gas sensors are introduced in detail.

8.2 Gas Sensing Applications

Monitoring of some gas species, which influence the air quality, is crucial in both work ambient and indoor well-being care. Carbon dioxide concentration, for example, is a good factor to evaluate air freshness [1]. Taking into account the security issues, detection of inflammable gases, such as methane and hydrogen sulfide, turns a protection tactic against explosion and fire incidents [2]. Detection of some organic gases such as NO in exhaled breath is an approved pathological method of airway inflammations including asthma [3]. Apart from the sensing of straight gases, commercial interests in the detection of complex super-molecular species are a driving force for an innovative technology called "electronic nose." This cutting-edge technology aims to mimic the olfactory system so that it is capable of measurement and characterization of volatile aromas [4]. A highly demanded application of this revolutionary technology is the detection of food spoilage in its initial stage [5].

8.3 Classic Gas Sensors

Gas detectors are in action right after the discovery of hazardous effects of some species such as carbon monoxide and hydrogen sulfide. Since the early years of industrial revolution, when fuel became a necessity, until today, methane detection inside coal mines is a security bottom line. The first methane detectors were volunteer workers who used to enter the mine shaft holding a flame torch and a wet blanket as a shield. When a methane pocket was found inside the mine, the flame would ignite it. Later, canaries substituted for these volunteer workers. If the canary stopped chirping, the workers knew that there might be a methane leak [6]. The first evidence of scientific logic applied to gas detection was the "flame light" monitoring inside the coal mines. An encapsulated flame inside a flame-arrestor shell may change the light level depending on the atmosphere composition. When the flame was extinguishing itself, it indicated a lake of oxygen in the mine's atmosphere. On the other hand, when the flame light would rise, it meant a possible methane pocket [7].

An approach called "catalytic-type sensors," which dates back to the 1800s, measures the concentration of combustible gases. A classic prototype counts on the thermal energy released as these gases burn due to catalytic sites on the surface of a heated (approximately 1000°C) platinum filament. The increase in the temperature of the filament is attributed to the volume percentage of combustible gases in the test atmosphere [8].

During the 1980s, sensors made from semiconductor materials gained a lot of interest [9]. They operate through absorption of the gas at the surface of a semiconductor, which is normally a thin metal oxide film (e.g. SnO_2 and ZnO). Once absorbed, the gas changes semiconductors' physical characteristics, for example, electrical resistance or dielectric constant. If these parameters were measured in real time, any changes in their steady state value indicate the presence of the sensed gas [10].

8.4 Gas-Sensitive Substances

Polycrystalline SnO_2 is a fundamental *n*-type semiconductor in gas sensing area principally for reducing gases such as H_2, CH_4, C_4H_{10}, CO, and H_2S. The main disadvantages of SnO_2 are the poor selectivity and stability of the sensor signal. Under ambient air conditions, $(O_2^-)_{ad}$ and $(O^-)_{ad}$ chemisorbed donors are present at the surface of SnO_2. These species are involved in reversible catalytic oxidation of reducing gases [11]. The x-ray photoelectron spectroscopy (XPS) spectra gained from a tin oxide sample after exposure to H_2S demonstrate an S(2s) core level peak at 161 eV binding energy. Consequently, the following reversible chemical reaction is proposed [12]:

$$SnO_{2-x}(s) + yH_2S(g) = SnO_{2-x-3y}(s) + yH_2O(g) + ySO_2(g) - 6ye^- \quad (8.1)$$

In a test starting with reduced tin oxide, oxygen deficiency (x in SnO_{2-x}) is modified via exposure to oxygen for different intervals. The results demonstrate reverse relationship between electrical resistance of the sample and x. In other words, oxygen vacancy as a defect in the lattice is responsible for changes in electrical conductivity of tin oxide. This observation is in agreement with the conductivity mechanism of "electron hopping between defect state" that is expected from n-type semiconductors [12].

Another interesting material is tungsten oxide, which is sensitive to hydrogen, and it is reported that WO_3 changes its color from greenish yellow to blue as it is exposed to hydrogen. More interestingly, WO_3 exposure to hydrogen causes modulations in optical properties such as transmittance, absorption, and refractive index [13].

As a hydrogen sensitive metal, Pd dissociates molecular hydrogen (H_2) to atomic (H), which undergoes a further diffusion process transforming the crystal structure of Pd:

$$(\alpha)Pd + \frac{k}{2}H_2 = (\beta)PdH_k \tag{8.2}$$

The variation of hydrogen concentration can modify the dielectric constant of Pd [13].

Graphene, a two-dimensional material composed of layers of carbon atoms in a honeycomb-shaped network, presents remarkable physicochemical behavior. Graphene, graphene oxide (GO), and reduced graphene oxide (rGO) are important materials for chemical- and bio-sensors. There is a series of NH_3 and humidity sensors developed based on the graphene interaction [14].

Fluorescence is another analytic method for detection of molecular compounds. The process involves a fluorophore substance that absorbs light and emits photons with less energy than initial exciting photons received by the fluorophore. In a phenomenon called "fluorescence quenching," the fluorescence behavior is changed due to the interaction with another chemical species. Among the compounds that present fluorescence quenching, some are sensitive to active gases. For example, a 2–[(3,5–dichloro–2–hydroxy–benzylidene)–amino]–2–hydroxymethyl–propane–1,3–diol Schiff base ligand, which during another process takes part in a Pb^{+2} complex, demonstrates fluorescence quenching when it is in contact with a nitroaromatic gas, especially dinitrotoluene [15].

Fluorophore quenching depends on several factors. However, the effect can be simplified through the Stern–Volmer equation [16]:

$$\frac{I_0}{I} = 1 + K_{SV}[Q] \tag{8.3}$$

where $K_{SV} = k_q\tau_o$

Q is the concentration of the quencher substance;

I_0 is the intensity in the absence of the quencher;

I is the intensity in the presence of a quencher;

τ_o is the fluorescence lifetime of the fluorophore; and

k_q is the quencher rate coefficient.

In the ideal case, a plot of I_0/I versus concentration is linear with a slope that corresponds to K_{SV} and an intercept of 1.

8.5 Optical Gas Sensors

Molecular absorption spectroscopy is a crucial optical analysis of gas species. Some compounds exhibit strong absorption in the UV/visible, near infrared, or mid-infrared regions of the electromagnetic spectrum. The absorption lines or bands are fingerprints of each species which form the basis for their detection and measurement [17]. Optical gas detection using absorption spectroscopy is based on the Beer-Lambert law [18]:

$$I = I_0 \exp(-\alpha l) \quad \textbf{(8.4)}$$

where:

I is the quantity of light transmitted through the test cell;
I_0 is the initial incident beam;
α is the absorption coefficient of the test sample; and
l is the distance between two test cell walls perpendicular to the incident beam.

The Beer-Lambert law is applied for monochromatic light sources. For low αl (small test samples or low gas concentrations), the law is conveniently linear.

Commercially available telecommunication optical fibers possess core diameters ranging from 5 to 60 μm. If the fiber were used as a part of a spectroscopy light transmission procedure, interference between the modes in a multimode fiber would cause undesirable effects on the signal. Consequently, single-mode fibers are desirable in high-resolution spectroscopy. In some cases, a conventional spectroscopy cell is connected to optical fibers to share certain light sources and reduce the cost of the sensor set. On the other hand, in some gas detectors, which are designated to monitor explosive gases, optical fibers offer more security than conventional electrically charged wires [19].

Partial removal of cladding enables the interaction between the light guided through an optical fiber and the surrounding media. The cladding can be tapered, etched, or even side-polished. In some cases, the sensing procedure functions are based on evanescent field (EF). As the proportion of light interacting with the surroundings depends on its refractive index, EF-operated sensors are more used with liquid samples (as liquids have more refractive index than gases due to their higher density) [20]. In a more advanced method, gas-sensitive thin films partially substitute the cladding, enabling the light to be influenced indirectly with the test gas [19].

Other than core-cladding structured fibers, some microstructured optical fibers have also gained interest in gas sensing as the holes in their structure may act as convenient repositories for sample gases [21].

8.6 POF-Associated Gas Sensors

Plastic optical fibers have a series of advantages such as ease of handling, low precision connectors (inexpensive), and mechanical resistance. POFs, however, have a low operating temperature (80°C–100°C) and are not resistant to humidity as they can absorb up to 1.5 wt.% of water [19].

A large number of POF sensors operate based on light intensity variation through transmission, reflection, spectroscopic, and evanescent field transduction mechanisms. The FBG POFs also are being investigated in the sensing methods and have gained major interest thanks to their low cost and flexibility. There are other techniques such as backscattering and optical frequency domain reflectometers that cause more interest in mechanical sensors than chemical ones [22].

A simple set of this optical sensor is a light source (LED), a POF, and a photodetector (or a photocell) at the end of the light terminal. This model can be divided into extrinsic and intrinsic configurations. In the former case, the optical fiber are only a transmission media for the optical signal and in the latter, the light does not leave the optical fiber (**Figure 8.1**).

Optical losses are a natural consequence of curved (bent) fibers. Bend loss has been usefully exploited as a transduction mechanism in fiber optic sensors. Factors affecting bend loss are its radius, fiber numerical aperture, core size, and the launching conditions of the input signal [23].

There are a vast number of reflectance studies at which POFs containing FBG, Fabry-Pérot cavities, and/or metal films for surface plasmon resonance (SPR) are exploited as sensors [22].

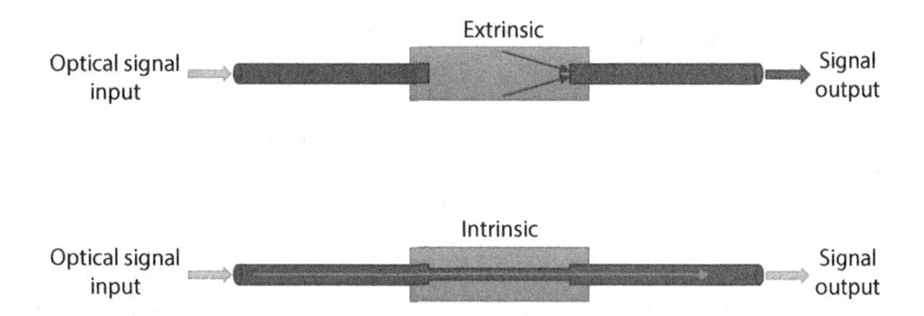

FIGURE 8.1 Extrinsic and intrinsic configurations of POF-assisted sensors.

8.6.1 Fluorescence-Quenching Aided Oxygen Sensor

Quantification of oxygen concentration is a fundamental issue in environment quality measurements, transportation, medicine, and agriculture. A wide range of O_2 sensors has been developed such as solid electrolyte based potentiometric, amperometric, and metal oxide-based semiconductors [24], as well as Clark electrodes for dissolved oxygen [25]. The drawback of actual sensors is response time and surface poisoning (i.e. chemical adsorption of molecules other than oxygen to the sensitive material). The use of fluorescent techniques offers some advantages such as non-invasive, extreme sensitivity, fast and reversible response, and low toxicity. The fluorescent materials can be coupled to an optical fiber set to provide effective fluorescence intensity monitoring [26].

There are two fluorescence measurement setups via optical fibers; in the first configuration, a fluorophore is attached (coated) to the fiber end. The exciting light inside the fiber goes all the way down to the tip. Meanwhile, the fluorescence light is back emitted and is guided to detectors. The second configuration relays on the interaction between evanescent wave (EW) and fluorophore deposited on the fiber. The weak point of such methods is poor signal to noise ratio (SNR), superimposition of the exciting light, and fluorescence and low penetration depth of the EW [16]. Side-illumination of the fluorophore immobilized on the fiber is proven to get better SNR [27]. The fluorescence is gathered by an optical fiber and is guided to detectors. The fibers that undergo a tapering procedure are able to get even better SNR [16].

Fluorophore immobilization at the fiber tip or on top of it is a critical issue. There are studies about materials which are able to perform immobilization such as zeolites [28], polymers like silicon rubber [29], sol-gels [30], and xerogel [31].

In a successful study, Ru (II) polypyridyl fluorophore immobilized on a locally tapered POF is explored to perform as an oxygen sensor [27]. The 980 μm core diameter PMMA fiber (SK-40 of Mitsubishi Rayon Co.) is tapered using a local heater and constant pulling in reverse directions forming a biconical shaped taper (**Figure 8.2**).

The next phase of the sensor fabrication is the immobilization of fluorophore (Ru (II) polypyridyl fluorophore) onto the tapered area. The procedure involves submerging the fiber

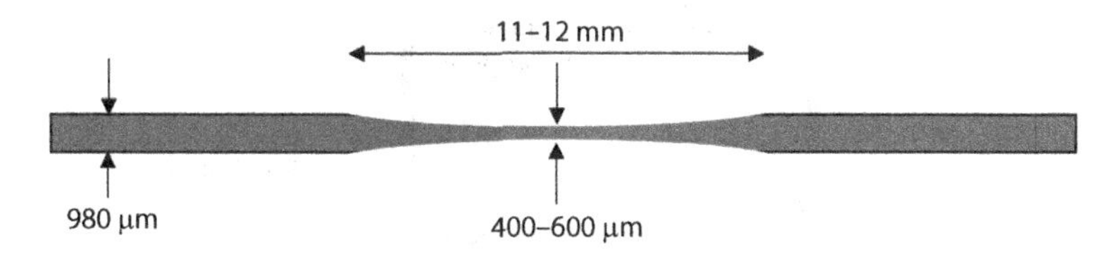

FIGURE 8.2 Schematic of the biconical taper formed upon POF via local heating and stretching method.

for 3 min in a 30% solution of acetone in water containing different concentrations of the fluorophore. Two solutions with different fluorophore concentrations (0.02 and 0.1 wt.%) were used.

A chamber containing gas inlet and outlet, a built-in LED (maximum emission wavelength of 421 nm), as well as an optical fiber connection is developed to test the current sensor. It is worth mentioning that the LED can be substituted by an optical fiber which delivers the necessary exciting light for the fluorophore. Therefore, the sensor probe can be constructed free of electric charge. The optical signal produced via fluorophore material is transmitted through the tapered polymer optical fiber and detected by an external photodetector. After being exposed to different concentrations of O_2 in pure nitrogen, changes in fluorescence caused by the quenching phenomenon are recorded by a data processor (**Figure 8.3**).

In a simple test, a biconically tapered POF immersed in a 0.1 wt.% fluorophore (Ru (II) polypyridyl fluorophore for 3 min is installed inside the test chamber. Then, the input oxygen concentration is changed from 0 to 100 vol.% with unit step of 25 vol.% and the fluorophore quenching output in photodetector is recorded as mV (**Figure 8.4**). The real-time monitoring of the sensor affirms the influence of oxygen concentration on the output signal in the form of electrical potential generated by the photodetector.

In order to prove the Stern–Volmer equation in the current case (see Section 10.4), the 591 nm wavelength (the peak of the fluorescence of Ru (II) polypyridyl fluorophore) is adopted and the light output intensity is measured upon different concentrations of oxygen. The results of the measurements are plotted and demonstrate a linear relationship between I_0/I and oxygen concentration (**Figure 8.5**).

As a conclusion, there are some important points about the studied sensor that follows:

The response time of the sensor is relatively long. However, it may be improved using other types of fluorophores or other tapering geometries [32].

It is possible to construct a completely electrical charge-free sensor by using a second fiber to transmit exciting light. Therefore, the sensor may offer extreme safety while installing in hazardous environments (i.e. explosion risks).

The signal to noise ratio (SNR) of the sensor is low and restricts its sensing limits to a minimum of about 10 vol.%. In order to elevate the SNR, it is recommended that the POF covered area with the flurophore be illuminated with exciting light from both sides. Additionally, more concentrated fluorophore solutions may be tested to gain high thickness.

According to the test represented in this work, the sensor was exposed to a simple mixture of nitrogen and oxygen. There are lots of other gas species in a real atmosphere, indeed. Consequently, the reported behavior of the sensor may not represent its response when it becomes exposed to a realistic air composition.

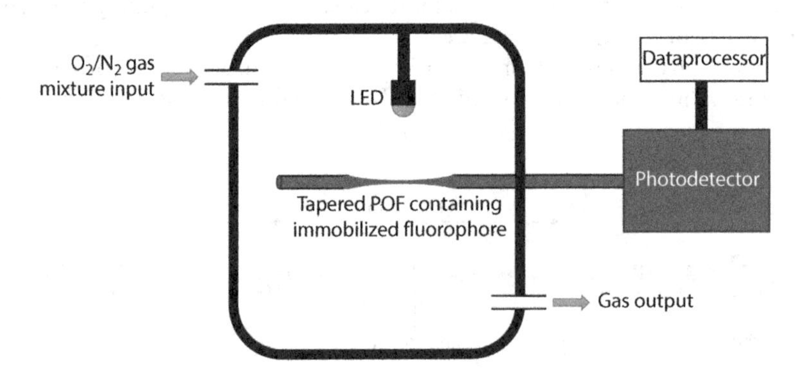

FIGURE 8.3 Schematic of the test chamber for fluorophore quenching oxygen sensor. (Modified from Pulido, C. and Esteban, Ó., *Sensor Actuat B*, 146, 190–194, 2010.)

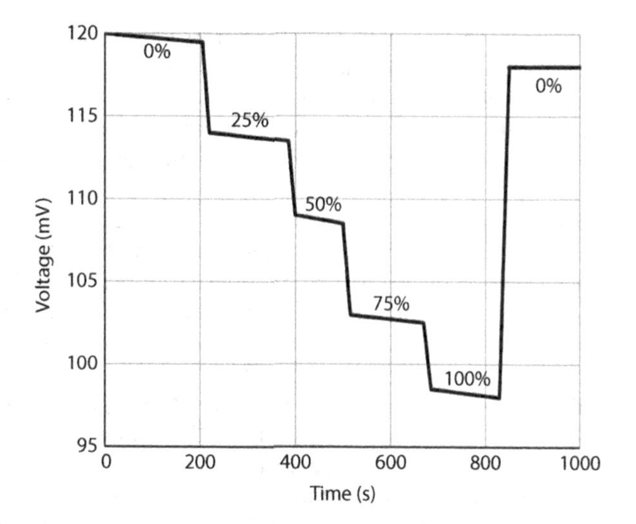

FIGURE 8.4 Schematic of the signal changes during exposure of the sensor do different oxygen concentrations (the numbers in %). (Modified from Pulido, C. and Esteban, Ó., *Sensor Actuat B*, 146, 190–194, 2010.)

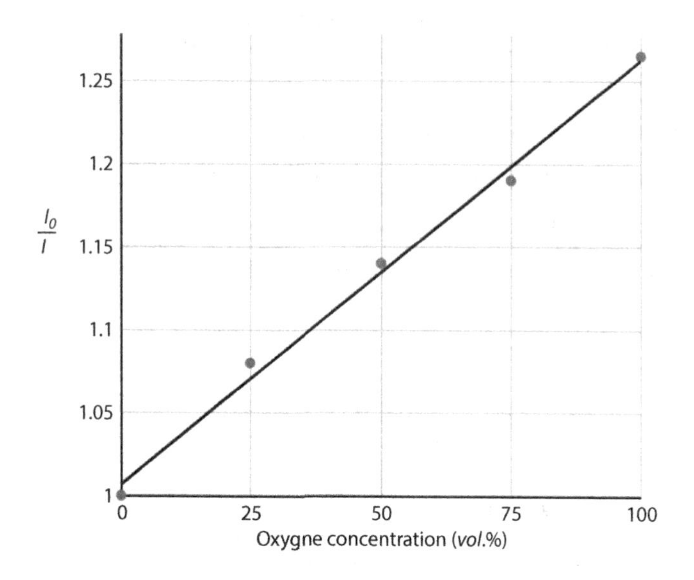

FIGURE 8.5 Combination of the sensor signal (circular points) with Stern–Volmer equation. The linear fit renders K_{SV} = 0.0026. (Modified from Pulido, C. and Esteban, Ó., *Sensor Actuat B*, 146, 190–194, 2010.)

8.6.2 Hydrogen Sensors Made Out of Mg-Ti Alloys

Hydrogen is considered the fuel of the future. Besides being clean and highly efficient, H_2 in gas form in concentrations of 4% is flammable, which demands high sensibility detectors. Catalytic resistors and electrochemical devices are some examples of hydrogen detectors. Their major drawback is being electrically charged. A palladium thin film is proven to demonstrate some changes in optical properties as it meets H_2. The phenomenon, however, shows poor signal and lack of selectivity [33,34] (see also Section 10.4).

Some metal (or alloy) layers that are intrinsically mirrors become transparent during their reaction with hydrogen. Magnesium-rare earth (Mg-RE) and magnesium-transition metals (Mg-T) are the most studied switchable (from mirror to transparent) alloys. Mg_yTi_{1-y} thin films demonstrate a remarkable reflection-transmission transformation, especially when receiving a nanometric Pd layer to protect the functional film from oxidation, as well as acting as a catalyzer of the hydrogen absorption [34,35].

A state-of-the-art extrinsic optical fiber hydrogen sensor is developed by deposition of a multilayer thin film on the cleaved end of the fiber (**Figure 8.6**). The first layer is a 60 nm-thick $Mg_{0.7}Ti_{0.3}$ thin film deposited via magnetron plasma sputtering co-deposition from both Mg and Ti targets. The intermediate layer is a 3 nm thickness thin film of palladium also formed through a plasma sputter deposition. The uppermost layer is a protective polytetrafluoroethylene (PTFE) thin film. It is mainly designated to protect the other thin films from condensed water [34,36].

A test chamber with a controlled atmosphere is prepared to expose the sensor to various H_2 concentrations. The optical fiber containing sensitive thin films ($Pd/Mg_{0.7}Ti_{0.3}$) is located inside the chamber through an appropriate opening. An LED with maximum emission at 635 nm wavelength and a photodetector (containing a built-in data processor) is connected to the optical fiber sensor via a bifurcator (**Figure 8.7**). The light emitted from the LED goes down to the final end of the optical fiber and is reflected (partially) yet inside the fiber and goes up to the photodetector. The quantity of reflected light is monitored during the experiments. Meanwhile, the chemical composition of flowing gases inside the chamber is adjusted by mass flow controllers. Consequently, changes in reflectance index of the sensitive thin films can be attributed to changes in the H_2 concentration in the test chamber.

In order to study the performance of the current sensor, the atmosphere of the chamber undergoes the following cycle:

Steady state flow of pure argon to clean up other gas species;

Injection of x vol.% of hydrogen in argon; (x ranges from 0.2 to 1);

Steady state flow of argon;

Injection of 20 vol.% of oxygen to recover the sensitive film (free from adsorbed hydrogen)

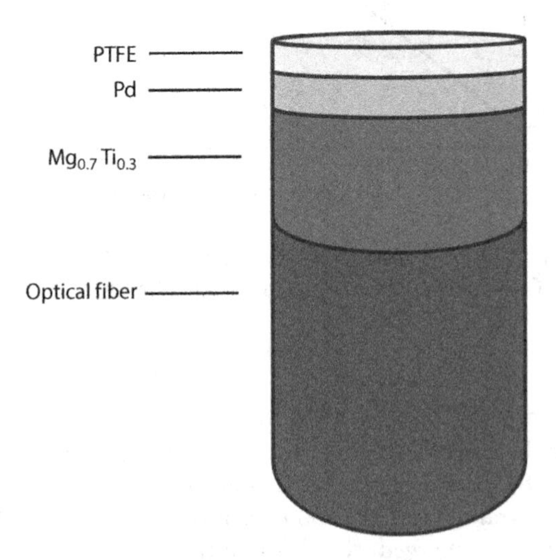

FIGURE 8.6 Schematic of the extrinsic optical fiber sensing head and the sequence of thin films. (Adapted from Mak, T. et al., *Sens. Actuat B-Chem.*, 190, 982–989, 2014.)

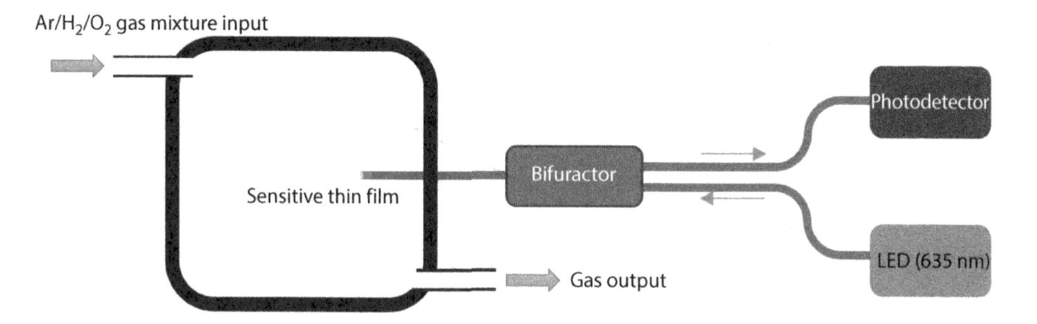

FIGURE 8.7 Schematic of the sensing apparatus coupled to a controlled atmosphere test chamber. (Adapted from Mak, T. et al., *Sens. Actuat B-Chem.*, 190, 982–989, 2014.)

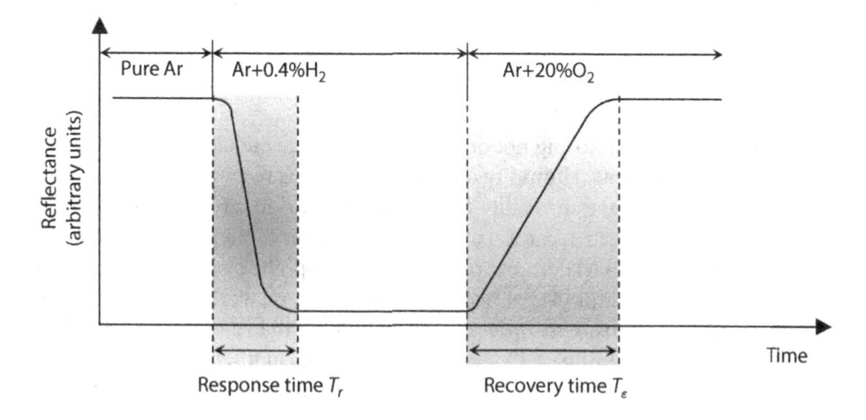

FIGURE 8.8 Typical thin film reflectance versus time behavior of the sensor. (Modified from Mak, T. et al., *Sens. Actuat B-Chem.*, 190, 982–989, 2014.)

A typical reflectance versus time plot in the current experiment demonstrates steady state, transition, saturated, and recovery phases (**Figure 8.8**). This regime is characteristic of a basic gas sensor.

The reaction between hydrogen and the sensitive thin film can be demonstrated as follows:

$$Mg_{0.7}Ti_{0.3}(s) + xH_2(g) = Mg_{0.7}Ti_{0.3}H_{2x}(s) \quad \textbf{(8.5)}$$

In order to study the effect of hydrogen concentration on the response time (T_r), the default cycle was run for different x values. A method developed to determine the controlling phase in the reaction kinetics [37,38] proposes to plot reverse time response versus reagent (hydrogen) concentration (**Figure 8.9**). If there is a linear relation between T_r^{-1} and H_2 concentration, the overall kinetics is surface-controlled.

As reversibility is a fundamental factor for a functional sensor, the exposure of the present sensor to H_2 and its refreshing with oxygen was performed in up to 50 cycles. The result guarantees complete reversibility of the developed method. However, it is worth noting that the sensitive thin film is not in contact with other aggressive gas species such as CO and H_2S. Further studies may be carried out using normal air instead of argon to improve the performance of the sensor in real working conditions.

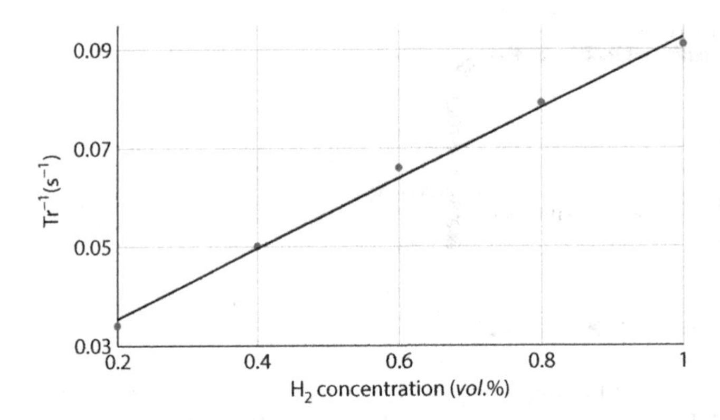

FIGURE 8.9 Hydrogen sensing response changes in different testing runs. The linear relationship addresses surface reaction between hydrogen, and the sensitive film is the controlling phase for the whole reaction rate as well as the sensor's response. (Adapted from Mak, T. et al., *Sens. Actuat B-Chem.*, 190, 982–989, 2014.)

8.6.3 U-Shaped POFs and Humidity Sensor

A comprehensive respiration monitoring not only measures its rate but also can retrieve useful information about the health of patients. Humidity content of a person's respiration is an important factor in medical diagnostics [39]. Indoor air quality (especially onboard in aircraft) is being monitored to maintain some essential factors such as carbon monoxide and water content at a comfortable level [40]. Poly-dimethylacrylamide (PDMAA) demonstrates strong hydrophilic characteristics, which makes it a good choice as a humidity indicator [41]. A bent optical fiber sensor, working based on absorption or evanescent field, can achieve a sensitivity order of 70 times higher than a straight form [42].

In a successful study, the cladding of a U-shaped plastic optical fiber is replaced by PDMAA to form a probe to detect and measure humidity concentration in air. As the PDMAA is sensitive to air humidity level, changes in the optical scattering properties of a cladding made from it can be attributed to humidity content in the air. A 1 mm diameter plastic optical fiber is bent in 180° in a 5 mm curve radius form. A cladding is formed upon curved area during the dip coating and UV radiation process as follows [43]:

First, a hydrogel based on poly-dimethylacrylamid containing UV active benzophenone (MABP) groups is deposited on the fiber through a dip coating procedure. Then, a UV radiation on the coated region provokes crosslinking by triggering a C, H insertion in MABP groups, as well as anchoring the cladding to the PMMA substrate (**Figure 8.10**).

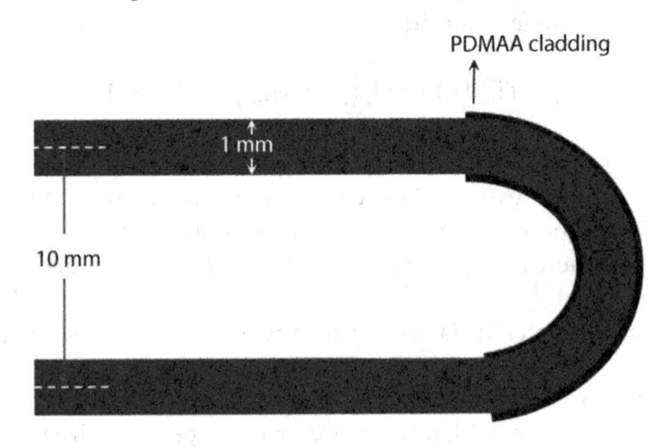

FIGURE 8.10 Schematic of the POF probe with a PDMAA cladding covering the U-shaped area. (Adapted from Kelb, C. et al., *Procedia Technol.*, 26, 530–536, 2016.)

An LED is fixed to one end (as a light emitting source) and a photodiode to another side of the fiber. The LED potential is kept constant, and the changes in optical power transmitted through the sensing probe are monitored through the photodiode (photodetector). In order to record any changes in the light energy transmitted caused by other factors than the PDMAA cladding (such as temperature), a similar probe without cladding is also prepared and installed next to the sensor. The couple is mounted in a specific chamber with an air inlet and outlet openings. A data processor records in real time the value of transmitted light counted by photodetectors in both fibers (**Figure 8.11**).

The humidity level inside the chamber is kept stable at 85.1% and at the seventh second of the experiment, the laboratory air, with 23.4% humidity, is injected into the chamber. Consequently, the sensor experiences a change in humidity in the form of the step function. The gathered data is projected as the real-time sensor response to the humidity change (**Figure 8.12**). The reference fiber set represents no changes during modification of humidity level. The experiment reveals that the response time of the sensor is extremely low (less than 5 seconds). Additionally, the

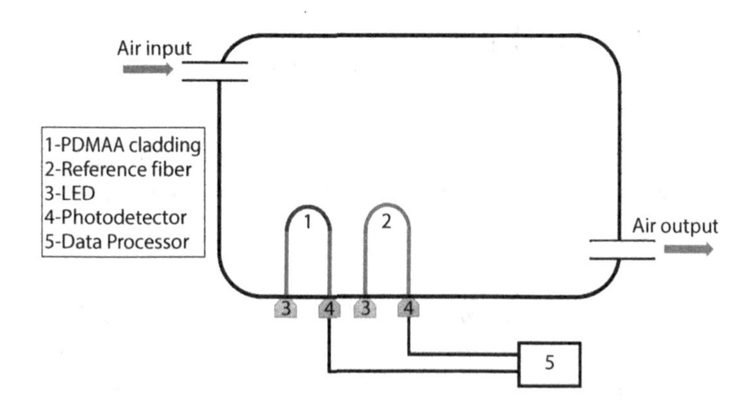

FIGURE 8.11 Schematic setup of the test chamber and optical fibers. (Modified from Keley, M.M. et al., Functionalized POPF H_2S sensor, *27th International Conference on Plastic Optical Fibers (POF2018)*, Seattle, Washington, WA, 2018.)

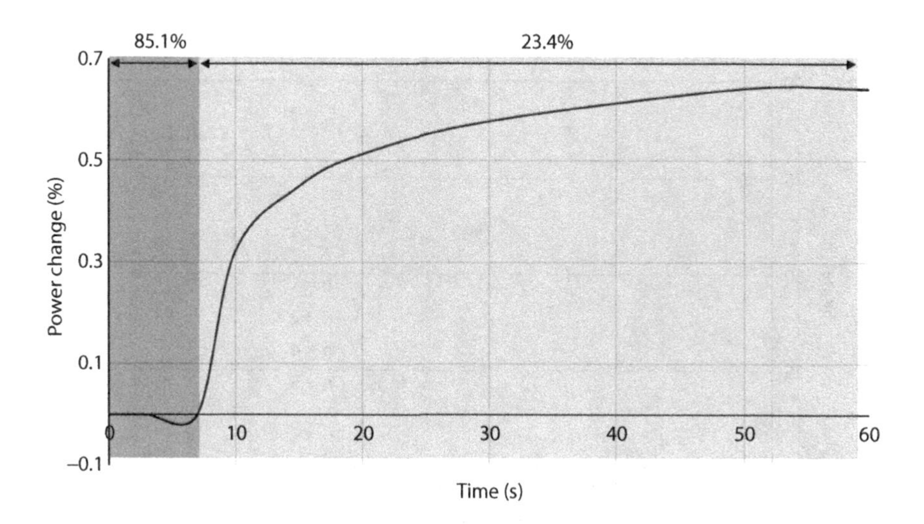

FIGURE 8.12 The response plot of the sensor subjected to a humidity change from 85.1% to 23.4% in humidity at time 7 sec. (Adapted from Keley, M.M. et al., Functionalized POPF H_2S sensor, *27th International Conference on Plastic Optical Fibers (POF2018)*, Seattle, Washington, WA, 2018.)

signal to noise ratio is satisfactory. In another test with human respiration as a test cycle, the sensor demonstrates reversibility with small recovery time [41].

As can be seen from the experiment, a remarkable change (74%) in humidity caused a poor signal of about 0.7%, which limits the detection range of the sensor to high values. The SNR in curved optical fiber sensors can be improved by giving more curves (such as W that possess three bending points). In other words, the light transmitting through the fiber engages with the thin film, either in the evanescent field or reflective index forms, in more than one point. Here comes a mathematical model that explains the phenomenon. Let x be the total incident optical energy entering an optical fiber with n bent (equally) points and y be the attenuation fraction caused in each point (**Figure 8.13**).

The following equation can be proposed:

$$Signal(\%)\frac{incident\,light - transmitted\,light}{incident\,light} \times 100 = 100\frac{\left(x - x(1-y)^n\right)}{x} = 100 \times \left(1-(1-y)^n\right) \tag{8.6}$$

The equation reveals that if $y > 0.1$, more bent points can improve remarkably the signal value. It also can be seen that in $y > 0.8$ more curves have a minor effect on the signal (**Figure 8.14**).

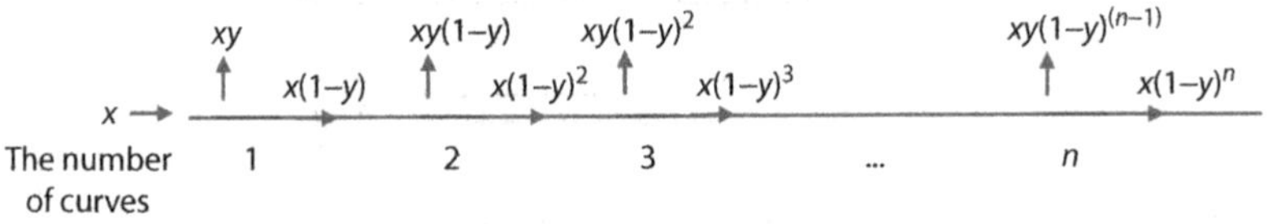

xy $\quad xy(1-y) \quad xy(1-y)^2 \quad xy(1-y)^{(n-1)}$

$x \rightarrow$ $\quad x(1-y) \quad x(1-y)^2 \quad x(1-y)^3 \quad x(1-y)^n$

The number of curves 1 2 3 ... n

FIGURE 8.13 The hypothetical distribution of light transmitted through a curved optical fiber in n points.

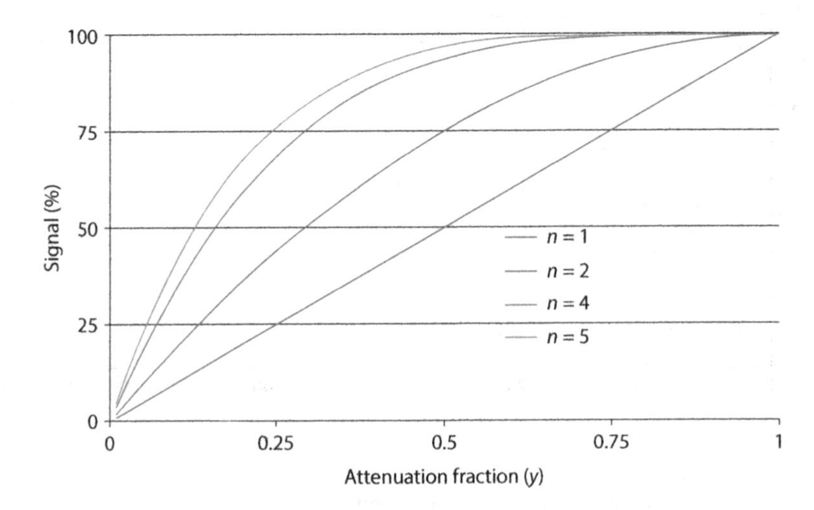

FIGURE 8.14 Signal versus attenuation plotted for different n values. Note that there is not a significant change in signal for curve numbers higher than four.

8.6.4 Functionalized POF H_2S Sensor

In addition to tin oxide, nickel and copper oxides also represent affinity to hydrogen sulfide (see Section 10.4). In a recent study, U-shaped POF received a copper oxide thin film through pulsed laser deposition (PLD) [44]. The laser power is maintained the same for all samples, and deposition time is varied from 2 to 30 min. The POF sensing probe is equipped with LED and photodetector (see Section 8.6.3).

In order to have a reference film morphology free of POF roughness and curvatures, a standard silicon wafer in <100> crystal growth direction is used as a substrate. The Si plate is cut in 1 × 1 cm² dimensions and subjected to CuO thin film PLD for 30 min. The sample then is analyzed using JPK® (NanoWizard®) atomic force microscopy (AFM) in intermittent contact mode and topography channel [45]. According to the results, PLD thin films produced a homogeneous infrastructure accompanied with distinguishable circular islands with diameter variety falling in the 100 to 500 nm range (**Figure 8.15**).

All sensors are exposed initially to pure nitrogen and then, as a step change in concentration, another mixture of nitrogen containing 200 ppm of H_2S. The responses of sensors per exposure time are recorded applying the following equation:

$$\text{Signal}(\%)\, y = 100 \times (y_0 - y_t)/y_0 \quad \textbf{(8.7)}$$

where:

y_0 is the initial light transmission yield through the POF in mV;
y_t is the yield in time t while the test is running.

It was observed that for thin films deposited for times longer than 30 min, a complete blocking of transmitted light takes place. In other words, the sensors are no longer functional. The signal in respect to test time is recorded and used to draw the response of sensors in respect to H_2S inlet and simplified in **Figure 8.16**. It can also be seen that the PLD duration presents a direct relation to the maximum signal. The phenomenon is attributed to covered fraction of the POF external surface with the thin film. The more time is given to deposition to take place, the more material is interacting with the evanescent field.

In order to investigate the reversibility of these sensors, they were subjected to fresh air and then to H_2S. However, it was confirmed that the sensors did not record the presence of the mentioned gas for the second time. At this phase of the current research, copper oxide, thin films

FIGURE 8.15 3D AFM image of a CuO thin film deposited on standard Si wafer. (Courtesy of Keley, M.M. et al., Functionalized POPF H_2S sensor, *27th International Conference on Plastic Optical Fibers (POF2018)*, Seattle, Washington, WA, 2018.)

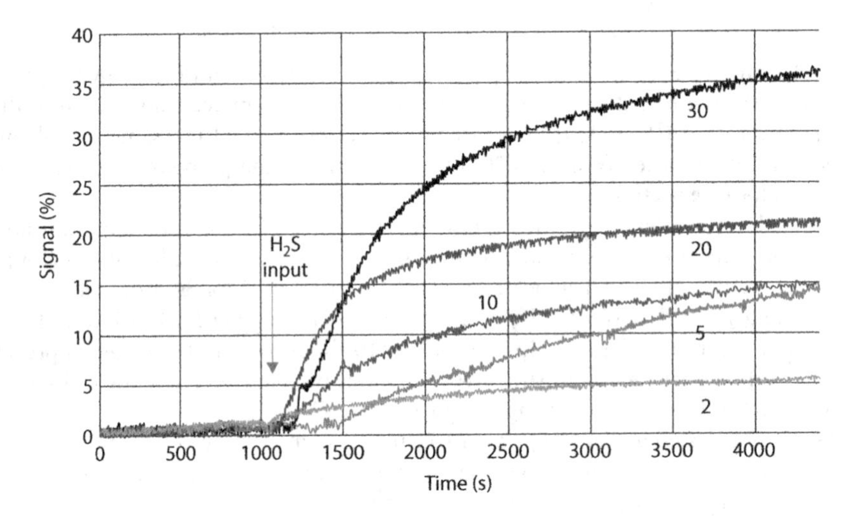

FIGURE 8.16 The signal in relation to time for 2 to 30 min depositions. H_2S in 200 ppm concentration is purged to test chamber in 1100 seconds from the initiation of the recording. (Courtesy of Keley, M.M. et al., Functionalized POPF H_2S sensor, *27th International Conference on Plastic Optical Fibers (POF2018)*, Seattle, Washington, WA, 2018.)

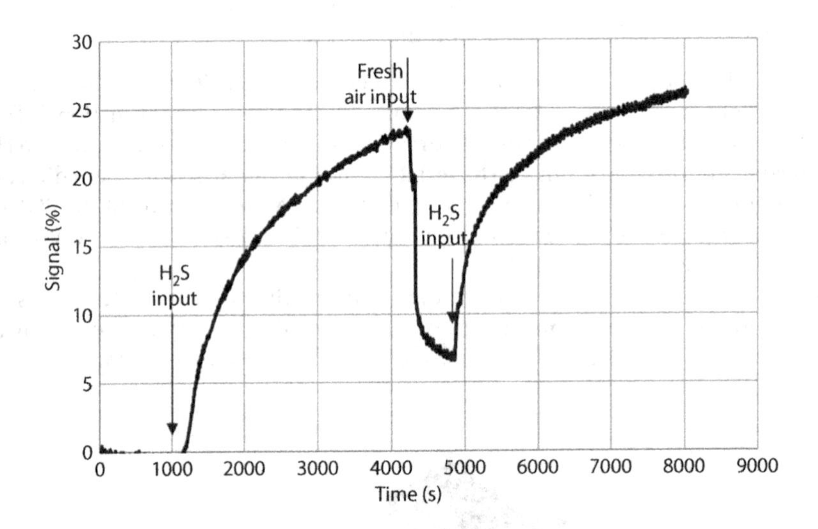

FIGURE 8.17 The behavior of a sensor that demonstrates reversibility to some extent. (Courtesy of Keley, M.M. et al., Functionalized POPF H_2S sensor, *27th International Conference on Plastic Optical Fibers (POF2018)*, Seattle, Washington, WA, 2018.)

are the best choice for detecting H_2S. However, they suffer from reversibility fault. Therefore, more PLD CuO thin films with different laser potentials are fabricated. It was recorded that potentials up to 90 mJ produce thin films that demonstrate partial reversibility (**Figure 8.17**).

In conclusion, it is worth mentioning that the sensor has a low operational temperature and the sensor is free of any heating apparatus that used to be mandatory in conventional H_2S setups. Its confirmed sensibility falls in the order of 100 ppm of the mentioned gas. The experiment revealed that CuO thin films undergo a remarkable change in optical properties as it is exposed to H_2S in ppm concentration orders. It also was proved that there is an optimum film thickness at which maximum signal takes part. Back to the reversibility issue, the best version of the studied sensor is demonstrating reversibility as it is exposed to extremely low concentrations of H_2S. In other words, the operational conditions of the sensor fall in the 10 ppm range. Otherwise, the sensor tends to react chemically with the sensed gas in an irreversible regime. More studies are needed to measure exact reversible exposure limit of the sensor. Additionally, it is recommended to study hybrid thin films composed of a metal oxide as matrix and another as precipitation/dopant [46,47]. This combination is proven to demonstrate improved sensibility and selectivity results.

References

1. A. Schieweck, E. Uhde, T. Salthammer, L. C. Salthammer, L. Morawska, M. Mazaheri, P. Kumar, "Smart home and the control of indoor air quality", *Renewable and Sustainable Energy Reviews* 94 (2018) 705–718.
2. K. V. D. Ambeth, "Human security from death defying gases using an intelligent sensor system", *Sensing and Bio-Sensing Research* 7 (2016) 107–114.
3. R. A. Dweik, P. B. Boggs, S. C. Erzurum, C. G. Irvin, M. W. Leigh, J. O. Lundberg, A. C. Olin, A. L. Plummer, D. R. Taylor, "An official ATS clinical practice guideline: Interpretation of exhaled nitric oxide levels (FENO) for clinical applications", *American Journal of Respiratory and Critical Care Medicine* 184(5) (2011) 602–615.
4. A. D. Wilson, M. Baietto, "Applications and advances in electronic-nose technologies", *Sensors (Basel)* 9(7) (2009) 5099–5148.
5. M. G. Varnamkhasti, C. Apetrei, J. Lozano, A. Anyogu, "Potential use of electronics noses, electronic tongues and biosensors as multisensors systems for spoilage examination in foods", *Trends in Food Science & Technology* 80 (2018) 71–92.
6. J. G. Murphy, R. S. Wright, "Social media posts and search engine queries as the canary in the coal mine for public health surveillance", *Mayo Clinic Proceedings* 93(9) (2018) 1155–1157.
7. N. S. Lawrence, "Analytical detection methodologies for methane and related hydrocarbons", *Talanta* 69 (2006) 385–392.
8. J. B. Miller, "Catalytic sensors for monitoring explosive atmospheres", *IEEE Sensors Journal* 1(1) (2001) 88–93.
9. N. Barsan, D. Koziej, U. Weimar, "Metal oxide-based gas sensor research: How to?" *Sensors and Actuators B* 121 (2007) 18–35.
10. A. Dey, "Semiconductor metal oxide gas sensors: A review", *Materials Science and Engineering B* 229 (2018) 206–217.
11. C. E. Simons, O. G. Florea, A. Stanoiu, "Gas sensing mechanism involved in H_2S detection with NiO loaded SnO_2 gas sensors", *Romanian Journal of Information Science and Technology* 20 (2017) 415–425.
12. T. W. Capehart, S. C. Chang, "The interaction of tin oxide films with O_2, H_2, H_2S", *Journal of Vacuum Science and Technology* 18 (1981) 393–397.
13. Y. Zhang, H. Peng, X. Qian, Y. Zhang, G. An, Y. Zhao, "Recent advancements in optical fiber hydrogen sensors", *Sensors and Actuators B: Chemical* 244 (2017) 393–416.
14. M. Yin, B. Gu, Q. An, C. Yang, Y. L. Guan, K. T. Yong, "Recent development of fiber-optic chemical sensors and biosensors: Mechanisms, materials, micro/nano-fabrications and applications", *Coordination Chemistry Reviews* 376 (2018) 348–392.
15. T. Chattopadhyay, S. Chatterjee, I. Majumder, S. Ghosh, S. Yoon, E. sim, "Fluorometric detection of nitromatics by fluorescent lead complexes: A spectroscopic assessment of detection mechanism", *Spectrochimica Acta Part A: Molecular and Biomolecular Spectroscopy* 194 (2018) 222–229.
16. P. A. S. Jorge, P. Caldas, C. C. Rosa, A. G. Oliva, J. L. Santos, "Optical fiber probes for fluorescence based oxygen sensing", *Sensors and Actuators B* 103 (2004) 290–299.
17. A. Paliwal, A. Sharma, M. Tomar, V. Gupta, "Carbon monoxide (CO) optical gas sensor based on ZnO thin films", *Sensors and Actuators B: Chemical* 250 (2017) 679–685.
18. W. S. Struve, *Fundamentals of Molecular Spectroscopy*, John Wiley & Sons, New York, 1989.
19. L. Bilro, N. Alberto, J. L. Pinto, R. Nogueira, "Optical sensors based on plastic fibers", *Sensors* (2012) 12184–12207.

20. H. Waechter, J. Litman, A. H. Cheung, J. A. Barnes, H. P. Loock, "Chemical sensing using fiber cavity ring-down spectroscopy", *Sensors (Basel)* 10(3) (2010) 1716–1742.
21. W. Jin, H. L. Ho, Y. C. Cao, J. Ju, L. F. Qi, "Gas detection with micro- and nano-engineered optical fibers", *Optical Fiber Technology* 19 (2013) 741–759.
22. H. H. Qazi, A. B. Mohammad, M. Akram, "Recent progress in optical chemical sensors", *Sensors* 12 (2012) 16522–16556.
23. A. A. P. Boechat, D. Su, D. R. Hall, J. D. C. Jones, "Bend loss in large core multimode optical fiber beam delivery system", *Applied Optics* 30(3) (1991) 321-327.
24. N. Izu, W. Shin, N. Murayama, S. Kazaki, "Resistive oxygen gas sensors based on CeO_2 fine powder prepared using mist pyrolysis", *Sensors and Actuators B* 87 (2002) 95–98.
25. R. Ramamoorthy, P. K. Dutta, S. A. Akbar, "Oxygen sensors: Materials, methods, designs and applications", *Journal of Materials Science* 38 (2003) 4271.
26. W. Zhong, P. Urayama, M. A. Mycek, "Imaging fluorescence lifetime modulation of a ruthenium-based dye in living cells: The potential for oxygen sensing", *Journal of Physics D: Applied Physics* 36 (2003) 1689-1695.
27. C. Pulido, Ó. Esteban, "Improved fluorescence signal with tapered polymer optical fibers under side-illumination", *Sensors and Actuators B* 146 (2010) 190–194.
28. P. Payra, P. K. Dutta, "Development of a dissolved oxygen sensor using tirs(bipyridyl) ruthenium (II) complexes entrapped in high siliceous zeolites", *Microporous and Mesoporous Materials* 64 (2003) 109–118.
29. E. V. Donckt, B. Camerman, R. Herne, R. Vandeloise, "Fibre-optic oxygen sensor based on luminescence quenching of a Pt(II) complex embedded in polymer matrices", *Sensors and Actuators B* 32 (1996) 121-127.
30. B. D. Mac Craith, G. O'Keeffe, C. McDonalds, A. K. McEvory, "LED-based fibre optic oxygen sensor using sol-gel coating", *Electronics Letters* 30(11) (1994) 888–889.
31. C. Chu, Y. Lo, "High-performance fibre-optic oxygen sensors based on fluorinated xerogels doped with Pt(II) complexes", *Sensors and Actuators B* 124 (2007) 376–382.
32. D. F. Merchant, P. J. Scully, N. F. Schmitt, "Chemical tapering of polymer optical fibre", *Sensors and Actuators* 79 (1999) 365–371.
33. S. Mao, H. Zhou, S. Wu, J. Yang, Z. Li, X. Wei, X. Wang, Z. Wang, J. Li, "High performance hydrogen sensor based on Pd/TiO_2 composite film", *International Journal of Hydrogen Energy* 43(2018) 22727–22732.
34. M. Slaman, B. Dam, M. Pasturel, D. M. Borsa, H. Schreuders, J. H. Rector, R. Griessen, "Fiber optic hydrogen detectors containing Mg-based metal hydrides", *Sensors and Actuators B: Chemical* 123 (2007) 538–545.
35. D. M. Borsa, A. Baldi, M. Pasturel, H. Schreuders, B. Dam, R. Griessen, "Mg-Ti-H thin films for smart solar collectors", *Applied Physics Letters* 88 (2009) 241910.
36. T. Mak, R. J. Westerwaal, M. Slaman, H. Schreuders, A. W. Van Vugt, M. Victoria, C. Boelsma, B. Dam, "Optical fiber sensor for the continuous monitoring of hydrogen in oil", *Sensors and Actuators B: Chemical* 190 (2014) 982–989.
37. F. Schweppe, M. Martin, E. Fromm, "Model on hydride formation describing surface control, diffusion control and transition regions", *Journal of Alloys and Compounds* 261 (1997) 254–258.
38. A. Borgschulte, R. J. Westerwaal, J. H. Rector, H. Schruders, B. Dam, R. Griessen, "Catalytic activity of noble metals promoting hydrogen uptake", *Journal of Catalysis* 239 (2006) 263–271.
39. P. Wolkoff, "Indoor air humidity, air quality, and health-An overview", *International Journal of Hygiene and Environmental Health* 221 (2018) 376–390.
40. Y. Pang, J. Jian, T. Tu, Z. Yang, J. Ling, Y. Li, X. Wang, Y. Qiao, Y. Yang, T. Ren, "Wearable humidity sensor based on porous grapheme network for respiration monitoring", *Biosensors and Bioelectronics* 116 (2018) 123–129.
41. C. Kelb, M. Körner, O. Prucker, J. Rühe, E. Reithmeier, B. Roth, "PDMAA hydrogel coated U-bend humidity sensor suited for mass-production", *Sensors* 17 (2017) 517–526.
42. S. Tao, L. Xu, J. C. Fanguy, "Optical fiber ammonia sensing probes using reagent immobilized porous silica coating as transducers", *Sensors and Actuators B* 115 (2006) 158–163.
43. C. Kelb, M. Körner, O. Prucker, J. Rühe, E. Reithmeier, B. Roth, "A Planar low-cost full-polymer optical humidity sensor", *Procedia Technology* 26 (2016) 530–536.

44. M. M. Keley, F. F. Borghi, A. Dante, C. Carvalho, R. Allil, F. Dutra, P. Cardoso et al., "U-shaped plastic optical fiber functionalized with metal oxides thin film for H_2S gas sensor applications", *IEEE I2MTC International Instrumentation and Measurement Technology Conference*, 2018, Houston, TX.
45. M. M. Keley, A. Dante, F. Dutra, R. C. S. B. Allil, C. C. Carvalho, M. M. Werneck, "Functionalized POPF H_2S sensor", *27th International Conference on Plastic Optical Fibers (POF2018)*, 2018, Seattle, Washington, WA.
46. J. Liu, X. Huang, G. Ye, W. Liu, Z. Jiao, W. Chao, Z. Zhou, Z. Yu, "H_2S detection sensing characteristics of CuO/SnO_2 sensor", *Sensors* 3 (2003) 110–118.
47. M. Kaur, B. K. Dadhich, R. Singh, D. Ganapathi, T. Bagwaiya, S. Bhattacharya, A. K. Debnath, K. P. Muthe, S. C. Gadkari, "RF sputtered SnO_2:NiO thin films as sub-ppm H_2S sensor operable at room temperature", *Sensors and Actuators B* 242 (2017) 389–403.

Biological Sensing

Regina Célia da Silva Barros Allil

Contents

9.1 Introduction

This chapter aims to study related to biological sensing with POF fibers as a proposal for the detection of microorganisms with low cost and fast response time.

Before beginning the approach on biological sensing, we must first review some phenomena and technologies such as evanescent field (EV), optical attenuation (by total reflection and the curvature of the fiber), surface plasmon resonance (SPR), and thin film (TF), with the goal of providing the reader with the key concepts used in the research and development of a biosensor.

Later, we will introduce the step-by-step development of a biosensor based on polymer optical fiber (POF), from a sensor for measuring and monitoring the refractive index (IR) associated with the mentioned technologies and phenomena, for the detection of *Escherichia coli.*

9.2 Light Propagation in Optical Fiber

The propagation of light within the optical fiber occurs according to Snell's Law (**Equation 9.1**) through the phenomenon of total internal reflection.

$$\frac{\text{Sen}\,\alpha}{\text{Sen}\,\beta} = \!\!\rightarrow n_1 \,\text{sen}\,\alpha = n_2 \,\text{sen}\,\beta \tag{9.1}$$

where

α = incident angle
β = refracted angle
n_1 = core refractive index
n_2 = cladding refractive index

Figure 9.1 shows an illustration of the operating principle of optical fiber based on the phenomenon of total internal reflection. Where core and cladding present different refractive indices and in this way, the light can be guided inside the fiber core.

However, a condition must be obeyed: the refractive index of the core material must be greater than the refractive index of the cladding material.

Where θ_{max}—maximum angle, according to the numerical aperture of the fiber, θ_1—light angle inside the fiber, α—incidence angle between core and cladding, β—refracted angle and $\theta_{critical}$—maximum angle for total reflection.

Figure 9.2 shows the light propagation in the fiber core, from the example of a given optical fiber with $n_{core} = 1.5$, $n_{cladding} = 1.46$, and critical angle = 76.7°.

Having reviewed of the phenomenon of total internal reflection in the optical fiber, we will now turn to the evanescent field phenomenon.

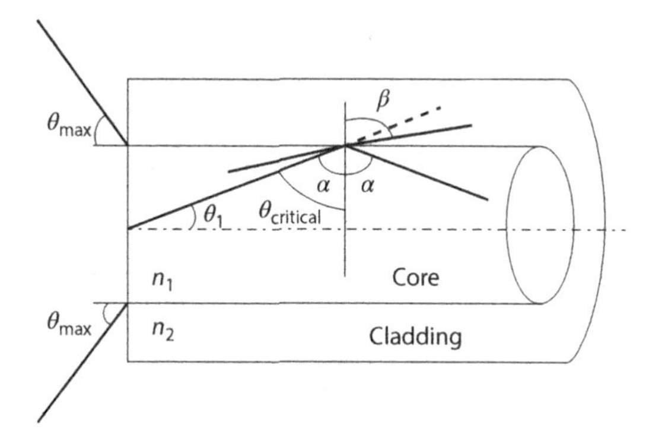

FIGURE 9.1 Representation of the guided light in the optical fiber. (Adapted from Leung, A. et al., *Sens. Actuators B*, 125, 688–703, 2007.)

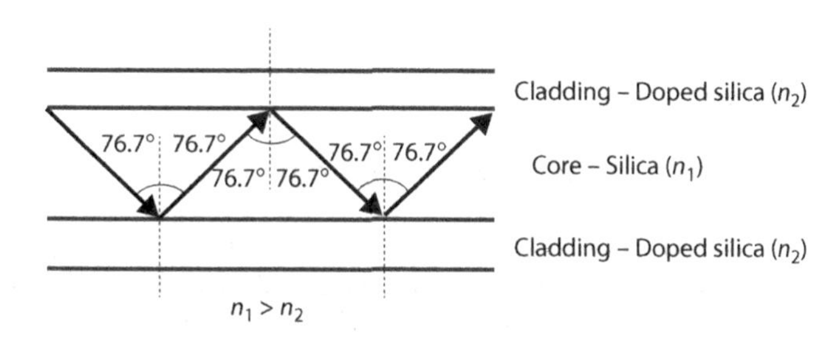

FIGURE 9.2 Demonstration of light propagation in the fiber core.

9.3 Evanescent Field

In-depth knowledge of the physical properties of fiber is indispensable for the advancement of sensor research. The study of the evanescent field, the relationship between the refractive indices of the core and the fiber cladding, and the effect of the curvature in the fiber are investigated in several studies [1].

There are several ways to increase the sensitivity of the sensor to the optical fiber, such as by reducing the diameter of the fiber in the sensing area, forming the so-called tapers, and through a curvature in the fiber to obtain the evanescent field.

The evanescent field approach begins by studying the electromagnetic radiation when crossing a flat interface to a medium with different refractive index from which it has been propagating. **Figure 9.3** shows the incident electromagnetic wave that propagates in the n_1 medium, following in the direction k_i, with angle of incidence θ_i; when finding an interface whose refraction index is different, part of this wave will be reflected in the direction k_r, with angle of reflection θ_r.

We can also observe that part of the electromagnetic wave is transmitted by the surface, propagated through the medium of refractive index n_2, following in the direction of propagation k_t with transmission angle θ_t [2]. Where, k_i represents direction of propagation of the incident electromagnetic wave, θ_i, angle of incidence, k_r the direction of propagation of the reflected electromagnetic wave, θ_r reflection angle, k_t, direction of propagation of the transmitted electromagnetic wave, θ_t transmission angle, n_1, refractive index of the core material, and n_2, index of refraction of the cladding material.

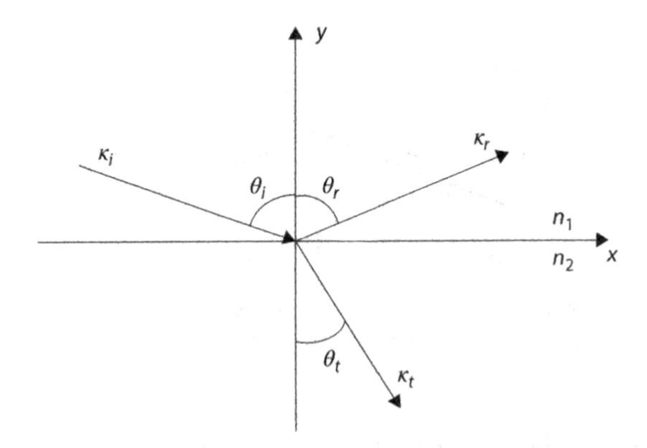

FIGURE 9.3 Geometry of the reflection and refraction of a light ray in an electromagnetic approach. (Adapted from Leung, A. et al., *Sens. Actuators B*, 125, 688–703, 2007.)

Thus, from the wave equation,

$$E = E_0 \exp i(k \cdot r - \mathrm{wt}) \tag{9.2}$$

where E_0 is the amplitude of the wave and k determines the direction of propagation. One can analyze the angles and amplitudes of the transmitted and reflected electric fields.

In this approach, we obtain the "Fresnel equations," which determine relations between the electric fields transmitted and reflected with the incident electric field.

The Fresnel equations for the bias in the plane of incidence are [3]:

$$E_{0_r} = \left(\frac{\alpha - \beta}{\alpha + \beta}\right) E_{0_i}, E_{0_t} = \left(\frac{2}{\alpha + \beta}\right) E_{0_i} \tag{9.3}$$

with

$$\alpha \equiv \frac{\cos \theta_t}{\cos \theta_i} \tag{9.4}$$

and

$$\beta \equiv \frac{\mu_1 v_1}{\mu_1 v_1} = \frac{\mu_1 n_2}{\mu_2 n_1} \tag{9.5}$$

where:

E_{0i}, E_{0r}, and E_{0t}—amplitude of the incident, reflected, and transmitted electric field, respectively.

θ_i and θ_t—angles of incidence and transmitted, respectively.

μ_1 and μ_2—permeability of media 1 and 2, respectively. One can consider $\mu_1 \approx \mu_2$.

v_l and v_2—wave velocity in means 1 and 2, respectively.

It is observed in **Equation (9.3)** that the amplitudes of the transmitted and the fields depend on α, which is a function of the angle of incidence (**Equation 9.4**), as well as the refractive indices and the permeability of the media (**Equation 9.5**).

By rewriting α as a function of the angle of incidence, we have:

$$\alpha = \frac{\sqrt{1-\text{sen}^2\theta_t}}{\cos\theta_i} = \frac{\sqrt{1-\left[\left(\frac{n_1}{n_2}\right)\text{sen}\,\theta_i\right]^2}}{\cos\theta_i} \tag{9.6}$$

Another important approach is the "Poynting vector" (S), which informs the energy per unit time, per unit area, which means the power per unit area, transported through a field, which propagates in the direction k and is then defined as:

$$\boldsymbol{S} \equiv \frac{1}{\mu}(\boldsymbol{E}\times\boldsymbol{B}) \tag{9.7}$$

The average power per unit area carried by the electromagnetic wave is called the intensity (I):

$$I \equiv \langle \boldsymbol{S} \rangle = \frac{1}{2} v\varepsilon E_0^2 \tag{9.8}$$

where ε is the permissiveness of the medium.

From the scalar product of S by the unit vector in the direction of propagation, we can extract the incident intensity by:

$$I_i = \frac{1}{2}\varepsilon_1 v_1 E_{0_i}^2 \cos\theta_i \tag{9.9}$$

Similarly, the intensities transmitted and reflected are:

$$I_r = \frac{1}{2}\varepsilon_1 v_1 E_{0_r}^2 \cos\theta_r \;\; e \;\; I_t = \frac{1}{2}\varepsilon_2 v_2 E_{0_t}^2 \cos\theta_t \tag{9.10}$$

The coefficients of reflection (R) and transmission (T) are then defined, according to the equations below. The transmission coefficient will be important in the study of the behavior of the electromagnetic wave in the curved fiber.

$$R \equiv \frac{I_r}{I_i} = \left(\frac{E_{0_r}}{E_{0_i}}\right)^2 = \left(\frac{\alpha-\beta}{\alpha+\beta}\right)^2 \tag{9.11}$$

$$T \equiv \frac{I_t}{I_i} \equiv \frac{\varepsilon_2 v_2}{\varepsilon_1 v_1}\left(\frac{E_{0_t}}{E_{0_i}}\right)^2 \frac{\cos\theta_t}{\cos\theta_i} = \alpha\beta\left(\frac{2}{\alpha+\beta}\right)^2 \tag{9.12}$$

By observing the total reflection process, it is expected that all light emitted will be reflected in the same medium. However, this is not exactly what happens. By treating the light through an electromagnetic approach, it is observed that by assuming null the transmitted electromagnetic wave, it becomes impossible to satisfy the boundary conditions using only the incident and reflected waves [4,5]. However, this transmitted wave cannot carry energy through the boundary created at the materials interface. One way of studying this phenomenon of the emergence of the evanescent field is by considering the transmitted wave, as we can see in **Figure 9.4** demonstrating the emergence of this phenomenon.

$$\boldsymbol{E}_t = \boldsymbol{E}_{0_t} \exp i(\boldsymbol{k}\cdot\boldsymbol{r} - \omega t) \tag{9.13}$$

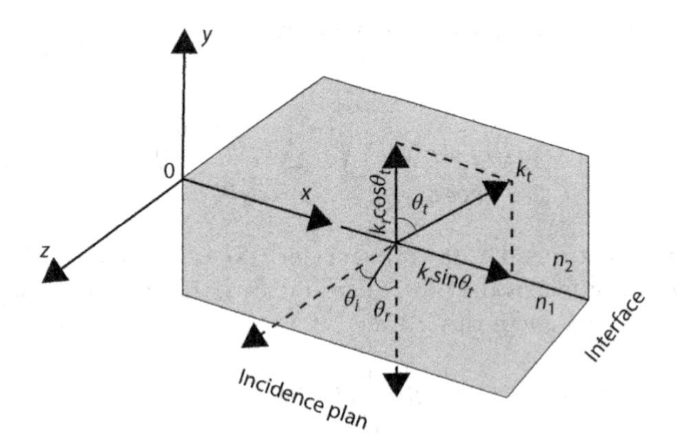

FIGURE 9.4 Emergence of the evanescent field during the passage of light in the waveguide. k_i—direction of propagation of the incident electromagnetic wave and θ_i angle of incidence; k_r—direction of propagation of the reflected electromagnetic wave and θ_r angle of reflection; k_t—direction of propagation of the transmitted electromagnetic wave and θ_t transmission angle; n_1—core index; and n_2—index of the cladding. (Adapted from Ferreira, A.P., *Bacteriosensor—Tecnologia de sensoriamento bacteriológico a fibra óptica*, Tese de Doutorado, Programa de Pós-Graduação de Engenharia COPPE/UFRJ, Rio de Janeiro, RJ, Brasil, 2000.)

where

$$\mathbf{k}\cdot\mathbf{r} = k_{tx}x + k_{ty}y \tag{9.14}$$

Note that there is no component z in k, but we have:

$$k_{tx} = k_t \operatorname{sen}\theta_t \tag{9.15}$$

$$k_{ty} = k_t \cos\theta_t \tag{9.16}$$

Using Snell's law (1), we find:

$$k_t\cos\theta_t = \pm k_t\sqrt{1 - \frac{n_1^2 \operatorname{sen}^2\theta_i}{n_2^2}} \tag{9.17}$$

Considering the case for total reflection, where $n_1 > n_2$, the radical is negative. Thus, we have:

$$k_{ty} = \pm i k_t \sqrt{\frac{n_1^2 \operatorname{sen}^2\theta_i}{n_2^2} - 1} = \pm i\delta \tag{9.18}$$

and

$$k_{tx} = \frac{n_1 k_t}{n_2}\operatorname{sen}\theta_i \tag{9.19}$$

Thus, substituting in (9.13) we have:

$$\mathbf{E}_t = \mathbf{E}_{0t} e^{\mp\delta y} e^{i\left(k_t x \frac{n_1 \operatorname{sen}\theta_i}{n_2} - \omega t\right)} \tag{9.20}$$

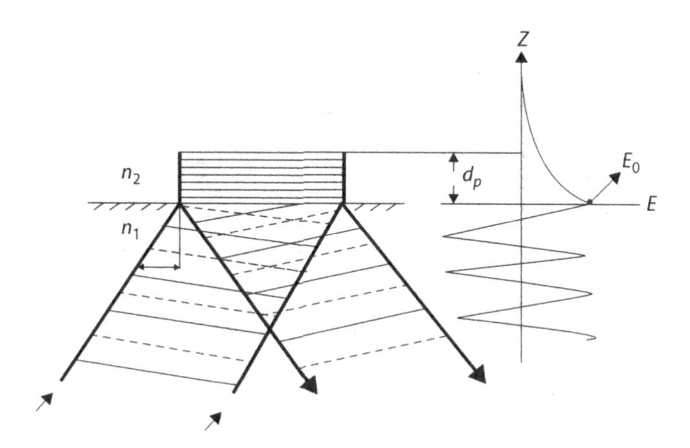

FIGURE 9.5 Penetration distance or length (d_p) determined by the drop of the electric field strength of 1/*e* and the value at the interface. E_0—amplitude of the electric field; n_1—core index; and n_2—index of the cladding. (Adapted from Hecht, E., *Optics*, 4rd ed, Ed. Addison Wesley, 2002.)

Since the positive exponential solution is physically impossible, it must be ignored. Thus, we have a wave whose amplitude decays exponentially as it penetrates the less dense medium. The penetration distance (dp), also known as penetration length, is defined by the value at which the field falls 1/e of the value at the interface, as shown in **Figure 9.5**.

This distance is given by **Equation (9.21)**, where θ_i is the incident angle at the interface and $\theta_i > \theta_c$, λ_0 is the wavelength of light in the vacuum. This equation can also be written as (9.22), where $k_0 = 2\pi/\lambda_0$ [7].

$$d_p = \frac{\lambda_0}{2\pi n_1 \sqrt{\text{sen}^2\theta_i - \text{sen}^2\theta_c}} \tag{9.21}$$

$$d_p = \frac{1}{k_0 \sqrt{n_1^2 \text{sen}^2\theta_i - n_2^2}} \tag{9.22}$$

In **Figure 9.6** we can see the magnetic field structure through the field distributions (modes $m = 0$, $m = 1$), and the light intensities that correspond in a given cross section of the guide (frontal and lateral view) are found not only in the core, but also in the cladding.

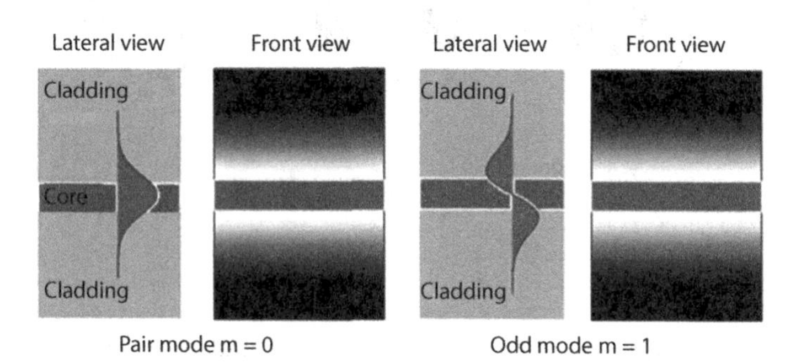

FIGURE 9.6 Field distribution in the transverse direction and intensity of light in the guide output for the even ($m = 0$) and odd ($m = 1$) modes. (Adapted from Zilio, S.C., *Optica Moderna, Fundamentos e Aplicações*, Fotônica, IFSC-USP, 2010; Nunes, F.D., *Fibras e Dispositivos para Comunicação Ópticas*, São Paulo, Brazil, Editora Renovarum Ltda, 2001.)

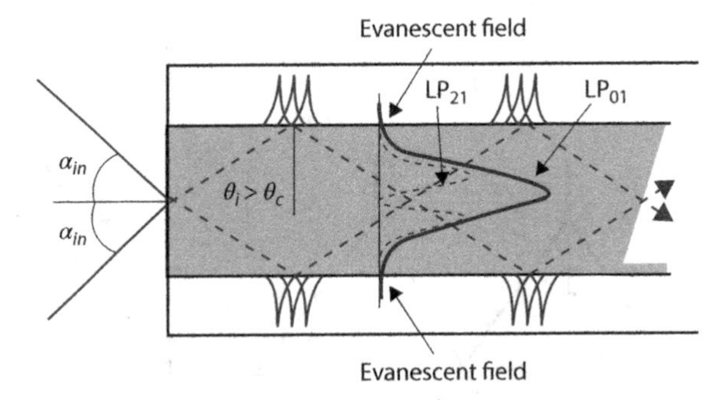

FIGURE 9.7 Demonstration of the evanescent wave. (Adapted from Kitano, C., Tecnologia de Fibras Ópticas, Universidade Estadual Paulista "Júlio De Mesquita Filho" Faculdade De Engenharia De Ilha Solteira, Agosto, https://docplayer.com.br/61011094–1-estrutura-geral-da-fibra-optica.html, 2017; Rodrigues, D.M.C., Desenvolvimento E Caracterização De Sensores A Fibra Óptica Plástica Para Refratometria Baseados Em Modulação De Amplitude, Tese Mestrado, Programa de Engenharia Elétrica, COPPE/UFRJ, Abril 2013.)

Another demonstration of the location of the evanescent field, from the leaked light of the core penetrating the shell, can be seen in **Figure 9.7**.

The evanescent field in the fiber optic shell region can be calculated using guided wave theory. According to this theory, there is more than one form of light propagating in the fiber, being able to cite the so-called fields of LP_{mn} modes.

These modes are discrete solutions of the Helmholtz wave equation and can be associated to the different angles from the propagating rays inside the nucleus, which is formed with the longitudinal axis [9,10].

The nomenclature of the LP_{mn} modes portrays the existence of maximum "*m*" in the azimuthal direction and at least "*n*" in the radial direction.

Figure 9.8 shows some examples of these modes. LP_{01} mode shows amplitude of a single maximum field, exactly at the central point of the nucleus, which decays radially towards the cladding and presents a bell-like profile. The LP_{21} mode displays two peaks (maximum) and

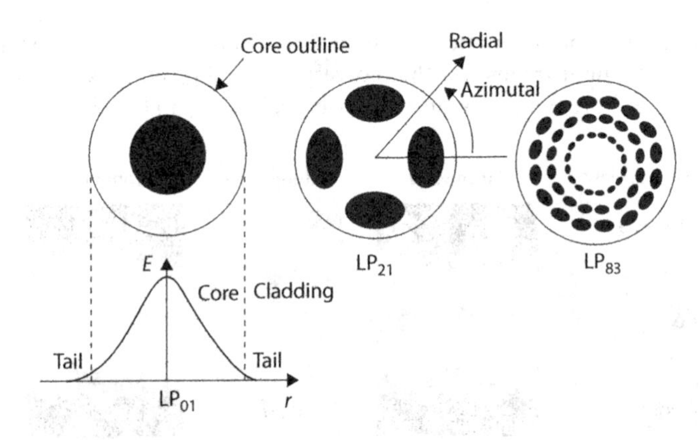

FIGURE 9.8 Profiles of modes LP_{01}, LP_{21}, and LP_{83} (side view). (Adapted from Kitano, C., Sensores de Amplitude em Fibras Ópticas, Universidade Estadual Paulista "Júlio De Mesquita Filho" Faculdade De Engenharia De Ilha Solteira, www.feis.unesp.br/Home/departamentos/engenhariaeletrica/optoeletronica/sensor-de-intensidade.pdf, 2017; Kitano, C., Tecnologia de Fibras Ópticas, Universidade Estadual Paulista "Júlio De Mesquita Filho" Faculdade De Engenharia De Ilha Solteira, https://docplayer.com.br/61011094–1-estrutura-geral-da-fibra-optica.html, 2017.)

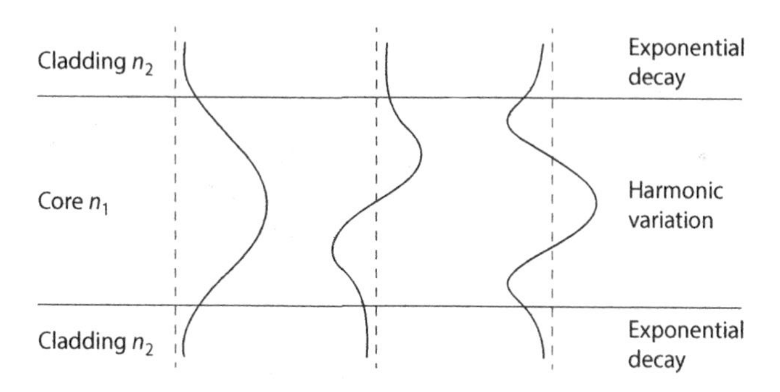

FIGURE 9.9 Illustration of the size of evanescent wave modes. (Adapted from Kitano, C., Sensores de Amplitude em Fibras Ópticas, Universidade Estadual Paulista "Júlio De Mesquita Filho" Faculdade De Engenharia De Ilha Solteira, www.feis.unesp.br/Home/departamentos/engenhari-aeletrica/optoeletronica/sensor-de-intensidade.pdf, 2017; Kitano, C., Tecnologia de Fibras Ópticas, Universidade Estadual Paulista "Júlio De Mesquita Filho" Faculdade De Engenharia De Ilha Solteira, https://docplayer.com.br/61011094–1-estrutura-geral-da-fibra-optica.html, 2017.)

two (minimum) valleys in the azimuthal direction, and at least one maximum in the radial direction. And the LP_{83} mode displays eight highs and eight lows in the azimuthal direction, and at least three highs in the radial direction.

In this way, the evanescent field satisfies the necessary boundary conditions at the core-cladding interface in the shell region, i.e., the evanescent wave is related to a specific extension that leaks from the core through a small tail of the field profile that penetrates the cladding [9,10].

When analyzing the fiber from the top, we can observe a profile in the circular form, due to the fact that the structure of the fiber and the size or extension of the tail will be larger, the more incident the field can reach the cladding, the greater the order of the mode in the fiber [9,10].

Figure 9.9 demonstrates the size of the evanescent wave through the following characteristics: it is smaller for LP_{01} mode, it grows into LP_{11} mode, and it is even larger for LP_{02} mode.

We can conclude this approach that, part of the light which travels in the core, also penetrates on the cladding of the optical fiber, through total internal reflection, where angles of incidence are higher than the critical angle, will be accompanied by an evanescent wave, in order to satisfy the boundary conditions at the interface [2,9,10].

The evanescent field can be defined as part of the light that affects the shell, in addition to its interface with the nucleus, called photonic tunneling, and that has an evanescent behavior, quantified by the exponential decay of the field strength in the shell of the guide [2].

9.4 Attenuation

9.4.1 Attenuation of the Output Power in the Optical Fiber by Total Reflection

One of the optical characteristics used in the biological sensor under study is the attenuation of the output power in the optical fiber by total internal reflection. Thus, it was necessary to study this behavior in the fiber.

The attenuation of the light guided in the nucleus depends on the fraction of the total power that is guided in the region of the evanescent field. This attenuation between the input (I_0) and output (I) powers shown in **Figure 9.10** is defined by **Equation (9.23)** [7,11].

$$I = I_0 \exp\left[-\left(r_f \alpha_m l C\right)\right] \tag{9.23}$$

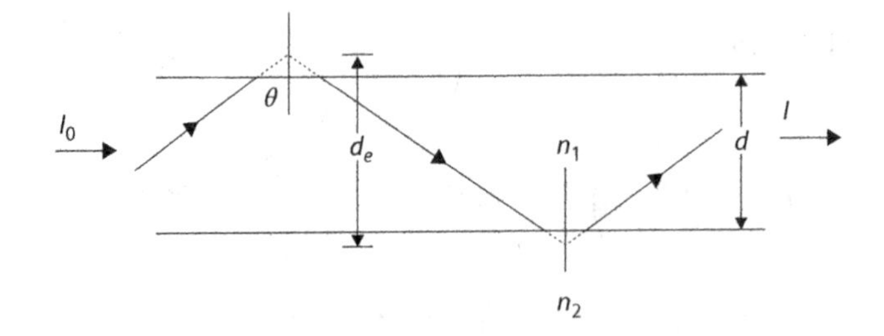

FIGURE 9.10 Loss of power along the waveguide as a function of fiber diameter and external refractive indexes. I_0—input power; I—output power; n_1—core index; n_2—bark index; d—effective diameter of the core. (Adapted from Culshaw, B. and Dakin, J., *Optical Fiber Sensors Components and Subsystems*, Norwood, MA, Ed. Artech House, 1996; Rodrigues, D.M.C., Desenvolvimento E Caracterização De Sensores A Fibra Óptica Plástica Para Refratometria Baseados Em Modulação De Amplitude, Tese Mestrado, Programa de Engenharia Elétrica, COPPE/UFRJ, 2013.)

where the parameters α_m and C are defined for the study of an absorber cell, α_m is the molar absorption coefficient, the cell length, C the chemical concentration of the cell and r_f represents the reduction factor ($r_f < 1$) in the coefficient of attenuation (α_m) when waveguide is used instead of a direct absorption cell.

Considering that l and C are fixed parameters, the attenuation can be analyzed by the variation of r_f. In general, r_f can be written as:

$$r_f = \frac{n_2}{n_e} f \tag{9.24}$$

Where,

n_2—absorber index.

n_e—effective index of the guided wave ($n_e = n_1 \text{sen}\, \theta$).

f—fraction of the total optical power that interacts with the absorber.

Since the field distribution in the waveguide is not known in the present case, the power fraction f cannot be calculated. Alternatively, one can calculate r_f for a flat guide through reflection loss and multiplying by the number of reflections per unit length. For a multimode fiber as in **Figure 9.10**, r_f is given by [12]:

$$r_f = \frac{n_r}{k_0 n_e d_e \sqrt{n_e^2 - n_2^2}} \left[\frac{n_1^2 - n_e^2}{n_1^2 - n_2^2} \right] \tag{9.25}$$

or,

$$r_f = \frac{n_2}{n_e} \left[\frac{n_1^2 - n_e^2}{n_1^2 - n_2^2} \right] \frac{1}{1 + \pi R \sqrt{n_e^2 - n_2^2}} \tag{9.26}$$

9.4.2 Attenuation of the Output Power in the Optical Fiber by the Curve

A study concerning the variation of power as a function of the evanescent field will be presented.

In the proposed sensor, the attenuation of the optical power by the curved fiber in relation to the variation of the refractive index of the external medium is also considered [11].

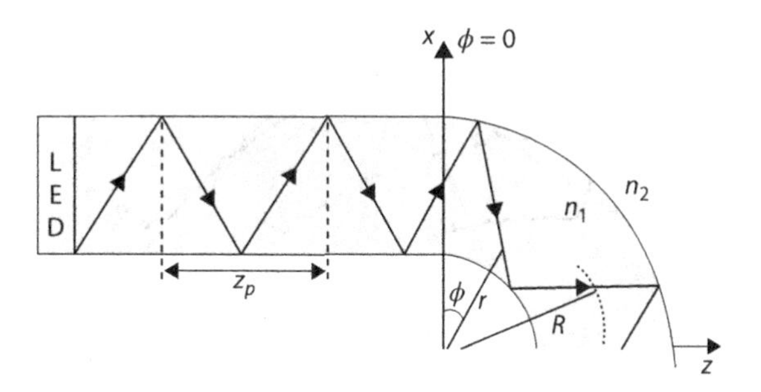

FIGURE 9.11 Curved fiber with angle variation (φ)—core radius; n_1—refractive index of the fiber; n_2—refractive index of the external environment; R—radius of the curve; r—distance between the center of the curve and the guided light; z_p—half-time distance between successive reflections in the stretched fiber. (Adapted from Rodrigues, D.M.C., Desenvolvimento E Caracterização De Sensores A Fibra Óptica Plástica Para Refratometria Baseados Em Modulação De Amplitude, Tese Mestrado, Programa de Engenharia Elétrica, COPPE/UFRJ, 2013.)

It is worth mentioning that the biological sensors that will be demonstrated are U-shaped, characterizing a fixed curve. However, it is important to know the behavior of the fiber in relation to the variation in the angle of the curve, where we will demonstrate the analysis developed for a curve fiber [5], as shown in **Figure 9.11**.

$$P(z) = P(0)\exp(-\gamma z) \tag{9.27}$$

where γ is the attenuation coefficient determined by the relation between the mean value of the transmission coefficient T, **Equation (9.12)**, and the half-period distance between successive reflections, called z_p [5].

$$\gamma = \frac{T}{z_p} \tag{9.28}$$

By analogy, the attenuation of the power along the curved fiber as a function of the angle of curvature φ is given by,

$$P(\phi) = P(0)\exp(-\gamma\phi) \tag{9.29}$$

where, also by analogy, the attenuation coefficient γ now relates to the angle φ_p between successive crests of the wave, or reflections, in which there is loss of power. **Figure 9.12** shows this analogy.

So,

$$\gamma = \frac{T}{\phi_p} \tag{9.30}$$

To simplify the analysis, we consider the reflections on the external surface of the curve. Thus, we have:

$$\phi_p = 2\theta_\phi \tag{9.31}$$

where

θ_φ—angle between the path of the light beam and the tangent of the external interface of the curve.

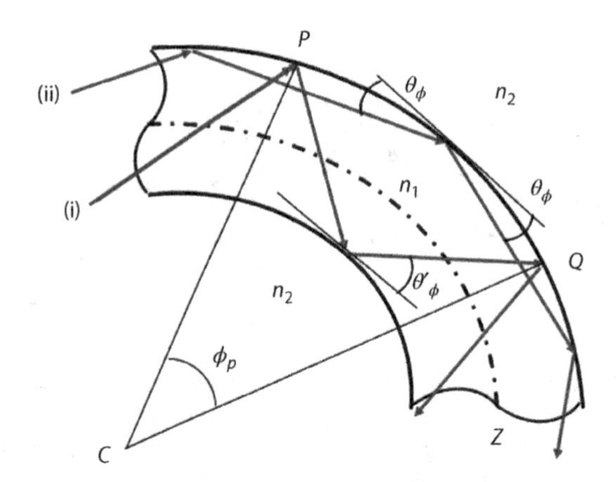

FIGURE 9.12 Highlight for the reflection angles inside the curved fiber. (i) Reflections alternating between the external and internal surfaces; (ii) reflections on the outer surface; φ_P—angle between two consecutive reflections on the outer surface; θ_φ—reflection angle on the external surface; θ'_φ—reflection angle alternating between the inner and outer surfaces; z—direction of propagation of light; n_1—internal refractive index; n_2—external refractive index. (Adapted from Snyder, W.A. and Love, J.D., *Optical Waveguide Theory*, New York, Ed. Chapman and Hall, 1983.)

The angle θ_φ relates to the angle of incidence θ_i through the expression,

$$\cos(\theta_\phi) = r_0 \cos(\theta_i)/(R+\rho) \tag{9.32}$$

Then,

$$\theta_\phi = \cos^{-1}\left(r_0 \cos(\theta_i)/R+\rho)\right) \tag{9.33}$$

where R is the radius of the curve, ρ is the radius of the nucleus, and r_0 is the value of r when $\varphi = 0$, according to **Figure 9.11**. Then, obtaining T and φ_p, one can calculate the attenuation coefficient γ of **Equation (9.30)**.

The total power P_s as a function of the curve is found by integrating all the radii:

$$P_s(\phi) = P(0) \int_{R-\rho}^{R+\rho} dr \int_{-\theta_C(r)}^{\theta_c(r)} \exp(-\gamma\phi)\, d\theta_i \tag{9.34}$$

where, in this case,

$$\theta_c(r) = \theta_c = \cos^{-1}\left(\frac{n_2}{n_1}\right) \tag{9.35}$$

It is observed from **Equation (9.34)** that the power output of light guided along a curved fiber changes as a function of the angle of inclination of the fiber.

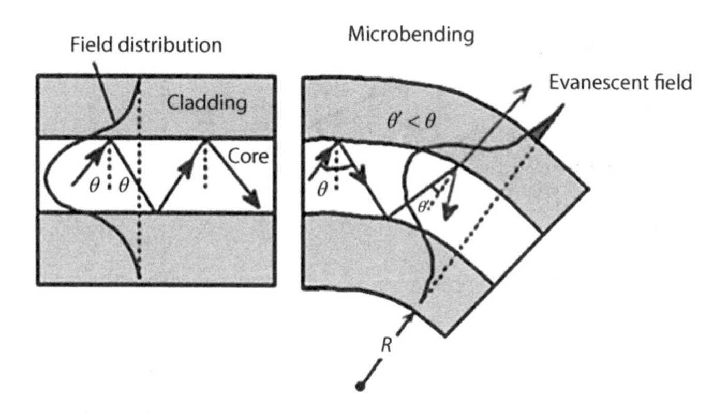

FIGURE 9.13 Demonstration of the attenuation from the curvature. (Adapted from Kitano, C., Sensores de Amplitude em Fibras Ópticas, Universidade Estadual Paulista "Júlio De Mesquita Filho" Faculdade De Engenharia De Ilha Solteira, www.feis.unesp.br/Home/departamentos/engenhariaeletrica/optoeletronica/sensor-de-intensidade.pdf, 2017; Kitano, C., Tecnologia de Fibras Ópticas, Universidade Estadual Paulista "Júlio De Mesquita Filho" Faculdade De Engenharia De Ilha Solteira, https://docplayer.com.br/61011094–1-estrutura-geral-da-fibra-optica.html, 2017.)

We can conclude this study by stating that in the presence of some kind of disturbance (curvature) in the fiber, attenuation will occur to the outer medium through the more elongated tail mode from the higher order mode. The higher order modes are less confined and the fundamental mode is the more confined mode. That is, a curvature in the fiber can cause a ray (upper mode) to exit the total internal reflection condition. In the region close to the curvature, the angle of incidence becomes smaller than the critical angle, the lightning refraction occurring and transmitting it to the outside of the fiber, resulting in loss of power to the outside.

Figure 9.13 shows the demonstration of optical power loss in fiber curvature.

9.4.3 Study and Analysis of Sensor Characteristics at the POF Curvature

In this section we will discuss the study and analysis of the characteristics of the biological sensor using a curved optical fiber with a fixed angle of 180°, aiming at the analysis of the relation between the output power as a function of the index of refraction of the external medium to the fiber.

9.4.3.1 Characteristic of Output Power at the Sensor: Attenuation across the Evanescent Field

Using **Equation (9.22)** of the penetration length (d_p) and knowing the parameters—fiber diameter (d), wavelength of the guided light (λ), and refractive index of the fiber core (n_1)—we can write the reduction factor in the attenuation coefficient r_f (reduction factor), described by **Equation (9.23)**, as a function of n_2.

To facilitate the analysis, a claddless fiber is considered, where the diameter of the plastic fiber is $d = 1$ mm and $\lambda = 880$ nm, which allows a fixed value of $R = d/\lambda$ in **Equation (9.24)**.

It is possible to draw the characteristic of the variation of r_f as a function of the variation of the external index (n_2), as shown in **Figure 9.14**, where the angle $\theta_i = 70°$ is used, since it has the highest normalized power loss within the range allowed by angle limit.

With the normalized input power for the water refractive index equal to 1.33, an attenuation of approximately 0.15% is observed for the variation of the refractive index up to 1.39.

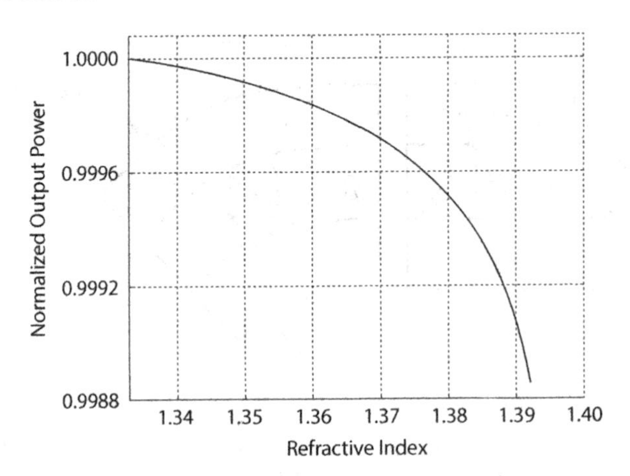

FIGURE 9.14 Characteristic curve of normalized output power drop, with variation of the refractive index as a function of the evanescent field in a fiber with a core diameter equal to 1 mm. (Adapted from Rodrigues, D.M.C., Desenvolvimento E Caracterização De Sensores A Fibra Óptica Plástica Para Refratometria Baseados Em Modulação De Amplitude, Tese Mestrado, Programa de Engenharia Elétrica, COPPE/UFRJ, 2013.)

9.4.3.2 Characteristics of the Output Power with the Variation of the Refractive Index in the Curve

The integral that defines the output power, **Equation (9.34)**, depends on the attenuation coefficient γ determined by the ratio between T and φ_p, according to **Equation (9.30)**.

Considering that in the case of a fiber without a cladding of 1 mm the ratio $r_0/(R + \rho)$ is approximately one, we have from **Equation (9.33)**:

$$\theta_\phi = \cos^{-1}\left(\cos\left(\theta_i\right)\right) \quad \textbf{(9.36)}$$

Using **Equation (9.21)** is θ_p.

The coefficient of transmission T is found from **Equation (9.12)**, considering that α is a function of the input angle. Integrating **Equation (9.34)**, one has:

$$P_s(\phi) = P(0)\left[R+\rho-R+\rho\right]\left[\theta_c+\theta_C\right]\exp(-\gamma\phi) \quad \textbf{(9.37)}$$

$$P_s(\phi) = 4P(0)\rho\theta_c \exp(-\gamma\phi) \quad \textbf{(9.38)}$$

Recalling that θ_c and γ depend on n_1 and n_2 and, in the present case, $\varphi = \text{cte} = \pi$, we can calculate the power as a function of the external index, $P\,(n_2)$ like this,

$$P_s(n_2) = 4P(0)\rho\theta_c(n_2)\exp\left[-\gamma(n_2)\pi\right] \quad \textbf{(9.39)}$$

It is also observed that, for a normalized output, it is necessary to calculate the initial power, that is, when n_2 is equal to its minimum value. In this case the output power is maximum, equal to one.

$$P(0) = P\left(n_{2\text{min}}\right) \quad \textbf{(9.40)}$$

$$1 = 4P\left(n_{2\text{min}}\right)\rho\theta_c\left(n_{2\text{min}}\right)\exp\left[-\gamma\left(n_{2\text{min}}\right)\pi\right] \quad \textbf{(9.41)}$$

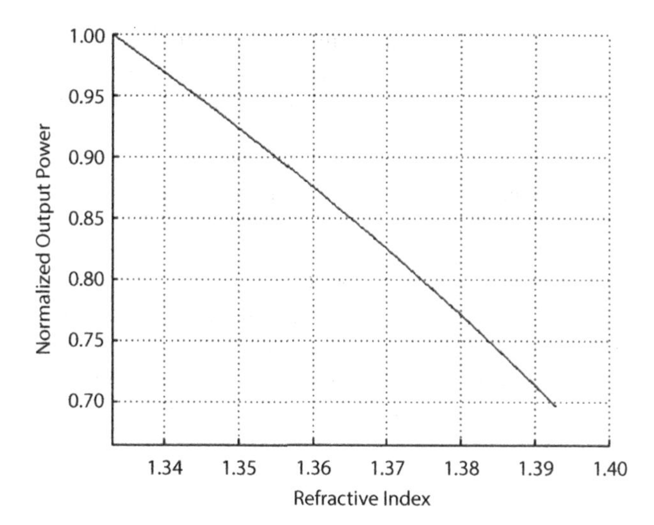

FIGURE 9.15 Normalized power characteristic at the output with the variation of the refractive index as a function of the curve in the fiber with a fixed angle of 180°.

$$P\left(n_{2_{\min}}\right)=\frac{1}{4\rho\theta_c\left(n_{2_{\min}}\right)\exp\left[-\gamma\left(n_{2_{\min}}\right)\pi\right]} \tag{9.42}$$

Figure 9.15 shows the power attenuation with the normalized refractive index for water (refractive index equal to 1.33). An attenuation of approximately 30% of the input for the variation of the index up to 1.39 is observed.

It is observed in this study that the attenuation of 30% in relation to the normalized input power using curved fiber (180°) is much more significant, than the attenuation of 0.15% in relation to the input power due to the change of the reduction factor (r_f) in the attenuation coefficient.

In this way, it is shown that the curved fiber characteristic presents a relevant contribution to the biological sensor approach, this attenuation being a quasi linear function, in addition to proving the proof of concept of the adopted methodology.

We can also observe the full functionality of the sensor in responding to variations in refractive indexes in the range of 1.33–1.39 and it can be applied to measure and monitor the index of refraction.

Once the review of the phenomena and concepts adopted in this specific study was completed, we are now beginning to approach the technologies used for the development of the biological sensor using POF based on SPR and thin film technologies.

9.5 SPR e LSPR Technologies

In this section we will study the main theoretical aspects of the techniques based on SPR-surface plasmon resonance (SPR) and localized surface plasmon resonance (LSPR), theoretical and experimental models, as well as the necessary conditions for the occurrence, the calculation of the evanescent field wave vector and the SPR wave vector, and finally, the proof of the existence of the phenomena arising from the respective technologies.

9.5.1 Surface Plasmon Resonance

The surface plasmon resonance (SPR) phenomenon has been studied for about 40 years and its theory is based on optical properties through a technique that does not require markers (radioactive or biochemical), which has a flexible experimental design, as well as automatic and fast response time. The use of SPR in biosensors and sensors is increasing significantly. Thousands of relevant publications are released, indicating enormous interest in this devoted research for an important subject [13,14].

Currently, the SPR technique has been widely used as an optical detection system, employing metal/dielectric surface phenomena aiming mainly at the measurement of the control and monitoring of refractive indices of liquid substances as a function of the variation of the angle of incidence of the light source [15–18].

Initially, it is necessary to define some terms, such as polaritons, phonons, and plasmons, which are fundamental for a perfect understanding of the phenomenon of superficial plasmon resonance.

Polaritons are mixed modes formed by two or more elemental excitations, such as, phonons, plasmons, and so on, where one of its components is the photon. Plasmons are charge densities (oscillations) from the coupling of the free electrons of a metal with the photons of a given light radiation [19–21].

9.5.2 Theoretical and Experimental Models Proposed for the Occurrence of SPR

The simplest theoretical model was proposed by Drude [22], which allowed prediction of the electrical and thermal conductivity of the metals through the imposition of two fundamental conditions. The first one portrays the field wavelength need, being larger than the free electron mean path of the free electrons of the metal. That is, making a critical but necessary choice of the metal to be used. The most suitable metals are the noble metals, such as silver, gold, copper, and aluminum, among which silver and gold are the metals that most attend the requirements and, consequently, are the most used [23]. The second condition is related to the frequency that a wave must have in order for the charge density to propagate. The natures of these charge densities, which are the plasmid oscillations themselves, can be understood very simply from the two experimental systems developed, described below.

For the occurrence of surface plasmon excitation, two different experimental systems were developed: one by Otto [24] and the other by Kretschmann [25]. However, the attenuated total reflectance configuration, developed by Kretschmann, is usually used in most SPR instruments for refractometry [16,26].

9.5.3 Conditions for Occurrence of Surface Plasmon Excitation by Kretschmann

The Kretschmann configuration, shown in **Figure 9.16**, is based on the phenomenon of total internal reflection. This phenomenon occurs when polarized light passes through an optical medium (glass) and reaches an interface between this medium and a lower optical density medium (air), and is reflected back into the denser medium. Although the incident light is fully reflected internally, a component of this radiation, wave, or photon penetrates the interface of the less dense medium, up to the limit distance of a specific wavelength. At a certain angle of incidence, when the plasmon wave vector is equal to the evanescent field wave vector ($K_{sp} = K_{ev}$), part of the radiation couples with the free oscillating electrons (plasmon) in the metal film, and the surface plasmon resonance occurs. As a result, there is a loss of energy from the incident light, resulting in the reduction of the light intensity reflected at that particular angle of incidence of the source. This attenuation can be detected with a photodetector [14].

In **Figure 9.16** the surface plasmon resonance is excited at the metal air/metal interface when the incident light angle (θ) is such that the evanescent component of the wave vector itself (K_{ev}) is equal to the wave vector of the surface plasma propagation (K_{sp}).

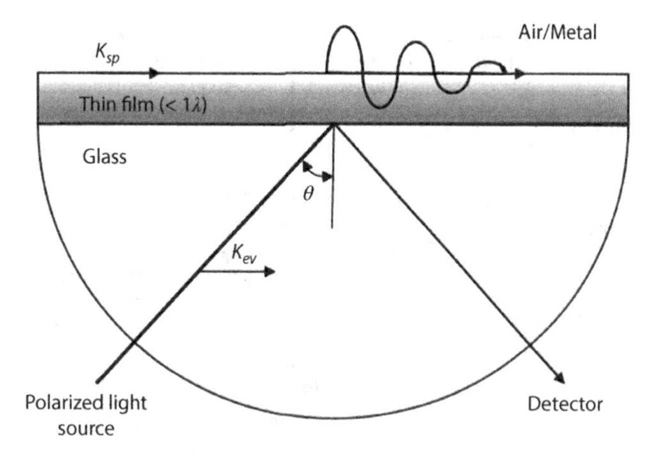

FIGURE 9.16 Basic configuration for SPR. (Adapted from Carvalho, R.M. et l., *Quím. Nova*, 26, 97–104, 2003.)

Then, we can summarize that, in order that the interaction between the light and the metallic surface generates the phenomena of plasmon resonance, the following conditions must be satisfied:

1. The metallic film must be between two dielectric media, and the refractive index of the dielectric through which the incident light is propagated must be greater than that of the dielectric on the other side of the film;
2. The light shall focus on the surface of the metallic film with an angle larger than the critical angle so that there is total internal reflection. In this condition, an evanescent field is generated perpendicular to the metal-dielectric interface that is responsible for exciting the surface plasmon;
3. Depending on the wavelength of the incident light, the plasmon resonates resulting in the maximum absorption of light;
4. The kind of metal is also important. Silver exhibits higher absorption [27]; however, gold is more used for its greater chemical stability and for not oxidizing as easily as silver;
5. The film thickness should be smaller than the range of the evanescent field.

Figure 9.17 shows the detailed Kretschmann configuration.

Looking at **Figure 9.17**, polychromatic light (a) impinges on a fixed angle on the interface of a prism and a metallic film, usually a thin film of gold (b) with 50 nm thickness. A beam of light excites coherent oscillations of the electrons of the valence band (c) causing an absorption of power of the light that is measured by the detector (d). If the incident angle is fixed and the light is polychromatic, resonance causes maximum absorption at only a certain wavelength. Through a spectrometer (e) it is possible to detect at what frequency the maximum absorption of reflected light occurs. By increasing the refractive index of the medium, the maximum absorption occurs at a longer wavelength, i.e., the minimum point in the transmission spectrum shifts to red. This is a method of frequency modulation detection [16,30].

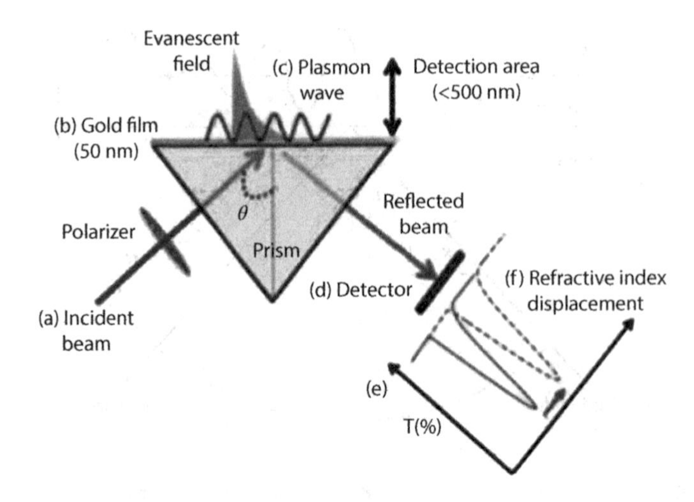

FIGURE 9.17 SPR sensing—Kretschmann configuration. (a) Light source; (b) gold film; (c) resonance produced; (d) detector; (e) spectrum; and (f) response of the spectrum to changes in the refractive index of the medium. (Adapted from Yanase, Y. et al., *Opt. Mater. Express*, 6, 1339, 2016; Arcas, A.S., Nanobiossensor A Fibra Óptica Revestido Com Filme Fino De Ouro Para Detecção Da Bactéria Escherichia Coli, Tese de Mestrado, Programa da Engenharia da Nanotecnologia, 2017.)

9.5.4 Calculation of the Evanescent Field Wave Vector and the Surface Plasmon Wave Vector

The wave vector of the evanescent field $\left(K_{ev}\right)$ is given by:

$$\left(K_{ev}\right) = (\omega_0 / c)\eta_g \operatorname{sen}\theta \tag{9.43}$$

where:

ω_0 is the frequency of incident light;
η_g is the refractive index of the dense medium (glass);
θ is the angle of incidence of the light;
c is the speed of light in vacuum.

The surface plasmon wave vector $\left(K_{sp}\right)$ can be approximated to:

$$K_{sp} = (\omega_0 / c)\left[(\varepsilon_m \eta_s^2)/(\varepsilon_m + \eta_s^2)\eta_s^2\right] \tag{9.44}$$

where ε_m is the dielectric constant of the metallic film and ε_m the refractive index of the dielectric medium.

The surface plasmon wave vector (K_{sp}) is dependent on the refractive index (**Equation 9.43**) of the water/air medium above the metal film, which can be monitored to the thickness of approximately 200 nm above the metal surface (thickness limit of detection of SPR).

If the index of refraction immediately above the surface of the metal undergoes some change, by changing from one substance to another substance with different refractive index, a change in the angle of incidence is required for the excitation of surface plasmon to occur. By monitoring this angle, resonance is observed. Variations around 0.5 milligrams corresponding to a resolution of approximately 1 x refractive index (for = 670 nm) can be detected [15,26], but the resolution depends on the wavelength of the incident light.

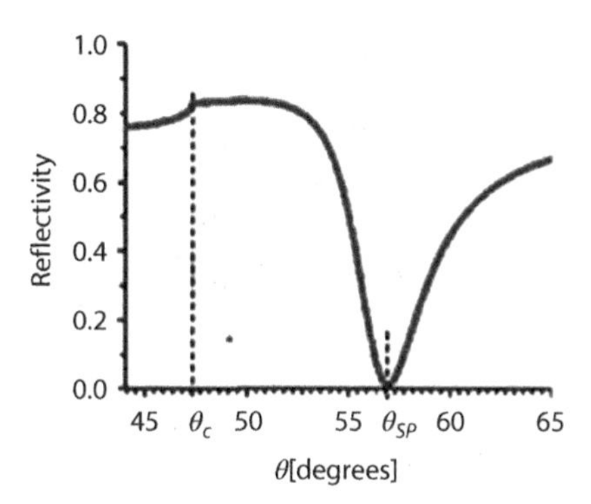

FIGURE 9.18 Characteristics of the plasmon resonance curve. (Adapted from Allil, R.C.S.B., Desenvolvimento de sensor óptico utilizando fibra plástica (POF) baseado em ressonância de plasmon superficial para aplicação em refratometria, Tese de Mestrado, Programa de Engenharia Biomédica, COPPE/UFRJ, 2004; Kretschmann, E., *Zeitschrift fur Physik*, 241, 313–324, 1971; Zhang, S. et al., *Biosens. Bioelectron.*, 15, 273–282, 2000.)

Surface plasmons can only be excited if at the interface the electric field has a normal component at the surface; then polarized light "*s*" is unable to produce surface plasmons. Such modes can only be excited by "*p*" polarized light.

9.5.5 Evidence of the Occurrence of the SPR Phenomenon

Experimentally this resonant coupling of SPR can only be observed by the high attenuation of the reflected optical intensity, as a function of the angle of incidence [31,32].

In **Figure 9.18** is shown the reflectivity curve, measured from a 50 nm Au film evaporated over a straight prism.

We observed some characteristics of the surface plasmon resonance curve [31,32]: For angles of incidence less than the critical reflection angle, the reflectivity is much higher. This is due to the fact that the metallic film acts partially as a mirror. The critical value is important for calibration of the measuring equipment.

For angles of incidence greater than $\theta_{critical}$, the reflectivity is less than the total internal reflection due to certain absorptions in the metal that remains, practically constant until it approaches the coupling angle of the surface plasmids θ_{SP}, where the reflectivity tends to decrease for very low values.

9.5.6 Comparison of Au and Ag Deposition as a Function of the Reflectivity Minimum

In **Figure 9.19** the reflectivity and electric field strength curves on the surface for Ag and Au deposition (with 50 nm thickness) [13,31,32] are shown.

It is observed that the minimum of reflectivity is more pronounced for silver, because silver is a more transparent metal than gold. However, the most commonly used metal for this purpose is gold, due to the enormous tendency to oxidize silver.

9.5.7 Localized Surface Plasmon Resonance (LSPR)

If the metal-dielectric interface is replaced by nanoparticles, the resonance phenomenon cannot propagate on the surface and remains localized. This phenomenon is called localized surface plasmon resonance (LSPR) and presents, as a great advantage, no dependence on the angle of incidence of light [33].

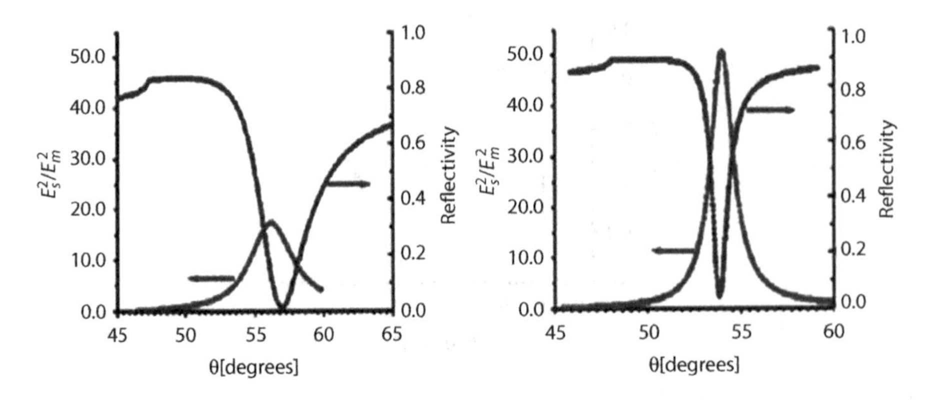

FIGURE 9.19 Comparison of the SPR resonance curve. (Adapted from Kretschmann, E., *Zeitschrift fur Physik*, 241, 313–324, 1971; Zhang, S. et al., *Biosens. Bioelectron.*, 15, 273–282, 2000.)

9.6 Thin Film Technology

In this section we will discuss the technology of thin films, including definitions and the main techniques that have been adopted and that have been contributing in the area of sensing.

9.6.1 Definition and Applications of Thin Films

Thin film technology is dedicated to the study of advanced materials in the form of thin films for surface analysis and interfaces [34]. This technique involves minute processes capable of producing films with layers of micra or angstrom sizes.

Fine films are coating layers applied to the surface of optical components (substrates), such as lenses, prisms, mirrors, and filters. They alter the light incident on the component and can perform several analyses involving surface phenomena and interfaces [34].

This meets the great demand in the field of optics for cutting-edge technological applications in optoelectronic devices, in the microelectronics industry, and in coatings of metals and metal alloys [35].

Basically, films are formed by the condensation (solidification) of atoms or molecules of a vapor on the substrate. The condensation process begins by the formation of small agglomerates of material, called nuclei, randomly spread over the surface of the substrate. Electrostatic attraction forces are responsible for fixing atoms to the surface. The fixation mechanism is called chemical adsorption, when the electron transfer occurs between the substrate material and the deposited particle and physical adsorption, if this does not occur. The binding energy associated with the chemical adsorption varies from 8 to 10 eV and that associated to the physical adsorption is approximately 0.25 eV. Adsorbed atoms migrate on the surface of the substrate interacting with other atoms to form the nuclei in a process called nucleation. As more atoms interact, the nuclei grow and consequently, they come in contact with other nuclei causing the coalescence that results in larger structures. The process has continuity in the formation of channels and holes in the exposed substrate, which are filled with new nuclei until the formation of a continuous thin film [35].

9.6.2 Mechanical Properties of Thin Films

Two fundamental mechanical properties in the production of a thin film are adhesion and tension.

The adhesion (greater contact area) of a thin film deposited on a substrate should be very good. However, poor adherence (detachment) can lead to faulty device behavior. Adhesion is related to two important factors, which are the cleaning procedures and the roughness of the substrate. A certain roughness can increase the adhesion. However, excessive roughness can lead to defects in coverage, thereby impairing adhesion. The adhesion can be, qualitatively, checked by sticking an adhesive tape onto the surface. When removing the tape the film should remain intact on the substrate. Another method is to scrape the surface of the film with a chrome-steel tip with varying tensions until the film is removed. This critical stress provides information on adherence [35].

The internal tension of a film can be classified as compression or expansion. Films with compressive stresses tend to expand, parallel to the surface of the substrate. In extreme cases they may form protrusions on the surface. And films with expansion tensions tend to contract parallel to the substrate, and may show cracks when they exceed their coefficient of elasticity.

The total voltage of a film results from the sum of three voltages: thermal voltage, which results from the different coefficients of expansion of the film and the substrate, external voltage, which can come, for example, from another film, and intrinsic voltage, which is related to the structure of the film and therefore is strongly dependent on parameters such as deposition temperature, thickness, deposition rate, process pressure, and type of substrate [35].

9.6.3 Equipment Used for the Production of Thin Films

Thin films are manufactured in special equipment called evaporators. This name comes from the fact that the films are produced from the high vacuum evaporation of dielectric materials through different deposition techniques. It is extremely sophisticated equipment, expensive, and requires a very high investment in installation, maintenance, and raw materials.

The properties of the thin films are directly related to and dependent on the deposition processes chosen. The miniature geometries of the devices must have a reliable thickness, good adhesion, and roughness [29,35].

The evaporators are systems composed of a stainless chamber, a diffusion pump, a mechanical pump whose purpose is to aid the diffusion pump, making the primary vacuum of the chamber, and a system of valves and pipes for vacuum configuration [34].

9.7 Biological Sensing

Finally, after the previous reviews of relevant issues, we can now address the development of a biological sensor.

We present the step-by-step technique for development POF biosensor based on evanescent field and on SPR and thin film technologies for *Escherichia coli* detection.

9.7.1 Working Principle

A bacterium presents an RI of about 1.39 at 400–800 nm [36–38], slightly higher than that of pure water (1.333). Nonetheless, even in heavily contaminated water, with 10^8 cells/mL, for instance, the bacteria concentration is not high enough to change the RI of the water substantially. Therefore, in order to detect bacteria by RI sensing, the detection system needs to concentrate the target bacteria around the fiber.

The technique known as immunocapture consists of covalently bonding the specific antibody on the sensor sensitive region in such a way that solely bacteria of specific species are captured and fixed on the sensor surface [39–41].

The proposed sensor works by the measurement principle of optical power loss in a U-shaped POF. The light conducted by the optical fiber is attenuated depending on the refractive index of sensor surrounding medium. In the shaped region, two effects lose part of the guided light: bending loss and surface plasmon resonance (SPR).

In a multi-mode optical fiber, the light propagation takes place by the principle of total internal reflection (TIR); as a consequence, the incidence angle of the traveling optical ray is required to be higher than that of the critical angle. In the core-cladding interface, an EW is generated that extends with a range from 50 to 1000 nm into the cladding depending on the wavelength, the refractive index, and the angle of incidence [42].

At a fiber bend, the higher order propagation modes (light rays reflecting at angles close to the critical angle) reach the core-cladding interface in an angle smaller than the critical one and refract into the cladding instead of reflecting by TIR. At the cladding interface with the surrounding medium there is another critical angle so that all cladding modes that reach this interface at smaller angles are lost forever into the surrounding medium. Thus, an increase in RI in the surrounding of a U-shaped POF increases the critical angle of the cladding interface, which results in the refraction loss of more propagation modes and, as a consequence, more attenuation of the light. The attenuation phenomenon by the fiber bend is known as bending loss. An illustration of the bare U-shaped POF probe operating by bending loss effect and with gold-coated operating by SPR effect in the presence of bacteria is shown in **Figure 9.20** [40,43–46].

Covering the U-shaped POF probes with gold film, if the layer is thin enough to be transparent to a certain light wavelength, the attenuation is still predominantly imposed by the effect of bending loss. In this case, the sensor can be used in the same way as the previous one, but it is endowed with a layer of gold for better immobilization of antibodies. However, for specific wavelengths, the SPR effect absorbs these wavelengths causing light attenuation. SPR is the resonant oscillation of conduction electrons at the interface between a noble metal and a dielectric material stimulated by incident light [47,48]. The SPR effect depends on the light wavelength, the incident angle, film thickness, and surrounding RI [49–51].

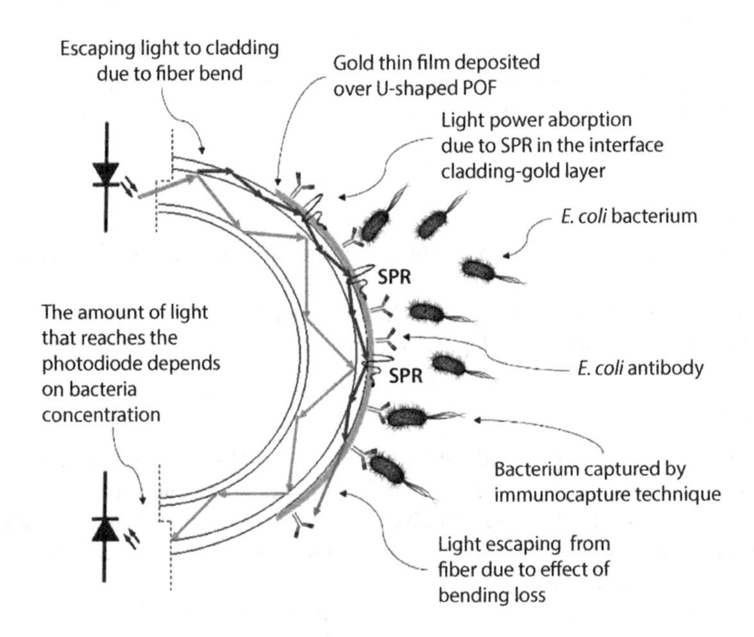

FIGURE 9.20 Working principle illustration of a U-shaped gold-coated biosensor.

9.7.2 Probe Fabrication

U-shaped probes were fabricated using Mitsubishi Rayon Eska GH 4001 multi-mode POF with 1 mm in diameter and cladding thickness of 10 μm. The core material is polymethyl methacrylate (PMMA) with RI = 1.49 in the visible range and the cladding material is fluorinated polymer with RI = 1.41. The fabrication process consisted in cutting the POF into about 10-centimeter-long section and both end surfaces were cleaved and polished for maximum light coupling. Then, the 10 cm-long sections were shaped and placed in a manually operated custom-made device (**Figure 9.21a**) for molding the U-shaped POF in a curve with 8 mm in diameter (**Figure 9.21c**). The fiber was heated by a hot-air blower gun (**Figure 9.21b**) for 15 s. During the molding procedure, the temperature was maintained below 70°C to not reach the fiber melting point.

A radio-frequency (RF) magnetron sputtering system (Aja International, USA) was used for gold thin film deposition on the U-shaped probe surface (**Figure 9.22**). A 40 W RF power was applied to ionize the argon gas (plasma) inside the vacuum sputtering chamber. The argon flow rate was maintained at 12 sccm (standard cubic centimeters per minute) at a pressure of 4×10^{-3} mbar. The gold target size was 25.4 mm in diameter and 0.3 mm in thickness.

After molding, the U-shaped probes were cleaned with isopropyl alcohol for two minutes (not exceeding this time in order to avoid fiber cracking), washed in ultrapure water, and dried with ultrapure nitrogen. Then they were attached to an acrylic support fixed in the substrate holder 10 cm above the gold target and tilted 45° from vertical, as shown in **Figure 9.22a**. The substrate holder was rotated at 20 rpm during the deposition process to improve film thickness uniformity. Under these conditions, the gold deposition rate is about 3.5 nm/min and layers of 3.5, 7, 10, 18, 30, 50, 70, and 100 nanometers were deposited in several U-shaped probes (**Figure 9.22b**).

After deposition, probe ends were cleaved with the device shown in **Figure 9.22c** and polished with a 3-micrometer polish film as illustrated in **Figure 9.22d**.

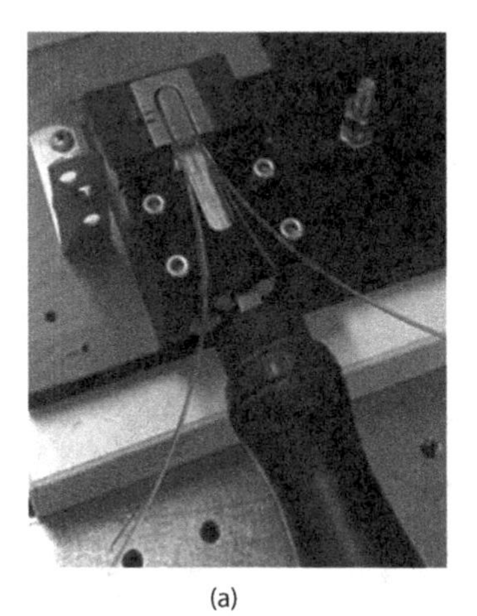

(a)

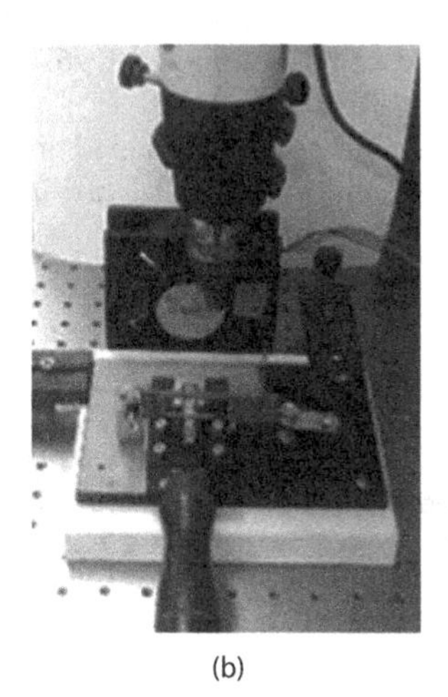

(b)

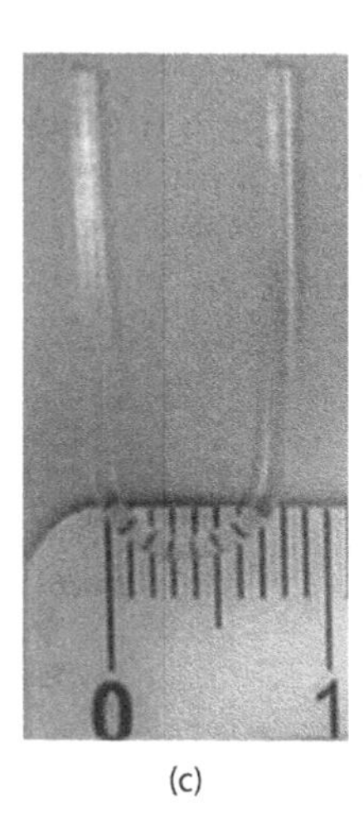

(c)

FIGURE 9.21 (a) Custom-made device used for U-shaped POF probe molding; (b) hot-air blower gun used for heating; and (c) the finished U-shaped POF fiber 8 mm in diameter.

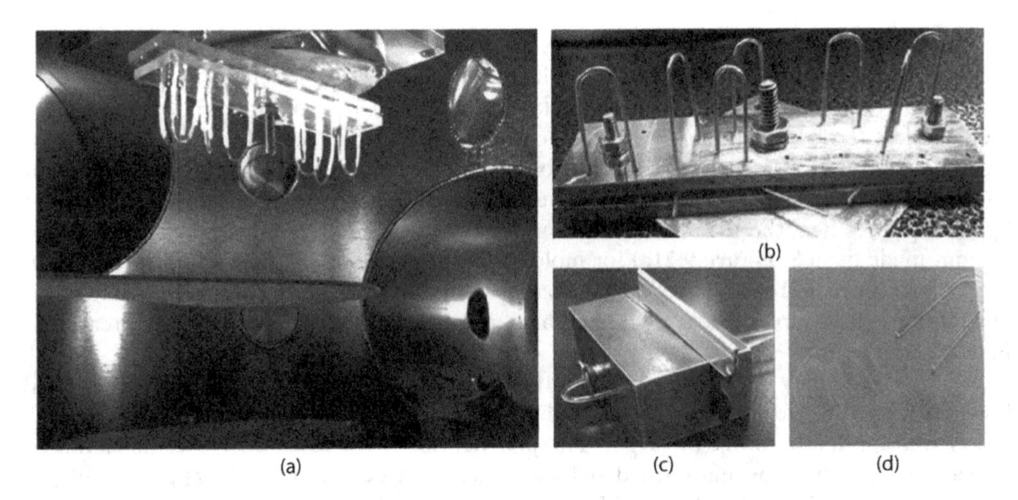

FIGURE 9.22 (a) Bare U-shaped POF probes installed inside the vacuum sputtering chamber 10 cm above gold target and tilted 45° from vertical; (b) U-shaped POF probes coated with 70 nm gold thin film; (c) custom-made device used to cleave the probe ends; and (d) end faces were polished with a 3-micrometer polish film in figure-of-eight movements.

9.7.3 Optoelectronic Setup

This setup is responsible for reading and acquiring biological sensor data through an optoelectronic instrumentation shown in **Figure 9.23**. The setup consists of a spectrometer and a white light source (HR-4000, HL-2000, Ocean Optics, USA, respectively) connected to both ends of the U-shaped probe. The block diagram and its picture are shown in **Figure 9.23** (left) and (right), respectively.

9.7.4 Sucrose Solutions for RI Testing

Aqueous sucrose solutions with RI varying from 1.33 to 1.39 were prepared in ultrapure water for evaluating sensor sensitivity and repeatability. The solutions refractive index was measured at 589 nm wavelength by an Abbe refractometer (Q767BD, Quimis, Brazil) with an uncertainty of 0.0002 refractive index unit (RIU).

In this experiment we used gold-coated U-shaped probes with 10 and 70 nm gold thicknesses than were tested in six sucrose solutions with RI varying from 1.33 to 1.39, randomly. For each concentration, 30 readouts were taken.

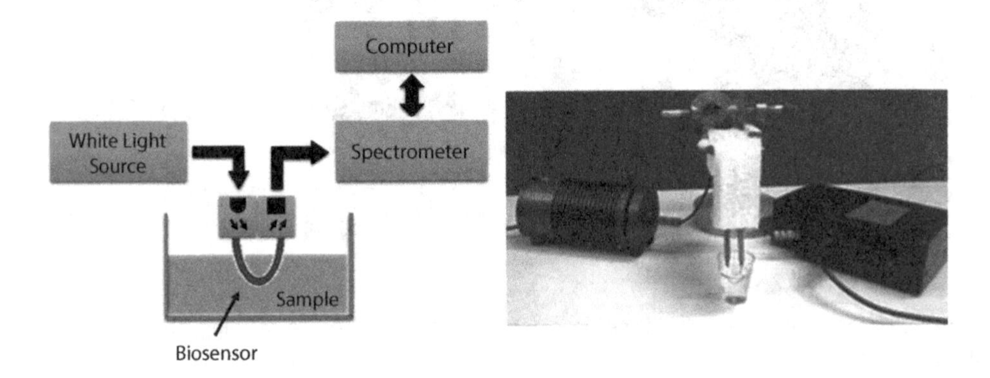

FIGURE 9.23 Left: Block diagram; Right: Picture of the setup. The white light source is at the left and the spectrometer at the right.

Straight lines were fitted to experimental data by linear regression for determining a best linear relationship for sensor response versus RI. The expanded uncertainties with 95% confidence level were obtained by the mean standard deviations multiplied by the student's t-distribution k-factors, referring to the degrees of freedom used in the tests.

9.7.5 Antibody Immobilization Protocol

The U-shaped POF probes were immobilized with rabbit anti-*E. coli* polyclonal IgG (Bio-Rad, Brazil). The antibody immobilization protocol was adapted from [52], which will be briefly described here. First, the probes were cleaned in isopropyl alcohol P.A. for 2 min and after that in ultrapure water. Then, they were immersed in a 4 mM cysteamine solution for 1.5 h that acts as a covalent bond between the gold surface and the antibody. After chemisorption of cysteamine on gold, the probes were washed in absolute ethyl alcohol and rinsed in ultrapure water twice and were ready for antibody immobilization.

The next step is the antibody immobilization. This protocol consists of the following steps: a solution of 100 mM 1-Ethyl-3-(3-dimethylaminopropyl) carbodiimide (EDAC) and 40 mM N-Hydroxysuccinimide (NHS) in phosphate-buffered saline (PBS) were prepared for activation of the carboxylic groups. Thereafter, an antibody concentration of 10 µg/mL was added in the solution and left to rest for 1 h. At the end of this period, the U-shaped POF probes were immersed in a bovine serum albumin (BSA) solution for 30 min. Subsequently, they were washed in PBS solution three times to remove unbound antibody molecules and kept immersed in PBS for 2 h (PBS, EDAC, and NHS supplied by Sigma-Aldrich and BSA supplied by Merck).

9.7.6 Bacterial Suspension and Testing

Cultures of *E. coli* O55 bacteria (Osvaldo Cruz Foundation, Fiocruz, Brazil) were grown in tryptone soya agar (TSA). For selectivity tests, cultures of *Enterobacter cloacae*, *Salmonella typhimurium* bacteria (Laboratory of Food Microbiology, UFRJ, Brazil), and *Bacillus subtilis* (Osvaldo Cruz Foundation, Fiocruz, Brazil) were grown in soybean triptych broth (TSB) and TSA.

Bacterial suspensions were prepared using ultrapure water and NaCl P.A. (Synth, Brazil). The concentrations were determined using the McFarland 0.5 turbidity standard by visual comparison which correspond to an *E. coli* concentration of 1.5×10^8 CFU/mL [53,54] and confirmed by absorbance measurements in the spectrophotometer (UV-1800, Pro-Tools). The McFarland 0.5 scale has an absorbance value between 0.08 and 0.1 for a wavelength of 625 nm and optical path of 1 cm [53,54]. All bacterial suspensions were prepared using a solution of 0.85% sodium chloride.

Bacteria suspensions with concentrations from 10^3–10^8 CFU/mL were prepared for testing the biosensors. We adopted just the gold-coated U-shaped probes with 70 nm thickness. They were immersed into a 5 mL beaker with bacteria suspension. Readouts were carried out at intervals of 5 min for up to 60 min.

After the bacterial suspensions testing, the sensors were submitted to safranin solution for 5 s for cell coloring. Then, they were washed with ultrapure water and let to dry at room temperature. After drying, the sensors were taken to an optical microscope with a 40x magnification. Images were produced with a 13-megapixel camera. The bacterial coverage area was estimated by the ImageJ software.

9.7.7 Results and Discussion

9.7.7.1 RI Measurements—Sucrose Solutions Testing

The setup, as shown in **Figure 9.23**, provides a transmittance spectrum as response. Tests with aqueous sucrose solutions were carried out with bare, 10–70 nm gold-coated probes. The 10 nm gold-coated probe, in which the effect of bending loss is more pronounced, was chosen because it did not compromise much the sensitivity, apart from needing a smaller amount of gold. The 70 nm gold-coated probe, in which the SPR effect is more pronounced, was chosen because it had a reasonable linear response for 1.33–1.38 RI range and used a smaller amount of gold than the 100 nm one.

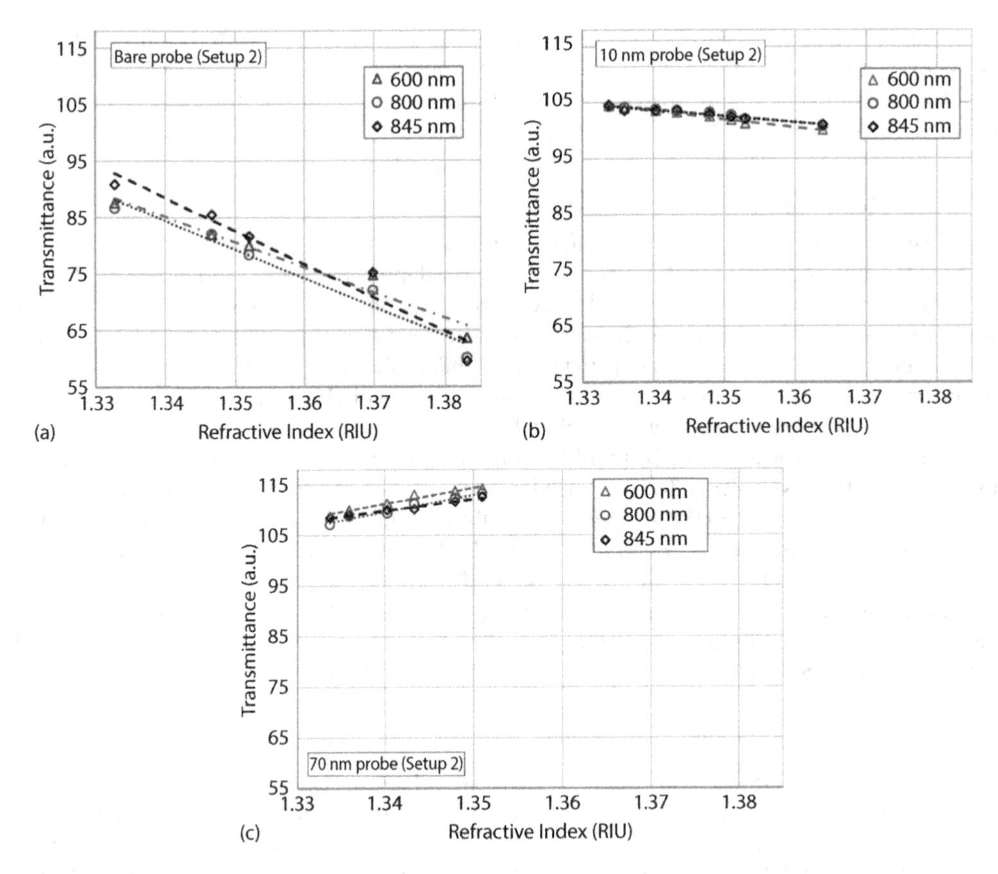

FIGURE 9.24 Sensor responses for RI variations using Setup 2 for wavelengths 600, 800 and 845 nm. (a) Bare U-shaped probe; (b) 10 nm gold-coated probe; and (c) 70 nm gold-coated probe.

Figure 9.24 shows the sensor responses for 600, 800, and 845 nm wavelengths, which presented highest sensitivity. By observing **Figure 9.24a**, it can be noticed that the bare probe presented a linear and indirectly proportional relationship between sensor response and RI changes, as expected. **Figure 9.24b** shows the 10 nm gold-coated sensor responses that present the same behavior of the bare probe sensor for a RI range of 1.33–1.38, however, with a lower sensitivity due to the gold layer absorption.

Figure 9.24c shows the 70 nm gold-coated sensor responses that presented linear and directly proportional at RI variations for the range of 1.33–1.352. The sensitivity was better than that of 10 nm gold-coated probe.

Table 9.1 shows a summary of the sensitivity and maximum expanded uncertainty (95% of confidence level) of the inverse regression for each gold thickness.

Table 9.1 Summary of the Biosensor Sensitivities and Maximum Expanded Uncertainties (95% Confidence Level) in the Setup 2

Gold Thickness (nm)	Sensitivity (1/RIU)	Uncertainty (RIU)
None	−591.78 (845 nm)	5.8×10^{-4}
10	−148.66 (600 nm)	5.1×10^{-4}
70	338.26 (800 nm)	5.6×10^{-4}

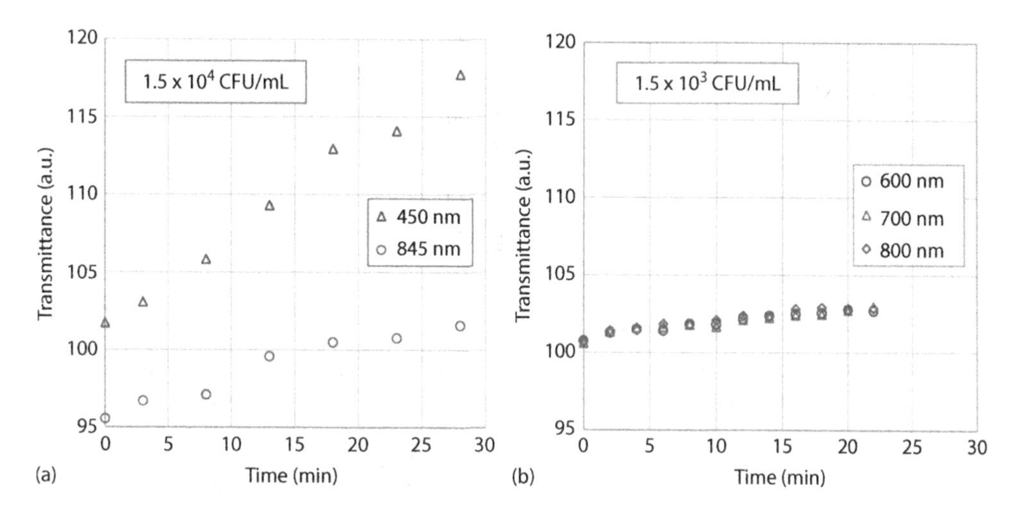

FIGURE 9.25 Sensor responses for bacterial suspensions using Setup 2. (a) 70 nm gold-coated biosensors in a 1.5×10^4 CFU/mL bacterial suspension; and (b) 70 nm gold-coated biosensors in a 1.5×10^3 CFU/mL bacterial suspension.

9.7.7.2 Bacterial Measurements

Probes immobilized with antibodies were tested in different bacterial suspensions in order to evaluate response time and sensitivity. For these tests the 10–70 nm gold-coated U-shaped probes were chosen.

9.7.7.3 Bacterial Suspensions Testing

For bacterial suspensions testing in Setup 2, the 70 nm gold-coated probes were chosen. **Figure 9.25a** and **b** show the sensor responses for bacteria concentration of 1.5×10^4 CFU/mL and 1.5×10^3 CFU/mL, respectively.

Notice in **Figure 9.25** that the sensor was able to detect a concentration of 1.5×10^3 CFU/mL. The sensor output voltage increased with time due to the increase in the surrounding RI, as expected. As the response was a transmittance spectrum, it was possible to choose the wavelength in which the sensitivity was higher. Tests with 880 nm were not possible as the wavelength range of the spectrometer ends at 850 nm. It has been shown that the system is capable of providing positive response to the bacterial concentration in less than 10 min. Notice that the higher the bacteria concentration, the more rapidly the output signal rose.

9.7.7.4 Cross Sensitivity Testing

In order to verify sensor selectivity, tests were carried out with the bacteria *Enterobacter cloacae*, *Salmonella typhimurium*, and *Bacillus subtilis*.

Figure 9.26a shows the results of the 70 nm probe tested in Setup 2 for *Salmonella typhimurium* suspensions at a concentration of 1.5×10^8 CFU/mL. The graphs show the results for 600–845 nm wavelengths. It is observed that, in this case, the sensor detected the presence of bacteria other than *E. coli*. It is assumed that this occurred due to the antibody being a polyclonal one rather than a monoclonal, which has high specificity and lower cross-reactivity.

Lastly, **Figure 9.26b** shows the results of the 70 nm U-shaped biosensor tested with *Bacillus subtilis* suspensions at a concentration of 1.5×10^8 CFU/mL. The graphs show the results for 600–845 nm wavelengths. It is observed that there is no tendency, that is, the sensor was not able to capture the bacteria even in high concentration.

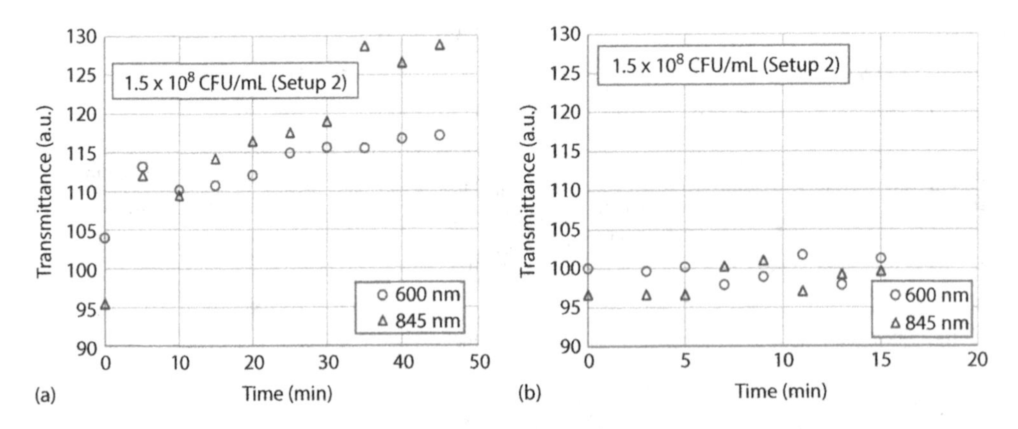

FIGURE 9.26 Sensor responses for 70 nm gold-coated probe for selectivity testing in a 1.5×10^8 CFU/mL bacteria concentration with Setup 2 for 600 and 845 nm wavelengths. (a) *Salmonella typhimurium*; and (b) *Bacillus subtilis*.

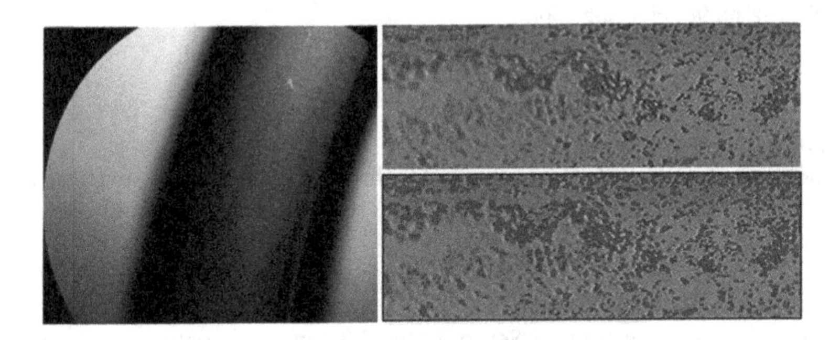

FIGURE 9.27 Optical microscope image (40x) of *E. coli* coverage over the 70 nm probe after immersion in a bacteria suspension at 1.5×10^8 CFU/mL concentration. At left, the probe microscope image. At upper right, a 40x magnification image of the bacteria adhered to the fiber surface colored by safranin solution. At bottom right, the coverage area in red estimated by the ImageJ. The coverage area estimated by ImageJ software was 16%.

9.7.7.5 Bacterial Adhesion on Sensor Surface

Figure 9.27 shows the image obtained by optical microscopy of the bacteria adhesion on the surface of the sensor tested under an *E. coli* concentration of 1.5×10^8 CFU/mL. The coverage area estimated by ImageJ software was 16%. This small coverage indicates that immunocapture could be improved to further reduce the detection limit.

9.8 Conclusions

A gold-coated U-shaped POF biosensor for *E. coli* detection has been presented. This technique allows using POF as a RI sensor even without removing the cladding.

U-shaped probes coated with gold layers up to 18 nm thick present the same behavior as bare U-shaped probes, in which the bending loss is the sensing predominant effect. Probes coated with 30–50 nm gold layer are not useful for RI sensing in the range of this application. In 70–100 nm gold-coated probes, the SPR effect is predominant and the biosensor was capable to detect bacteria concentrations as low as of 1.5×10^3 CFU/mL.

The results demonstrated that the proposed biosensor can be an efficient, low cost, and portable tool for routine analysis of drinkable and bathable water and food quality. This technique can be applied to other types of bacteria by simply immobilizing the antibody of the bacteria to be detected.

In future work, efforts should be devoted to improving the bacteria immobilization and thereby expanding the coverage area on the fiber; a new high-resolution optoelectronic setup operating with an LED whose wavelength has greater sensitivity should be developed; and the selectivity could be improved by using monoclonal instead of polyclonal antibodies.

References

1. Leung, A., Shankar, P. M., Mutharasan, R., "A review of fiber-optic biosensors," *Sensors and Actuators B*, 125, pp. 688–703, 2007.
2. Zilio, S. C., *Optica Moderna, Fundamentos e Aplicações,* Fotônica, IFSC-USP, 2010.
3. Griffiths, D. J., *Introduction to Electrodynamics*, 3rd ed, Upper Saddle River, New Jersey, NJ, Ed. Prentice Hall, 1999.
4. Hecht, E., *Optics*, 4th ed, Ed. New York, Addison Wesley, 2002.
5. Snyder, W. A., Love, J.D., *Optical Waveguide Theory*, New York, Ed. Chapman and Hall, 1983.
6. Ferreira, A. P. *Bacteriosensor—Tecnologia de sensoriamento bacteriológico a fibra óptica.* Tese de Doutorado, Programa de Pós-Graduação de Engenharia COPPE/UFRJ, Rio de Janeiro, Brazil, 2000.
7. Culshaw, B., Dakin, J., *Optical Fiber Sensors Components and Subsystems.* Vol. 3. Norwood, MA, Ed. Artech House, 1996.
8. Nunes, F. D, *Fibras e Dispositivos para Comunicação Ópticas*, São Paulo, Brazil, Editora Renovarum Ltda, 2001.
9. Kitano, C., Sensores de Amplitude em Fibras Ópticas, Universidade Estadual Paulista "Júlio De Mesquita Filho" Faculdade De Engenharia De Ilha Solteira, Agosto 2017. (www.feis.unesp.br/Home/departamentos/engenhariaeletrica/optoeletronica/sensor-de-intensidade.pdf)
10. Kitano, C., Tecnologia de Fibras Ópticas, Universidade Estadual Paulista "Júlio De Mesquita Filho" Faculdade De Engenharia De Ilha Solteira, Agosto 2017. (https://docplayer.com.br/61011094-1-estrutura-geral-da-fibra-optica.html).
11. Rodrigues, D. M. C., Desenvolvimento E Caracterização De Sensores A Fibra Óptica Plástica Para Refratometria Baseados Em Modulação De Amplitude, Tese Mestrado, Programa de Engenharia Elétrica, COPPE/UFRJ, Abril 2013.
12. Stewart, G., Culshaw, B., "Optical waveguide modelling and design for evanescent field chemical sensors," *Optical and Quantum Electronics*, 26, pp. 249–259, 1994.
13. Allil, R. C. S. B., Desenvolvimento de sensor óptico utilizando fibra plástica (POF) baseado em ressonância de plasmon superficial para aplicação em refratometria, Tese de Mestrado, Programa de Engenharia Biomédica, COPPE/UFRJ, Março 2004.
14. Carvalho, R. M., Rath, S., Kubota, L.T., "SPR—Uma nova ferramenta para Biossensores," *Química Nova*, 26(1), pp. 97–104, 2003.
15. Jonsson, U., Fagerstam, L.I., Jonsson, B. et al., "Real-time biospecific interaction analysis using surface plasmon resonance and a sensor chip technology," *Biotechniques*, 11, p. 620, 1991.
16. Homola, J., Yee, S. S., Gauglitz, G., "Surface plasmon resonance sensors: Review," *Sensors Actuators B*, 54, pp. 3–15, 1999.
17. O' Brien, M. J., Pérez-Luma, V. H., Brueck, S. R. et al., "A surface plasmon resonance array biosensor based on spectroscopic imaging," *Biosensors Bioelectronics*, 16, p. 97, 2001.
18. Rich, R. L., Myszka, D. G., "Survey of the 1999 surface plasmon resonance biosensor literature," *Journal of Molecular Recognition*, 13, pp. 388–407, 2000.
19. Raether, H., *Surface Plasmon on Smooth and Rough Surfaces and on Gratings*, Berlin, Germany, Springer, 1988.
20. Agranovich, V. M., *Surface Polaritons*, North Holland, Amsterdam, the Netherlands, 1982.
21. Yeh, P. *Optical Waves in Layered Media*, New York, John Wiley & Sons, 1988.
22. Rich, L. R., Myszka, D. G., "Advances in surface plasmon resonance biosensor analysis," *Current Opinion in Biotechnology*, 11, p. 54, 2000.

23. Steiner, G., Sablinskas, V., Hubner, A. et al., "Surface plasmon resonance imaging of microstructured monolayers," *Journal of Molecular Structure*, 509, p. 265, 1999.
24. Otto, A., "Excitation of surface plasma waves in silver by the method of frustrated total reflection," *Zeitschrift für Physik A Hadrons and nuclei*, 216, pp. 398–410, 1968.
25. Kretschmann, E., Kroger, E., "Reflection and transmission of light by rough surfaces, including results for surface-plasmon effects," *Journal of the Optical Society of America*, 65, p. 150, 1975.
26. Green, R. J., Frazier, R. A, Shakesheff, K. M. et al., "Surface plasmon resonance analysis of dynamic biological interactions with biomaterials," *Biomaterials*, 21, p. 1823, 2000.
27. Trouillet, A., Ronot-Trioli, C., Veillas, C., Gagnaire, H., "Chemical sensing by surface plasmon resonance in a multimode optical fibre," *Pure and Applied Optics: Journal of the European Optical Society Part A*, 5, pp. 227–237, 1996.
28. Yanase, Y., Sakamoro, K., Kobayashi, K., Hide, M., "Diagnosis of immediate type allergy using surface plasmon resonance," *Optical Materials Express*, 6(4), p. 1339, 2016.
29. Arcas, A. S., Nanobiossensor A Fibra Óptica Revestido Com Filme Fino De Ouro Para Detecção Da Bactéria *Escherichia coli*. Tese de Mestrado, Programa da Engenharia da Nanotecnologia, Maio 2017.
30. Khanna, V. K., *Nanosensors: Physical, Chemical, and Biological*, CRC Press, Boca Raton, London, New York, 2011.
31. Kretschmann, E., "Die Bestimmung Optischer Konstanten von Metallen Durch Anregung von Berflachenplasmaschwingungen," *Zeitschrift fur Physik*, 241, pp. 313–324, 1971.
32. Zhang, S., Wright, G., Yang, Y., "Materials and techniques for electrochemical biosensor design and construction," *Biosensors and Bioelectronics*, 15, pp. 273–282, 2000.
33. Camara, A. R., Gouvêa, P. M, Dias, A. C., Braga, A. M., Dutra, R. F., De Araújo, R. E, Carvalho, I. C., "Dengue immunoassay with an LSPR fiber optic sensor," *Optics Express*, 21(22), pp. 27023–27031, 2013.
34. Santos, A. L. N., Fabio, L., Deposição de Filmes de Ti02 pelo método de Evaporação por feixe de elétrons, Monografia de fim de curso, Instituto Militar de Engenharia-IME, Rio de Janeiro, 2002.
35. Maissel e GLANG, Handbook of Thin Film Technology, Mc Graw-Hill Book Company, NY, 1970.
36. Beres, C., Nazaré, F., Souza, N., Miguel, M., Werneck, M., "Tapered plastic optical fiber-based biosensor—Tests and application," Biosensors Bioelectronics, 30, pp. 328–332, 2011. doi:0.1016/j.bios.2011.09.024.
37. Balaev, A. E., Dvoretski, K. N., Doubrovski, V. A., Refractive index of Escherichia coli cells. In *Proceedings of SPIE*, 2002, Vol. 4707, Saratov Fall Meeting 2001, Optical Technologies in Biophysics and Medicine III, Saratov, Russian Federation. doi:10.1117/12.475627.
38. Bryant, F. D., Seiber, B. A., Latimer, P., "Absolute optical cross sections of cells and chloroplasts," *Archives of Biochemistry and Biophysics*, 135, pp. 97–108, 1969.
39. Wandermur, G., Rodrigues, D., Allil, R., Queiroz, V., Peixoto, R., Werneck, M., Miguel, M., "Plastic optical fiber-based biosensor platform for rapid cell detection," *Biosensors Bioelectronics*, 54, pp. 661–666, 2014. doi:10.1016/j.bios.2013.11.030.
40. Werneck, M., Lopes, R., Costa, G., Rodrigues, D., Arcas, A., Dutra, F., Queiroz, V., Allil, R., "POF Biosensors Based on Refractive Index and Immunocapture Effect." In *Fiber Optic Sensors*; Matias, I. R., Ikezawa, S., Corres, J., Eds.; Springer International Publishing: Cham, Switzerland, 21, pp. 69–93, 2017.
41. Baccar, H., Mejri, M. B., Adams, C. P., Aouini, M., Obare, S. O., Abdelghani, A., "Functionalized gold nanoparticles for biosensors application," *Sensors Letters*, 9, pp. 2336–2338, 2011. doi:10.1166/sl.2011.1744.
42. Gouveia, C. A. J., Baptista, J. M., Jorge, P. A. S., "Refractometric Optical Fiber Platforms for Label Free Sensing." In *Current Developments in Optical Fiber Technology*; Harun, S. W., and Arof H., Ed.; InTech, Rijeka, Croatia, 2013. doi:10.5772/55376.
43. Chen, C.-H., Tsao, T.-C., Tang, J.-L., Wu, W.-T., "A multi-D-shaped optical fiber for refractive index sensing," *Sensors*, 10, pp. 4794–4804, 2010. doi:10.3390/s100504794.
44. Arrue, J., Zubia, J., "Analysis of the decrease in attenuation achieved by properly bending plastic optical fibres," *IEE Proccedings—Optoelectron*, 143, p. 135, 1996. doi:10.1049/ip-opt:19960181.
45. Durana, G., Zubia, J., Arrue, J., Aldabaldetreku, G., Mateo, J., "Dependence of bending losses on cladding thickness in plastic optical fibers," *Applied Optical*, 42, pp. 997–1002, 2003.

46. Gloge, D., "Bending loss in multimode fibers with graded and ungraded core index," *Applied Optical*, 11, pp. 2506–2513, 1972.
47. De Carvalho, R. M., Rath, S., Kubota, L. T., "SPR-Uma nova ferramenta para biossensores," *Quimica Nova*, 26, pp. 97–104, 2003.
48. Navarrete, M.-C., Díaz-Herrera, N., González-Cano, A., Esteban, Ó. A., "Polarization-independent SPR fiber sensor," *Plasmonics*, 5, pp. 7–12, 2010. doi:10.1007/s11468-009-9108-0.
49. Christopher, C., Subrahmanyam, A., Sai, V. V. R. Gold Sputtered U-Bent Plastic Optical Fiber Probes as SPR- and LSPR-Based Compact Plasmonic Sensors. Plasmonics 2017. doi:10.1007/s11468-017-0535-z.
50. Mitsushio, M., Miyashita, K., Higo, M., "Sensor properties and surface characterization of the metal-deposited SPR optical fiber sensors with Au, Ag, Cu, and Al," *Sensors Actuators Physics*, 125, pp. 296–303, 2006. doi:10.1016/j.sna.2005.08.019.
51. Cennamo, N., Massarotti, D., Conte, L., Zeni, L., "Low cost sensors based on SPR in a plastic optical fiber for biosensor implementation," *Sensors*, 11, pp. 11752–11760, 2011, doi:10.3390/s111211752.
52. Celikkol-Aydin, S., Suo, Z., Yang, X., Ince, B., Avci, R., "Sharp transition in the immunoimmobilization of *E. coli* O157:H7," *Langmuir*, 30, 7755–7761, 2014. doi:10.1021/la501545n.
53. Bd McFarland Turbidity Standard No 5 2010.
54. Zibaii, M. I., Latifi, H., Saeedian, Z., Chenari, Z., "Nonadiabatic tapered optical fiber sensor for measurement of antimicrobial activity of silver nanoparticles against *Escherichia coli*," *Journal of Photochemistry and Photobiology B: Biology*, 135, pp. 55–64, 2014. doi:10.1016/j.jphotobiol.2014.03.017.

10

POF Displacement Sensors

Joseba Zubia

Contents

10.1 Introduction

In recent years, a variety of optical sensors for distance measurements have been reported in the literature [1]. Non-contact displacement sensors are a key element in a large amount of industrial applications, such as micromechanics, process control, quality assessment, and prototyping. For that reason, high-quality optical sensors with high resolution and wide dynamic range at a minimum cost have been requested.

Two kinds of optical sensors are commonly adopted, namely interferometric optical sensors and intensity modulation-based sensors. Interferometric optical sensors rely on fringe counting and offer high resolution and stability, but their response depends on the wavelength of light [2].

Fiber optic sensors (FOS) represent a valuable alternative to interferometric optical sensors, thanks to their advantageous properties: immunity to electromagnetic interference, intrinsically high sensitivity, lightweight, small size, fast response, and high precision non-contact measurement, along with their fireproof and remote operation capabilities [3]. Because of that, companies like Omron, Keyence, Banner, MTI instruments, Philtec, and Althen, among others, early on recognized the versatility and importance of FOS.

Plastic optical fibers (POFs) are specially suited for displacement sensors due to their facility for handling, large diameter, numerical aperture (*NA*), and flexibility [4]. These valuable qualities make them convenient for design and manufacturing as well as stable and reliable in performance. POF displacement sensors (POFDS) were demonstrated using an intensity modulation technique [5–9]. Comparison of the reflected or transmitted light intensity with the launched light provides information on the displacement between the probe and the target. The displacement is related to the optical power which can be easily measured using very low-cost components. POFDS operate with LED sources, have an extremely straightforward coupling, and work at visible wavelengths with cheap photodetectors, and all this at a reasonable resolution, long operating range, wide frequency capability, and extremely low displacement detection limit [10].

As secondary transducers, POFDS correlate the amount of displacement to a broad range of parameters such as temperature [11], vibration [12], pressure or sound [13], acceleration [14], surface profile [15], monitoring cracks in building walls [16–18], wind speed [19], rail deflections [20], gasoline [21], and wear detection of milling tools [22], just to name a few. POFDS can also be used for medical applications such as spine bending monitoring [23] or stain formation on teeth [24,25].

This chapter is structured as follows. First, a mathematical analysis based on both ray and wave optics is developed in order to predict the response function of the sensor. This analysis will be useful to understand how different parameters acting on the sensor affect its response. In particular, the next section will describe the influence that several parameters, such as fiber radius asymmetry, separation between transmitting and receiving fibers, angle of the fiber tip and between fibers, and tilting and roughness of the reflecting surface, have on the sensor response. Finally, an application of a POFDS for the measurement of the tip clearance of an aeronautical turbine will be shown.

10.2 Theoretical Model

The basic structure of the POFDS consists of a light source, usually emitting in the visible wavelength range, a transmitting fiber (TF), a receiving fiber (RF), and a photodetector with a transimpedance amplifier to achieve high gain and broad bandwidth. The light, after exiting the TF, reflects on the target surface to which distance has to be measured. The RF that collects the reflected light can be either the TF itself or a different fiber mounted close to it. At the RF, the light intensity is measured with a suitable optical receiver, which is responsible for transforming light power into electric voltage. The amount of receiving light intensity is a function of the distance between the reflecting surface and the TF tips, so the displacement can be obtained through measuring the receiving light intensity [26].

The analysis and response of POFDS can be addressed from two different points of view: geometrical and wave-like. We will present both viewpoints and show how they predict similar results.

There are two key differences between these approaches. Whereas in the geometrical approach the light intensity is constant and restricted to the light cone defined by the NA of the TF, in the wave approach the light intensity is neither homogeneous in the cross-section nor restricted to the NA.

10.2.1 Geometrical Approach

We will assume that the irradiance remains constant for all points inside the light cone and that no irradiance occurs outside it (see **Figure 10.1a**):

$$I = \begin{cases} \dfrac{P_o}{\pi d^2} & \text{Inside the cone} \\ 0 & \text{Outside the cone} \end{cases} \tag{10.1}$$

P_o represents the optical power emitted by the TF and d the radius of the reflected light cone at a distance z from the RF: $d = r_T + z \cdot \tan\theta_c$, where $\theta_c = \sin^{-1}(NA/n)$ is the acceptance angle of the TF, n the refractive index of the surrounding medium (usually air), and r_T is the radius of the TF core.

In the transmission mode configuration (**Figure 10.1a**), only a fraction of the light from the TF will be intercepted by the RF [27]. The fraction of optical power coupled into the RF is given by the ratio of the cross-section area of the RF (πr_R^2, where r_R is the core radius of the RF) to the area over which the emitted light power is distributed at a distance z from the RF ($\pi \cdot d^2$).

The transmitted power P_T for an offset joint z between two identical step-index POFs is found to be:

$$P_T = P_o\left[\frac{r_R}{d}\right]^2 = P_o\left[\frac{r_R}{r_T + z \cdot \tan\theta_C}\right]^2 = P_o\left[\frac{r_R}{r_T + z \cdot \tan\left(\sin^{-1}\left(\dfrac{\text{NA}}{n}\right)\right)}\right]^2. \tag{10.2}$$

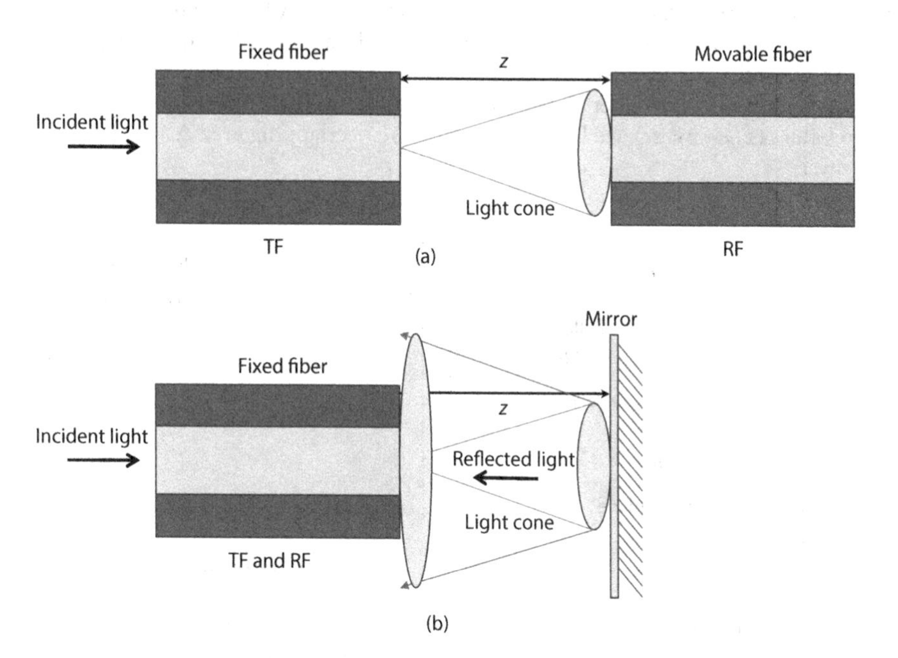

FIGURE 10.1 Illustration of a basic POFDS in (a) transmission and (b) reflection mode.

If the fiber core radius is the same for the TF and RF $\left(r_T = r_R = a\right)$, **Eq. 10.2** simplifies to:

$$P_T = P_o\left[1+\frac{z}{a}\cdot\tan\left(\sin^{-1}\left(\frac{\text{NA}}{n}\right)\right)\right]^{-2}. \tag{10.3}$$

Figure 10.2b shows the response of the sensor ($P_N = P_T/P_o$) according to **Eq. 10.3**. When the fiber tips are in contact ($z = 0$), the transmitted power equals the received power. However, as the gap between both fibers increases, the received optical power decreases according to the inverse of the square of the distance z. A deeper approach can be found in references [28,29] where fibers terminated with lenses are also studied.

Instead of recording transmitted light levels (transmission mode configuration, **Figure 10.1a**), a sensor is usually configured to measure changes in reflected light (reflection mode configuration, **Figure 10.1b**). In this case, a mirror or a target surface located at a distance z from the fiber end reflects the light. A percentage of the reflected light is coupled into the RF and then transmitted to a photodetector that records the optical power level.

The analysis set out for the transmission mode configuration (Eqs. 10.2 and 10.3) can also be applied to the reflection mode configuration. In the latter case, however, symmetry considerations allow us to place the RF in the image plane at a distance $2z$ from the TF (image fiber). This situation is depicted in **Figure 10.2a**. Then, the fraction of reflected power coupled into the RF is given by:

$$P(z) = \frac{P_o\Gamma\Delta A}{\pi\left(r_T + z\cdot\tan\theta_c\right)^2} \tag{10.4}$$

where Γ is the reflectivity of the target surface and ΔA is the overlapping area of the emitted light cone with the RF core.

If we denote the lateral offset between the TF and RF axes as s, and assuming that the refractive index of the medium between the fibers and the mirror is $n = 1$ (air), the RF is completely illuminated for distances larger than $z_s = \frac{s - r_T + r_R}{2\cdot\tan\theta_c}$, and therefore $\Delta A = \pi r_R^2$ and the received power decreases as z^{-2}. On the contrary, for distances smaller than $z_{off} = \frac{s - r_T - r_R}{2\cdot\tan\theta_c}$ no light is coupled into the RF and we will have an offset in the measurement that could disappear if the separation $s = r_T + r_R$, which means that fibers have no cladding (see **Figure 10.3**). For distances between these two values ($z_{off} < z < z_s$), we have to compute the overlapping area ΔA to get the received optical power.

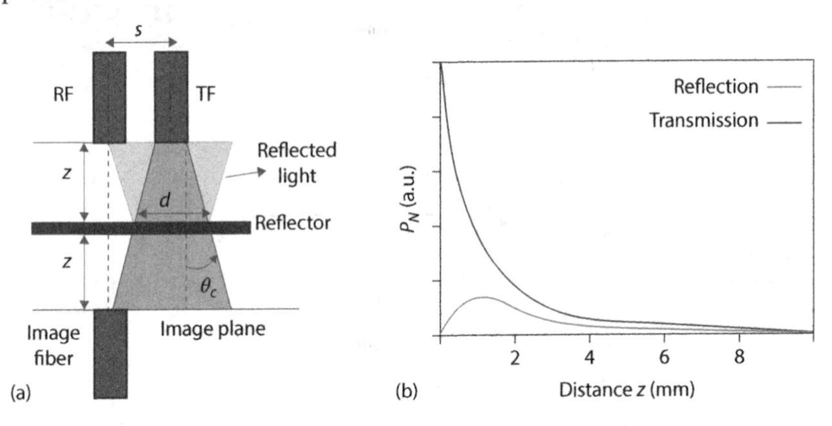

FIGURE 10.2 (a) Illustration showing the meaning of the image fiber in the reflection configuration. (b) Typical normalized response on the sensor in transmission (red curve) and reflection mode (green curve) according to the geometrical approach.

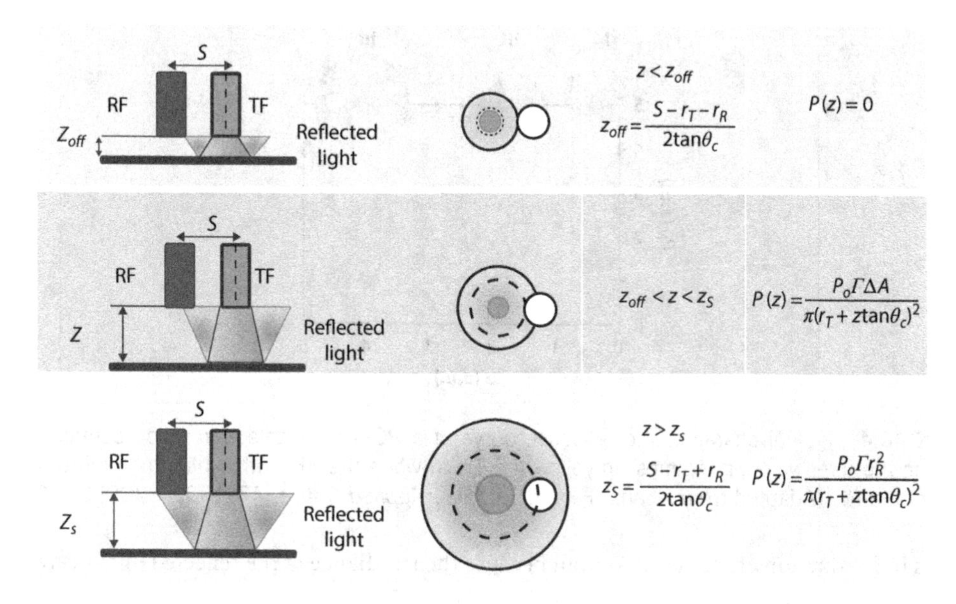

FIGURE 10.3 The three regimes of a bifurcated POFDS with the definitions of z_{off} and z_s.

The overlapping area ΔA depends on geometrical parameters according to the following expressions [30–35]:

$$\Delta A = a^2\left[k_c^2\varphi + \sin^{-1}(k_c\sin\varphi) - mk_c\sin\varphi\right],\ k_c \le \sqrt{m^2+1} \tag{10.5a}$$

$$\Delta A = a^2\left[k_c^2\varphi + \pi - \sin^{-1}(k_c\sin\varphi) - mk_c\sin\varphi\right], \sqrt{m^2+1} < k_c < m+1 \tag{10.5b}$$

$$\varphi = \cos^{-1}\left[\frac{k_c^2+m^2-1}{2mk_c}\right];\ k_c = 1 + \frac{2z}{a}\tan\theta_c \tag{10.5c}$$

where $m = \frac{s}{a}$ is the dimensionless distance between the axes of adjacent fibers, assuming that the TF and RF have the same core radius a.

It is clear from **Figure 10.2b** that the received irradiance is higher for POFDS in transmission mode configuration. Even then, the reflection mode configuration is preferable since the POFDS can be realized in a more compact way arranging all fibers at the same side.

In **Figure 10.3** we have plotted the three regimes of a bifurcated POFDS and in **Figure 10.4** the response of the POFDS according to Eqs. 10.4 and 10.5 [30,31]. At small distances z (region I), the RF is located outside of the light cone, resulting in a "dead zone" or offset, in the displacement-power curve ($z < z_{off}$). Having passed the offset (region II), the sensor output starts to increase steadily with the target distance. This is because increasing the target distance produces an enlarged light cone that intersects a bigger area of the RF core. As a result of that, the quasi-linear region with positive slope is created where the sensor output increases monotonously with the target distance. Therefore, the response curve is determined by the interplay of two conflicting mechanisms as z grows: the increase of the intersecting area and the decrease of the irradiance.

The sensor output reaches its maximum value when the center of the receiving fiber core coincides with the boundary of the light cone ($z < z_s$). Further increase of the target distance

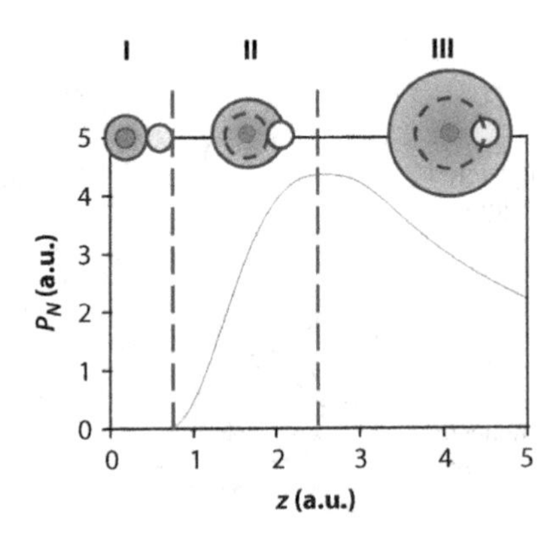

FIGURE 10.4 Typical characteristic response curve of a POFDS in reflection mode configuration showing the three working regimes. In yellow the TF, in white the RF, and in blue the emitted light cone at the RF. (Adapted from Huang, H. and Tata, U., *Applied Optics*, 47, 1302–1309, 2008.)

results in a reduction of the sensor output because the irradiance of the reflected light decreases as z^{-2} while the light collecting area remains constant: πr_R^2 (region III).

10.2.2 Gaussian Beam Approach

The Gaussian beam approach is more realistic than the geometrical approach. Based on experimental data some authors such as Libo et al. and Faria [36–42] have found that the light distribution at the end of the fiber can be fit to a Gaussian-like function for any kind of light source. The irradiance of the emitted light from the TF decreases radially over the beam cross section according to:

$$I(\rho,z)=\frac{2P_o}{\pi q^2(z)}\exp\left[-\frac{2\rho^2}{q^2(z)}\right] \tag{10.6a}$$

where the effective radius $q(z)$ is:

$$q(z)=\sigma r_T\left[1+\xi\left(\frac{z}{r_T}\right)^{\eta}\tan\left(\arcsin\left(NA\right)\right)\right]. \tag{10.6b}$$

In these expressions ρ is the radial coordinate from the axis of the TF, and η, σ and ξ are three regulating parameters of light intensity distribution relative to the characteristics of the light source and TF. σ is the characterization parameter of the optical fiber refractive index and ξ is the modulation parameter related to the fiber coupling conditions (**Figure 10.5**).

As before, the light collected by the RF is the same as the light intensity received from a mirror-image fiber located at twice the distance from the TF to the target surface:

$$I_R(\rho,z)=\Gamma I(\rho,2z)=\frac{2\Gamma P_o}{\pi q^2(2z)}\exp\left[-\frac{2\rho^2}{q^2(2z)}\right] \tag{10.7a}$$

$$I_R(0,0)=\frac{2P_o\Gamma}{\pi(\sigma r_T)^2} \tag{10.7b}$$

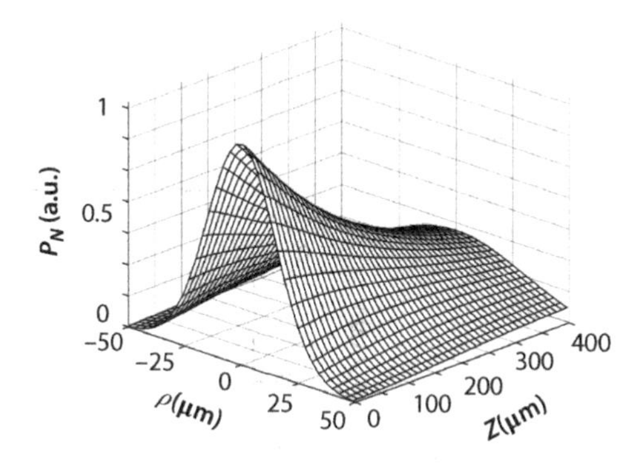

FIGURE 10.5 Gaussian-like function according to Eq. 10.6.

To obtain the received power, the irradiance must be integrated over the RF core. Assuming that the light irradiance is approximately constant across the receiving core and equal to its value at the center of the RF:

$$P(z) = I_R(\rho, 2z)\cdot A = \frac{2r_R^2 \Gamma P_o}{q^2(2z)} \exp\left[-\frac{2s^2}{q^2(2z)}\right] = \frac{2r_R^2 \Gamma P_o}{q^2(2z)} \exp\left[-\frac{8r_R^2}{q^2(2z)}\right] \tag{10.8}$$

where $A = \pi r_R^2$ is the RF core area and $\rho = s = 2\cdot r_R$, i.e. fibers with no cladding. This is the power in reception which can be further simplified defining a new variable $\zeta^2 = \frac{q^2(2z)}{r_R^2}$:

$$P(z) = \frac{2\Gamma P_o}{\zeta^2} \exp\left[-\frac{8}{\zeta^2}\right] \tag{10.9}$$

Defining $f = -\frac{1}{\zeta^2}\exp\left[-\frac{8}{\zeta^2}\right]$ as the modulation characteristic function of a fiber pair, **Eq. 10.9** can be simplified to

$$P(z) = 2\Gamma P_o f(r_T, r_R, NA, \rho, z). \tag{10.10}$$

We can perform a better calculation computing the light power at the RF from the intersection of the reflected light distribution with the RF core as plotted in **Figure 10.6a**. The total received power is the integration of the light intensity over the light intersecting area A_i (see **Figure 10.6a**).

$$P_R(z) = \int_{A_i} I(A)dA = \int_{A_i} 2\beta I(\rho)\rho d\rho \tag{10.11a}$$

$$P_R(z) = \frac{4\Gamma P_o}{\pi q^2(2z)} \int_{s-r_R}^{s+r_R} \arccos\left(\frac{\rho^2 - s^2 - r_R^2}{2s\rho}\right) \exp\left[-\frac{2\rho^2}{q^2(2z)}\right] \rho d\rho \tag{10.11b}$$

This equation describes the relation between the sensor output and the distance z to the target surface. A typical response curve of the sensor is depicted in **Figure 10.6b**, which closely resembles the result found before using the geometrical approach.

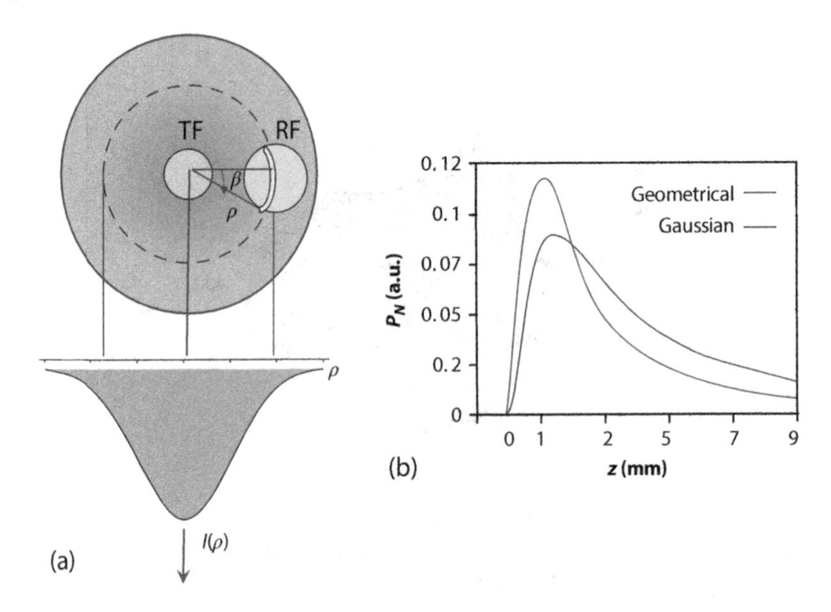

FIGURE 10.6 (a) Front view of the reflected Gaussian pattern with integration parameters. (b) Comparison between the geometrical and the Gaussian model.

The geometrical approach overestimates the maximum value of the sensor response, around 30%, and it has steeper front and rear slopes. In the Gaussian model, in addition, when the reflecting target is close to the fiber tips, the received power is not exactly null as it happens in the framework of the geometrical approach.

10.2.3 Sensor Sensitivity

The sensor sensitivity S can be calculated by deriving $P(z)$ with respect to ζ:

$$S = \frac{\partial P(z)}{\partial \zeta} = \frac{4\Gamma P_o \exp\left(-\frac{8}{\zeta^2}\right)}{\zeta^3}\left[\frac{8}{\zeta^2} - 1\right] \tag{10.12}$$

The sensitivity goes to zero at $\zeta^2 = 8$, which means that the received irradiance is maximum at this value:

$$8 = \sigma^2\left[1 + \xi\left(\frac{z}{r_T}\right)^{\eta} \tan\left(\arcsin(NA)\right)\right]. \tag{10.13}$$

Solving for z yields

$$z_{peak} = r_T\left[\frac{\frac{8}{\sigma^2} - 1}{\xi \cdot \tan\left(\arcsin(AN)\right)}\right]^{\frac{1}{\eta}} \tag{10.14}$$

z_{peak} is the distance at which the response of the sensor is maximum. It depends on the characteristics of both the light source and on the TF. In POFDS (see **Figure 10.6b**), the front slope is highly sensitive and useful for close distance targets and the back slope is less sensitive and useful for long distance measurements. As we will see through the next section, the obtained experimental response curve agrees well with the theoretical one, which verifies the successful modeling of the POFDS. There are no typical values for the sensitivity as they depend on the choice of the fiber parameters, photodetectors, and on the associated electronics. We can identify configurations ranging from some μV/μm to mV/μm.

10.3 Parameters Affecting the Response of the Sensor

POFDS are very sensitive to geometrical parameters of both the sensing head and the target. In this section we provide a qualitative estimation of the sensor performance, and its tolerance to geometrical parameters (fiber geometry and target rotation, *NA*, distance between fibers, etc.) as well as to target non-idealities (non-uniform reflectivity and diffusive surface) [30]. In the following section we will assume a bifurcated POFDS with one TF and one RF placed at the same plane as shown in **Figure 10.2a**.

Unless otherwise stated, we will assume that the RF and TF core radii r_R and r_T are the same and equal to a.

10.3.1 Effect of Numerical Aperture

In **Figure 10.7** we show the response of the POFDS for different *NA* of the TF. For TFs with higher *NA* the two linear regions are much steeper, resulting in raised displacement sensitivity. On the contrary, it results in a shorter dynamic range as well as a smaller working distance. This can be explained as follows. A higher *NA* reduces the irradiance, that is to say, the same light power is distributed across a wider light cone of radius $d = r_T + z \cdot \tan\theta_C$. For instance, if the RF is covered with light (region III, beyond the maximum), a higher *NA* means less irradiance and so less power into the RF [43]. Due to the increase in the *NA* most of the light is collected by the RF at smaller z values, and hence the peak position is shifted toward smaller distances, which in turn increases the sensitivity and sharpens the peak. Finally, it is worthy of mention that the front quasi-linear operating range decreases as total overlap occurs at smaller distances.

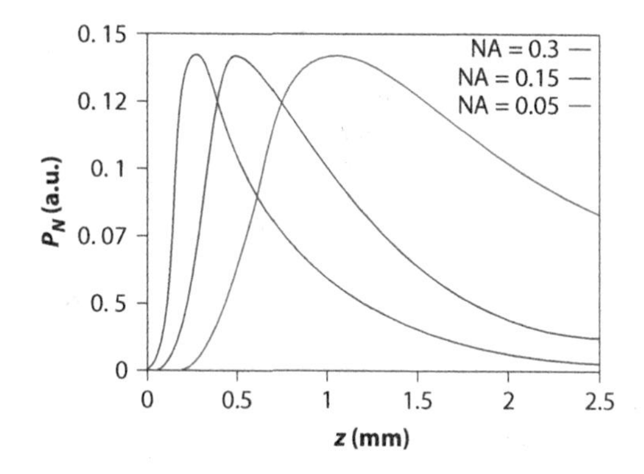

FIGURE 10.7 Effect of the *NA* of the TF on the characteristic curve of the bifurcated sensor. (Adapted from Huang, H. and Tata, U., *Appl. Opt.*, 47, 1302–1309, 2008 and Patil, S.S. and Shaligram, A.D., *Int. J. Sci. Eng. Res.*, 4, 2013.)

10.3.2 Effect of the Separation *s* between Emitting and Receiving Fibers

Ordinarily a POF consists of two layers: the core, where the light propagates through it by total internal reflection, and the cladding, a thin layer of lower refractive index surrounding the core. In the case of a 50 mm POF, the core diameter is 980 μm and the remaining 20 μm corresponds to the cladding layer. As we can infer from **Eq. 10.8**, the characteristic curve depends on the center-to-center separation between the TF and RF (from now on *offset*). The response curve of the POFDS for various *s* values is shown in **Figure 10.8**.

A comparison of the characteristic response curves reveals three important features. First, the reflected signal is smaller for longer offset values, and the corresponding peak value both decreases and shifts to longer distances. Second, the dynamic range is higher for long offsets. Finally, the sensitivity diminishes and z_{off} grows up with increasing values of *s*. The explanation for that characteristic behavior relies on the fact that for large values of *s*, the light cone needs to travel a longer distance in order to intercept the RF, causing the power density as well as the irradiance coupled to the RF to diminish. The separation *s* between the axis of the emitting and receiving fibers includes not only the cladding and coating but also any disturbance due to an improper assembly of the fibers [43,44].

10.3.3 Effect of the RF's Core Radius

We have previously assumed that the core radii of the TF and RF were the same: *a*. Let us think now that we have a POFDS with a fixed TF, but that the RF core is variable in size. **Figure 10.9** replies to that conjecture [26]. The response curves corresponding to this new case are shown in **Figure 10.9** [26].

Not surprisingly, the amount of light that is coupled into the RF is proportional to the core area of the RF as long as we keep a constant offset. Hence, doubling the RF core diameter will increase fourfold the response of the sensor, as **Eq. 10.4** confirms.

10.3.4 Effect of the Cutting Angle of the Emitting and Receiving Fiber Tips

Hadjiloucas et al. [45] presented a radiometric analysis of a bifurcated POFDS with optimally cut optical fiber tips (see **Figure 10.10**). Uncut fiber sensors show the largest range but a smaller responsivity. Single cut sensors exhibit an improved responsivity (a factor of 4) at the expense of a smaller dynamic range. An additional increase in responsivity (a factor of 64) as well as a reduction in the operational range is achieved when a double cut sensor configuration is implemented.

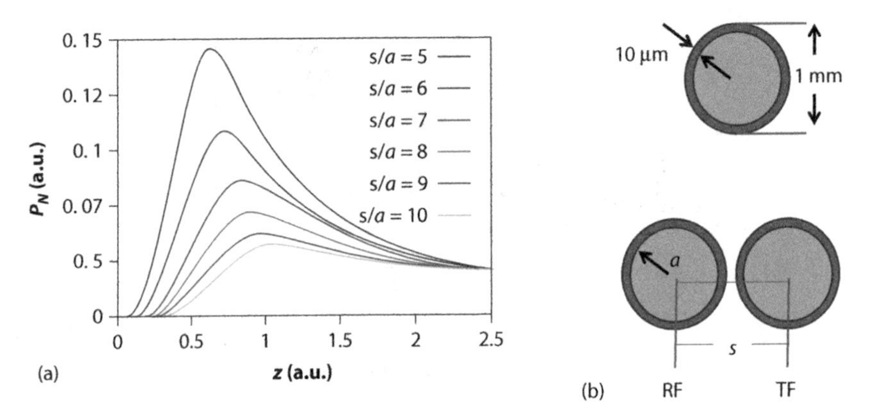

FIGURE 10.8 (a) Characteristic curve of a bifurcated sensor as a function of the separation *s* between TF and RF. (Adapted from Huang, H. and Tata, U., *Appl. Opt.*, 47, 1302–1309, 2008 and Jafari, R. and Golnabi, H., *Opt. Laser Technol.*, 43, 814–819, 2011.) (b) Geometry of the fibers.

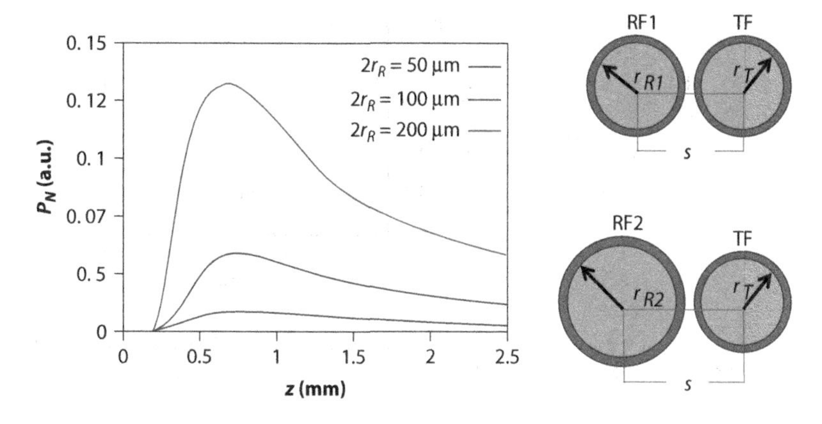

FIGURE 10.9 Characteristic curves of a bifurcated sensor as a function of the radius r_R of the RF core.

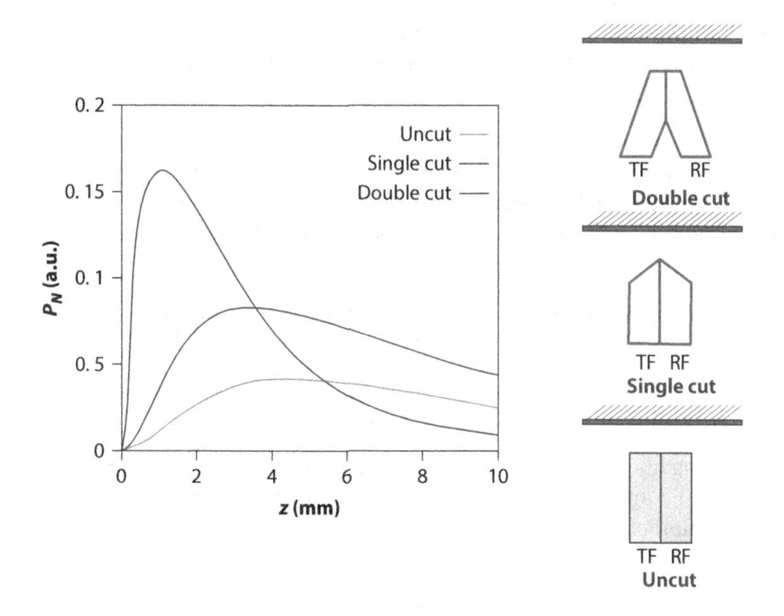

FIGURE 10.10 Simulated response curves for three configurations of a bifurcated POFDS: uncut, single cut, and double cut. (Adapted from Hadjiloucas, S. et al., *Rev. Sci. Instr.*, 71, 2000.)

10.3.5 Effect of the Offset between Transmitting and Receiving Fiber Ends

Due to the difficulty aligning the fiber tips of the emitter and receiving fibers, their distance to the target is different. It has been demonstrated [30] that an offset between the two fiber ends changes the response curve as plotted in **Figure 10.11**. Placing several RFs around a single TF at different offset distances enables an increase of the linear range of the response curve. If one looks at **Figure 10.10**, it can be observed that the linear distance range of the three-fiber POFDS sensor is doubled if the output of the sensor is switched from RF1 to RF2. The whole range of the POFDS is a concatenation of the linear ranges of the RFs.

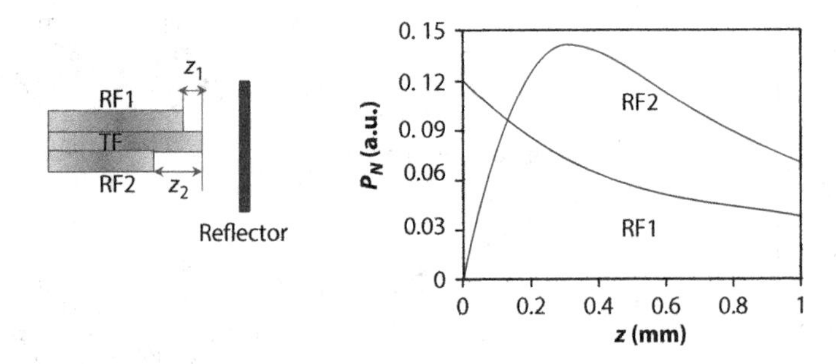

FIGURE 10.11 Characteristic curve of a bifurcated sensor as a function of the offset between the TF and RF ends. (Adapted from Huang, H. and Tata, U., *Appl. Opt.*, 47, 1302–1309, 2008.)

10.3.6 Effect of Angle

The results presented so far only hold when the targeting surface is set perpendicular to the fibers axes. Accidental perturbations could give rise to the tilting of the mirror or target surface. In this situation, the transfer function of the sensors must be modified [46–50].

Figure 10.12a shows the geometry of the fibers. The TF and RF are placed at an angle θ to the reflecting surface. **Figure 10.12b** shows the variation of the sensor response with distance for different angles between TF and RF. From a close inspection of the results we can infer the following:

1. We observe a decrease in offset region with angle θ. As the angle between fibers increases, the size d of the reflected light cone also increases ($d = r_R + 2z\cos\theta + 2r_T\sin 2\theta$). That means that more light power is coupled to the RF. Hence, even at the lesser distance RF is able to receive reflected light.
2. The peak maximum power $P_{N,max}$ moves up with increasing tilt although the critical distance z_{peak} at which the maximum power is reached moves left at shorter distances.
3. The slope of the response curve (sensitivity) grows with the angle θ between fibers. On the contrary, the linear range of the response curve narrows.

Consequently, tilting fibers provide a means of optimizing the response of the sensor. Increasing the angle between TF and RF enhances both sensitivity and the overall received signal at the expense of a reduction of the linear range.

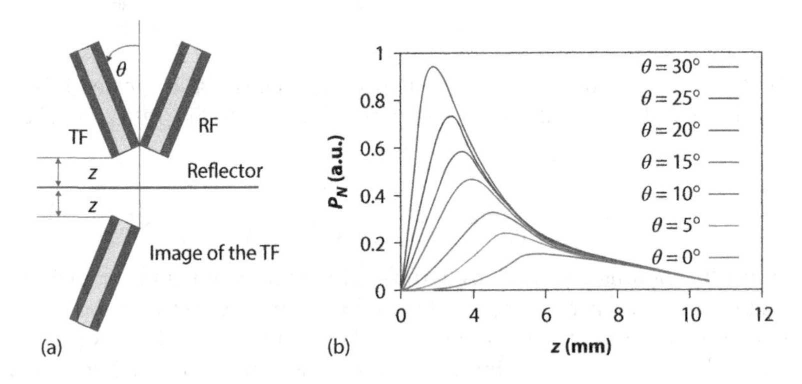

FIGURE 10.12 (a) Geometry of an inclined bifurcated sensor: θ is the angle between the fiber tips and the normal to the reflector. (b) Sensor response for different angle values. (Adapted from Buchade, P.B. and Shaligram, A.D., *Sens. Actuators A*, 136, 199–204, 2007.)

Faria et al. considered this effect from the wavelike approach [37]. They reached the same conclusions. Positive mirror tilting (increasing the angle with the RF) brings the image fiber further inside the light cone where radiation intensity is stronger.

It was found by Kleiza et al. that, when the angle between the TF and RF changed from 0° to 30°, the sensitivity increased between 16 and 30 times and the linear range decreased to 33%. The blind region, i.e. the region where the response of the sensor is null, decreases with the angle between TF and RF and reduces to zero at 25° [51–53].

A novel differential variation of this sensor was proposed by Jia et al. [54]. That sensor can directly measure the angular and axial linear displacements of a flat surface. The structure of the sensor probe is shown in **Figure 10.13**. It is composed of a double ring optical fiber bundle perpendicular to the surface target (RF_1 fibers of the first ring and RF_2 fibers of the second ring) (see **Figure 10.13**). Besides, a left optical fiber bundle (RF_{Left}) and a right optical fiber bundle (RF_{Right}) are arranged at an angle α with the central bundle. As we will see later on, the distance from the central bundle to the reflector can be measured by taking the ratio of the received light power V_{RF1}/V_{RF2} in rings RF_1 and RF_2, (light power is usually proportional to the voltage at the photodetector). It is the difference of the voltage $V_{Right}-V_{Left}$ at the photodetectors plugged to RF_{Right} and RF_{Left}, that allows inferring the angular deflection of the surface γ. The sensitivity increases with the angle α. The main drawback of these sensors is that the size of the fiber head increases.

A similar principle also serves to measure slide [41,42,55]. The sensing probe consists of five fiber optics as shown in **Figure 10.14**. The light from the reflecting surface is received by four fibers (fibers 1–4) arranged around the illuminating fiber (fiber 5). The rotation angles can be inferred from the ratio of the irradiances I_i of the RFs.

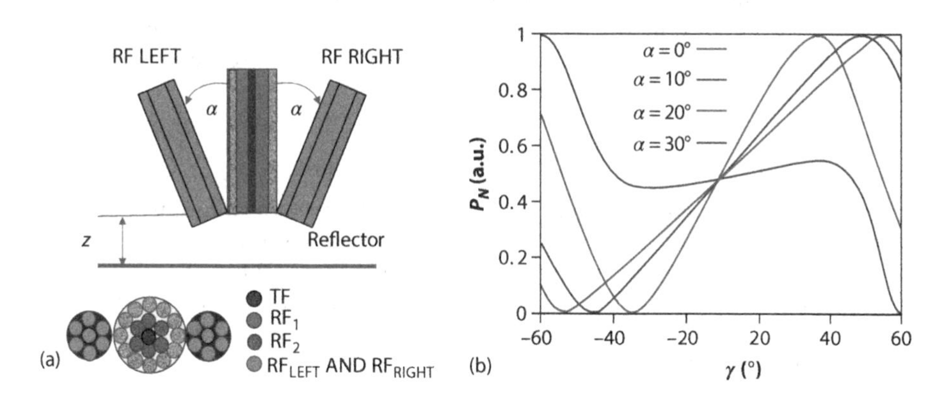

FIGURE 10.13 (a) Structure of the differential reflective optical fiber displacement sensor. (b) Normalized characteristic curve $V_{Right}-V_{Left}$ when the angle γ changes. (Adapted from Jia, B. et al., *Sensors*, 16, 1508, 2016.)

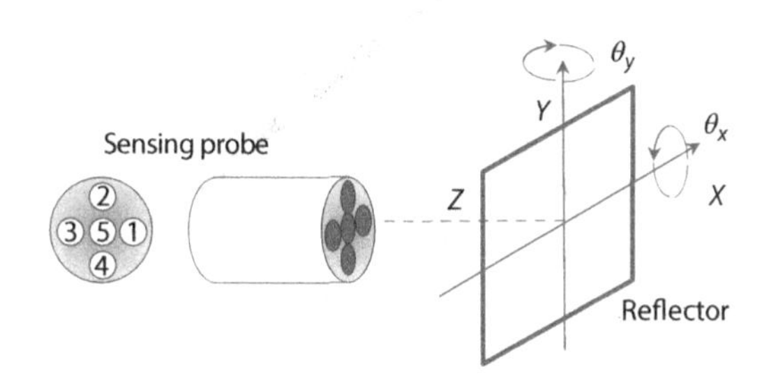

FIGURE 10.14 Five-fiber-optic sensing probe for slide sensing. (Adapted from Yuan, L., *Opt. Fiber Technol.*, 7, 340–349, 2001.)

Under very realistic conditions Yuan et al. [41] and Sagrario et al. [55] demonstrated that the angles θ_x and θ_y can be deduced from the irradiances collected in fibers 1–4. Specifically,

$$\log\left(\frac{I_1}{I_3}\right) = S(z,s)\theta_y \tag{10.15a}$$

$$\log\left(\frac{I_2}{I_4}\right) = S(z,s)\theta_x \tag{10.15b}$$

$$S(z,s) = \frac{8zs}{a^2\left[1+\xi(2z/a)^{3/2}\tan\theta_c\right]^2} \tag{10.15c}$$

where z is as usual the distance from the illuminating fiber to the reflecting surface and s is the distance between the fiber axes of any of the RFs and the illuminating fiber. The sensitivity $S(z, s)$, as outlined in **Figure 10.15**, is more dependent on the s than on the distance z. This approach follows a linear dependence of the irradiances ratios on the angles at least up to 0.1 rad.

Other authors use a bifurcated POFDS with twin RF at both sides of the TF [56]. The response of the sensor is obtained through the differential subtraction of responses P_{RF1}–P_{RF2}. Another approaches make use of variable fiber geometries, with an asymmetrically inclined fiber optic sensor configuration [57]. A positive lens can also be utilized to achieve a high sensitivity fiber optic angular displacement sensor [58] which even can be upgraded to detect ultrasounds [59]. Similar results were obtained by Khiat et al. using 1TF and 4 RF [60]. They enhanced the sensitivity by a factor of two with a GRIN microlens. Ko and Wu have also worked on resembling issues [61,62].

10.3.7 Effect of Surface Roughness and Surface Reflectivity

Previously we have assumed that the reflection is purely specular which means that the irradiance distribution is the same before and after the reflection at the surface.

Reflecting surfaces are neither specular nor Lambertian ones. When the light is reflected from the target surface, the random characteristics of its roughness will make the light scatter at the incident point. Based on the scattering theory, the intensity distribution of reflection from the coarse target can be easily introduced in the calculus [63–65].

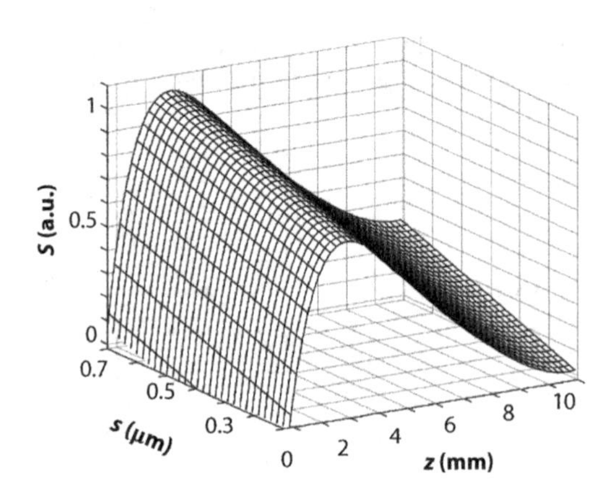

FIGURE 10.15 The sensitivity $S(z, s)$ according to Eq. 10.15c, versus z and s. (Adapted from Yuan, L., *Opt. Fiber Technol.*, 7, 340–349, 2001.)

Nan-Nan et al. [66] used a multiwavelength fiber sensor for measuring surface roughness. They used a modified version of **Eq. 10.7a**. According to Beckmann's scattering theory [63], when the light impinges perpendicularly on the rough surface the average scattered power P_s is:

$$P_s = I_o \exp\left[-2\left(\frac{4\pi\sigma}{\lambda}\right)^2\right] \quad \textbf{(10.16)}$$

where λ is the light wavelength, I_o is the intensity of the incident light, and σ is the root mean square value. The received optical power will be:

$$P_R = \Gamma P_s = I_o \Gamma \exp\left[-2\left(\frac{4\pi\sigma}{\lambda}\right)^2\right] \quad \textbf{(10.17)}$$

By using two lasers at central wavelengths λ_1 and λ_2:

$$P_R(\lambda_1) = I_o(\lambda_1)\cdot\Gamma\cdot\exp\left[-2\left(\frac{4\pi\sigma}{\lambda_1}\right)^2\right] \quad \textbf{(10.18)}$$

$$P_R(\lambda_2) = I_o(\lambda_2)\cdot\Gamma\cdot\exp\left[-2\left(\frac{4\pi\sigma}{\lambda_2}\right)^2\right] \quad \textbf{(10.19)}$$

Taking the ratio between both optical powers,

$$\frac{P_R(\lambda_1)}{P_R(\lambda_2)} = \frac{I_o(\lambda_1)}{I_o(\lambda_2)}\exp\left[-32\pi^2\sigma^2\left[\frac{1}{\lambda_1^2} - \frac{1}{\lambda_2^2}\right]\right], \quad \textbf{(10.20)}$$

the root mean square σ can be calculated as below:

$$\sigma^2 = \frac{\lambda_1^2\lambda_2^2}{32\pi^2\left(\lambda_1^2 - \lambda_2^2\right)}\log\left[\frac{P_R(\lambda_1)}{P_R(\lambda_2)}\frac{I_o(\lambda_2)}{I_o(\lambda_1)}\right] \quad \textbf{(10.21)}$$

Finally, the surface roughness R_a is given by [67]:

$$R_a = \frac{4}{5}\sigma \quad \textbf{(10.22)}$$

Within this approach authors found a sensor response ten times higher than that obtained with just one wavelength.

Yang proposed the sensor head shown in **Figure 10.16a**. This fiber bundle allows to measure simultaneously both distance to the surface and surface roughness [64]. Defining different normalized quantities with the irradiances received at different RFs, they were capable of measuring target surface inclinations from −50° to 50°. According to these authors gap distances can be achieved from the normalized quantity G_p:

$$G_p = \frac{\sum_{j=1}^{4} I_{1j} - \sum_{j=1}^{4} I_{2j}}{\sum_{i=1}^{2}\sum_{j=1}^{4} I_{ij}} \quad (10.23)$$

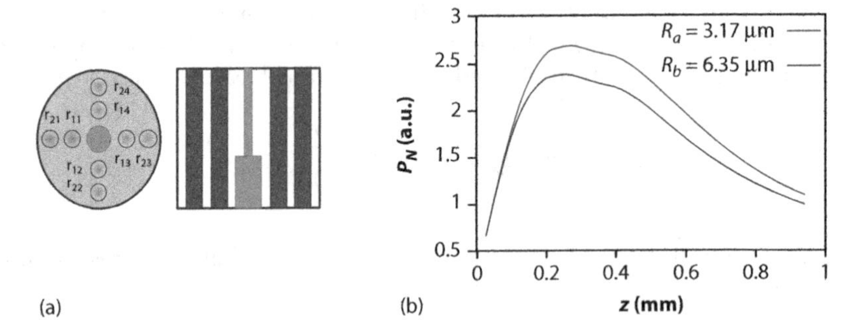

FIGURE 10.16 (a) Lateral and cross section of the fiber probe of the sensor. (b) Response of the sensor for different values of the surface roughness. (Adapted from Yang, Y. et al., *Precis. Eng.*, 24, 32–40, 2000.)

where I_{ij} represents the light intensity detected by each of the receiving fibers RF_{ij}. Other measurements such as the surface inclination angle and surface roughness can be accomplished from similar expressions. The experimental results are shown in **Figure 10.16b**, where the scattering provoked by roughness causes the response to diminish [67–70].

Another issue that deserves attention is the composition of the target surface. The reflectivity depends not only on the roughness but also on the material composition. Yasin et al. tested different surface materials (see **Figure 10.17**) [71,72]. The highest and lowest sensitivities were obtained using a mirror and a plastic target respectively. Higher reflectivity surfaces show a higher sensitivity. However, for the back slope, the lower reflectivity surfaces have the largest linear range, which in this case corresponds to plastic and aluminum. The slope is relatively shallower at small powers, which contributes to a larger linearity range.

We have just seen that, in general, the response of a POFDS is modulated by the change in displacement and the surface target nature. Abdullah et al. [73] took advantage of this to measure the refractive index of liquids. In their approach the surface target is just the limiting surface of a liquid. In that case, **Eq. 10.8** reads as follows:

$$P_R(z)=\left(\frac{n_1-n_o}{n_1+n_o}\right)^2\frac{2r_R^2I_o}{q^2(2z)}\exp\left[-\frac{2s^2}{q^2(2z)}\right] \tag{10.24}$$

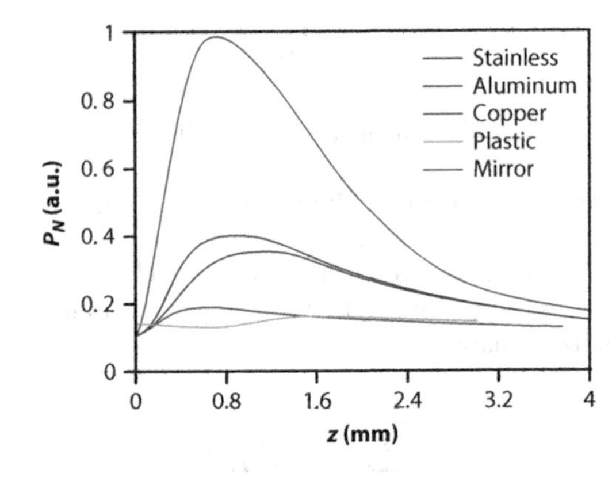

FIGURE 10.17 Variation of the characteristic response curve with the type of target surface. (Adapted from Yasin, M. et al., *Laser Phys. Lett.*, 5, 55–58, 2008.)

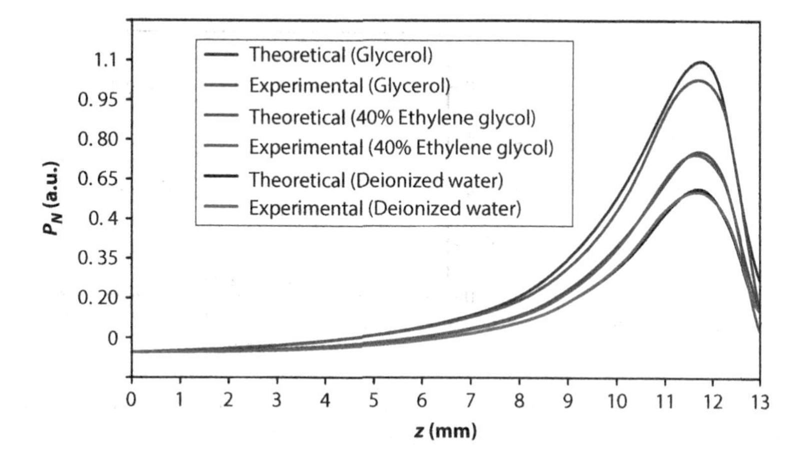

FIGURE 10.18 Theoretical output curve and experimental output curve for deionized water, 40% ethylene glycol, and glycerol. (Adapted from Abdullah, M. et al., *Laser Phys.*, 28, 035102, 2018.)

Thus, the reflected optical power onto the RF by the surface of the liquid medium n_l is the same as that from surface with a reflectivity $\Gamma = (\frac{n_l - n_o}{n_l + n_o})^2$. n_o is the refractive index of air, and the polarization has been disregarded. The liquid's refractive index modifies the response curve. Higher refractive indices of n_1 lead to higher values of the front and back slopes, as well as higher values of the peak of the characteristic curve, as shown in **Figure 10.18**. By correlating these three parameters with the refractive index, a mirrorless refractometer with high accuracy can be built.

Harun [48,74], Govindan [75], and Keliza [53] employed a different approach with a POFDS made of two asymmetrical inclined fibers. The liquid filled up the gap between probe and reflector. The liquid changes the light numerical aperture and so the emission angle θ, which in turn, modulates the amount of received power (**Eq. 10.6b**). They successfully tested the sensor with water, chlorobenzene, glycerol, alcohol, and methanol.

10.3.8 Effect of Surface Shape

In the preceding sections we have assumed that the reflecting surface was flat. Some authors have proposed analytical models for curved reflectors. Gaikwad et al. designed a bifurcated POFDS with a convex reflector [76]. The radius of the light cone is re-evaluated, and it was found that it depends, among other parameters, on the focal length of the convex mirror. The mathematical model shows that the offset or insensitive region of the response curve is minimized if the plane reflector is substituted by a convex one. Such a of sensor can be utilized when a high sensitivity at small distances is required (see **Figure 10.19**).

A POFDS with a concave mirror was proposed by Yang et al. to enhance flexibility in sensitivity selection and linear range [77–79].

The response curve of the sensor is rather complicated, and it depends both on the focal length of the concave mirror and on its diameter as shown in **Figure 10.20**. In the near displacement range (<4 mm), the characteristic curve shares similar features with the flat mirror POFDS. As the sensor head moves further, the response of this sensor deviates strongly from the response of the flat sensor. The characteristic curve reaches another two maxima located at both sides of $2f$, f meaning the focal length of the mirror. As the focal length increases, the height of the second and third maxima decreases. On the other hand, the larger the mirror diameter, the nearer the normalized power of the second and third peaks to 1. The response curve is not monotonic anymore, but it has several regions of steep slopes, i.e. of large sensitivities. Moreover, the response

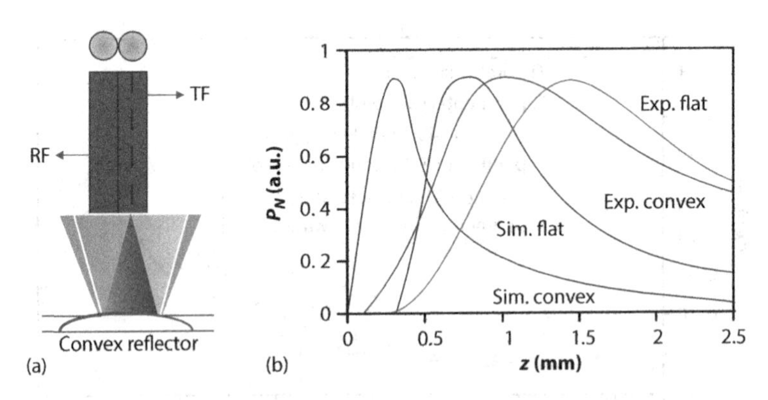

FIGURE 10.19 (a) Differences of the light cone reflected on a flat (blue) or a convex (orange) reflector. (b) Experimental and simulated responses (geometrical approach) of a POFDS with plane and convex reflectors in air. Blue–simulated response for a convex reflector, Red–simulated response for a plane reflector, Green–measured response for a convex reflector, Orange–measured response for a plane reflector. PMMA, 1 mm diameter fibers with $NA = 0.5$. (Adapted from Gaikwad, A.D. et al., *Int. J. Adv. Res. Electr. Electron. Instrum. Eng.*, 1, 29–35, 2012.)

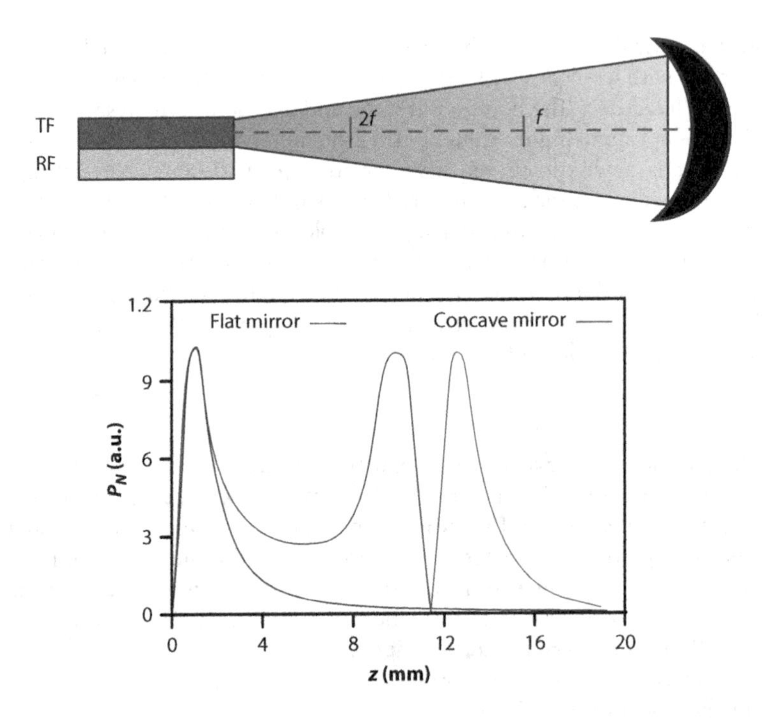

FIGURE 10.20 Experimental results for displacement responses of flat and concave mirrors with $2f = 11$ mm. (Adapted from Yang, H.Z. et al., *Sens. Actuators A*, 162, 8–12, 2010.)

is very high even at large distances, and by a cleaver choice of the focal length and diameter of the mirror we can get a suitable sensibility for the displacement region of interest.

Zhao et al. also used a concave mirror with an optical head composed of eight symmetrically cross-arranged RFs fibers and one centered TF. That sensor can accomplish distance measurement with high vertical resolution of 0.1 μm [80].

10.4 Fiber Bundles

Once the single fiber pair model is clearly defined, the model can be upgraded for a bundle with more than one TF or RF. Fiber bundles are well suited not only for displacement sensors but also for gauging and surface assessment. When only one fiber is used to launch the light and only one fiber is used to collect the light, the maximum measuring range is very short. Soon, a new structure of the POFDS was proposed with two RF to extend the dynamic range.

Compared with the conventional sensor with only one RF, the improved sensor has a better linearity range because it has a larger area to amass the reflected light from the target. However, the sensitivity of both sensors is almost similar (see **Figure 10.21**). The linearity ranges are enhanced by about 44% compared with the conventional sensor [81].

Figure 10.22a shows a classic example of hexagonal packing for a probe consisting of two rings of collecting fibers. In the shown arrangement the central fiber is the light delivery channel. In **Figure 10.22b** we draw the corresponding characteristic curves. The response of the inner ring is narrower, higher, of smaller range, and with a peak at a smaller distance z.

Instead of having a single pair of fibers interacting with each other, now we have a number of TFs (n_T) and RFs (n_R) interacting, so the total power received will be the summation of each interaction. In this case it is assumed that the same power is emitted from each TF.

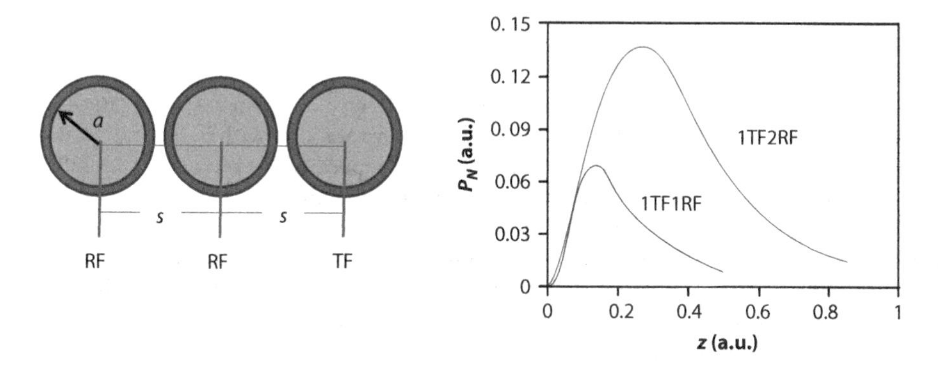

FIGURE 10.21 Response curve of a POFDS with one TF and two RFs located at both sides of the TF. (Adapted from Harun, S.W. et al., *Microw. Opt. Technol. Lett.*, 52, 373–376, 2010.)

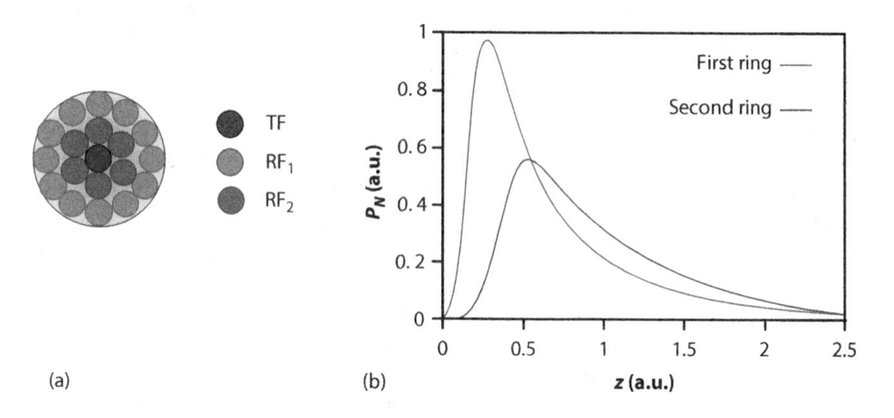

FIGURE 10.22 (a) Sketch of a fiber bundle with two rings of receiving fibers. (b) Response of the POFDS.

The theory for sensor response of POFDS based on the geometrical approach was first developed by Cook [82] and modified by Cuomo [33–35], Nevshupa [83], Shimamoto [84,85], and Zhao [86], among others. However, in the following we will take the wave-optics approach.

Starting from **Eq. 10.10**, the total received power is:

$$
\begin{aligned}
P(z) &= \exp\left[-\sum_k \eta_k r_k\right] \sum_{i=1}^{nT} \sum_{j=1}^{nR} P_{ij}(z) \\
&= \exp\left[-\sum_k \eta_k r_k\right] \Gamma P_o \sum_{i=1}^{nT} \sum_{j=1}^{nR} f\left(r, NA, s_{ij}, z\right) \\
&= \exp\left[-\sum_k \eta_k r_k\right] \Gamma P_o\, F\left(n_T, n_R, a, NA, s_{ij}, z\right)
\end{aligned}
\tag{10.25}
$$

where $F = \sum_{i=1}^{nT} \sum_{j=1}^{nR} f\left(a, NA, s_{ij}, z\right)$ is defined as the modulation characteristic function of a fiber bundle and $\exp\left[-\sum_k \eta_k r_k\right]$ are the additional losses caused by the receiving fibers' microbends [36,87]. We have also disregarded, among other things, Fresnel and fiber absorption losses. Keeping in mind that $\exp\left[-\sum_k \eta_k r\right]$ (no vibrations), Γ, P_o, a, and NA are all constant, the modulation function will only depend on the difference between fiber axes s_{ij}, which simplifies the mathematical model [88].

Using different combinations of optical fibers, we will have different response curves. In the following figure we have plotted the response of two sensors, one with 1 TF and 16 RFs surrounding it, and another one with 1000 TF and 1000 RFs separated in two semicircles [28,89,90].

On one side, the 16RF configuration will respond immediately to the changes in the displacement due to the coaxial configuration of the RF. Small increase in the size of the reflected light cone will significantly increase the amount of light received by the RF's. On the other hand, the hemispheric side-by-side configuration reacts slowly. The curve for 1000 RF is shifted toward larger displacement because the effect of the reflected Gaussian beam becomes more significant at larger distances (**Figure 10.23**).

Many bundle structures for POFDS have been proposed. **Figure 10.24** is a summary of the response curve of POFDS with different fiber probes: single fiber, bifurcated optical fiber, coaxial and semi-circular, and random [88,91].

According to the applications of the sensor, some of the features of the curves should be strengthened, such as the dead zone or offset, the slope of the curve before and after the maximum, (i.e. the sensitivity for small or long distances), and the linear and dynamic range. **Figure 10.24** summarizes some of the fiber arrangements frequently used in bundles to measure distance together with their response curves. A random distribution of TFs and RFs gives the steepest response curve while a semicircular distribution of TF and RF gives the smoothest one. On the other hand, a single fiber POFDS has a response curve with a unique and a steep decreasing slope but a narrow range. Random configuration offers a high front slope sensitivity with a small linear range just the opposite to what happened on the back slope in which the linear range is big and the sensitivity low. For the hemispheric configuration, the peak positions shift for longer distances which results in the extension of linear operating range. It has a big linear range and low sensitivity on the front slope and middle linear range and middle slope sensitivity on the back slope. The random configuration displays an average performance.

Moro et al. [92] proposed an optimization approach that employs a genetic algorithm to design the bundle sensor architecture. From the converged output of the optimization routine, a bundled displacement sensor configuration was designed and tested, offering linear performance with high sensitivity over a variety of axial displacement ranges, and with an acceptable error level.

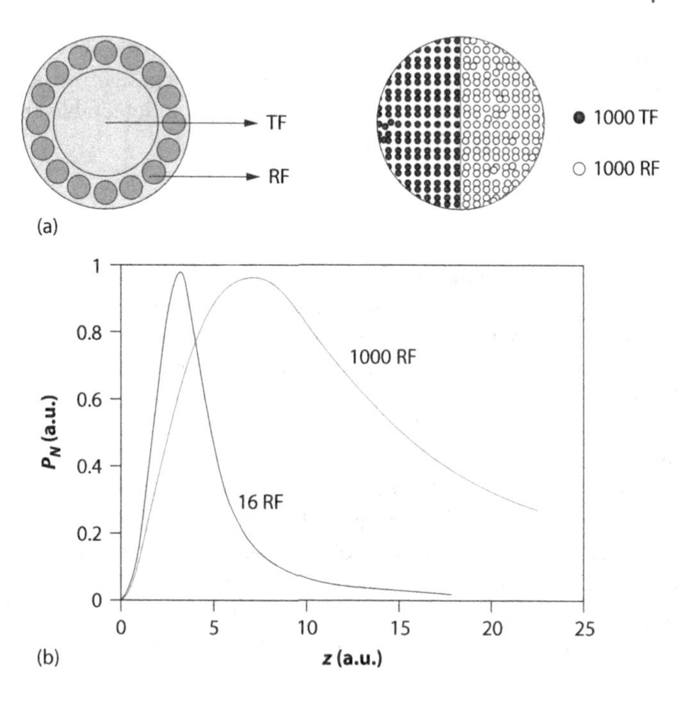

FIGURE 10.23 (a) Structure of two bundles with one TF and 16 RFs and another with 1000 TFs and 1000 RFs. (b) Characteristic curves corresponding to these two fiber probe structures. (Adapted from Abdullah, M. et al., *IEEE Sens. J.*, 13, 4522–4526, 2013.)

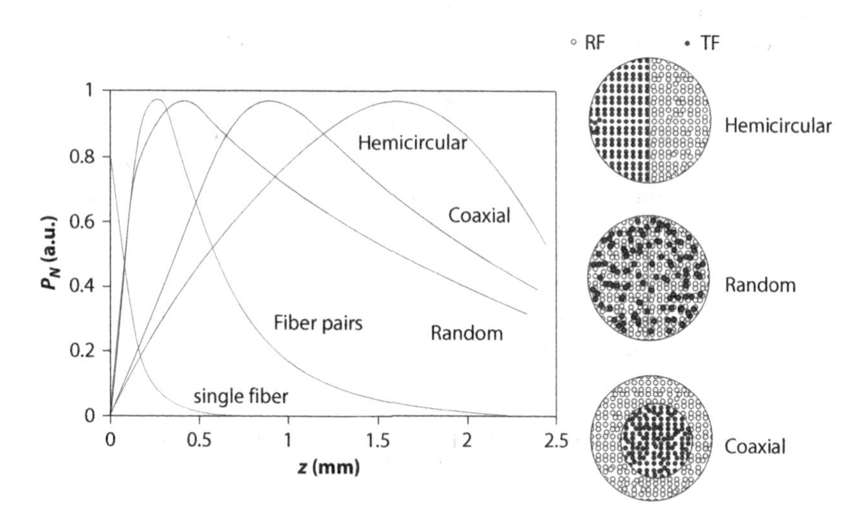

FIGURE 10.24 Characteristic curves corresponding to several fiber probe structures.

10.4.1 Dynamic Range

The dynamic range of a POFDS in dB is defined as:

$$DR = 20\log\frac{Z}{\Delta Z_{min}} \tag{10.26}$$

where Z is the maximum linear response range in which the sensitivity S is almost constant, i.e. the range where the output voltage is proportional to the change in the displacement z. ΔZ_{min} is the displacement detection limit, that is, the minimum detectable change in z which depends on the detector and its associated electronics.

10.4.2 Compensation and Autoreference

For the reflective intensity modulated POFDS, changes of displacement are correlated with changes of the collected light intensity. One of the drawbacks of the early intensity modulated POFDS was its sensitivity to fluctuations of light source power, vibrations, bending of the POFs, and changes of other characteristics of system components. Besides, as we have just seen, the modulation characteristic of the sensor does not only fall on the target distance but also on the roughness of the surface, the relative inclination to the fibers, the optical fiber parameters, the bundle configuration, etc. For instance, different surface reflectivities and light source instabilities greatly impair displacement measurements based on absolute light power.

Bifurcated POFDS depends strongly on target reflectivity [6–83], which varies over time due to external light exposure, aging, or deposition of superficial dust or dirt, all conditions that can occur frequently in harsh industrial environments. Another issue is the output drift of a system when ambient temperature changes. The drift deteriorates the total resolution of the system, and it is mainly generated from two sources: the drift of electronics and the fluctuation of the light source power.

Therefore, light intensity POFDS is prone to change in the transduction process. A lack of a compensation mechanism leads usually to a wrong and noisy measurement. Accordingly, when a higher measurement accuracy and stability is required, we must adopt some compensation technique to reduce or eliminate the effect brought from the above factors.

Some smart techniques have been developed to provide compensation without adding excessive complexity or cost. As shown by Libo et al. sensor schemes with two or more receiving fibers are enough to compensate not only light source intensity variations but also the losses of fiber bends. Suganuma et al. [83] use a trifurcated fiber bundle which consists of one TF and two rings of RFs surrounding the TF as shown in **Figure 10.22a**. The use of a trifurcated fiber bundle not only compensates the light fluctuations originated in different elements of the systems but also extends the linear range of the sensor response [8,80]. By taking the ratio of the output signals from the inner and output rings of RFs or what is better the ratio of the difference to the sum, we can eliminate the variation of the sensitivity of the sensor to the reflectivity of the target, and the working distance can be extended by four times compared with other conventional approaches [42,93,94].

The output voltage of the photodetector depends on its responsivity $R(\lambda)$, the transimpedance of the electronic amplifier circuit, and the received optical power $P(z)$ [95,96]:

$$V_{out} = R(\lambda) * \text{Transimpedance} * \exp\left[-\sum_k \eta_k r_k\right] \Gamma P_o \, F\left(n_T, n_R, a, NA, s_{ij}, z\right) \tag{10.27}$$

For a fiber bundle which has a single TF and two RF rings arranged concentrically at different distances from the center of the bundle:

$$\frac{V_2}{V_1} = \frac{R_2(\lambda) * \text{Transimpedance}(PD2) * F_2\left(n_T, n_R, a, NA, s_{ij}, z\right)}{R_1(\lambda) * \text{Transimpedance}(PD1) * F_1\left(n_T, n_R, a, NA, s_{ij}, z\right)} \tag{10.28}$$

where the magnitudes with subindex 2 are related to the light that has been received by the outer ring of RFs, and the ones with a 1 are related to the light received by the inner circle of RFs. By using this model, it is guaranteed that the final results are independent not only of Γ and P_0 but also from bending losses. Besides, if detectors with variable gain are used, we can extend the range of the sensor as well as the slope of its response curve [96].

10.5 Other Approaches

In this section we will summarize other approaches that have been reported in literature for displacement measurements. Li et al. use a grating panel with a periodic variation of its refractive index to modulate the reflected light intensity. Successful experiments were carried out for the measurement of the lateral displacement in harsh working environments. This approach has two strengths: long distance measurement range (>30 cm), and sense and value of speed [97].

The sensor design suggested by Liu et al. [98] is completely different. Their proposal is founded on the macro-bend coupling effect which causes power coupling between two twisted bending POFs. The coupling power changes with the bending radius of the fibers. Although the characteristic curve of the POFDS is quite linear in the range between 0 and 150 mm (0.12 mm resolution), and it is not influenced by external light sources, its temperature dependence degrades the operation of the sensor. From 25°C to 75°C the temperature drift is above 100%.

Sastikumar et al. uses the displacement sensor to measure the thickness of a transparent plate [99]. The output characteristic of the sensor shows that the peak position is dependent on the thickness and it varies linearly with it. Moreover, the sensitivity increases with the increase in the NA although it does not vary with the RF and TF diameters. The accuracy of identification of peak position is higher when the NA is higher.

10.6 Tip Clearance Measurement

Finally, we will briefly explain the results we have obtained with a POFDS for the measurement of the tip clearance (TC) of aeronautical turbines [95,96,100–102]. TC consists of measuring the distance from the casing of the engine to the tip of the blade (see **Figure 10.25**), which is a critical measure since the efficiency and the health of the engine are directly related to it. As TC values get lower, the blade will be closer to the casing and the engine will be more efficient due to a better sealing and the reduction of unwanted air and gas leaks. In addition, having the blades monitored allows preventive actions such as substituting damaged blades. Considering the engines rotate at high speeds, data acquisition timing becomes critical since it is mandatory to acquire the signal from all blades simultaneously. Blades are exposed to different vibrations: tangential, rotational, and axial, which can affect the measurement and the TC. For that reason, a high-resolution acquisition system offers the chance of characterizing completely the vibrational behavior of the engine apart from TC measurements.

The optical head consisted of a trifurcated bundle as shown in **Figure 10.26** in which the fibers are arranged concentrically. The common leg with the fiber configuration depicted in

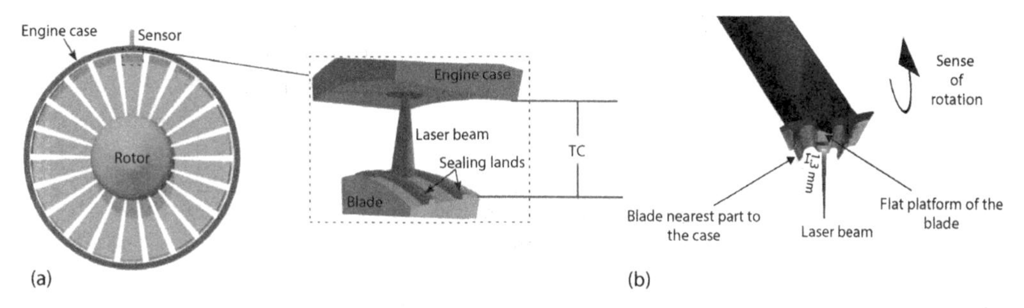

FIGURE 10.25 Schematic of (a) the tip clearance and (b) a detail of a blade.

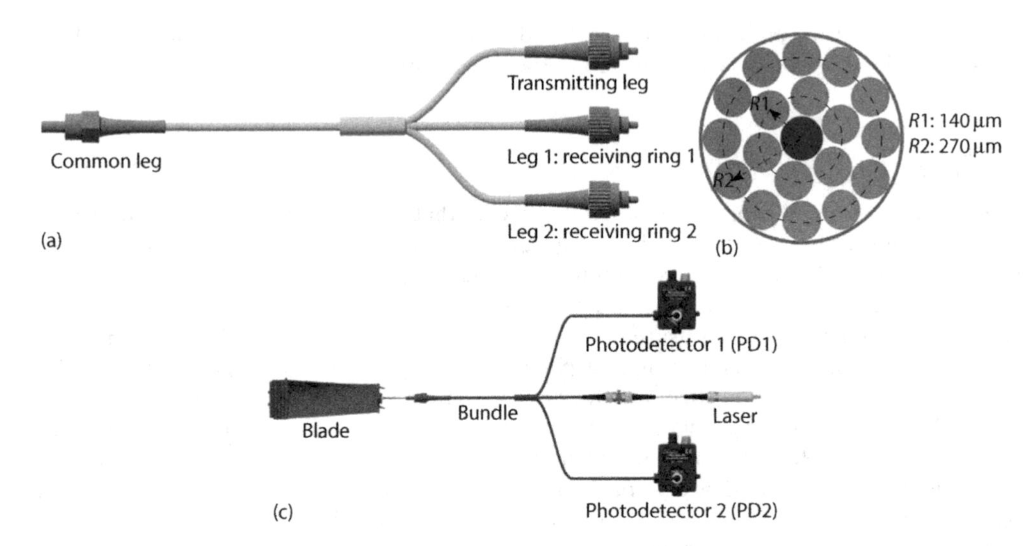

FIGURE 10.26 Schematic of (a) the manufactured bundle, (b) the common leg's cross-section, and (c) its connection to the optical devices.

Figure 10.26 is pointing to the target. Legs 1 and 2 correspond to the RFs of the inner and outer rings respectively and are connected to the photodetectors. The transmitting leg with only one fiber is attached to the laser [103].

Both photodetectors are connected to a DAQ system that takes samples at 2 MS/s per channel simultaneously, so a possible source of error for taking the samples at different moments is avoided. This setting was tested in first place at the laboratory. The calibration curve is shown in **Figure 10.27**. The calibration curve has been obtained by approaching as much as possible the blade to the tip of the bundle and then moving it away in steps of 10 microns. Since TC values in compressors vary from 2 to 4 mm, this bundle suits for these type of measurements. The range of the calibration curve that is used for the measurement goes from approximately 3.5 to 5.5 mm. To accommodate the range of the TC and the linear range of the calibration curve, the bundle was fixed at a distance of 1.3 mm inside the case.

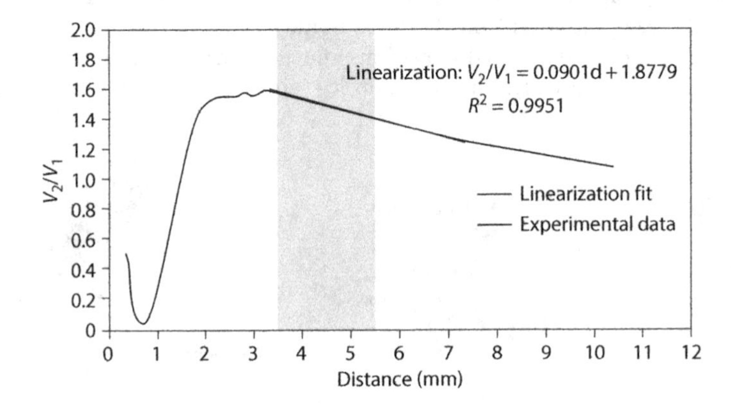

FIGURE 10.27 Calibration curve and its linearization for the expected TC region. (Adapted from García, I. et al., *IEEE J. Lightw. Technol.*, 33, 2663–2669, 2015.)

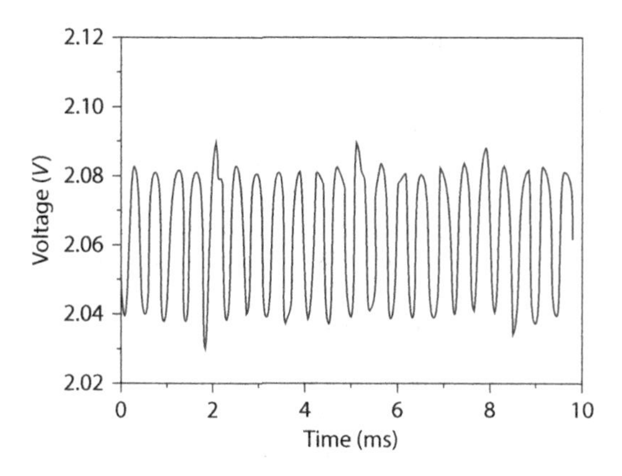

FIGURE 10.28 Typical trace of a measurement of the TC, showing several blades. (Adapted from García, I. et al., *Sensors*, 17, 165, 2017.)

In **Figure 10.28** we can see the response of the sensor for several blades. Within this approach the resolution of the system is around 15 microns, well below the requirements for this kind of application [104].

References

1. Optical methods for distance and displacement measurements, G. Berkovic and E. Shafir, *Advances in Optics and Photonics*, Vol. 4, 2012, 441–471.
2. Fiber-optic displacement sensor with 0.02 mm resolution by white-light interferometry, A. Koch and R. Ulrich, *Sensors and Actuators A: Physical*, Vol. 25, No. 1–3, 1990, 201–207.
3. *Handbook of Optical Fiber Sensors*, J. M. López-Higuera, Wiley, Chichester, England, 2002.
4. Plastic optical fibers: An introduction to their technological processes and applications, J. Zubia and J. Arrue, *Optical Fiber Technology*, Vol. 7, No. 2, April 2001, 101–140.
5. Micro-displacement sensors based on plastic photonic bandgap Bragg fibers, H. Qu, T. Brastaviceanu, F. Bergeron, J. Olesik, and M. Skorobogatiy, *Optical Sensors* 2013, Rio Grande, Puerto Rico United States, 14–17 July 2013.
6. Optical sensors based on plastic fibers, L. Bilro, N. Alberto, J. L. Pinto, and R. Nogueira, *Sensors*, Vol. 12, 2012, 12184–12207.
7. Displacement sensing with polymer fibre optic probe, S. Binu and J. George, *Frontiers in Sensors (FS)*, Vol. 1, No. 3, July 2013, 49–52.
8. A review of recent developed and applications of plastic fiber optic displacement sensors, H. Z. Yang, X. G. Qiao, D. Luo, K. S. Lim, W. Y. Chong, S. W. Harun, *Measurement*, Vol. 48, 2014, 333–345.
9. Fiber optic displacement sensor for industrial applications, D. H. A. Munap, N. Bidin, S. Islam, M. Abdullah, F. M. Marsin, and M. Yasin, *IEEE Sensors Journal*, Vol. 15, No. 9, September 2015.
10. Optical fiber bundle displacement sensor using an ac-modulated light source with subnanometer resolution and low thermal drift, A. Shimamoto and K. Tanaka, *Applied Optics*, Vol. 34, No. 25, 1 September 1995, 5854–5860.
11. Fiber optic displacement sensor for temperature measurement, H. A. Rahman, S. W. Harun, N. Saidin, M. Yasin, and H. Ahmad, *IEEE Sensors Journal*, Vol. 12, No. 5, May 2012.
12. Fiber optic displacement sensors for the measurement of a vibrating object, C.-S. Lin, R.-S. Chang, *Precision Engineering*, Vol. 16, No. 4, October 1994, 302–306.
13. Fiber-optic diaphragm-curvature pressure transducer, C. M. Lawson and V. J. Tekippe, *Optics Letters*, Vol. 8, No. 5, May 1983, 286–288.
14. Plastic optical fiber displacement sensor for study of the dynamic response of a solid exposed to an intense pulsed electron beam, R. B. Salins, *Review of Scientific Instruments*, Vol. 46, 1975, 879.

15. Compensation of an optical fibre reflective sensor, C. P. Cockshott and S. J. Pacaud, *Sensors and Actuators*, Vol. 17, 1989, 167–171.
16. Methodology for the design, production, and test of plastic optical displacement sensors, M. Rahlves, C. Kelb, E. Reithmeier, and B. Roth, *Advanced Optical Technologies*, Vol. 5, No. 4, 1–10.
17. Wide range fiber displacement sensor based on bending loss, J. Zhao, T. Bao, and T. Kundu, *Journal of Sensors*, Vol. 2016, 1–5.
18. Two-dimensional displacement sensor based on plastic optical fibers, A. Vallan, Member, IEEE, M. L. Casalicchio, M. Olivero, G. Perrone, *IEEE Transactions on Instrumentation and Measurement*, Vol. 62, No. 5, May 2013, 1233–1240.
19. Barrier sensor based on plastic optical fiber to determine the wind speed at a wind generator, J. Zubia, O. Aresti, J. Arrúe, and M. López-Amo, *IEEE Journal on Selected Topics in Quantum Electronics*, Vol. 6, No. 5, September/October 2000, 773–779.
20. Modified fiber optic system for monitoring the railgun rail deflections, V. Kleiza and J. Verkelis, *Acta Physica Polonica A*, Vol. 119, No. 2, 2011, 271–273.
21. Gasoline level sensor based on displacement sensor using fiber coupler, Samian, G. Y. Y. Yhosep, A. H. Zaidan, and H. Wibowo, Measurement, Vol. 58, December 2014, 342–348.
22. On-line wear detection of milling tools using a displacement fiber optic sensor, E. Castillo-Castañeda, *Journal of Applied Research and Technology*, Vol. 1, No. 2, México 2003.
23. Plastic optical fibre sensor for spine bending monitoring with power fluctuation compensation, M. A. Zawawi, S. O'Keeffe and E. Lewis, *Sensors*, Vol. 13, 2013, 14466–14483.
24. Detection of stain formation on teeth by oral antiseptic solution using fiber optic displacement sensor, H. A. Rahman, H. R. A. Rahim, S. W. Harun, M. Yasin, R. Apsari, H. Ahmad, W. A. B. Wan Abas, *Optics & Laser Technology*, Vol. 45, 2013, 336–341.
25. Fiber optic displacement sensor using multimode plastic fiber probe and tooth surface, H. A. Rahman, S. W. Harun, M. Batumalay, F. A. Muttalib, and H. Ahmad, *IEEE Sensors Journal*, Vol. 13, No. 1, January 2013.
26. Fiber optic lever displacement transducer, *Applied Optics*, Vol. 18, No. 19, 1 October 1979, 3230–3241.
27. M. L. Casalicchio, Innovative plastic optical fiber sensors. PhD thesis, 2012.
28. Design and operation of different optical fiber sensors for displacement measurements, H. Golnabi, *Review of Scientific Instruments*, Vol. 70, 1999, 2875.
29. Lateral and axial displacements measurement using fiber optic sensor based on beam-through technique, M. Yasin, S. W. Harun, W. A. Fawzi, Kusminarto, Karyono, and H. Ahmad, *Microwave and Optical Technology Letters*, Vol. 51, No. 9, September 2009.
30. Simulation, implementation, and analysis of an optical fiber bundle distance sensor with single mode illumination, H. Huang and U. Tata, *Applied Optics*, Vol. 47, No. 9, 20 March 2008, 1302–1309.
31. Force measuring system for high-precision surface characterization under extreme conditions, R. Nevshupa and M. Conte, *New Trends and Developments in Metrology*, edited by L. Cocco, IntechOpen, London, UK, July 20, 2016.
32. Measurement uncertainty of a fibre-optic displacement sensor, R. Nevshupa, M. Conte, and C. van Rijn, *Measurement Science and Technology*, Vol. 24, 2013, 035104.
33. Displacement response, detection limit, and dynamic range of fiber-optic lever sensors, G. He and F. W. Cuomo, *Journal of Lightwave Technology*, Vol. 9, No. 11, November 1991.
34. The analysis of a three-fiber lever transducer, *Proceedings of the SPIE*, 478, F. W. Cuomo, 1984, 28–32.
35. A light intensity function suitable for multimode fiber optic sensors, G. He and F. W. Cuomo, *Journal of Lightwave Technology*, Vol. 9, 1991, 545–551.
36. Analysis of the compensation mechanism of a fiber-optic displacement sensor, Y. Libo, P. Jian, Y. Tao and H. Guochen, *Sensors and Actuators A*, Vol. 36, 1993, 177–182.
37. Automatic characterization of a bifurcated optical fiber bundle displacement sensor taking into account reflector tilting perturbation effects, J. Brandao Faria, O. Postolache, J. D. Pereira, and P. S. Girao, *Microwave and Optical Technology Letters*, Vol. 26, No. 4, August 20, 2000, 242–247.
38. Modeling the y-branched optical fiber bundle displacement sensor using a quasi-Gaussian beam approach, J. A. Brandao Faria, *Microwave and Optical Technology Letters*, Vol. 25, No. 2, April 20, 2000.
39. A theoretical analysis of the bifurcated fiber bundle displacement sensor, J. Brandao Faria, *IEEE Transactions on Instrumentation and Measurement*, Vol. 47, 1998, 742–747.

40. Fiber-optic diaphragm pressure sensor with automatic intensity compensation, Y. Libo and Q. Anping, *Sensors and Actuators A: Physical*, Vol. 28, No. 1, June 1991, 29–33.
41. A novel fiber-optic slide-sensing scheme, L. Yuan, *Optical Fiber Technology*, Vol. 7, No. 4, October 2001, 340–349.
42. Automatic-compensated two-dimensional fiber-optic sensor, L. Yuan, *Optical Fiber Technology*, Vol. 4, 1998, 490–498.
43. Analytical study of performance variations of fiber optic micro-displacement sensor configurations using mathematical modeling and an experimental test jig, S. S. Patil and A. D. Shaligram, *International Journal of Scientific & Engineering Research*, Vol. 4, No. 11, November 2013, 533–540.
44. Fibre position effects on the operation of opto-pair fibre displacements sensors, R. Jafari and H. Golnabi, *Optics & Laser Technology*, Vol. 43, 2011, 814–819.
45. Radiometric analysis of the light coupled by optimally cut plastic optical fiber amplitude modulating reflectance displacement sensors, S. Hadjiloucas, J. J. Irvine, and J. W. Bowen, *Review of Scientific Instruments*, Vol. 71, No. 8, August 2000.
46. Regularities of signal and sensitivity variation of a reflection fiber optopair sensor dependent on the angle between axes of fiber tips, V. Kleiza and J. Verkelis, *Nonlinear Analysis: Modelling and Control*, Vol. 14, No. 1, 2009, 41–49.
47. Simulation and experimental studies of inclined two fiber displacement sensor, P. B. Buchade and A. D. Shaligram, *Sensors and Actuators A*, Vol. 128, 2006, 312–316.
48. Theoretical and experimental studies on liquid refractive index sensor based on bundle fiber, S. W. Harun, H. Z. Yang, and H. Ahmad, *Sensor Review*, Vol. 31, 2011, 173–177.
49. Mathematical modeling of intensity-modulated bent-tip optical fiber displacement sensors, P. Puangmali, K. Althoefer, and L. D. Seneviratne, *IEEE Transactions on Instrumentation and Measurement*, Vol. 59, No. 2, February 2010, 283.
50. Influence of fiber geometry on the performance of two-fiber displacement sensor, P. B. Buchade and A. D. Shaligram, *Sensors and Actuators A: Physical*, Vol. 136, No. 1, 1 May 2007, 199–204.
51. An analytical analysis of a fiber-optic reflective sensor, V. Kleiza and J. Verkelis, Elektronika ir Elektrotechnika, Vol. 62, No. 6, 2005, 77–81.
52. Modelling light transmission in a fiber-optical reflection system, V. Kleiza, J. Paukšt, and J. Verkelis, *Nonlinear Analysis: Modelling and Control*, Vol. 9, No. 1, 2004, 55–63.
53. Some advanced fiber-optic amplitude modulated reflection displacement and refractive index sensors, V. Kleiza and J. Verkelis, *Nonlinear Analysis: Modelling and Control*, Vol. 12, No. 2, 2007, 213–225.
54. A differential reflective intensity optical fiber angular displacement sensor, B. Jia, L. He, G. Yan and Y. Feng, *Sensors*, Vol. 16, 2016, 1508.
55. Axial and angular displacement fiber-optic sensor, D. Sagrario and P. Mead, *Applied Optics*, Vol. 37, No. 28, October 1998, 6748–6754.
56. Differential reflective fiber-optic angular displacement sensor, M. Shan, R. Min, Z. Zhongn, Y. Wang, and Y. Zhang, *Optics & Laser Technology*, Vol. 68, 2015, 124–128.
57. Modeling and experimental studies on retro-reflective fiber optic micro-displacement sensor with variable geometrical properties, S. S. Patil and A. D. Shaligram, *Sensors and Actuators A: Physical*, Vol. 172, No. 2, December 2011, 428–433.
58. Geometrical parameter analysis of the high sensitivity fiber optic angular displacement sensor, J. M. S. Sakamoto, G. M. Pacheco, C. Kitano, and B. R. Tittmann, *Applied Optics*, Vol. 53, No. 36, 20, December 2014.
59. High sensitivity fiber optic angular displacement sensor and its application for detection of ultrasound, J. Marcos, S. Sakamoto, C. Kitano, G. M. Pacheco, and B. R. Tittmann, *Applied Optics*, Vol. 51, No. 20, 10 July 2012, 4841.
60. High-resolution fibre-optic sensor for angular displacement measurements, A. Khiat, F. Lamarque, C. Prelle, N. Bencheikh, and E. Dupont, *Measurement Science & Technology*, Vol. 21, 2010, 025306.
61. A fiber-optic reflective displacement micrometer, W. H. Ko, K.-M. Chang, and G. J. Hwang, *Sensors and Actuators A*, Vol. 49, 1995, 51–55.
62. Fiber optic angular displacement sensor, C. Wu, *Review of Scientific Instruments*, 66, 1995, 3672.
63. *The Scattering of Electromagnetic Waves from Rough Surfaces*, P. Beckmann and A. Spizzichino, Artech House, Norwood, MA, 1987.

64. Fiber optic surface topography measurement sensor and its design study, Y. Yang, K. Yamazaki, H. Aoyama, and S. Matsumiya, *Precision Engineering*, Vol. 24, 2000, 32–40.
65. Design, development and experimental study of novel configuration of reflectivity invariant retro reflective fiber optic sensor for measuring of surface roughness, S. S. Patila, A. D. Shaligramb, *Journal of Institute of Smart Structures and Systems (ISSS) JISSS*, Vol. 5, No. 1, March–September 2016, 9–13.
66. Surface roughness measurement based on fiber optic sensor, Z. Nan-nan and Z. Jun, *Measurement*, Vol. 86, 2016, 239–245.
67. Theoretical and experimental investigation on the optical in-process measurement of surface characteristics, G. Jiguang, *Journal of Optoelectronics Laser*, Vol. 21, 7, July 2010, 1040–1043.
68. The performance of a fiber optic displacement sensor for different types of probes and targets, M. Yasin, S. W. Harun, H. A. Abdul-Rashid, Kusminarto, Karyono, and H. Ahmad, *Laser Physics Letters*, Vol. 5, No. 1, 2008, 55–58.
69. LED-based fibre-optic sensor for measurement of surface roughness, B. Cahill and M. A. El Baradie, *Journal of Materials Processing Technology*, Vol. 119, No. 1-3, 20 December 2001, 299–306.
70. Design and operation of a double-fiber displacement sensor, H. Golnabi and P. Azimi, *Optics Communications*, Vol. 281, 2008, 614–620.
71. Estimation of metal surface roughness Using fiber optic displacement sensor, S. W. Harun, M. Yasin, H. Z. Yang, Kusminarto, Karyono, and H. Ahmadm SSN 1054660X, *Laser Physics*, Vol. 20, No. 4, 2010, 904–909.
72. Fiber-optic sensor to estimate surface roughness of corroded metals, G. Gobi, D. Sastikumar, A. B. Ganesh, and T. Radhakrishnan, *Optica Applicata*, Vol. XXXIX, No. 1, 2009, 5–11.
73. Theoretical and experimental study of mirrorless fiber optics refractometer based on quasi-Gaussian approach, M. Abdullah, G. Krishnan, T. Saliman, M. F. S. Ahmad, and N. Bidin, *Laser Physics*, Vol. 28, 2018, 035102.
74. Fiber optic displacement and liquid refractive index sensors with two asymmetrical inclined fibers, H. Z. Yang, S. W. Harun, and H. Ahmad, *Sensors & Transducers Journal*, Vol. 108, No. 9, September 2009, 80–88.
75. Measurement of refractive index of liquids using fiber optic displacement sensors, G. Govindan, S. G. Raj, and D. Sastikumar, *Journal of American Science*, Vol. 5, No. 2, 2009, 13–17.
76. An intensity modulated optical fiber displacement sensor with convex reflector, A. D. Gaikwad, J. P. Gawande, A. K. Joshi, and R. H. Chile, *International Journal of Advanced Research in Electrical, Electronics and Instrumentation Engineering*, Vol. 1, No. 1, July 2012, 29–35.
77. Environment-independent liquid level sensing based on fiber-optic displacement sensors, H. Z. Yang, S. W. Harun, H. Arof, and H. Ahmad, *Microwave and Optical Technology Letters*, Vol. 53, No. 11, November 2011, 2451–2453.
78. Enhanced bundle fiber displacement sensor based on concave mirror, H. Z. Yang, K. S. Lim, S. W. Harun, K. Dimyati, and H. Ahmad, *Sensors and Actuators A*, Vol. 162, 2010, 8–12.
79. Theoretical and experimental studies on concave mirror-based fiber optic displacement sensor, H. Z. Yang, S. W. Harun, and H. Ahmad, *Sensor Review*, Vol. 31, No. 1, 2011, 65–69.
80. A novel fiber optic sensor used for small internal curved surface measurement, Y. Zhao, P. Li, C. Wang, and Z. Pu, *Sensors and Actuators*, Vol. 86, 2000, 211–215.
81. Theoretical and experimental study on the fiber optic displacement sensor with two receiving fibers, S. W. Harun, H. Z. Yang, M. Yasin, and H. Ahmad, *Microwave and Optical Technology Letters*, Vol. 52, No. 2, February 2010, 373–376.
82. Fiber optic lever displacement transducer, R. O. Cook and C. W. Hamm, *Applied Optics*, Vol. 18, No. 19, 1979, 3230–3241.
83. Force measuring system for high-precision surface characterization under extreme conditions, R. Nevshupa and M. Conte, chapter 6, *New Trends and Developments in Metrology*, Edited by L. Cocco, IntechOpen, London, UK, July 20, 2016.
84. Development of a differential optical-fiber displacement sensor, F. Suganuma, A. Shimamoto, and K. Tanaka, *Applied Optics*, Vol. 38, No. 7, 1999, 1103–1109.
85. Geometrical analysis of an optical fiber bundle displacement sensor, A. Shimamoto and K. Tanaka, *Applied Optics*, Vol. 35, No. 34, December 1996, 6767–6774.
86. Modulation function of a reflective fiber sensor with random fiber arrangement based on a pair model, Z. Zhao, W. S. Lau, A. C. K. Choi, and Y. Y. Shan, *Optical Engineering*, Vol. 34, No. 10, 1995, 3055.

87. Research on displacement sensor of two-circle reflective coaxial fiber bundle, X. Zhang and L. Yang, *Proceedings of the 2008 IEEE/ASME International Conference on Advanced Intelligent Mechatronics*, July 2–5, 2008, Xi'an, China.
88. Theoretical and experimental study on the optical fiber bundle displacement sensors, H. Cao, Y. Chen, Z. Zhou, and G. Zhang, *Sensors and Actuators A*, Vol. 136, 2007, 580–587.
89. Performance of a new bundle fiber sensor of 1000 RF in comparison with 16 RF probe, M. Abdullah, M. Yasin, and N. Bidin, *IEEE Sensors Journal*, Vol. 13, No. 11, November 2013, 4522–4526.
90. https://www.mtiinstruments.com/technology-principles/fiber-optic-sensors/
91. Theoretical modeling, simulation and experimental studies of fiber optic bundle displacement sensor, S. S. Patil and P. B. Buchade, *Sensors and Actuators A: Physical*, Vol. 201, 15 October 2013, 79–85.
92. Using a validated transmission model for the optimization of bundled fiber optic displacement sensors, E. A. Moro, M. D. Todd, and A. D. Puckett, *Applied Optics*, Vol. 50, No. 35, 2011, 6526–6535.
93. Fiber optic displacement sensor with new reflectivity compensation method, A. Wego, and G. Geske, *Journal of Sensor Technology*, Vol. 3, 2013, 21–24.
94. Study on an intelligent optical fibre displacement sensor, J. Zhao, X. Zhang, and Y. Wang, *Procedia Engineering*, Vol. 15, 2011, 989–993.
95. An optical fiber bundle sensor for tip clearance and tip timing measurements in a turbine rig, I. García, J. Beloki, J. Zubia, G. Aldabaldetreku, M. A. Illarramendi, and F. Jiménez, *Sensors*, Vol. 13, No. 6, 2013, 7385–7398.
96. Different configurations of a reflective intensity-modulated optical sensor to avoid modal noise in tip-clearance measurement, I. García, J. Zubia, A. Berganza, J. Beloki, J. Arrue, M. A. Illarramendi, J. Mateo, and C. Vazquez, *IEEE Journal of Lightwave Technology*, Vol. 33, No. 12, June 15, 2015, 2663–2669.
97. An optical fiber lateral displacement measurement method and experiments based on reflective grating panel, Y. Li, K. Guan, Z. Hu and Y. Chen, *Sensors*, Vol. 16, No. 6, 2016, 808.
98. A wide-range displacement sensor based on plastic fiber macro-bend coupling, J. Liu, Y. Hou, H. Zhang, P. Jia, S. Su, G. Fang, W. Liu, and J. Xiong, *Sensors*, Vol. 17, 2017, 196.
99. Determination of the thickness of a transparent plate using a reflective fiber optic displacement sensor, D. Sastikumar, G. Gobi, and B. Renganathan, *Optics & Laser Technology*, Vol. 42, 2010, 911–917.
100. Tip-clearance measurement in the first stage of the compressor of an aircraft engine, I. García, R. Przysowa, J. Amorebieta, and J. Zubia, *Sensors*, Vol. 16, 2016, 1897.
101. Numerical analysis of the blade tip timing signal of a fiber bundle sensor probe, H. Guo, F. Duan, Z. Cheng, *Optical Engineering*, Vol. 54, No. 3, 2015, 034103.
102. Theoretical and experimental study on wide range optical fiber turbine flow sensor, Y. Du and Y. Guo, *Sensors*, Vol. 16, No. 7, 2016, 1095.
103. Optical fiber sensors for aircraft structural health monitoring, I. García, J. Zubia, G. Durana, G. Aldabaldetreku, M. A. Illarramendi, & J. Villatoro, *Sensors*, Vol. 15, No. 7, 2015, 15494–15519.
104. Optical tip clearance measurements as tool for rotating disk characterization, I. García, J. Zubia, J. Beloki, J. Arrue, G. Durana, & G. Aldabaldetreku, *Sensors*, Vol. 17, No. 1, 2017, 165.

Chemical Sensing with POF

Filipa Sequeira, Rogério N. Nogueira, and Lúcia Bilro

Contents

11.1 Introduction

What is a sensor? This may seem a meaningless question although this is an important aspect that's worth clarification. Sensors are present in everyday life without even being noticed, as they are so common nowadays. A sensor is expected to give a reliable output of a certain parameter that is aimed to be detected and/or measured, in a repeatable way and in real-time, without interference of external media. When we dive into optical chemical sensors, also known as optodes or optrodes, this means that a sensing device based on light can be used to determine the presence (on/off) or the properties of a specific chemical specie over time, i.e. concentration. Chemical sensors are expected to be small, act reversibly, be free from sample treatment, work in complex samples (selectivity), and give a response in short time. The lifetime of the sensor is also an important parameter to be addressed. In on/off sensing the disposable sensors can be of interest (although less sustainable and cheap solutions), where reversibility and lifetime (in operation) are no longer an issue. Single-use disposable sensors are known as probes.

Chemical sensing with optical fibers allows fast and real-time detection, with the possibility of low-cost equipment and procedures. Moreover, it is possible to develop several in-line remote sensing systems with real-time monitoring through specific computational platforms. Chemical analysis is usually time consuming and performed in a laboratory by a skilled operator using expensive equipment or complex procedures, critical aspects that can be overcome with optical fiber sensing. Nowadays, smaller and miniaturized components are being developed with high quality, which allows increased interest, development, and performance of fiber optic chemical sensors (FOCS).

The basic design of an optical sensing system includes a light source, a medium or waveguide where the light can propagate, optical detection, and data acquisition; see **Figure 11.1**. The changes in the light propagation can therefore be monitored as a consequence of the interaction with the external medium. A transduction mechanism is necessary in chemical sensing, responsible for the conversion of the desired quantity to be detected in a variation that can be measured, namely optical properties (through absorption, fluorescence, diffraction, scattering, etc.).

POF technology is a good solution when short lengths of sensing elements are foreseen, due to the high attenuation of light while traveling in the POF in comparison with glass optical fibers (GOFs). This attenuation is dependent on the length and material of the fiber as well as the wavelength of the light source. Usually the selection of the light source, and respective wavelength, has to take into account the window of lower attenuation of the selected POF. Therefore, wavelengths around 600 nm are commonly used in sensing schemes with POF. Nevertheless, POFs are especially advantageous due to their excellent flexibility, easy handling, and high numerical aperture which makes them suitable to use with low-cost light emitting diodes (LEDs) and large diameters allowing easier and cheaper alignment and interconnection. Computers or microprocessors are used to control the optical instrumentation and are employed to analyze the output signals.

Chemical sensing with POFs means that a chemical specie will be detected and/or measured through optical detection; therefore, two basic operation principles can be present: label-free (an intrinsic property of the analyte allows for its detection without the use of labels or indicators) or label based (when the analyte does not possess an intrinsic property that can be used for sensing and so there is the need of labels or indicators). Furthermore, the POF can act only as a waveguide for light propagation (extrinsic sensing) or can also act as the substrate/sensing platform (intrinsic sensing) together with the sensing material/selective layer, which can be immobilized on the surface or end-face of the fiber.

Intrinsic fiber optic chemical sensors have been reported in the scientific literature since 1946 (Lieberman, 1993). Hesse described an oxygen sensor in 1974 (Baldini, Chester, Homola, & Martellucci, 2006) and, in 1975, Lübbers and Opitz developed a carbon dioxide (CO_2) and oxygen (O_2) sensor used in biological fluids, giving rise to optical fiber-based biosensors (Biran, Yu, & Walt, 2008). The first POF chemical sensor is believed to have been reported by Zhou et al. in 1991. These POF chemical sensors were developed by covalently bonded selective chemical indicators systems with the POF, allowing the detection of various chemical parameters, both in gas and liquid media (Zhou, Tabacco, & Rosenblum, 1991).

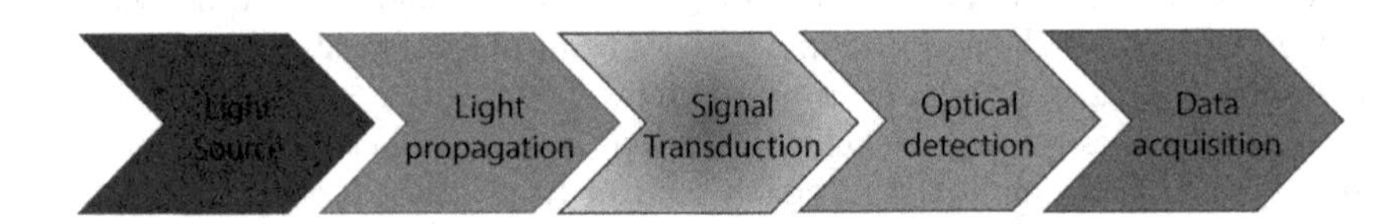

FIGURE 11.1 Schematic representation of the basic design of an optical sensing system.

11.2 Light-Matter Interaction

Photonics is the "science of light," the technology that enables us to generate, control, and detect light waves and photons, the particles that constitute light. The light that reaches the planet comes from our closest star, the Sun. The colors of the rainbow, visible with the naked eye, are only a small part of the electromagnetic spectrum of light, called visible light; see **Figure 11.2**. The energy of a photon depends on the frequency (f) of the light and consequently on its wavelength (λ); higher frequency and lower wavelength mean higher energy:

$$E = h.\,f = \frac{h.\,c}{\lambda} \tag{11.1}$$

where h is the Planck's constant (6.626×10^{-34} J.s) and c is the velocity of light in vacuum (2.998×10^{8} m/s).

The light that reaches the planet interacts with all the particles, ions, and molecules that constitute the atmosphere, giving rise to different visual effects visible on Earth. The interaction of light with matter and chemical species is dependent on their characteristics and also on the wavelength of the incident light. These light–matter characteristic interactions allow the so-called spectroscopic sensing techniques: reflection, refraction, scattering, diffraction, absorption, emission, etc.

The velocity of the light that propagates in a medium will decrease depending on the refractive index of the medium, n; the higher the refractive index, the lower the velocity, v, at which the light travels in the medium:

$$v = \frac{c}{n} \tag{11.2}$$

When light is incident on an interface between two mediums it can be reflected, transmitted and refracted, scattered, and/or absorbed. As depicted in **Figure 11.3** the angle of reflection (θ_r) is equal to the incident angle (θ_1) according to the law of reflection, and the angle of refraction (θ_2) is dependent on the refractive index of the mediums according to the Snell's law:

$$n_1.\sin\theta_1 = n_2.\sin\theta_2 \tag{11.3}$$

Smooth and flat surfaces allow specular (mirror-like) reflections, while rough and irregular surfaces give rise to diffuse reflections, as depicted in **Figure 11.4**.

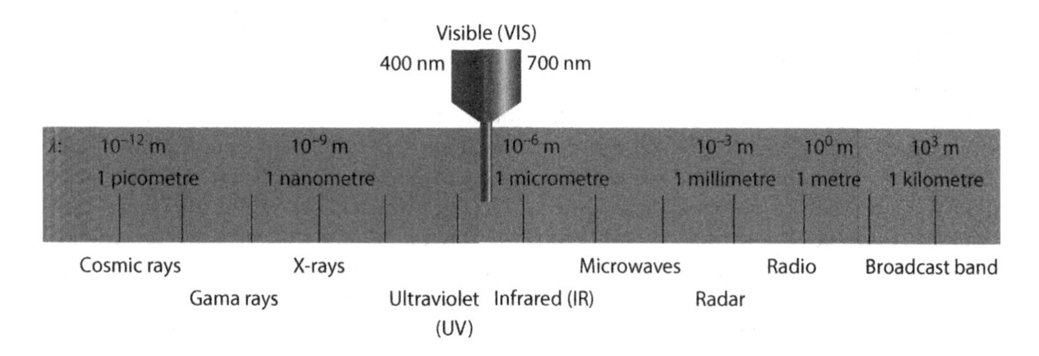

FIGURE 11.2 Electromagnetic spectrum.

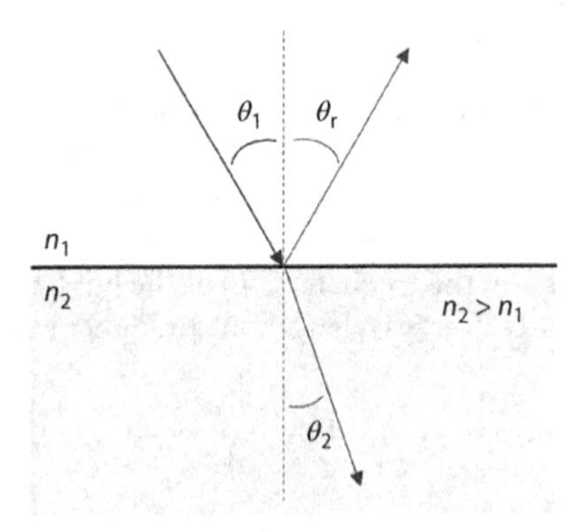

FIGURE 11.3 Light ray reflected and refracted after incidence on an interface between two mediums.

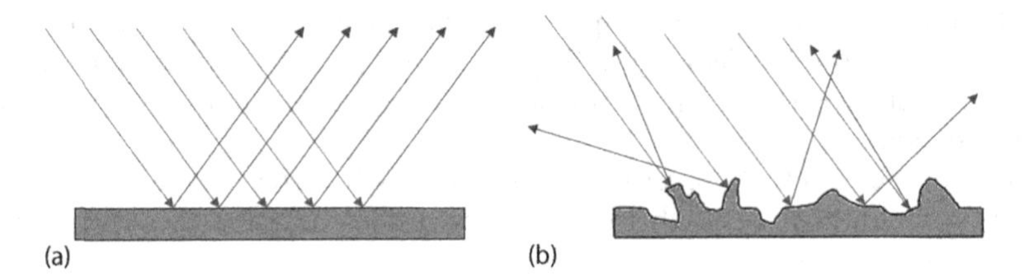

FIGURE 11.4 Representation of (a) specular reflection and (b) diffuse reflection, after incidence on a smooth flat surface and on a rough and irregular surface, respectively.

When light interacts with matter, a redirection of the light may occur, with the same or higher wavelength (lower energy), and it may have different polarization. When the scatterer has the same dimensions as the wavelength of the incident light, or higher, all wavelengths are equally scattered, known as Mie scattering. When light from the Sun interacts with the water droplets in a cloud, all the wavelengths of the visible light are scattered in a similar way, making clouds white. When the dimensions of the scatterer are much smaller than the wavelength of the light, the light can be absorbed and reemitted in a different direction with the same (Rayleigh scattering) or longer (Raman scattering) wavelength. In the last case, the molecule absorbs light and gets into an excited state, energy that is reemitted later when the molecule returns to the ground state (lower energy). The air molecules like di-oxygen (O_2) and di-nitrogen (N_2) are much smaller than water droplets and more effectively scatter shorter wavelengths, like blue and violet, which gives us the impression that the sky is blue (Rayleigh scattering).

Absorption is very common. When visible light hits an object, its color will be the set of the various wavelengths that are reflected, which corresponds to the so-called complementary color of the wavelength(s) that is/are absorbed. Absorption is therefore an optical property that can also be used for sensing, from microwaves (MW) to the infrared (IR), visible (VIS), and ultraviolet (UV) (Menzies & Chahine, 1974; Barringer & Davies, 1977). However, when sensing with POFs the windows of transmission (low attenuation) are usually centered in the visible part of the light spectrum, from 400 to 800 nm, depending on the characteristics of the POF. In order to obtain a transmission of light with low losses in other regions of the electromagnetic spectrum, POFs can be doped or produced with special materials and geometries.

Absorption is a transfer of energy from the light wave to the atoms or molecules of the medium. Due to the absorption of energy, the electrons that constitute the atoms can be transferred to higher energy states (excited state) and, in the case of molecules, vibration or rotational states can be present. Infrared (IR) radiation is not energetic enough to excite electrons, but this radiation is absorbed generally by all organic molecules causing an excitation in vibrational energies. The IR absorption spectra of compounds are a unique signature of their molecular structure. The IR absorption in specific bands (IR absorptiometry) allow the development of sensors for gases and pollutants (CO_2, CH_4, SO_2, NO_x) and other hydrocarbons while the ultraviolet (UV) absorption allow the monitoring of nitrate, nitrite (with absorption bands around 300 and 350 nm, respectively) (Moo, Matjafri, Lim, & Tan, 2016) or PAHs (polycyclic aromatic hydrocarbons) (Axelsson et al., 1995) in water as well as different organic compounds such as potassium hydrogen phthalate (KHP, acidic salt compound) (Kim, Eom, Jung, & Ji, 2016).

Simple optical configurations can be used in absorption studies; a light source and an optical detector are basic instruments for performing the analysis. As the incident light (I_0) is absorbed, less light is transmitted (I) and a decrease in the light that reaches the detector is obtained; see **Figure 11.5**. The transmittance (T) and absorbance (A) can be easily calculated through the following equations:

$$T = \frac{I}{I_0} \text{ and } A = -\log(T) \quad \textbf{(11.4)}$$

When the detection and monitoring of chemical species are foreseen, a finer analysis must be performed. If the target analyte has a characteristic absorption at a specific wavelength, the broad spectrum of the light that is transmitted will have a valley at the absorbed wavelength. A simple example will be given in order to clarify the procedures that can be used. Rhodamine B is commonly used as an indicator due to the characteristic absorption in the visible region and the fluorescence emission depending on the form that is used. In order to obtain the transmitted spectra after passing through a liquid sample, a white light source and a spectrometer can be used placing the sample in between; see **Figure 11.6**. The transmission spectra of water and a

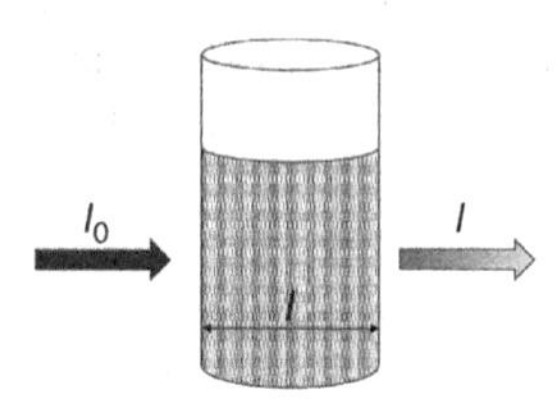

FIGURE 11.5 Light absorption by chemical species in a sample ($I < I_0$) when the light travels through an absorbing medium with length l.

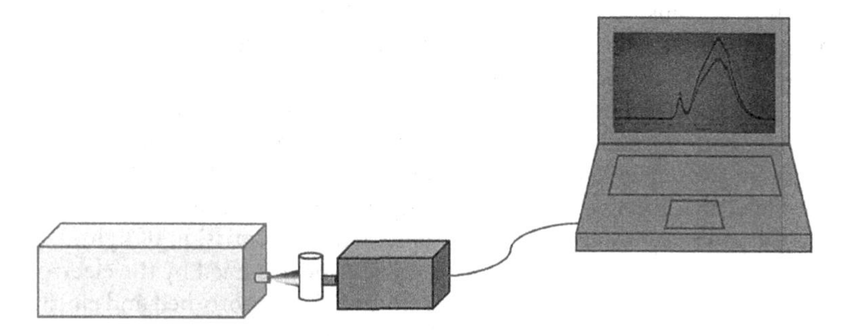

FIGURE 11.6 White light source, sample, and spectrometer connected to a laptop—simple optical setup that can be used to obtain the transmitted spectrum of different samples.

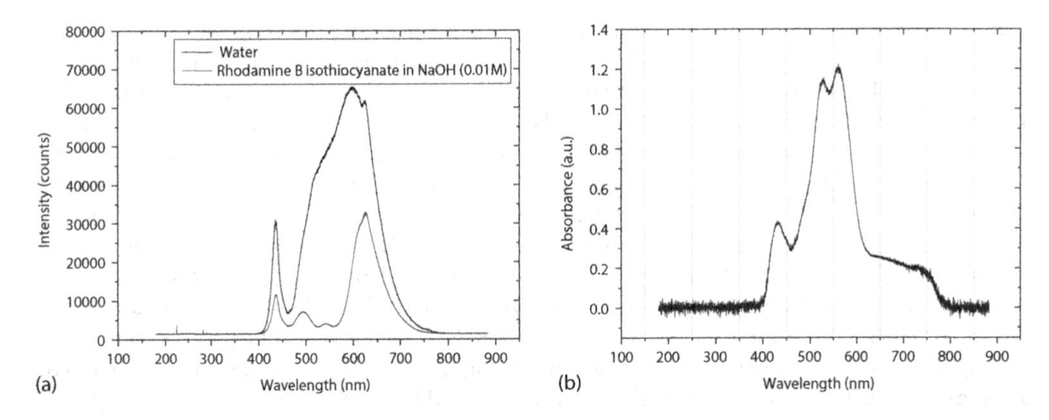

FIGURE 11.7 (a) Transmission spectra—water and solution of rhodamine B isothiocyanate and (b) obtained absorbance of the rhodamine B solution.

solution of rhodamine B isothiocyanate (ITC) in sodium hydroxide (NaOH, 0.01M) are depicted in **Figure 11.7** as well as the respective calculated absorbance according to **Equation 11.4**. From the manufacturer it was expected a maximum absorption (λ_{max}) at 555 nm.

The absorption of light by an absorbing specie present in a medium is directly related to its concentration; the higher the concentration, the higher the absorption of light that passes through the medium, a relation known as the Beer-Lambert law, which can be expressed as:

$$A = \varepsilon . l . c \tag{11.5}$$

where A is the absorbance, ε is the molar absorptivity (dependent on the wavelength of the light and characteristic of the chemical specie), l is the distance that light travels in the medium (in cm), and c is the molar concentration (mol/L) of the chemical specie in the medium. The Beer-Lambert law can be securely used when the light source is monochromatic (the molar absorptivity is wavelength dependent) and the concentration (c) is actually the concentration of the absorbing species (Swinehart, 1962). Furthermore, the Beer-Lambert law is adequate for concentration studies related with light absorption when the concentration of the specie is low (<0.01 M) and in clear solutions, or else electrostatic interactions and refractive index changes may occur as well as scattering due to the presence of particles in the sample. The sample and chemical species that are present should not have fluorescence or phosphorescence properties.

Fluorescence and phosphorescence are two examples of luminescence (emission of radiation), which can be visible or not and are a consequence of light absorption. A fluorescent specie emits light while absorbing light, and the emission stops immediately as the source of light is no longer present, as for example in traffic signs or fluorescent clothes. On the contrary, a phosphorescent specie will maintain the emission of light even after the source of light is no longer present. Chemiluminescence occurs when the emission of light is a result of a chemical reaction, and when present in glowing animals like jellyfish or some microorganisms, it is known as bioluminescence. When absorption of light occurs, the electrons that constitute the atoms and molecules will occupy higher energy states (excited state), after which they will decay again to the ground state (fundamental state) with the emission of light. Depending on the molecule and atoms, only discrete energy levels can be occupied by the electrons since the energy is quantized, which means only certain amounts can be absorbed and emitted. The necessary energy for excitation corresponds to the exact difference between the energy levels. The imaging of algae in the open sea is possible through the characteristic red fluorescence of chlorophyll.

Diffraction of the light that constitutes the visible part of the spectrum can be observed in a rainbow or when light hits the surface of a CD at a certain angle. Diffraction is also present in photonic crystals where color changes due to the change of the distance of nanoparticles in an ordered structure, for example as observed in butterflies, where the color observed (known as structural color) is due to the diffraction and interference of light. Diffraction and interference are also involved in the production and visualization of holograms, where, basically, a diffraction pattern is recorded in a photosensitive medium, and all the information about the object is recreated when light is incident in the same angle of recording. In this case Bragg's law can be used, the law of diffraction for constructive interference. The diffraction of light allows the development of optical sensors based on diffraction gratings and photonic crystals since the 1980s (Taylor, 1987; Morey, Dunphy, & Meltz, 1991; Lukosz, 1995; Asher & Holtz, 1997; Asher, Holtz, Weissman, & Pan, 1998; Homola, Koudela, & Yee, 1999).

11.3 Light Sensing with POF

Fiber optic sensors allow distributed and remote sensing in harsh environments and inaccessible sites at long distances. Improvements in the transparency and fabrication process of polymeric optical fibers (POFs), which started several years ago, have encouraged their use for optical fiber sensing (Koike & Koike, 2011).

Although the use of POFs for sensing has inherent advantages, one drawback is still present—the part of the spectrum of light that can be used for sensing applications. The attenuation and losses are dependent on the material of the POF, its chemical composition, and imperfections. The chemical composition and physical structure of the fiber are considered intrinsic losses, whereas the presence of contaminants or imperfections due to the production of the fiber are considered extrinsic losses. Nevertheless, the surface properties of the POFs can be more easily manipulated when compared to silica fibers, using simple and wide-ranging chemistry techniques to achieve the desired sensing application (Tow et al., 2017). More detailed information about the optical properties, transparency, and attenuation characteristics of POFs can be found in Chapter 2 ("Principles of Polymer Optical Fibers").

POFs are waveguides that allow the propagation of light through attenuated total reflection (ATR), and the development of POF sensors is based on the properties of light and its interactions with matter as described in the previous section. Light, when injected in one side of the fiber, will propagate in the core through multiple reflections in the core-cladding interface, due to the difference between the refractive index of the core (n_{co}) and cladding (n_{clad}) of the fiber. When the light reaches this interface between the two media, part of the light will be reflected to the core and part will be refracted into the cladding; see **Figure 11.8**. According to Snell's law (**Equation 11.3**), the refracted light will travel parallel in the cladding (90° with the normal of the interface) when the incident angle reaches a certain value, known as the critical angle (φ_c, defined in **Equation 11.6**),

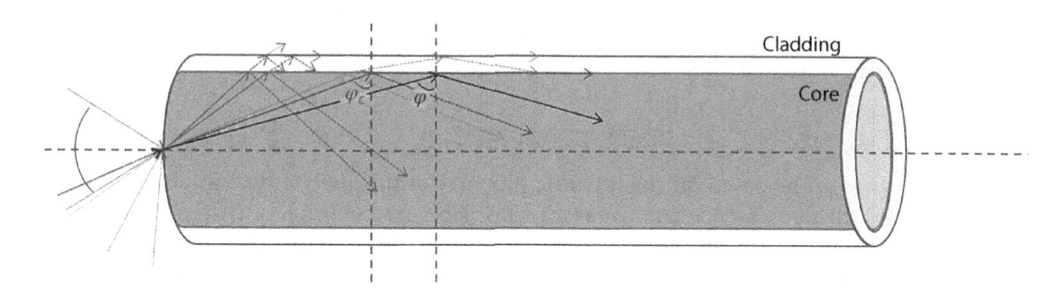

FIGURE 11.8 Schematic representation of light propagation in a step index multimode POF.

after which the light will be totally reflected in the interface between the two media (when $\varphi > \varphi_c$). The acceptance cone illustrates the incident light rays that will undergo total reflection inside the fiber core.

$$\varphi_c = \arcsin\left(\frac{n_{clad}}{n_{co}}\right) \tag{11.6}$$

Nonetheless, the totally reflected light evanesces into the cladding with penetration depth around the wavelength of the light ($\approx \lambda$). Several sensing schemes are based in this interaction between the light and this region of fiber, known as evanescent field sensing. In a straight configuration, the interaction of light with variations that may occur in the cladding are negligible; therefore, when POF sensors are developed, this interaction must increase. Some possible configurations in order to increase the interaction with the external medium is tapering or bending of the fiber in specific regions (sensing region). In the first case, tapered-POFs, the fiber is stretched (with temperature) in order to decrease the waist diameter, and in the second case a bending is permanently formed—U-bent POFs—allowing more light to leak out from the fiber and interact with the external medium. The cladding of the fiber can also be removed through etching (chemical dissolution of the cladding) and/or polishing (lateral polished or D-shaped-POFs).

For the development of POF chemical sensors that are sensitive and specific for a target analyte, the POF can be coated with active chemical layers in order to allow the interaction or binding of the analyte with the active layer making possible the transduction into a measurable signal of the property that is aimed to be measured. POF chemical sensors can be considered extrinsic or intrinsic, depending on if the POF is only the waveguide where the light travels (bringing light from and to a chemically-active coating or sensing medium) or if the POF is actively interacting with the external medium/selective layer acting also as a transducer and sensing element. The use of photonic crystal fibers (PCF), also known as microstructured POFs (mPOFs), allow liquid or gas samples to propagate inside the holes of the fiber which can promote the introduction of dopants in the fiber as well as the interaction between light and chemical or biochemical species over significant lengths (Webb, 2010).

Extrinsic and intrinsic configurations can also be based on reflection, where the light is reflected back and enters in the same or different fiber or in transmission, where the light propagates and interacts with the external medium/sensitive layer and propagates always in the same fiber or enters in a different one. When using reflection-based sensing schemes, it is mandatory to have a POF coupler connecting the POF to the optical source and detector, while in transmission schemes, the POF can be connected directly to the source and detector.

The optical fiber sensing principles are well explained and detailed in Chapter 3 ("Optical Fiber Sensing Principles"); sensing configurations can be based on intensity, wavelength, interference, and polarization through the development of sensing platforms based on intrinsic or extrinsic sensors. Intensity-based configurations allow low-cost experimental setups with LEDs and photodetectors as well as to monitor the variation of the transmitted/reflected light through spectroscopic principles, while other sensing techniques normally allow obtaining sensors with higher sensitivities and resolution, although more expensive and complex equipment is usually needed.

11.3.1 Direct (Label-Free) and Indirect (Label-Based) Sensing Methods

Chemical sensing can be based on the intrinsic property of the analyte such as characteristic absorption bands or fluorescence, known as direct or label-free sensing, or through the use of labels, indicators, probes, etc., in the case of species that cannot be sensed directly. Absorption, fluorescence, surface plasmon resonance, evanescent field, and photonic crystal-based sensing are considered label-free methods for sensing. In the case of indirect sensing, the labels or indicator probes used are supposed to remain inert (label and indicator-based sensing methods), as, for example, a fluorescent label that can be used as a marker by binding with a specific target and allow the detection of that target through the characteristic fluorescence of the label. Nowadays, labels are easily found as they are commercially available with different characteristics. Unlike labels, molecular indicators are supposed to react and respond to a chemical parameter (pH, chemical species, etc.).

Absorption studies, already described in the previous section, can also be easily monitored with POFs. The analysis can be performed through an extrinsic configuration and the light propagation in the POF is directed to and from the sensing medium (see **Figure 11.9a**) or based on intrinsic configurations where the light propagates continuously in the POF and is absorbed through the medium (see **Figure 11.9b**). If the concentration of the absorbing specie must be determined, is important that the concentration of the specie is low and the solution is clear, so that changes in the medium's refractive index or turbidity are not present and do not interfere with the measurements. More detailed information about the possibility to determine the concentration of an absorbing specie through absorption studies (and the Beer-Lambert law) can be found in the previous section.

When dealing with absorption spectroscopy, the matrix where the target is aimed to be sensed should be well-known in order to ensure that no other species will absorb in the same wavelength.

The same configurations can be used in the case of fluorescence studies. POFs can be doped with fluorescent dyes or also coated with specific layers that include a fluorescent specie that binds to the target analyte and allows label-based sensing applications. In both cases, the monitoring of light absorption and fluorescence or phosphorescence can be performed using simple and low-cost equipment such as LEDs and photodetectors (by monitoring the output optical power that reaches the detector when the matrix is known and only the target will absorb and emit light) or can be performed with a white light source and a spectrometer, allowing more detailed information, such as the wavelength at which light is absorbed and/or emitted by the specie.

In refractometric studies based on intensity sensing schemes, the variation of the medium refractive index will be responsible for the variation in the light output that reaches the detector. In extrinsic configurations, the light cone will decrease with the increase of refractive index and more light will re-enter in the POF and reach the detector. In intrinsic configurations, geometrical bending or deformation of the fibers, as well as increasing the roughness of the sensing region by polishing, can allow increase of the penetration depth of the evanescent field, which results in an increase of the sensitivity. When using an unclad straight or U-bent fiber, the increase in the refractive index will cause a decrease in the output power (when approaching the cladding refractive index), as more light is leaked out, a phenomenon that is favored with the decrease of the bending radius (see **Figure 11.10a**); with polished fibers, in straight or D-shaped POFs the output power increases as the refractive index of the medium increases, mostly due to roughness on the surface (Sequeira et al., 2019; Sequeira et al., 2016), and when a bending is applied in the POF, the sensitivity to refractive index decreases, see **Figure 11.10b**.

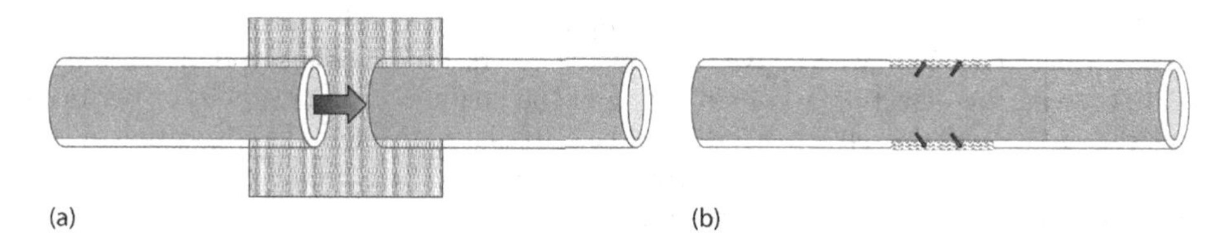

FIGURE 11.9 Schematic representation of light absorption by a specie in (a) extrinsic and (b) intrinsic configurations.

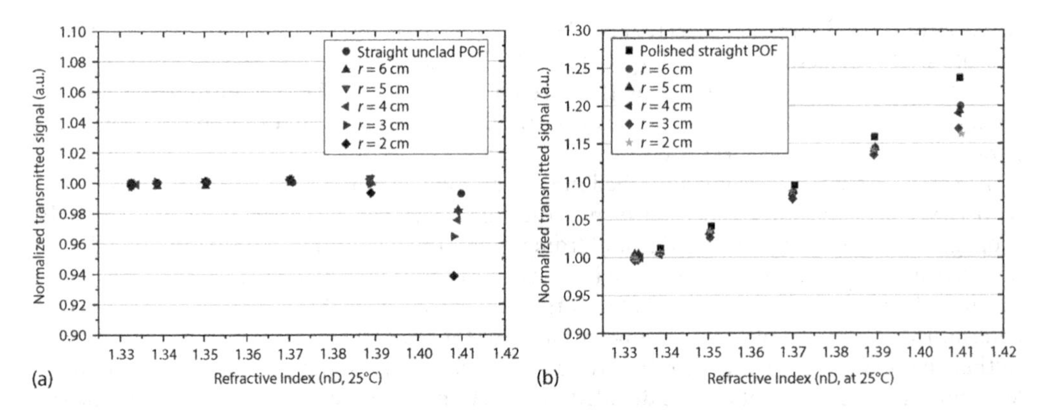

FIGURE 11.10 Normalized transmitted signal with increasing RI and decreasing bending radius: (a) unclad POF and (b) polished POF.

Although the above examples cannot be considered as chemical sensors, when only one species is present that will affect the medium refractive index, the concentration of the species can be related with the medium refractive index in a known matrix and temperature, through sensors calibration, i.e., the determination of sugar content in a liquid. Memon et al. reported in 2018 the development of a U-bent POF for the measurement of ultralow concentrations of ethanol in water, corresponding to the bioethanol production rate by cyanobacteria (0.1–0.5 $gL^{-1}day^{-1}$); the sensing method was based in the evanescent field, by the detection of the transmitted light that reaches the detector when the concentration of ethanol in water changes in the external medium (Memon, Ali, Pembroke, Chowdhry, & Lewis, 2018). The results obtained show linearity, repeatability, sensitivity of 817.76 O.D/RIU (O.D refers to optical density, unit of absorbance) with a limit of detection (LOD) of 9.2×10^{-7} RIU (predicted refractive index). Being a U-bent POF, this sensor does not present a selective property; therefore, it can be used for the monitoring of a chemical species when the matrix is well known and does not affect the results obtained (interferers are not present, or can be isolated).

Surface plasmons are collective charge oscillations that occur at the interface between conductors and dielectrics, ranging from freely propagating electron density waves along metal surfaces (surface plasmon resonance, SPR) to localized electron oscillations in metal nanoparticles (localized surface plasmon resonance, LSPR). SPR and LSPR are widely used as sensing principles in optical fiber technology for chemical and biochemical detection and monitoring through the deposition of sensitive and selective layers in the optical fiber above the metallic layer, or incorporated with metal nanoparticles; see **Figure 11.11a** and **b**. At a certain angle, light is not reflected but absorbed by the coated layer, and this angle depends on the refractive index of the coating. When a binding or adsorption is present between the target analyte and the sensitive layer, the refractive index at the interface changes, giving rise to a detectable change in the resonance wavelength (see **Figure 11.11c**) through the use of a proper interrogation system. (Usually, a common white light source and spectrometer can be used.)

For chemical sensing with POF, the above examples are of extreme importance. When coating the tip or surface of the POF with active and selective layers, if neither absorption nor luminescence are present, the detection and monitoring of the analytes of interest will be performed through the variation of the refractive index on the active layer, through the adsorption or binding of the analyte. The variation of the refractive index can therefore be correlated with the target concentration through calibrations studies.

POFs are not sensitive active elements, and the use of active layers is therefore commonly used for the development of POF chemical sensors. Tow et al. demonstrated as proof-of-concept the use of spider silk as a chemically-active optical fiber for chemical sensing. In this case, the natural

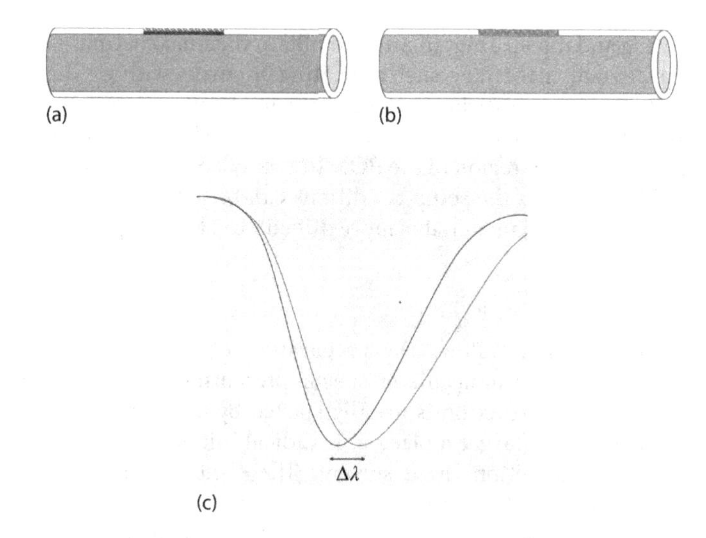

FIGURE 11.11 (a) POF with a selective layer deposited above a metallic layer, in the sensing region (SPR); (b) POF with a selective layer with incorporated nanoparticles (LSPR); and (c) variation on the resonance wavelength due to refractive index change.

polymer-based optical fiber was used to measure ambient relative humidity (RH) level, trough a polymeric setup with linearly polarized light and a polarization analyzer (Tow et al., 2017). Although POFs also can absorb water and be used as humidity sensors, the silk-based optical fiber is reported to be a chemically active element and can be used to detect specific chemical species (in controlled environments, as the element should not be selective). Neverthless, this communication is an open door for the research evolution in terms of the development of chemically-active POFs, increasing significantly the sensitivity and ability of these waveguides to act as low-cost chemical sensors.

11.4 Active and Selective Layers

Chemical sensors can be developed through the use of active layers that are selective and sensitive to specific analytes and that can be coated or bonded on the surface of the POF (lateral surface or tip). First, the sensitivity of the POF to the external medium or changes in the sensing region should be addressed, as the interaction of light with the active layer must occur in order to monitor the changes due to the presence of the analyte (target molecule).

Several deposition methods can be used such as sputtering, spin coating, dip coating, and functionalization through chemical reactions. When modifications of the fiber's surface are made through chemical reactions, the coating is chemically bonded to the fiber, with the possibility of high stability and chemical resistance.

Spin coating consists of the rotation at controlled speed and acceleration of a small quantity of the mixture that is dropped in the surface of the substrate (flat section of the POF, usually made in D-shaped POFs) in vacuum (performed in commercial spin coater machines). The thickness of the coating and its homogeneity are dependent on the spin coating parameters, including rotation time and speed as well as the density and viscosity of the mixture that is dropped on the POF's surface. Dip coating consists of the immersion and pulling of the POF in the prepared mixture. Depending on the characteristics of the desired coating, the fiber can stay immersed for some time before being pulled from the mixture preparation. The velocity of pulling and the viscosity of the mixture are important factors that will determine the thickness of the coating. Normally, the fiber is left for some time in a proper holder for draining and solvent evaporation.

Both in spin coating and dip coating, in order to obtain the final coating, a further curing or post-treatment step is usually necessary, such as heating or curing with UV light.

Sputtering of metallic layers or carbon can be performed with sputtering machines. In this case, all the exposed POF will be coated; therefore, some masks may need to be arranged in order to selectively coat a certain region of the POF. In case of coating of a flat surface, it is easier to obtain reproducible results, as the setup conditions can be repeated; on the contrary, when coating the POF in its circular diameter, it is more difficult to obtain reproducible results.

11.4.1 Molecularly Imprinted Polymers

Molecular imprinting technique allows the preparation of functionalized polymers with specific binding sites for a target molecule or specie, presenting good chemical stability and selectivity. The polymerization reaction is usually started by heating or under UV radiation, in the presence of the template (target molecule), a radical initiator, cross-linker monomer(s), functional monomer(s), and optionally, a solvent. These molecularly imprinted polymers (MIPs) can be grafted onto the POF's surface (or tip) through surface modification or coated using techniques such as spin coating or dip coating. After polymerization, the target molecule is extracted and the polymer retains a memory of the formed binding sites, allowing it to rebind selectively when the target molecule is present in a mixture of different compounds and even from similar ones.

When MIPs are grafted or coated directly on the POF's surface, intensity-based detection schemes are commonly used in order to monitor the changes due to the binding of the analyte (Cennamo et al., 2018b; Ton, Acha, Bonomi, Tse Sum Bui, & Haupt, 2015; Foguel et al., 2015; Cennamo et al., 2017), although wavelength-based configurations can also be used when absorbance or fluorescence are present. When MIPs are deposited above a metallic layer or contain metallic nanoparticles, the monitoring is usually performed through the variation of the resonance wavelength due to the binding of the analyte, surface plasmon resonance (SPR). The combination of D-shaped POFs and MIP receptors is an effective way to obtain highly selective and sensitive POF-SPR-MIP sensors, especially suitable for chemical sensing (Cennamo et al., 2016). Cennamo et al. reported in 2018 the development of a D-shaped POF-SPR-MIP chemical sensor for the detection of perfluorinated alkylated substances (PFAs) in water (Cennamo et al., 2018a) and a D-shaped POF-SPR biosensor based on a specific antibody for PFOA/PFOS (Cennamo et al., 2018c). In comparison, the sensor with the MIP receptor is low-cost, presents a better stability out of the native environment, has good reproducibility, and avoids the need of functionalizing procedures. In comparison with bio-receptors, MIPs also allow an easier and fast preparation, longer storage stability, higher mechanical strength and resistance to heat and pressure, and the possibility of sensing in remote and harsh environments (Yan & Ho Row, 2006).

Besides all the described advantages and high selectivity, the development of POF chemical sensors with MIPs as selective layers is still is a long and time-consuming process that needs to be constantly optimized due to the large numbers of variables, finding the balance between repeatability and selectivity, for example, as the development of highly efficient molecularly imprinted polymers entails the mastery of complex chemistry (Whitcombe, Kirsch, & Nicholls, 2014; Fu, Chen, Li, & Zhang, 2015; Derazshamshir & Yavuz, 2017). Batch binding studies of bulk polymers are usually performed in order to evaluate the properties of the MIP, such as selectivity, time of binding, and specificity by comparison with the non-imprinted polymer (NIP), that is prepared in the same way but without the presence of the analyte.

MIPs can also be produced with fluorescent monomers, enabling the emission/quenching of light due to the binding between the template molecule and the MIP (Ton et al., 2015; Foguel et al., 2015). The selection of the fluorophore is very important when designing a sensor and, generally, absorption and fluorescence in the visible range (or longer wavelength) are preferred to reduce absorption and scattering by the matrix (Wan, Wagner, & Rurack, 2016).

Ton et al. developed disposable POF-MIP sensors containing a fluorescent monomer (FIM) for the selective detection of the mycotoxin citrinin and the herbicide 2,4-dichlorophenoxyacetic acid (2,4-D), whose concentration could be monitored by evanescent wave sensing based in a reflection configuration (a mirror was placed in the end of the fiber), with fluorescent enhancement when the binding between the target and the MIP is present. MIPs were prepared through precipitation polymerization and dip-coated in the fiber. In-situ photopolymerization was also performed using the same 2,4-D MIP formulation (except for the amount of solvent and the initiator) (Ton et al., 2015).

The same sensor platform was coated with a different MIP for the textile dye basic red 9 (BR9), reported by Foguel et al. (2015), by dip-coating the fiber in the polymer suspension. The sensor was characterized with a spectrometer and, upon binding between the analyte and the MIP, an increase of absorption is observed, due to the characteristic absorption of the BR9 at 545 nm. The detection in the range of the μM makes this sensor a very promising solution for the detection of this compound in effluents.

11.4.2 Other Functionalized Coatings

Azkune et al. reported in 2018 the development of a U-bent POF probe through the functionalization of the core with phenylboronic groups, which have high affinity to the diol groups of the glucose allowing its detection. The sensing scheme was based on the variation of absorption in the presence of glucose, through a transmission-based intrinsic configuration. The characterization of the U-bent POF probe to glucose concentration was performed in three different physiological media, revealing good expectations in the development of low-cost glucose sensors (Azkune et al., 2017).

A simple POF sensor head for the detection and measurement of nitrogen dioxide (NO_2) concentration at temperatures below 50°C in a dry atmosphere was developed by Maciak et al. (Maciak, Sufa, & Stolarczyk, 2014). The POF tip was first subjected to thermoforming, after which was coated with a functionalized polymer by dip-coating technique. The detection was performed with a reflection-based setup, while the concentration of nitrogen dioxide (0–10 ppm) was changed inside a gas chamber where the POF sensor head was placed. The reflected optical signal that reaches the detector carries information about the changes that occur in the sensing layer due to NO_2 adsorption, resulting in the variation of the reflectance between 750 and 770 nm. The obtained sensitivities are below 1 ppm of NO_2, although the sensor reproducibility still needs to be improved and the selectivity addressed (Maciak et al., 2014).

A fluorescent dip-probe POF sensor was developed by Riviera et al. for the selective monitoring of PB^{2+}, based on an extrinsic reflection configuration. The surface of the dip-probe sensor was chemically activated with a lead-selective plasticized PVC (polyvinylchloride) membrane with commercial and synthesized ionophores, revealing a reversible response with a detection limit of 7×10^{-6}M (Rivera et al., 2009).

11.4.3 Functionalized Coatings Based on Non-reversible Chemical Reactions

Sultangazin et al. reported a handled and low-cost POF sensor integrated into a smartphone for hydrogen sulfide (H_2S) detection (Sultangazin et al., 2017). Hydrogen sulfide is a highly toxic, inflammable, and corrosive gas, the petrochemical industries, coal mines, and sewage treatment plants being its main sources. The POF sensing region was functionalized and coated with silver (Ag), which allows the detection of the hydrogen sulfide (H_2S) through the non-reversible chemical reaction:

$$2Ag^{+} + S^{-} \rightarrow Ag_2S + 2e^{-}$$

The silver sulfide (Ag_2S) product of this reaction decreases the reflectivity of the silver coating, which results in the decrease of the transmitted light that travels through the POF's sensing region and reaches the detector (smartphone camera) (Sultangazin et al., 2017). The all-smartphone fiber sensing system proposed opens the way to new portable, low-cost, and effective POF chemical sensors for environmental monitoring. The reaction that takes place in the sensitive layer and allows gas detection is not reversible, which is a drawback in this sensor.

Grassini et al. reported in 2015 a declad POF sensor using the same principle of detection through the sputtering of a thin film of silver in the POF's core. In this case, the sensor was developed to measure the cumulative response in indoor atmospheres for cultural heritage applications, where long exposition to hydrogen sulfide leads to tarnishing of metallic artifacts (Grassini et al., 2015). Another unclad POF sensor was also proposed, based on the coating of the unclad POF with a glass (SiO_2)-like coating through plasma-enhanced chemical vapor deposition (PECVD). The coating reacts with hydrogen fluoride vapors (HF), which leads to the formation of silicon tetrafluoride (SiF_4) also through a non-reversible chemical reaction:

$$SiO_2 + 4HF \rightarrow SiF_4 + 2H_2O$$

and allows monitoring of the cumulative response to HF vapors exposition inside the RPC (resistive plate chamber) muon detector of the Compact Muon Solenoid (CMS) experiment at CERN in Geneva, which can prevent filter degradation and schedule their maintenance. Both sensors were based on intrinsic transmission schemes, where the variation in the transmitted light could be related with the concentration of the target specie (Grassini et al., 2015).

11.5 Conclusions

Plastic optical fibers allow the development of chemical sensors with high stability and selectivity. POFs can be considered as simple platforms that can easily be modified, through several physical and chemical techniques that allow this waveguide to work as a reliable sensor. Nevertheless, the sensors that are found in the literature are mostly characterized in controlled samples, and the challenge of chemical sensing is the complex matrixes where the sensor must be able to work selectively.

Nowadays the interdisciplinary knowledge involved in the production of POF chemical sensors demands that researchers and scientists from different areas come together to share knowledge and work together, since so many areas can be involved in sensor development, such as physics, chemistry, biology, materials science, environment, ecology, toxicology, medicine, sea science, electronics, and many others.

The quality of air and water is a global concern, and the monitoring in-situ and in-real-time of important parameters, such as contaminants, pollutants, neurotoxins, hormones, and so many others can be foreseen with POF chemical sensors. For that, the next step is allowing for cheap and simple solutions that could be installed in large scale and also used by the average citizen.

Acknowledgements

This work is funded by FCT/MEC through national funds and, when applicable, co-funded by the FEDER–PT2020 partnership agreement under the projects AQUATICsens - POCI-01-0145-FEDER-032057 and INITIATE-IF/FCT-IF/01664/2014/CP1257/CT0002, investigator grant IF/01664/2014 and PhD fellowship (Filipa Sequeira: SFRH/BD/88899/2012).

References

Asher, S. A., & Holtz, J. H. (1997). Polymerized colloidal crystal hydrogel film as intelligent chemical sensing materials. *Nature*, *389*(6653), 829–832. doi:10.1038/39834.

Asher, S. A., Holtz, J., Weissman, J., & Pan, G. (1998). Mesoscopically periodic photonic-crystal materials for linear and nonlinear optics and chemical sensing. *MRS Bulletin*, *23*(10), 44–50. doi:10.1557/S0883769400029596.

Axelsson, H., Eilard, A., Emanuelsson, A., Galle, B., Edner, H., Ragnarson, P., & Kloo, H. (1995). Measurement of aromatic hydrocarbons with the DOAS technique. *Applied Spectroscopy, 49*(9), 1254–1260. doi:10.1366/0003702953965254.

Azkune, M., Ruiz-Rubio, L., Aldabaldetreku, G., Arrospide, E., Pérez-Álvarez, L., Bikandi, I., … Vilas-Vilela, J. (2017). U-shaped and surface functionalized polymer optical fiber probe for glucose detection. *Sensors, 18*(1), 34. doi:10.3390/s18010034.

Baldini, F., Chester, A. N., Homola, J., & Martellucci, S. (Eds.). (2006). *Optical Chemical Sensors (Proceedings of ASCOS 2014). Series II: Mathematics, Physics and Chemistry-Vol 224.* Springer & NATO Public Diplomacy Division. doi:10.1007/1-4020-4611-1.

Barringer, A. R., & Davies, J. H. (1977). Satellite monitoring of atmospheric cases. *Journal of the British Interplanetary Society, 30*(5), 178–183.

Biran, I., Yu, X., & Walt, D. R. (2008). Optrode-based fiber optic biosensors (bio-optrode). In F. S. Ligler & C. R. Taitt (Eds.), *Optical Biosensors: Today and Tomorrow* (2nd edition, pp. 3–82). Elsevier B.V. doi:10.1016/B978-044453125-4.50003-6.

Cennamo, N., D'Agostino, G., Porto, G., Biasiolo, A., Perri, C., Arcadio, F., & Zeni, L. (2018a). A molecularly imprinted polymer on a plasmonic plastic optical fiber to detect perfluorinated compounds in water. *Sensors (Switzerland), 18*(6), 1–11. doi:10.3390/s18061836.

Cennamo, N., D'Agostino, G., Sequeira, F., Mattiello, F., Porto, G., Biasiolo, A., Nogueira, R., Bilro, L., & Zeni, L. (2018b). A simple and low-cost optical fiber intensity-based configuration for perfluorinated compounds in water solution. *Sensors (Switzerland), 18*(9). doi:10.3390/s18093009.

Cennamo, N., Pesavento, M., Profumo, A., Merli, D., De Maria, L., Chemelli, C., & Zeni, L. (2016). Chemical sensors based on surface plasmon resonance in a plastic optical fiber for multianalyte detection in oil-filled power transformer. In B. Andò, F. Baldini, C. Di Natale, G. Marrazza, & P. Siciliano (Eds.), *Sensors. CNS 2016. Lecture Notes in Electrical Engineering* (Vol. 431, pp. 128–134). Cham, Switzerland: Springer. doi:10.1007/978-3-319-55077-0_17.

Cennamo, N., Testa, G., Marchetti, S., De Maria, L., Bernini, R., Zeni, L., & Pesavento, M. (2017). Intensity-based plastic optical fiber sensor with molecularly imprinted polymer sensitive layer. *Sensors and Actuators B: Chemical, 241*, 534–540. doi:10.1016/j.snb.2016.10.104.

Cennamo, N., Zeni, L., Tortora, P., Regonesi, M. E., Giusti, A., Staiano, M., … Varriale, A. (2018c). A High Sensitivity Biosensor to detect the presence of perfluorinated compounds in environment. *Talanta, 178*(October 2017), 955–961. doi:10.1016/j.talanta.2017.10.034.

Derazshamshir, A., & Yavuz, H. (2017). Molecular imprinting of macromolecules for sensor applications. *Sensors, 17*, 898. doi:10.3390/s17040898.

Foguel, M. V., Ton, X. A., Zanoni, M. V. B., Sotomayor, M. D. P. T., Haupt, K., & Tse Sum Bui, B. (2015). A molecularly imprinted polymer-based evanescent wave fiber optic sensor for the detection of basic red 9 dye. *Sensors and Actuators, B: Chemical, 218*, 222–228. doi:10.1016/j.snb.2015.05.007.

Fu, J., Chen, L., Li, J., & Zhang, Z. (2015). Current status and challenges of ion imprinting. *Journal of Materials Chemistry A, 3*(26), 13598–13627. doi:10.1039/C5TA02421H.

Grassini, S., Ishtaiwi, M., Parvis, M., & Vallan, A. (2015). Design and deployment of low-cost plastic optical fiber sensors for gas monitoring. *Sensors, 15*, 485–498. doi:10.3390/s150100485.

Homola, J., Koudela, I., & Yee, S. S. (1999). Surface plasmon resonance sensors based on diffraction gratings and prism couplers: Sensitivity comparison. *Sensors and Actuators, B: Chemical, 54*(1), 16–24. doi:10.1016/S0925-4005(98)00322-0.

Kim, C., Eom, J. B., Jung, S., & Ji, T. (2016). Detection of organic compounds in water by an optical absorbance method. *Sensors, 16*(1), 61. doi:10.3390/s16010061.

Koike, Y., & Koike, K. (2011). Progress in low-loss and high-bandwidth plastic optical fibers. *Journal of Polymer Science, Part B: Polymer Physics, 49*(1), 2–17. doi:10.1002/polb.22170.

Lieberman, R. A. (1993). Recent progress in intrinsic fiber-optic chemical sensing II. *Sensors and Actuators B: Chemical, 11*(1–3), 43–55. doi:10.1016/0925-4005(93)85237-5.

Lukosz, W. (1995). Integrated optical chemical and direct biochemical sensor. *Sensors and Actuators, B:Chemical, 29*, 37–50. doi:10.1016/0925-4005(95)01661-9.

Maciak, E., Sufa, P., & Stolarczyk, A. (2014). A low temperature operated NO_2 gas POF sensor based on conducting graft polymer. *Photonics Letters of Poland, 6*(4), 124–126. doi:10.4302/plp.2014.4.04.

Memon, S. F., Ali, M. M., Pembroke, J. T., Chowdhry, B. S., & Lewis, E. (2018). Measurement of ultralow level bioethanol concentration for production using evanescent wave based optical fiber sensor. *IEEE Transactions on Instrumentation and Measurement, 67*(4), 780–788. doi:10.1109/TIM.2017.2761618.

Menzies, R. T., & Chahine, M. T. (1974). Remote atmospheric sensing with an airborne laser absorption spectrometer. *Applied Optics, 13*(12), 2840–2849. doi:10.1364/AO.13.002840.

Moo, Y. C., Matjafri, M. Z., Lim, H. S., & Tan, C. H. (2016). New development of optical fibre sensor for determination of nitrate and nitrite in water. *Optik-International Journal for Light and Electron Optics, 127*(3), 1312–1319. doi:10.1016/j.ijleo.2015.09.072.

Morey, W. W., Dunphy, J. R., & Meltz, G. (1991). Multiplexing fiber Bragg grating sensors. *Fiber and Integrated Optics, 10*(4), 351–360. doi:10.1080/01468039108201715.

Rivera, L., Izquierdo, D., Garcés, I., Salinas, I., Alonso, J., & Puyol, M. (2009). Simple dip-probe fluorescence setup sensor for in situ environmental determinations. *Sensors and Actuators, B: Chemical, 137*(2), 420–425. doi:10.1016/j.snb.2009.01.064.

Sequeira, F., Cennamo, N., Rudnitskaya, A., Nogueira, R., Zeni, L., & Bilro, L. (2019). D-Shaped POF Sensors for Refractive Index Sensing—The Importance of Surface Roughness. *Sensors*, 19(11), 2476.

Sequeira, F., Duarte, D., Bilro, L., Rudnitskaya, A., Pesavento, M., Zeni, L., & Cennamo, N. (2016). Refractive Index Sensing with D-Shaped Plastic Optical Fibers for Chemical and Biochemical Applications. *Sensors, 16*(12), 2119. doi:10.3390/s16122119.

Sultangazin, A., Kusmangaliyev, J., Aitkulov, A., Akilbekova, D., Olivero, M., & Tosi, D. (2017). Design of a smartphone plastic optical fiber chemical sensor for hydrogen sulfide detection. *IEEE Sensors Journal, 17*(21), 6935–6940. doi:10.1109/JSEn.2017.2752717.

Swinehart, D. F. (1962). The Beer-Lambert law. *Journal of Chemical Education, 39*(7), 333–335. doi:10.1021/ed039p333.

Taylor, E. W. (1987). Optical waveguide diffraction grating sensor. In *Instrumentation in the Aerospace Industry* (Vol. 33, pp. 143–147).

Ton, X. A., Acha, V., Bonomi, P., Tse Sum Bui, B., & Haupt, K. (2015). A disposable evanescent wave fiber optic sensor coated with a molecularly imprinted polymer as a selective fluorescence probe. *Biosensors and Bioelectronics, 64*, 359–366. doi:10.1016/j.bios.2014.09.017.

Tow, K. H., Chow, D. M., Vollrath, F., Dicaire, I., Gheysens, T., & Thévenaz, L. (2017). Towards a new generation of fibre optic chemical sensors based on spider silk threads, 103231E. doi:10.1117/12.2264438.

Wan, W., Wagner, S., & Rurack, K. (2016). Fluorescent monomers: "bricks" that make a molecularly imprinted polymer "bright." *Analytical and Bioanalytical Chemistry, 408*(7), 1753–1771. doi:10.1007/s00216-015-9174-4.

Webb, D. J. (2010). Polymer photonic crystal fibre for sensor applications. In *SPIE 7726, Optical Sensing and Detection* (Vol. 7726, p. 77260Q). doi:10.1117/12.859090.

Whitcombe, M. J., Kirsch, N., & Nicholls, I. A. (2014). Molecular imprinting science and technology: A survey of the literature for the years 2004–2011. *Journal of Molecular Recognition, 27*. doi:10.1002/jmr.2347.

Yan, H., & Ho Row, K. (2006). Characteristic and synthetic approach of molecularly imprinted polymer. *International Journal of Molecular Sciences, 7*, 155–178. doi:10.3390/i7050155.

Zhou, Q., Tabacco, M. B., & Rosenblum, K. W. (1991). Development of chemical sensors using plastic optical fiber. In *Proceedings volume 1592, Plastic Optical Fibers*. doi:10.1117/12.50998.

12

POF Sensors for Structural Health Monitoring

Aleksander Wosniok

Contents

12.1 Sensing Techniques for Structural Health Monitoring

The ever more ambitious strategic goals of meeting the requirements in ensuring technical safety and security of civil structures have resulted in flourishing development of innovative structural health monitoring (SHM) technologies for early damage diagnosis and prognosis. At the same time, implementing SHM systems provides tangible economic benefits derived from lower life-cycle costs associated with reduction in the maintenance, repair, and insurance expenses. Due to the large size and harsh environmental conditions common to most civil structures, the broad range of favorable physical-mechanical properties of POFs allow for customized monitoring solutions for a wide variety of applications.

In addition to common SHM-related advantages of optical fibers including their electromagnetic immunity, small size, lightweight, and spark-free and non-conductive characteristics, POFs offer better bending and fracture resistance than their glass counterparts. Particularly, the improved robustness of POFs, their ease of handling, low Young's Modulus, and high elastic limit of 10% compared to 1% in silica glass [1] are relevant to practical applications. Depending on the composition, dopants, drawing process, and geometry [2], strain measurement up to 45% [3] or even above 100% [4,5] has been demonstrated with standard POFs. Therefore, the dominant market expected for POF sensors includes monitoring of high-strain-rate deformations in earthwork structures, crack detection in concrete and masonry structures [6], or overstressing in high-rise steel structures exposed to moisture, corrosion, leakage, fatigue, vibration, fire, overflow, earthquake, and intentional damage. For fracture monitoring within concrete structures, the sensory usage of POFs becomes especially favorable since the extremely alkaline environment of concrete mixtures is well known to be corrosive to standard silica glass optical fibers (GOFs) [1,7].

Most advanced distributed sensing techniques are commonly based on Rayleigh backscatter reflectometry using commercially available multimode (MM) POFs. Such typical MM POFs range from a step-index (SI) poly(methyl methacrylate) (PMMA) POF having a core diameter of 1 mm to a low-loss graded-index (GI) perfluorinated (PF) POF based on poly(perfluorobutenylvinylether), also known as CYTOP [8] with a 50 μm core diameter. The relatively low optical attenuation value of 30 dB/km at 1.3 μm [9,10] makes PFGI POFs also interesting for distributed Brillouin sensing [11–14]. Compared to GOFs, PFGI POFs offer better potential for temperature measurement and have comparably low theoretical attenuation limit [15]. Therefore, POF-based distributed Brillouin sensing is expected to play an important role in the future of SHM, especially at high-strain ranges. The significance of the Brillouin measurement technique can be also enhanced by further development of the single-mode (SM) POFs which are still subject of research and are used for coherent detection techniques [6]. Furthermore, SM PMMA POFs have been characterized in a Mach-Zehnder interferometer setup for strain values up to 15.8% [16,17].

The current development of SM perfluorinated and microstructured POFs (mPOFs) represents an immense promise for quasi-distributed dynamic measurement at high strain levels based on fiber Bragg grating (FBG) technology. While SM mPOFs with optical losses of about 1 dB/m can be fabricated [6], the SM PF POFs presented by Zhou et al. feature low attenuation of even less than 0.2 dB/m in the wavelength range of 1.41 to 1.55 μm [18]. At the same time, the SM PF POFs have the potential for improved thermal stability compared to their PMMA counterparts [19].

This whole chapter provides a comprehensive overview on current POF-based sensing principles and SHM technologies, highlighting their diverse applications in civil engineering structures. In the application-related context, close attention is paid to the development of smart sensor-based geotextiles and geogrids. Such geosynthetics-integrated distributed POF sensors have proven to be a promising solution for two- or even three-dimensional monitoring of critical high mechanical deformations in both geotechnical and masonry structures. Moreover, geosynthetics in the form of nonwoven geotextiles as well as polymer-based geogrids used as carrier materials for POF sensors enable optimized load transfer from the monitored structure to the measuring fiber without losing their original functionality. In other words, smart geosynthetics provide a cost-efficient dual solution for, on the one hand, well-established increase of structure stability and decrease of erosion effects, and on the other hand, early-warning and detection capabilities in the prevention and elimination of potential hazards and lasting damages.

Completing recent advances in the POF-based sensor research and application, the development of fiber optic sensor technologies for aircraft monitoring should also be mentioned at this point. These technologies based on POF-based sensors can be reliably used for monitoring both the structures of wing surfaces or fuselage sections [20,21] and turbine parameters like tip clearance as well as tip timing [22–24].

12.2 POF-Based Distributed Monitoring Systems

Distributed measurement techniques provide the advantage of complete spatial acquisition of measurands along the whole sensor fiber. Such spatially resolved recording of strain, temperature, and humidity is realized within the spatial resolution which, depending on the measurement method, typically ranges from a few cm to about 1 m. Here, the spatially resolved information about the extrinsic influences mentioned above is obtained by means of a modulation of the laser signals launched into the sensor fiber.

The most established measuring principles are based on the optical time domain reflectometry (OTDR) using pulse modulation of laser light. In such measuring methods, the runtime of light provides information on position of the light's interaction with the measured variable, while the spatial resolution is directly determined by the used pulse width.

Due to the time-frequency duality given by the Fourier transform for linear and time-invariant systems, the POF-based distributed measurement can be also realized in the frequency domain. The measuring procedure via the frequency-domain transfer function can be figuratively treated as a decomposition of the laser pulses into their spectral components. These components are single-frequency harmonic signals corresponding to the modulation frequencies of the performed intensity modulation of continuous wave pump laser light. The influence of modulation settings on the spatial resolution characteristics is explained in detail in Sections 12.2.2 and 12.2.3, respectively.

All distributed measurement techniques are physically based on backscattering effects in optical fibers. For POF-based sensor systems, the elastic Rayleigh scattering process plays the predominant role in SHM applications. Compared to their silica glass counterpart, POFs exhibit a significant Rayleigh-backscattered power dependence on all elementary quantities, i.e. strain, temperature, and humidity. This behavior is caused by the strong inhomogeneity of the refractive index along the POFs resulting in high susceptibility to mechanical, thermal, and humidity impact. In Rayleigh-based OTDR, two different MM fiber types are generally used as distributed sensors. Due to relatively high optical attenuation, PMMA POFs allow measurement only along limited fiber lengths up to about 100 m [4]. On the other hand, they offer the ease of connection due to their large core diameters and high numerical apertures. Using the second type of PF GI POFs, the measurement range can be increased up to 500 m. The development of low-loss PF GI POFs provided also the basis for implementation of distributed techniques based on both Brillouin backscattering [12,13,22] and interferometric detection.

12.2.1 Optical Time Domain Reflectometry (OTDR)

The Rayleigh-based OTDR is a common testing technique widespread in telecommunication industry for in-depth fiber characterization and fault analysis and has been recently utilized for SHM applications as well [4,23]. Here, a series of short optical pulses is launched into one end of the fiber. A backscattered signal resulting from Rayleigh scattering occurring at each position along the fiber is recorded as a function of time which can be converted to the spatial position along the fiber. Such backscattered signal is furthermore affected by perturbations (such as fiber kinking and micro- or macrobendings), resulting in either optical loss or reflection peaks in the backscatter signal at the location of the disturbance [1]. From the sensory point of view, longitudinal strain and external temperature gradients can also be seen as certain types of perturbations affecting spatial distribution of backscattered traces. This was shown for the first time by Husdi et al. and Nakamura et al. in two different MM PMMA SI POFs [24,25]. The predominant effect of backscatter intensity increase upon straining has been also proven by Liehr et al. in MM PF GI POFs [4]. In addition to the increase of the measurement range by a factor of five as mentioned above, the perfluorinated POFs allow sensing with a high spatial resolution of 10 cm due to their negligible material and modal dispersion. However, PF GI POFs can be used in a lower strain range than PMMA POFs, which remain the most suitable sensor for medium and large strain up to 45% [4,26]. In addition, compared to perfluorinated fibers, PMMA POFs exhibit only insignificant temperature cross-sensitivity [3,27]. Generally, for standard PMMA POFs, a positive temperature

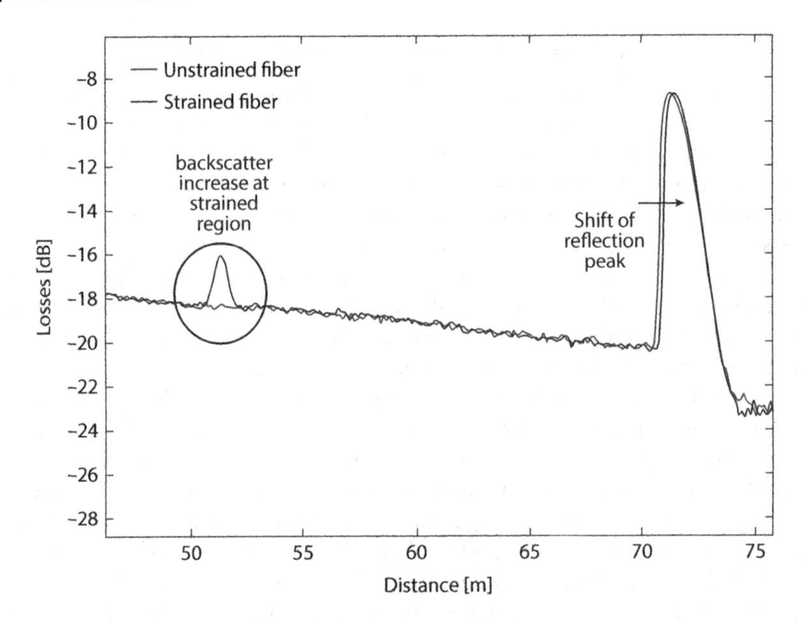

FIGURE 12.1 OTDR traces of standard MM PMMA POFs—principle of POF-based distributed strain measurements.

gradient of 5 K induces the same increase of backscattered signal as strain of 0.1% [3,4]. The sensing principle based on OTDR is presented in **Figure 12.1** in case of strain measurements using MM PMMA POFs. Obviously, apart from strain-induced local backscatter increase, the spatial shift of Fresnel reflections at the fiber end can also be used for strain evaluation.

By the use of high-resolution photon counting OTDR, the measurement time is between 10 s and several minutes, mainly depending on the demanded accuracy. On the other hand, the accuracy can be deteriorated by relaxation processes occurring in the strained fiber section, which lead to an asymptotic decrease of the backscatter level over time. In spite of this, as shown in **Figure 12.2**, the strain step changes in the range of 1% are generally well resolvable. Moreover, the strain resolution under laboratory conditions can be as high as 0.1% [3,4].

The greatest potential for applications in SHM using Rayleigh-based OTDR is expected for high strain measurements in the geotechnical field for detecting deformation in earthwork structures or crack openings in concrete and masonry structures. For the best possible transfer of quantities to be measured (mainly longitudinal strain), the POFs can be integrated in various support structures used as carrier materials for fiber optic sensors. Thereby, the POF-based smart geosynthetics represent a particularly flourishing research and application branch whose development and application examples are presented in the separate Section 12.4.

For some applications in SHM, the measurement of moisture gains in importance. What is meant here is the determination of moisture content in concrete and screed in building foundation, moisture ingress detection in sealed structured such as bridges, and monitoring of seepage line and water leakage in dams and dykes [28].

CYTOP is hydrophobic and absorbs only minute amounts of water [29]. Therefore, only PMMA POFs can be used to measure small—compared to strain-related effects—wavelength-dependent changes of Rayleigh-backscattered light with the relative humidity. The dependence on wavelength can be traced back to additional attenuation along the fiber due to the OH vibrational absorption, which is negligible at 500 nm but increasing with rising relative humidity at 650 nm [28]. Thus, Rayleigh-based OTDR measurements at the two wavelengths can provide a way to separate the impact of humidity from the other factors (strain, temperature) in the backscattered signals.

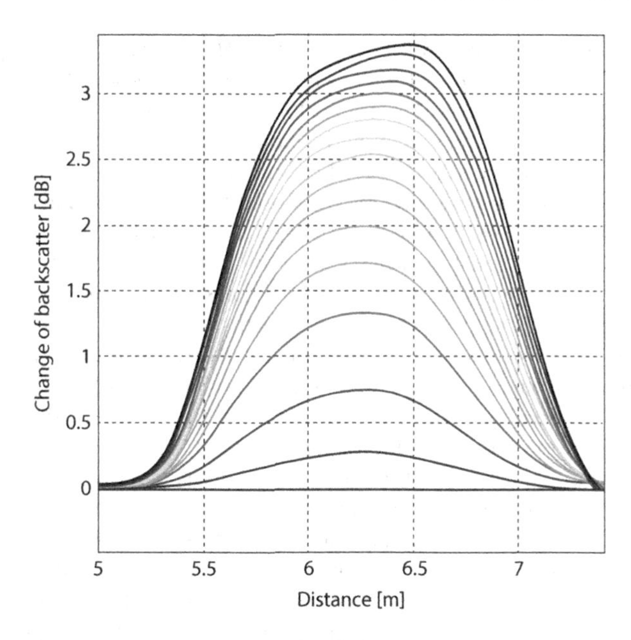

FIGURE 12.2 Strain-backscatter dependency for 1.5 m long standard MM PMMA fiber section gradually strained at 6.3 m from 0% up to 16% with the step of 1%.

With regard to possible expansion of application, improvement of a slow hours-long-response of PMMA-based distributed sensor to humidity gradients is crucial. However, the response times under one hour are possible by decreasing the fiber core diameter below 250 µm [30–32].

12.2.2 Incoherent Optical Frequency Domain Reflectometry (I-OFDR)

The frequency domain techniques basically use a Fourier pulse decomposition given by a progressive amplitude modulation of a continuous-wave (cw) light source at different modulation frequencies using an electro-optic modulator (EOM). A stepwise sinusoidal amplitude modulation of cw pump light replaces the pulsed modulation in time domain method described in Section 12.2.1. In the basic I-OFDR configuration shown in **Figure 12.3**, the laser light backscattered in the fiber under test (FUT) adopts the amplitude modulation at the same modulation frequency f_m as the pump light. The backscattered light monitored by the vector network analyzer (VNA) is used to determine the fiber transfer function $H(f_m)$. For the distributed measurement, the modulation frequency f_m is swept by an RF signal of the vector network analyzer (VNA) with M equidistant frequency

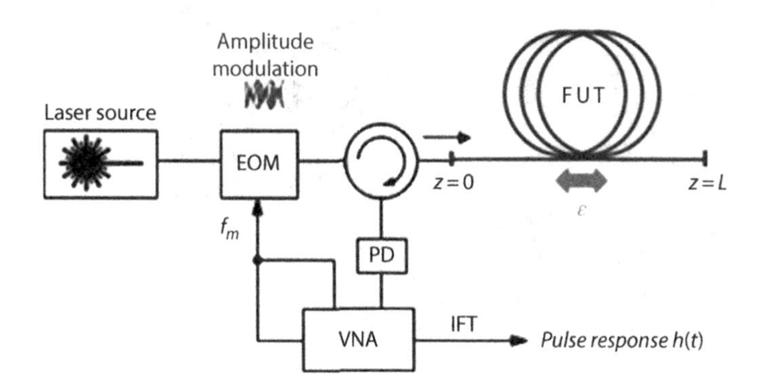

FIGURE 12.3 Rayleigh-based I-OFDR basic configuration.

steps Δf_m, so that the complete complex transfer function H can be recorded. Applying the inverse Fourier transform (IFT) to the recorded function H, the pulse response in time domain is obtained giving signal traces similar to **Figure 12.1**. The I-OFDR signal traces feature significant improved spatial resolution, signal stability and increased dynamic range at shorter interrogation times compared to OTDR [33,34]. Furthermore, the spatial resolution of I-OFDR is inversely proportional to the modulation frequency range generated by VNA used for pump signal amplitude modulation. For example, a technically achievable sweep over a frequency range from a few kHz to 3 GHz leads to the spatial resolution better than 4 cm when using MM PF GI POFs as distributed sensors.

The commercially available optical components of the Rayleigh-based I-OTDR setup shown in **Figure 12.3** can be easily adapted to the wavelengths in the second near-infrared spectral window around 1.3 µm. This wavelength range coincides with the lowest optical attenuation of PF GI POFs, which makes them most suitable for use in the I-OFDR approach. Moreover, their gradient reflective index structure and very low material dispersion facilitate the metrological potential for high-resolution measurement for sensor lengths up to 500 m.

The typical backscatter signal of PF GI POFs exhibits strong fluctuations due to a great number of randomly distributed scattering centers [33]. Such characteristic fingerprints distributed along the whole fiber length can be used to determine strain-induced local displacements by applying a cross-correlation approach. In this method, the scatter profiles from the two data sets (two separate measurements) are cross-correlated along the defined fiber segments representing individual sensing elements. In this way, changes in the scatter signal occurred in each sensing element can be determined, which can accordingly be converted into local strains.

Strongly reflecting events needed for the measurement of length changes based on the cross-correlation method can be also intentionally positioned at well-defined reference points in the fiber. Any physical contact connectors or Fresnel reflections on fiber end faces can be used for this purpose. Depending on the chosen measurement parameters, the quasi-distributed measurement of the length changes between reflection points using the I-OFDR technique enables dynamic deformation detection with a high repetition rate up to 2 kHz [35].

The potential of such dynamic strain analysis is presented by the example of earthquake tests schematically shown in **Figure 12.4**.

Here, two different fiber types in pre-strained condition, MM GOFs and a PF GI POF cable, were integrated onto a two-story masonry building constructed on a seismic shaker table [6]. The seismic load was applied during five tests with increasing acceleration. Each test lasted about 45 seconds. The seismic acceleration was always kept parallel to the sensor 1 (MM PF GI POF) and perpendicular to the sensors 2 and 3 (both MM GOFs). **Figure 12.5** presents the measured dynamic length changes of all three sensors during an earthquake test performed under pre-damage conditions of the building obtained by earlier seismic tests.

FIGURE 12.4 Earthquake test configuration with three sensors installed. Seismic load applied in direction of sensor 1.

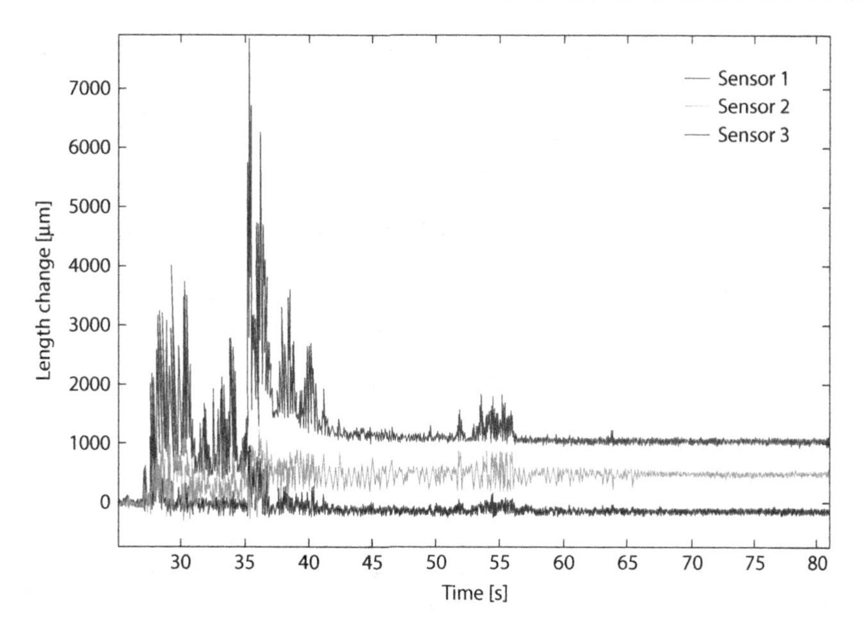

FIGURE 12.5 Dynamic length changes during the earthquake test measured with a repetition rate of 160 Hz.

Due to the crack openings in the walls, sensor 1 and 3 exhibit a permanent deformation.

In summary, the I-OFDR approach provides the possibility of fully distributed strain measurement based on the measurement of local increase of backscattered signal which can be converted into strain analogically to the OTDR method. The I-OFDR offers benefit of high signal stability allowing simultaneous detection of deformation events in the range of a few cm, which are non-resolvable using OTDR. Additionally, the inclusion of cross-correlation between backscattering peaks or Fresnel reflections along the sensor fiber allows highly-precise quasi-distributed measurement of dynamic length changes with spatial resolution below 1 µm [35]. The potential of high peak-to-noise dynamic range of I-OFDR in the range of 40 dB can be even increased by more than 10 dB by replacing the expensive VNA with a cost-effective digital circuit [36].

12.2.3 Brillouin Sensing

The Rayleigh-based methods for fully distributed deformation measurement presented above feature limited strain resolution and accuracy as already shown in **Figure 12.2**. The reason for this lies in the relatively low strain dependency of Rayleigh backscatter. The current research concerning implementation of POF-based distributed Brillouin sensing can provide a solution for significant improvement of measurement accuracy while simultaneously increasing the measurable strain range compared to GOFs. Presenting Brillouin optical correlation-domain reflectometry using POFs as distributed sensors, Hayashi et al. have already reported on achieved strain measurement errors down to 200 µm/m [13]. Considering further development of measurement setups, accuracies slightly below 100 µm/m are to be expected in the future. However, the improvement is physically limited by the multimodal nature [9,37] and material properties of POFs leading to three times wider stimulated Brillouin gain spectra with lower signal-to-noise ratio compared to their silica glass counterpart [22,38,39].

For the purpose of the successful establishment of POF-based distributed Brillouin sensing for high-strain detection, special attention should be paid to further investigation of signal non-linearities which are already occurring at strain values above 1.2% [40,41].

Unlike the Rayleigh-based methods, the investigated behavior of Brillouin frequency shift (BFS) as a quantity to be measured shows a distinct temperature dependence [11,38]. Concerning the relative ratio of their temperature versus strain BFS sensitivity, POFs are more than fifteen times more temperature sensitive than their silica glass counterparts. Therefore, POFs are very promising for implementation of Brillouin-based temperature sensing. However, as already mentioned above, the full applicability of POF-based distributed Brillouin sensing for structural health monitoring requires further development of measurement set-ups with the aim of increasing the measurement range and accuracy. Several techniques for achieving this aim have been demonstrated for short PF GI POF sections relying on analysis of stimulated Brillouin scattering amplification in frequency [39], correlation [13,38], or time domain [42]. Among them, the narrow-band measurement in frequency domain (Brillouin optical frequency-domain analysis; BOFDA), shown schematically in **Figure 12.6**, offers an essential low-cost potential [43].

To adapt the wavelength of pump and probe laser sources to the lowest optical attenuation of PF GI POFs most suitable for use in BOFDA, Nd:YAG lasers emitting light with a wavelength of 1.319 μm should be used. Such laser sources are indeed more expensive than laser diodes used by Minardo et al. [39] emitting light in the third near-infrared spectral window at 1.55 μm. On the other hand, they balance the cost increase by improving the measurement range by the factor of five. Furthermore, thanks to the implementation of the sideband technique [44], only one laser source is needed to generate both the pump and low-frequency probe signals.

According to **Figure 12.6**, the measuring procedures of BOFDA is based on a progressive sinusoidal amplitude modulation of the pump wave using EOM by sweeping the modulation frequency for a selected constant laser frequency difference f_D of counterpropagating probe signals. Furthermore, the frequency difference f_D between pump and probe signals should be swept in the range of characteristic BFS during the measurement process. In such a way, the fiber transfer function can be recorded providing a BFS distribution over the whole fiber length using inverse Fourier transform.

As presented in [43] and detailed description in [45], the complex transfer function does not necessarily have to be measured by the VNA. The VNA is clearly overdesigned for recording the single scattering parameter S_{21} corresponding to the transfer function in the BOFDA setup shown in **Figure 12.6**. Furthermore, for this measurement task a digital processing technique based on standard and thus also cost-efficient electronics devices can be used. Since the complete digital processing can be performed off-line, the digital BOFDA features a significant advantage of data acquisition time reduction while simultaneously increasing the dynamic range.

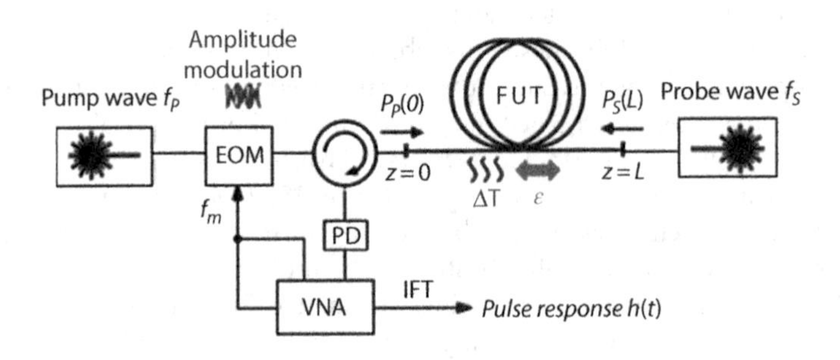

FIGURE 12.6 BOFDA basic configuration.

12.2.4 Interferometric Sensors

Interferometric sensing principles are characterized by their high-resolution potential. Such sensor systems are based on effects requiring single-mode operation for optimal results. As the suitable SM POFs with acceptable low-loss are not available on the market, the use of the POF-based interferometric techniques for SHM remains an important issue for future development. However, some research has been done in this area regarding highly-resolved strain and temperature sensing for short distances.

The sensitivity of SM POFs was investigated for the first time by Silva-Lopez et al. within a limited temperature range and for low strains using the Mach-Zehnder technique [46]. Here, the number of typical interference fringes can be measured as a function of the fiber displacement. Further investigations for higher strain values up to 15.8% performed by Kiesel et al. [16,17] have shown that photoelastic nonlinearities cannot be neglected for realization of high-strain interferometric fiber optic sensors. Favorably, no hysteresis has been detected for cyclic elongation up to 4%. Due to the low elastic modulus and beneficial acoustic properties, compared to glass fibers, POFs allow a significant improvement of sensitivity using a Mach-Zehnder interferometer arrangement for ultrasound and acoustic sensing applications [47].

Another existing interferometric technique for POF-based high-resolution distributed strain and temperature measurement is based on the swept-wavelength interferometry (SWI). This technique, well-known as optical backscatter reflectometry (OBR) [48], has been established on the market for use with SM GOFs over the last decade. As schematically illustrated in **Figure 12.7**, a tunable laser source is coupled into a fiber-based Mach-Zehnder interferometer built up of a fiber reference arm (internal delay path) and the sensor FUT itself as a second interference arm. The detected interference term therefore gives a complex scalar response of the fiber including both the phase and amplitude information. This allows the data to be presented both in the time- and frequency-domain related to one another by a discrete Fourier transform. Any relative local changes are detected in the form of frequency shifts and then converted into the strain/temperature distribution in small cross-correlated sections along the sensor fiber. The small section chosen for the cross-correlation algorithm are typically in the 5–20 mm range, thus determining the spatial resolution of strain and temperature measurements.

Achieving high accuracy of temperature- or strain-induced frequency shift measurement requires low-dispersion fiber characteristics. Hence, the measurement accuracy for MM POFs quickly degrades with increasing fiber length due to the modal dispersion and the strong mode coupling of these fibers. In order to achieve satisfying measurement length using POF-based OBR, low-loss SM POFs are to be developed. However, Kreger et al. demonstrated that OBR can be used for distributed strain and temperature sensing for 2–3 m long commercially available

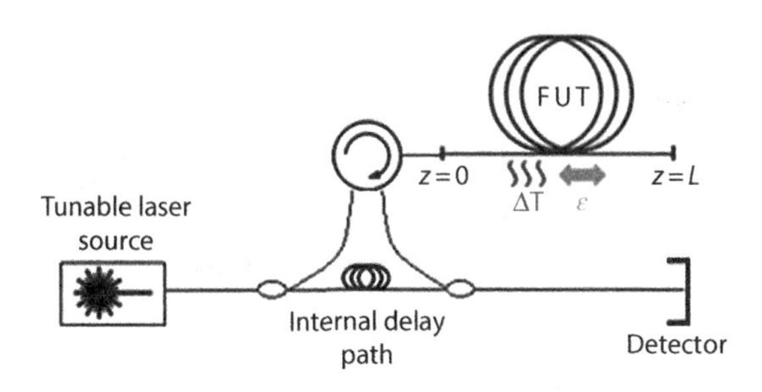

FIGURE 12.7 Basic diagram of fiber-based OBR.

PF GI POFs with relatively high measurement resolution better than 4 µm/m and 1°C [49]. Moreover, the 50 micron core PF GI POF tested by Kreger et al. exhibited 20% higher strain sensitivity than standard SM silica glass fibers. The temperature response proved to be about four times larger than for the silica fibers.

12.3 Point Sensors

In application-specific practical cases, to limit both the system costs and large quantities of distributed measurement data to be processed, a wide range of point sensors for hot-spot monitoring can be used. Some of these pointwise sensor solutions are based on simple and therefore low-cost principles comprising a single light source, commercially available optical fibers, common transducer and detector. Such cost-effective loss-based techniques allow for precise dynamic measurements of local strain and vibration. Additionally, the application of grating-based solutions can contribute to an increase in achievable sampling rates for highly sensitive multipoint detection.

12.3.1 Cost-Effective Optical Power Change Technique

The most basic fiber optic sensors are based on well-established concepts of optical power change measurements for detection of mechanical stress, displacement, temperature, and pressure. Such measurements can be realized in reflection or in transmission. Two prevalent examples of numerous sensors based on the transmission measurement for crack detection are presented in **Figure 12.8**. The widely-used extrinsic sensor solutions often require external transducer inducing optical power change, for example due to axial offset in a fiber system as shown in **Figure 12.8** (left). The optical power change can also result from an offset in radial direction, external pressure, or fiber bends. In addition, a specific structuring of the sensor fiber by a local side-polishing or removal of the cladding and even core material as proposed by Kuang et al. [50] can be used.

The overall approach for detection of the transmitted or reflected optical power using a photodetector can be complemented by an additional reference mechanism for compensation of source power fluctuations and connector deterioration, respectively. Due to the compensation, for example, a high-precision crack evolution measurement can be realized.

The application of fiber optic sensors based on optical power measurement provides a cost-effective solution for an early damage detection or a dynamic failure analysis as an alternative to electrical strain gauges at high strain ranges. However, these sensors can be applied only for pointwise detection making quasi-distributed interrogation impossible, which indeed complicates on-site installation when numerous hot spots are to be monitored.

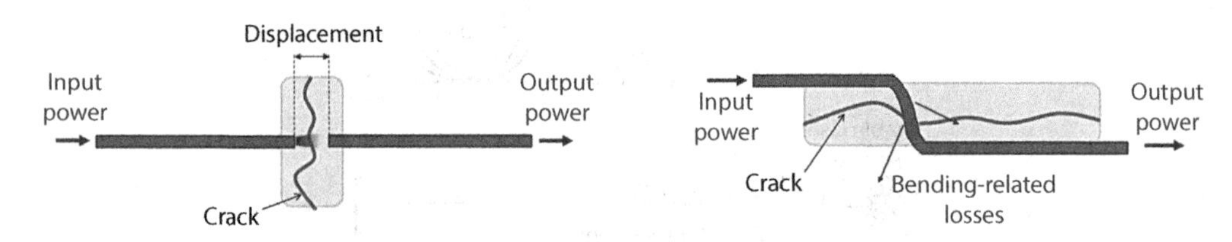

FIGURE 12.8 Example of POF-based extrinsic (left) and intrinsic (right) power change sensor implementations.

12.3.2 Fiber Bragg Gratings

The progress in the application of POF-based fiber Bragg gratings (POFBGs) for SHM remains closely coupled with the development and fabrication of low-loss SM POFs. During the last few years, some significant achievements relating to the long-term stability of POFBGs have already been reached. The attraction for development of POFBGs lies primarily in their potential for high-strain sensing combined with their lower stiffness and higher elasticity compared to GOFs [51,52]. POFBGs are predominantly written in SM mPOFs using either phase mask technique with high-power UV laser or point-by-point inscription with focused femtosecond laser. In addition to strain sensing, POFBGs can be also utilized for temperature und relative humidity monitoring [31,53,54]. As shown in [55,56], the femtosecond laser inscription enables the production of FBGs in MM PF GI POFs characterized by low attenuation and easy fiber coupling. According to novel research results published by Ishikawa et al., the MM PF POF-based FBGs are even suitable for fast-response high-sensitivity pressure sensing. The determined pressure-dependence coefficient of 1.3 nm/MPa is over five times higher than for other types of POF-based FBGs [57].

12.3.3 Long Period Gratings

Periodic changes of the fiber refractive index profile can be also realized using techniques based only on mechanical stress applied on heated fiber, which in turn simplifies the grating fabrication. The gating period fabricated in this way is much larger than that of typical POFBGs. Most of such long period gratings (LPGs) are inscribed in mPOFs, whose cross-sectional microstructure is representatively shown in **Figure 12.9** (left) [58]. As presented in **Figure 12.9** (right), the mPOF-based LPGs exhibit dips in their transmission spectra at wavelength positions dependent on the LPG period. These transmission dips arise from mode coupling from the core into the cladding occurring in the forward direction. The dips' spectral position is strain dependent as the LPG period changes when the fiber is strained. The strain sensitivity is typically about 1.1–1.4 pm/μm/m [59,60].

For structural engineering applications, such as the integration of mPOF-based LPGs into composite materials presented by de Oliveira et al. [61], existing cross-sensitivities to temperature and moisture should not be neglected for precise deformation measurements. In particular, the temperature sensitivity can reach the range of 0.2–0.3 nm/°C [60], whereas the increase of relative humidity causes a shift of transmission dip up to 0.3 nm/10% rh.

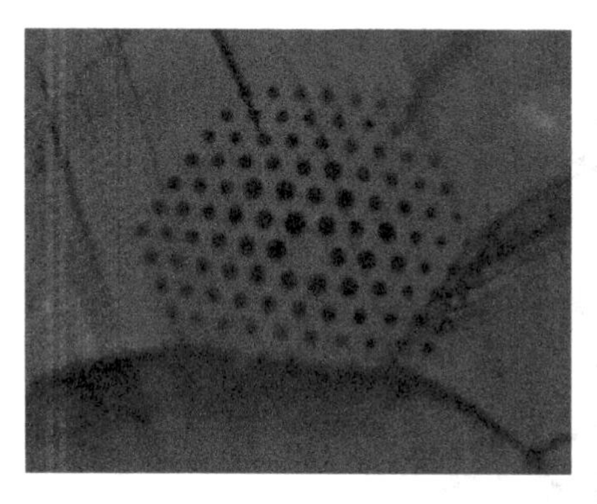

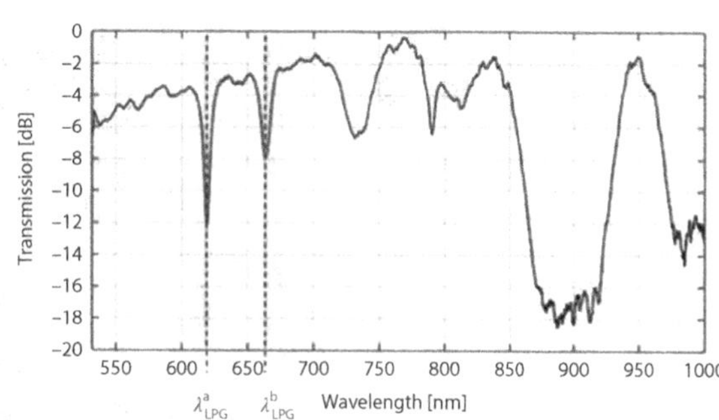

FIGURE 12.9 Development of LPGs in mPOFs: (left) cross section of an SM mPOF and (right) transmission spectrum of an LPG inscribed in the SM mPOF.

12.4 Smart Geosynthetics with Integrated POF Sensors

The large-scale monitoring of objects and facilities can be realized by integrating distributed fiber optic sensor elements in geosynthetic carrier structures. The eligibility of using geosynthetics as sensor carriers can be directly derived from the rapidly growing fields of application of these materials. Due to technical and economic advantages over conventional building materials, geosynthetics can be found in many civil engineering construction works and major projects. For example, they serve as building materials in coastal defense and flood protection, water traffic engineering, track and embankment constructions for landfills, and dam engineering. An additional integration of the fiber optic sensor cables into such geosynthetics allows creating multifunctional sensitive structures that have enormous potential for future solutions to many questions of large-scale monitoring for structural and civil engineering. Furthermore, the geosynthetics-integrated distributed fiber optic sensors are excellent tools to provide information about soil displacement, erosion control, critical mechanical deformations, and cracking in geotechnical, concrete, and masonry structures.

In the last decade, several laboratory and field tests both with GOF- [45,62–65] and POF-integrated [3,4,23,26,66] smart geosynthetics in the form of non-woven textiles and stiffer geogrids have been conducted. During the development of fiber-sensor-based geosynthetics for geotechnical applications and retrofitting of masonry structures, the ruggedness and the large core diameter of POFs have proven to be advantageous for fiber integration into the geosynthetic materials. While utilization of POFs allows easy sensor integration into the carrier materials, fabrication of GOF-based geotextiles and geogrids requires several optimization steps to avoid the prejudicial pre-strain and the occurrence of bending effects along the fiber. The latter can lead to significant increase in optical losses, which in turn restrict the possible measurement range.

Figure 12.10 illustrates the installation of a PMMA POF sensor-equipped geogrid placed perpendicular to the tear-off edge of a creeping slope at an open brown coal pit near Belchatow, Poland [26]. The cleft formation observed after installation, as shown on the right side of **Figure 12.10**, was monitored in irregular intervals using OTDR technique.

According to the results compiled in **Figure 12.11**, strain of more than 10% was measured here at the creeping location by the POF-based OTDR sensor. From the data depicted in **Figure 12.11**, the average creep velocity can be calculated to be around 2.5 mm per day.

As a result of the successful demonstration of POF-based OTDR sensors in the field boosted by the huge interest of the geotechnical industry, the first product based on distributed PMMA POF sensors integrated in geosynthetics, called GEDISE, has been commercially available on the market for several years [66].

FIGURE 12.10 (Left) Installation of geogrid with integrated POF sensor at a creeping slope (Belchatow, Poland) and (right) creeping effects leading to local strain of the integrated sensor-based geogrid.

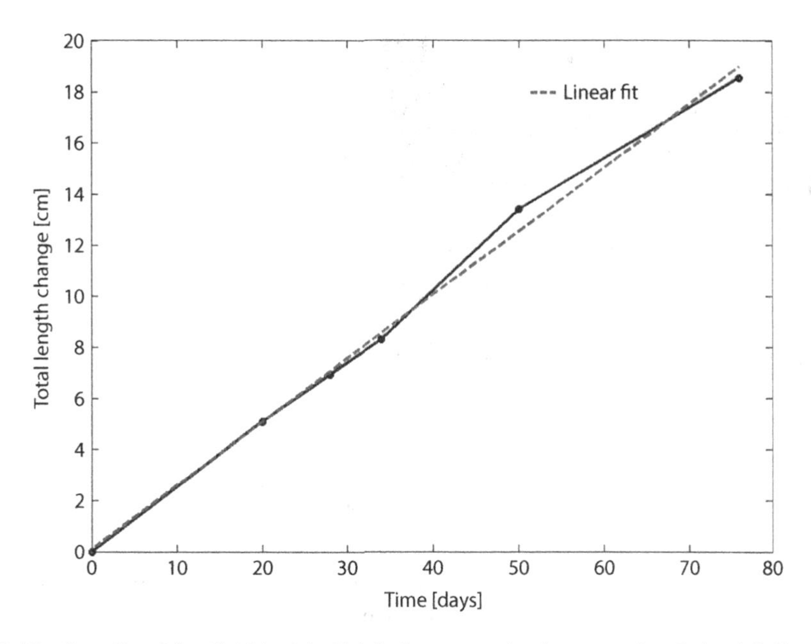

FIGURE 12.11 Results of the field test in Belchatow—constant creep rate of about 2.5 mm/day.

Another application of POF-integrated technical textiles is based on their direct embedding into concrete and masonry structures. In this case, typical damages and failures can be detected as crack formation and growth. For example, dynamic crack detection for earthquake protection is described in Section 12.2.2.

Figure 12.12 illustrates another series of tests that has been performed to prove the ability of diverse PMMA POF-integrated geogrids applied directly to masonry structures to detect mechanical deformation [3,4]. Here, by applying a load to the upper side of each specimen, a crack opened perpendicular to the sensor fiber on the bottom side of each sample. In this way of step-wise mechanical loading, crack formation below 1 mm could be detected by OTDR measurements. The direct fiber bonding to the structure provides optimized integration of the sensor. Thanks to that, short gauge lengths were achieved resulting in turn in high local strains in the fiber with crack widths up to 20 mm. In order to avoid fiber fracture at high crack widths, a trade-off between integration-related high-sensitivity crack detection and the range of crack widths to be detected need to be ensured.

The sensitivity characteristics of POF-integrated geosynthetics and its dependence on sensor embedding method can be investigated in the laboratory as shown in **Figure 12.13** [3]. Here, using the test setup at Saxon Textile Research Institute (STFI), various sensor-based

FIGURE 12.12 Crack opening test on a masonry specimen with integrated POF-based geotextile.

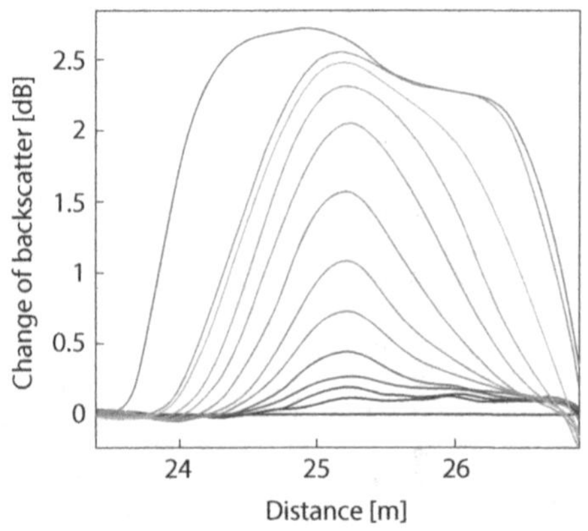

FIGURE 12.13 Load test on a POF-based geogrid performed at Saxon Textile Research Institute—the higher the mechanical load, the larger the change of backscatter.

geosynthetics specimens can be clamped on the two ends. Forcing a lateral displacement by the use of the hydraulic cylinder, the entire transfer behavior from the geosynthetics to the sensor fibers can be determined.

The test results presented for example on the right side of **Figure 12.13** in the form of local OTDR traces showed possibility of measurement of high strain values.

The ability of application of different geosynthetics configurations for efficient force transmission from the sensor environment to the measurement fibers creates new possibilities for widespread two-dimensional SHM of large-scale geotechnical and masonry structures.

12.5 Conclusion

This chapter provides a compact overview of fundamental POF-based sensor principles for structural health monitoring applications with the focus on distributed monitoring systems. The advantages and limitations of each individual fully distributed and multi-point sensor system were presented in detail. Although most of the measurement methods have not yet been commercially established, the ongoing POF material development toward SM and low-loss solutions can completely change this market situation. The crucial advantages of POF sensors like high strain measurement range, sensitivity to moisture, ruggedness, low Young's Modulus, low cost, and ease of handling are the key factors for their market-oriented breakthrough expected in the near future. Particularly, the use of low-cost extrinsic loss-based sensors for precise dynamic high-strain and acceleration measurements is expected to spread. The great capabilities for monitoring of extended structures, like dikes or railway embankments, using smart geosynthetics containing POFs for sensitivity-optimized multidimensional detection and localization of potential damages are also to be emphasized.

References

1. Kuang, K. S. C., Quek, S. T., Koh, C. G., Cantwell, W. J., & Scully, P. J. (2009). Plastic optical fibre sensors for structural health monitoring: A review of recent progress. *Journal of Sensors*, 2009.
2. Ishigure, T., Hirai, M., Sato, M., & Koike, Y. (2004). Graded-index plastic optical fiber with high mechanical properties enabling easy network installations. I. *Journal of Applied Polymer Science*, 91(1), 404–409.

3. Liehr, S., Lenke, P., Krebber, K., Seeger, M., Thiele, E., Metschies, H., Gebreselassie, B., Münch, J. C., & Stempniewski, L. (2008, April). Distributed strain measurement with polymer optical fibers integrated into multifunctional geotextiles. In *Proceedings of SPIE* (Vol. 7003, pp. 700302-700302).
4. Liehr, S., Lenke, P., Wendt, M., Krebber, K., Seeger, M., Thiele, E., Metschies, H., Gebreselassie, B., & Munich, J. C. (2009). Polymer optical fiber sensors for distributed strain measurement and application in structural health monitoring. *IEEE Sensors Journal*, 9(11), 1330–1338.
5. Liehr, S. (2010). Quasi-distributed and dynamic length change measurement in polymer opical fibers, *19th International Conference on Plastic Optical Fibers*, Yokohama, Japan.
6. Liehr, S. (2011). Polymer optical fiber sensors in structural health monitoring. In *New Developments in Sensing Technology for Structural Health Monitoring* (pp. 297–333). Berlin, Germany: Springer.
7. Murphy, K., Zhang, S., & Karbhari, V. M. (1999). Effect of concrete based alkaline solutions on short term response of composites. *Society for the Advancement of Material and Process Engineering, Evolving and Revolutionary Technologies for the New Millenium*, 44, 2222–2230.
8. Yoshihara, N. (1998, February). Low-loss, high-bandwidth fluorinated POF for visible to 1.3 μm wavelengths. In *Optical Fiber Communication Conference and Exhibit*, 1998, Technical Digest (p. 308). IEEE.
9. Aleksander, W., & Schreier, A. (2017). Toward investigation of Brillouin scattering in multimode polymer and silica optical fibers. Prague, Czech Republic: SPIE Optics + Optoelectronics.
10. Ziemann, O., Krauser, J., Zamzow, P. E., & Daum, W. (2008). *POF Handbook*. Berlin, Germany: Springer.
11. Mizuno, Y., & Nakamura, K. (2010). Potential of Brillouin scattering in polymer optical fiber for strain-insensitive high-accuracy temperature sensing. *Optics Letters*, 35(23), 3985–3987.
12. Minardo, A., Bernini, R., & Zeni, L. (2014). Distributed temperature sensing in polymer optical fiber by BOFDA. *IEEE Photonics Technology Letters*, 26(4), 387–390.
13. Hayashi, N., Mizuno, Y., & Nakamura, K. (2015). Simplified Brillouin optical correlation-domain reflectometry using polymer optical fiber. *IEEE Photonics Journal*, 7(1), 1–7.
14. Minakawa, K., Koike, K., Hayashi, N., Koike, Y., Mizuno, Y., & Nakamura, K. (2015). Brillouin frequency shift dependence on water absorption in plastic optical fibers. In *24th International Conference on Plastic Optical Fibers*.
15. Koike, Y., & Asai, M. (2009). The future of plastic optical fiber. *NPG Asia Materials*, 1(1), 22–28.
16. Kiesel, S., Peters, K., Hassan, T., & Kowalsky, M. (2008). Large deformation in-fiber polymer optical fiber sensor. *IEEE Photonics Technology Letters*, 20(6), 416–418.
17. Kiesel, S., Peters, K., Hassan, T., & Kowalsky, M. (2009). Calibration of a single-mode polymer optical fiber large-strain sensor. *Measurement Science and Technology*, 20(3), 034016.
18. Zhou, G., Pun, C. F. J., Tam, H. Y., Wong, A. C., Lu, C., & Wai, P. K. A. (2010). Single-mode perfluorinated polymer optical fibers with refractive index of 1.34 for biomedical applications. *IEEE Photonics Technology Letters*, 22(2), 106–108.
19. Liu, H. Y., Peng, G. D., & Chu, P. L. (2002). Thermal stability of gratings in PMMA and CYTOP polymer fibers. *Optics Communications*, 204(1), 151–156.
20. Gomez, J., Zubia, J., Aranguren, G., Arrue, J., Poisel, H., & Saez, I. (2009). Comparing polymer optical fiber, fiber Bragg grating, and traditional strain gauge for aircraft structural health monitoring. *Applied Optics*, 48(8), 1436–1443.
21. Durana, G., Kirchhof, M., Luber, M., de Ocáriz, I. S., Poisel, H., Zubia, J., & Vázquez, C. (2009). Use of a novel fiber optical strain sensor for monitoring the vertical deflection of an aircraft flap. *IEEE Sensors Journal*, 9(10), 1219–1225.
22. Mizuno, Y., & Nakamura, K. (2011). Brillouin scattering in polymer optical fibers: Fundamental properties and potential use in sensors. *Polymers*, 3(2), 886–898.
23. Lenke, P., Liehr, S., Krebber, K., Weigand, F., & Thiele, E. (2007). Distributed strain measurement with polymer optical fiber integrated in technical textiles using the optical time domain reflectometry technique, In *Proceedings of the III ECCOMAS Thematic Conference on Smart Structures and Materials*, Gdansk, Poland.
24. Husdi, I. R., Nakamura, K., & Ueha, S. (2004). Sensing characteristics of plastic optical fibres measured by optical time-domain reflectometry. *Measurement Science and Technology*, 15(8), 1553.
25. Nakamura, K., Husdi, I., & Ueha, S. (2004). Memory effect of POF distributed strain sensor. In *Proceedings of SPIE* (Vol. 5502, p. 145).

26. Liehr, S., & Krebber, K. (2016). Applications and prospects for distributed sensing using polymer optical fibres. *International Conference on Smart Structures and Construction*, Cambridge, UK.
27. Liehr, S., Lenke, P., Wendt, M., & Krebber, K. (2008). Perfluorinated graded-index polymer optical fibers for distributed measurement of strain. *17th International Conference on Plastic Optical Fibers*.
28. Liehr, S., Breithaupt, M., & Krebber, K. (2017). Distributed humidity sensing in PMMA optical fibers at 500 nm and 650 nm wavelengths. *Sensors*, 17(4), 738.
29. Technical information sheet, AGCCE Chemical Europe, Ltd. www.agcce.com/cytop-technical-information.
30. Zhang, C., Chen, X. F., Webb, D. J., & Peng, G. D. (2009). Water detection in jet fuel using a polymer optical fibre Bragg grating. In *Proceedings of SPIE* (Vol. 7503).
31. Zhang, W., Webb, D. J., & Peng, G. D. (2012). Investigation into time response of polymer fiber Bragg grating based humidity sensors. *Journal of Lightwave Technology*, 30(8), 1090–1096.
32. Rajan, G., Noor, Y. M., Liu, B., Ambikairaja, E., Webb, D. J., & Peng, G. D. (2013). A fast response intrinsic humidity sensor based on an etched single mode polymer fiber Bragg grating. *Sensors and Actuators A: Physical*, 203, 107–111.
33. Liehr, S., Wendt, M., & Krebber, K. (2010). Distributed strain measurement in perfluorinated polymer optical fibres using optical frequency domain reflectometry. *Measurement Science and Technology*, 21(9), 094023.
34. Liehr, S., Nöther, N., & Krebber, K. (2009). Incoherent optical frequency domain reflectometry and distributed strain detection in polymer optical fibers. *Measurement Science and Technology*, 21(1), 017001.
35. Liehr, S. (2015). Fibre optic sensing techniques based on incoherent optical frequency domain reflectometry. Ph.D. Thesis: Bundesanstalt für Materialforschung und -prüfung (BAM).
36. Liehr, S., Nöther, N., Steffen, M., Gili, O., & Krebber, K. (2014). Performance of digital incoherent OFDR and prospects for optical fiber sensing applications. In *Proceedings of SPIE* (Vol. 9157, pp. 915737-1).
37. Minardo, A., Bernini, R., & Zeni, L. (2014). Analysis of the Brillouin gain spectrum in a graded-index multimode fiber. In *Photonics Conference, 2014 Third Mediterranean* (pp. 1–3), Trani, Italy.
38. Mizuno, Y., Hayashi, N., & Nakamura, K. (2014). Distributed strain and temperature sensing based on Brillouin scattering in plastic optical fibers. In *OFS2014 23rd International Conference on Optical Fiber Sensors* (pp. 91573V–91573V). International Society for Optics and Photonics, Santander, Spain.
39. Minardo, A., Bernini, R., & Zeni, L. (2014, June). Brillouin optical frequency domain analysis in polymer optical fiber. In *OFS2014 23rd International Conference on Optical Fiber Sensors* (pp. 91576V–91576V). International Society for Optics and Photonics, Santander, Spain.
40. Minakawa, K., Mizuno, Y., & Nakamura, K. (2017). Cross effect of strain and temperature on Brillouin frequency shift in polymer optical fibers. *Journal of Lightwave Technology*, 35(12), 2481–2486.
41. Hayashi, N., Minakawa, K., Mizuno, Y., & Nakamura, K. (2014). Brillouin frequency shift hopping in polymer optical fiber. *Applied Physics Letters*, 105(9), 091113.
42. Dong, Y., Xu, P., Zhang, H., Lu, Z., Chen, L., & Bao, X. (2014). Characterization of evolution of mode coupling in a graded-index polymer optical fiber by using Brillouin optical time-domain analysis. *Optics Express*, 22(22), 26510–26516.
43. Wosniok, A., Nöther, N., & Krebber, K. (2009). Distributed fibre optic sensor system for temperature and strain monitoring based on Brillouin optical-fibre frequency-domain analysis. *Procedia Chemistry*, 1(1), 397–400.
44. Nikles, M., Thevenaz, L., & Robert, P. A. (1997). Brillouin gain spectrum characterization in single-mode optical fibers. *Journal of Lightwave Technology*, 15(10), 1842–1851.
45. Nöther, N. (2010). Distributed fiber sensors in river embankments: Advancing and implementing the Brillouin optical frequency domain analysis. Ph.D. Thesis: Bundesanstalt für Materialforschung und -prüfung (BAM).
46. Silva-López, M., Fender, A., MacPherson, W. N., Barton, J. S., Jones, J. D., Zhao, D., Dobb, H., Webb, J. D., Zhang, L., & Bennion, I. (2005). Strain and temperature sensitivity of a single-mode polymer optical fiber. *Optics Letters*, 30(23), 3129–3131.
47. Gallego, D., & Lamela, H. (2009). High-sensitivity ultrasound interferometric single-mode polymer optical fiber sensors for biomedical applications. *Optics Letters*, 34(12), 1807–1809.

48. Samiec, D. (2012). Distributed fibre-optic temperature and strain measurement with extremely high spatial resolution. *Photonic International*, 6, 10–13.
49. Kreger, S. T., Sang, A. K., Gifford, D. K., & Froggatt, M. E. (2009). Distributed strain and temperature sensing in plastic optical fiber using Rayleigh scatter. In *Proceedings of SPIE* (Vol. 7316, p. 73160A).
50. Kuang, K. S., Cantwell, W. J., & Scully, P. J. (2002). An evaluation of a novel plastic optical fibre sensor for axial strain and bend measurements. *Measurement Science and Technology*, 13(10), 1523.
51. Liu, H. Y., Liu, H., & Peng, G. D. (2005). Strain sensing characterization of polymer optical fibre Bragg gratings. In *Proceedings of SPIE* (Vol. 5855, pp. 663–666), Bruges, Belgium.
52. Liu, H. Y., Liu, H. B., & Peng, G. D. (2005). Tensile strain characterization of polymer optical fibre Bragg gratings. *Optics Communications*, 251(1), 37–43.
53. Zhang, C., Zhang, W., Webb, D. J., & Peng, G. D. (2010). Optical fibre temperature and humidity sensor. *Electronics Letters*, 46(9), 643–644.
54. Woyessa, G., Nielsen, K., Stefani, A., Markos, C., & Bang, O. (2016). Temperature insensitive hysteresis free highly sensitive polymer optical fiber Bragg grating humidity sensor. *Optics Express*, 24(2), 1206–1213.
55. Lacraz, A., Polis, M., Theodosiou, A., Koutsides, C., & Kalli, K. (2015). Femtosecond laser inscribed Bragg gratings in low loss CYTOP polymer optical fiber. *IEEE Photonics Technology Letters*, 27(7), 693–696.
56. Theodosiou, A., Lacraz, A., Polis, M., Kalli, K., Tsangari, M., Stassis, A., & Komodromos, M. (2016). Modified fs-laser inscribed FBG array for rapid mode shape capture of free-free vibrating beams. *IEEE Photonics Technology Letters*, 28(14), 1509–1512.
57. Ishikawa, R., Lee, H., Lacraz, A., Theodosiou, A., Kalli, K., Mizuno, Y., & Nakamura, K. (2017). Pressure dependence of fiber Bragg grating inscribed in perfluorinated polymer fiber. *IEEE Photonics Technology Letters*, 29(24), 2167–2170.
58. Witt, J., Breithaupt, M., Erdmann, J., & Krebber, K. (2011). Humidity sensing based on microstructured POF long period gratings. In *Proceedings of the International Conference on Plastic Optical Fibres* (pp. 409–414).
59. Witt, J., Steffen, M., Schukar, M., & Krebber, K. (2010, May). Investigation of sensing properties of microstructured polymer optical fibres. In *Photonic Crystal Fibers IV* (Vol. 7714, p. 77140F). International Society for Optics and Photonics.
60. de Oliveira, R. (2010). Advanced composite materials with embedded POF sensors for structural health monitoring. In *Proceedings of the 19th International conference on plastic optical fibers*, Yokohama, Japan.
61. de Oliveira, R., Schukar, M., Krebber, K., & Michaud, V. (2010). Development of adaptive composites with embedded plastic optic fibres. In *Proceedings of the 3rd International conference on recent advances in composite materials*, Limoges, France.
62. Nöther, N., Wosniok, A., Krebber, K., & Thiele, E. (2007). Dike monitoring using fiber sensor-based geosynthetics, In *Proceedings of the III ECCOMAS Thematic Conference on Smart Structures and Materials*, Gdansk, Poland.
63. Nöther, N., Wosniok, A., Krebber, K., & Thiele, E. (2009). A distributed fiber-optic sensing system for monitoring of large geotechnical structures. *Proceedings of SHMII-4*, 2009, Zurich, Switzerland.
64. Nöther, N., Wang, S., Wosniok, A., Glötzl, R., & Schneider-Glötzl, J. (2013). Distributed Brillouin sensing in optical fibers: Soil displacement monitoring using sensor-equipped geogrids, In *Proceedings of SHMII-6*, Hong Kong, China.
65. Wosniok, A., & Krebber, K. (2015). Smart geosynthetics for structural health monitoring using fully distributed fiber optic sensors, In *Proceedings of 6th International Technical Textiles Congress*, Izmir, Turkey.
66. Krebber, K., Lenke, P., Liehr, S., Schukar, M., Wendt, M., & Witt, J. (2010). Distributed POF sensors: Recent progress and new challenges. In Invited Paper, *International Conference of Plastic Optical Fiber (ICPOF)*, Yokohama, Japan.

13

POF and Radiation Sensing

Pavol Stajanca

Contents

13.1 Interaction of Optical Fibers with Ionizing Radiation

In general, three different effects can occur in optical fibers under irradiation:

- Decrease of fiber's transmission due to radiation-induced attenuation (RIA)
- Generation of UV, VIS, or NIR light through radiation-induced emission (RIE)
- Compaction of fiber material

Magnitude and kinetics of these radiation-induced effects depend on the irradiated fiber, nature and properties of interacting radiation, and irradiation conditions. As a result of that, radiation response of optical fibers remains a complex and not fully understood issue even after decades of extensive research. The interest in studying fiber radiation response is usually twofold. For fiber optic systems that ought to be deployed and operated in radiation environments, radiation-induced changes can be viewed as detrimental effects degrading the fiber performance. As such, they need to be characterized and understood in order to perform system optimization and minimize the negative impacts of irradiation. On the other hand, these changes can be exploited for direct radiation detection or monitoring. In this case, knowledge and optimization of fiber's radiation response is crucial for obtaining efficient optical fiber radiation sensor.

Ionizing radiation detection, monitoring, or dosimetry is of interest in many different fields including the nuclear industry, high-energy physics, medicine, sterilization, recycling, and the food industry. Compared to other radiation monitoring techniques, optical fiber radiation sensors (OFRSs) can offer several advantages inherent to fiber optic technology [1,2]. Optical fibers are electrically passive, immune to electromagnetic interference, lightweight, compact, and provide low-loss transmission over extended distances. As a result, OFRSs can be used for remote and real-time radiation monitoring even in hazardous, harsh, and difficult-to-access locations. In addition, certain OFRSs can be multiplexed on a single optical fiber link [3], or even have potential for fully-distributed measurement [4]. Realization of OFRS has been suggested based on all above mentioned interaction mechanisms, i.e. RIA, RIE, and material compaction, or better said, refractive index change. In the following, we will introduce these effects and how they can be exploited for radiation sensing. In addition to intrinsic sensors based on direct radiation-fiber interaction, optical fibers are also often employed in extrinsic OFRSs as a convenient interconnection between a sensor's electronics and sensing medium. These cases will not be discussed here. Instead, several notable examples relevant to POF technology will be addressed in Section 13.4.

13.1.1 Radiation-Induced Attenuation (RIA)

Radiation-induced attenuation has been studied extensively for silica-based optical fibers [5], mostly as a detrimental effect limiting the performance and lifetime of fiber systems operated in harsh radiation environments. The origin of RIA is an increase of material attenuation through additional absorption on radiation-induced defects, so-called color centers [6,7]. Nature of the defects depends on the fiber material. Different defects, in turn, absorb at different wavelengths. Therefore, RIA is strongly material and wavelength dependent [5]. Depending on fiber composition, both highly radiation hard (tolerant) as well as highly radiation sensitive fibers can be identified. Examples of the former are pure-silica core [8] or nitrogen-doped fibers [9], while phosphorus-doped [10] or lead-doped fibers [11] are representatives of the latter fiber category. Regarding RIA wavelength dependence, fiber's RIA sensitivity typically increases from the NIR toward UV wavelengths.

Magnitude of RIA in a fiber can be correlated to the total dose that the fiber has been subjected to. In this way, RIA monitoring in a fiber with suitable radiation response can be used for radiation sensing [12]. The utilization of fiber RIA for radiation dosimetry purposes dates back to the 1970s [13]. Since then, numerous different fibers and arrangements have been investigated. In the simplest case, fiber's RIA can be measured in a transmission mode when a light source and a detector are connected to the opposite ends of the sensor fiber [13]. Alternatively, a light source and a detector can be connected to the same fiber end with the help of a 3 dB coupler and a reflecting component placed on the remote end of the sensing fiber [14]. In this case, the effective length of the sensor (along with its sensitivity) is doubled as light travels through the sensing fiber two times. In addition, there are certain techniques, such as optical time domain reflectometry (OTDR) [15], that enable distributed measurement of fiber attenuation profile along its entire length. Monitoring of fiber's RIA with the help of such a technique can be used for distributed radiation measurement [4,16].

Over the years, it was shown that, besides fiber composition and monitoring wavelength, fiber's RIA response is influenced by a number of additional parameters. These include dose, dose-rate, temperature, or radiation energy dependence [5]. The complexity of RIA response dependent on many operation-related parameters limits the practical potential of RIA-based OFRSs. Therefore, emission-based monitoring techniques are nowadays prevailing for the majority of applications. While the research on the RIA in silica-based optical fibers is quite extensive, only a few studies were devoted to the RIA in POFs. These will be reviewed in Section 13.2.

13.1.2 Radiation-Induced Emission (RIE)

Interaction of ionizing radiation with optical fibers can lead to generation of light in the fiber material, i.e. radiation-induced emission. There are two main mechanisms responsible for RIE in optical fibers: scintillation, also known as radiation-induced luminescence (RIL), and Cerenkov radiation. Scintillation is one of the oldest and most popular techniques for detection and measurement of ionizing radiation [17,18]. Practically all standard telecommunication fibers, both silica- and polymer-based ones, yield RIL under irradiation. In the case of silica-based fibers, RIL comes from pre-existing or radiation-induced point defects in the glass structure that are excited by the incoming radiation. For POFs, excitation of the host material on the molecular level is responsible for the luminescence. Nevertheless, intensity of RIL in standard fibers is typically too low to be practical for radiation monitoring applications. Optical fibers are then usually combined with materials with higher RIL yield, known as scintillators. This can be done either by coupling of the optical fiber with a separate external scintillator or by direct incorporation of scintillation-promoting compound into the fiber via doping. The former arrangement gives rise to an extrinsic OFRS, while in the latter case optical fiber itself becomes the scintillator and main sensing element.

In general, the intensity of scintillator's RIL is proportional to the irradiation dose rate and therefore can be used for simultaneous real-time measurement of the instant dose rate and the total dose accumulated over time. Scintillators can be broadly divided into two categories: organic and inorganic. Both types have their own merits and limitations [18]. Inorganic scintillators usually come in the form of crystals or powders and have higher light yield and longer scintillation decay times. Organic scintillators are mostly plastic or liquid and have typically lower light yield and shorter decay times. Organic plastic scintillators (OPSs) can be relatively easily processed into almost arbitrary shape, which eventually led to the creation of scintillating polymer optical fibers. Scintillating POFs represent OPSs which can not only produce RIL but also guide generated light, and thus increase the light collection efficiency of the sensing system or cover extended lengths. Many scintillation-based extrinsic OFRSs have been investigated, and some notable examples relevant to POF technology will be discussed in Section 13.4. However, the key focus will be devoted to the scintillating POFs that will be presented in more detail in Section 13.3.

Cerenkov radiation is light which is generated when a charged particle, most commonly an electron, passes through a material at a speed which is faster than the speed of light in that material [19]. It is commonly viewed as an electromagnetic analogy to a sonic boom. Electrons responsible for Cerenkov light generation could be either primary, i.e. originating directly from high-energy beams, or secondary electrons set in motion through material ionization. The particle energy threshold for Cerenkov light generation is dependent on the material's refraction index n and decreases with increasing n value. As most of the optical materials used for optical fiber fabrication have a fairly high refractive index (RI), threshold for Cerenkov light generation is relatively low; e.g. 175 keV for a medium with $n = 1.5$ [20]. Cerenkov emission is broadband with a peak in the blue-violet spectral region and then drops quickly as λ^{-3} toward the longer wavelengths. Compared to RIL yield of common scintillators, the emission yield of Cerenkov radiation in optical fibers is up to two orders of magnitude lower. Despite that, Cerenkov radiation is typically viewed as an undesirable parasitic effect compromising the read-out of scintillation-based OFRSs [21]. Different correction techniques have been developed to cope with the issue. These will be briefly addressed in Section 13.3.3. Few works investigated utilization of Cerenkov emission in silica-based fibers for radiation monitoring purposes as well [22,23]. Recently, RIE in a bare polymethyl methacrylate (PMMA) POF has been suggested as an effective means for depth-dose measurement in proton therapy [24,25]. The RIE was first attributed to Cerenkov light generation; however, later studies used spectral analysis and Monte-Carlo simulations to show that this RIE is actually not Cerenkov light, but most likely comes from fluorescence of the fiber material itself [26,27]. In addition, the practical potential of Cerenkov-based OFRS is generally limited by rather low light yield and dependence on irradiation geometry. No other works on Cerenkov-based polymer optical fiber radiation sensors are known to the author and no further space will be devoted to the topic.

13.1.3 Refractive Index Change

Subjecting optical material to a high-energy irradiation can lead to a change of its refractive index via alteration of its internal structure [28,29]. In general, RI changes can come either from the RIA or material density changes. The relation between RI changes and the RIA or material density is guided by Kramers-Kronig [30,31] or Lorentz-Lorenz formula [32,33], respectively. For optical fibers, perhaps the most common and investigated manifestation of the radiation-induced RI change is the wavelength shift of fiber Bragg gratings (FBGs) or long period gratings (LPGs) [34]. In the case of silica-based fiber gratings, the effect has been studied extensively, mostly in quest for radiation hard gratings for strain or temperature sensing applications in radiation environments [35]. Use of radiation-induced FBG wavelength shift for large dose measurement has been previously suggested as well [36]. However, complex response, low inherent sensitivity and problems with strain and temperature cross-sensitivity limit the practical potential of fiber grating technology for radiation monitoring applications. Very little attention has been devoted to this topic since.

Virtually no information is available on the radiation response of polymer-based fiber gratings. Witt et al. measured response of LPG in PMMA-based microstructured POF to synchrotron irradiation with average energy of around 12 keV [37]. Hamdalla et al. investigated Bragg wavelength shift of PMMA-based polymer optical fiber Bragg grating under low dose neutron irradiation [38]. Due to a lack of information and low potential for radiation monitoring, this topic will not be discussed further in this chapter.

13.2 Radiation Monitoring with RIA-Based POF Sensors

Unlike for silica-based fibers, information on the RIA in polymer optical fibers is rather limited. Nevertheless, POFs can be an interesting alternative for certain radiation monitoring applications, especially as cheap, easy-to-use, disposable sensors. This is not only thanks to the lower direct cost of POFs, but also thanks to the ease of handling and connectorization they offer. This helps to minimize expenses for high-tech equipment and highly trained personnel. With regard to radiation monitoring applications, RIA in two different types of commercial POFs has been studied so far. O'Keeffe et al. investigated the RIA in standard PMMA POFs [39–42], while Stajanca et al. characterized the RIA response of commercial perfluorinated POFs [43–48]. Few other works have been published investigating the RIA in standard PMMA POF [49–51], however, not in radiation sensing context. The RIA in scintillating and other specialty POFs has been also investigated in terms of their radiation hardness for scintillation-based OFRS [52–56].

13.2.1 PMMA-Based POFs

O'Keeffe et al. suggested monitoring of RIA in a standard 1 mm multimode PMMA POF for a real-time *in situ* dose measurement in radiation processing industries, e.g. in radiation sterilization. They measured PMMA POF RIA response up to 50 kGy under gamma irradiation from spent fuel elements with dominant contribution from ^{137}Cs [39]. Later, an analogical study with similar results using a ^{60}Co source was performed as well [40]. They used simple transmission setup employing a broadband light source and a CCD-based spectrometer to measure the RIA in the entire transmission window of PMMA POF. Similarly as in the case of silica-based fibers, RIA in the studied POF grows monotonically with increasing irradiation dose. The rate of this growth is wavelength dependent, and the RIA grows faster at lower wavelengths.

In case the relation between the RIA and the dose is linear, fiber's RIA sensitivity at a given wavelength can be determined as a slope of the respective RIA versus dose growth curve. Along with sensitivity, each dosimetry system is also characterized by its monitoring range determined by the smallest and largest dose that the system is able to measure. Due to the strong wavelength dependence of the RIA in the PMMA fiber, O'Keeffe et al. argued that a large range of sensitivities and monitoring ranges can be achieved with the POF-based system by a suitable selection of monitoring wavelength. They demonstrated the RIA-based dose measurement in the range between 30 and 50 kGy with sensitivity as high as 0.6 dBm^{-1}/kGy at 525 nm [40]. The same group also investigated the possibility of extending the measurement principle to a low-dose measurement for medical applications [41]. The PMMA POF's response to 6 MV radiotherapeutic X-rays was measured up to 12 Gy. Even though a distinct RIA increase trend could be observed, the measurement noise was too large to perform a reliable dose quantification. This indicates that the sensitivity of PMMA POFs is not high enough for the low-dose measurement required in medical applications.

13.2.2 Perfluorinated POFs

More recently, Stajanca et al. characterized RIA of commercial perfluorinated POFs (PF-POFs) under gamma irradiation from a ^{60}Co source. Compared to PMMA fibers, PF-POFs based on perfluorinated polymer Cytop have much broader low-loss transmission window covering the VIS and NIR wavelengths [57]. Two different types of PF-POFs with slightly different characteristics are commercially available: fibers drawn from a preform and co-extruded ones [58]. The initial general studies of irradiation impacts on PF-POFs revealed rather high radiation sensitivity in both of these fiber types, especially in the VIS spectral region [43,45]. Later comparative studies showed that while the spectral shape of the RIA is very similar in both PF-POF types, co-extruded fibers offer more advantageous performance for radiation monitoring applications in terms of better linearity [46] or lower dose-rate dependence of their RIA response (**Figure 13.1**) [47].

In the latest study, a complex performance of RIA-based radiation sensing with PF-POFs was analyzed [47]. Dose-rate dependence, temperature dependence, annealing, and repeatability of the RIA measurement was investigated. It was shown that, with certain limitations, RIA monitoring

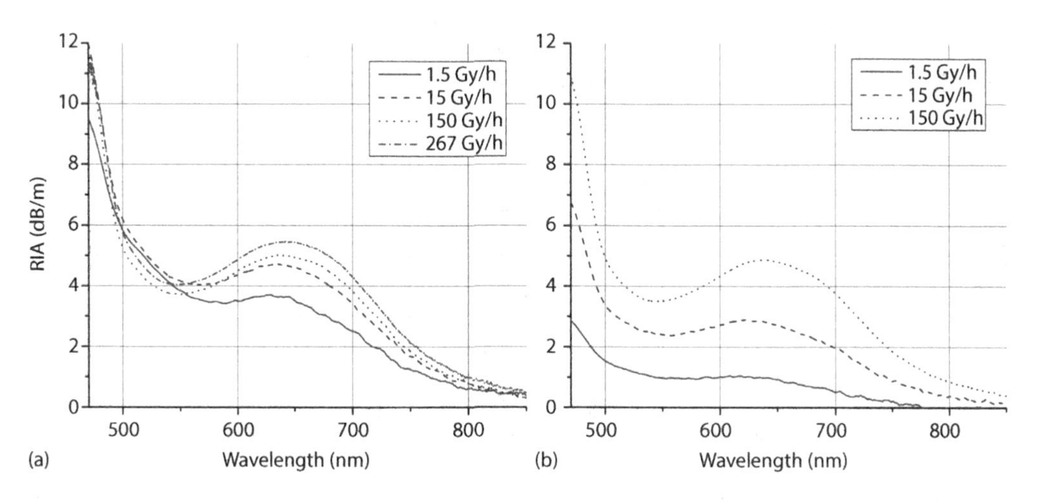

FIGURE 13.1 The spectral shape of RIA in (a) co-extruded PF-POF (GigaPOF-50S, Chromis Fiberoptics) and (b) PF-POF drawn from a preform (Fontex, Asahi Glass Company) after gamma irradiation to 100 Gy at different dose rates. (Reprinted from Stajanca, P. and Krebber, K., *Sensors*, 17, 1959, 2017. Creative Commons BY 4.0 [https://creativecommons.org/licenses/by/4.0/].)

Table 13.1 Comparison of RIA Sensitivity of the PMMA POF [40], and the Co-extruded PF-POF [46] for Irradiation with Gamma Rays

PMMA [40]		PF-POF [46]	
Wavelength (nm)	**Sensitivity (dBm^{-1}/kGy)**	**Wavelength (nm)**	**Sensitivity (dBm^{-1}/kGy)**
525	0.6	460	135.8 ± 0.6
570	0.3	600	47.2 ± 0.1
594	0.2	750	19.81 ± 0.03
650	0.06	830	5.67 ± 0.01
		890	1.99 ± 0.01

in this type of fiber holds the potential for high sensitivity radiation measurement with good reproducibility. **Table 13.1** compares the RIA sensitivity of the PMMA POF and co-extruded PF-POF at different wavelengths. It is clear that the PF-POF is considerably more sensitive than the PMMA fiber. The maximal demonstrated sensitivity of 135.8 ± 0.6 dBm^{-1}/kGy at 460 nm is more than 200 times higher than the maximal demonstrated sensitivity of the PMMA POF. Due to the higher sensitivity, PF-POFs can be used even for a low-dose measurement with sub-Gray resolution [48], which can be potentially interesting for certain medical applications.

In addition to characterization of PF-POFs' RIA using simple transmission setup, Stajanca et al. also explored the possibility of distributed radiation monitoring using OTDR-based RIA measurement [46,47]. Monitoring at different wavelengths, namely 650, 1060, and 1310 nm, was investigated. Among the tested wavelengths, 650 nm was finally found as the most suitable. It represents a compromise between a high RIA sensitivity required for efficient radiation detection and a low inherent attenuation required for monitoring of extended fiber lengths. Due to the temperature and dose-rate dependence of fiber's RIA response at 650 nm, this technique was deemed not suitable for a real quantitative dose measurement (dosimetry). However, the approach can be interesting for monitoring applications of a more qualitative nature, such as distributed radiation leak detection or radiation field profiling. Possibility of distributed detection of gamma radiation for doses as small as 20 Gy and length of irradiated fiber section down to 0.25 m was demonstrated using the co-extruded PF-POF and OTDR measurement at 650 nm (**Figure 13.2**) [47].

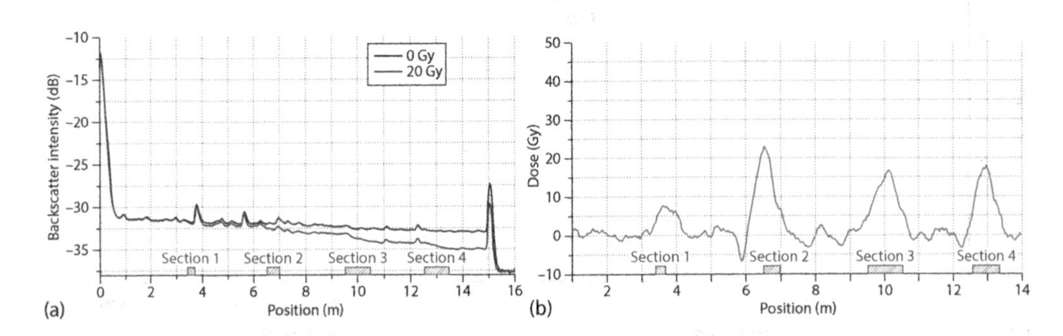

FIGURE 13.2 (a) Backscattering traces of co-extruded PF-POF (GigaPOF-50SR, Chromis Fiberoptics) before and after irradiation to 20 Gy and (b) reconstructed dose distribution along the fiber. Four separate fiber sections of different lengths (0.25, 0.5, 1, 0.75 m), indicated by the red bars on the X-axis, were irradiated simultaneously under same conditions and to the same total dose level of 20 Gy. (Reprinted from Stajanca, P. and Krebber, K. *Sensors*, 17, 1959, 2017. Creative Commons BY 4.0 [https://creativecommons.org/licenses/by/4.0/].)

13.3 Scintillating POFs

Scintillating polymer optical fibers (S-POFs), commonly denoted also as plastic scintillating fibers, can be considered an extension of organic plastic scintillators (OPSs). The possibility of drawing of OPS into thin filaments was demonstrated as early as 1957 [59]. However, addition of cladding layer to produce the first modern S-POFs was suggested only later by Borenstein et al. in the early 1980s [60–62]. The core of the early S-POFs was based on polyvinyltoluene (PVT), which remains the most common base material for the bulk OPS up to now. In the case of S-POFs, however, PVT was soon replaced by polystyrene (PS) as the base fiber core material thanks to its higher refractive index and more favorable drawing behavior [63,64]. Polyvinyl acetate (PVA) was used as a cladding material in these early fibers, but was replaced by PMMA later on [63–65]. Considerable effort has been devoted to the development of S-POF in the 1980s and 1990s, mainly for applications in particle physics. As a result of the extensive research, S-POF is today considered a mature, commercially available technology. Perhaps the two most common suppliers of S-POFs are Kuraray and Saint-Gobain Crystals. Today, double-cladding fibers with PS-based core, PMMA primary cladding, and secondary cladding from fluorinated methyl methacrylate (MMA) are the state of the art of S-POF technology.

Development of the first plastic scintillating fiber was motivated by its use in a particle detector for proton-proton collider ISABELLE (USA) [61,62]. Even though the collider itself was never fully constructed, S-POFs attracted considerable attention in the particle physics community and were the subject of extensive R&D activities ever since. The main motivation behind this research was construction of reliable and fast particle detectors with high spatial resolution. Compared to their silica-based counterparts, S-POFs offer shorter decay times, higher scintillation yield, and easier handling. Therefore, they became the superior choice for the majority of the fiber-based sensor applications in particle physics. On the other hand, S-POFs have rather poor radiation tolerance, typically around 10 kGy. Consequently, their use is limited to applications with relatively low overall irradiation levels. Number of different applications of scintillation-based OFRSs have been explored over the years. Nevertheless, high-energy particle physics remains the most traditional application field for modern S-POFs. In the last 15 years, a considerable growth of interest in scintillation-based OFRSs for medical applications has been observed. The main driving force in this case is realization of real-time, small-field, and *in vivo* dosimetry. In addition to the aforementioned benefits of S-POF technology, the key advantage of polymer fibers over other inorganic-based technologies is their water equivalence.

In this section we will first introduce main principles and properties of S-POFs and S-POF-based detectors. The discussion of their applications in particle physics and the medical industry will follow.

13.3.1 Basic Principles and Properties

The role of S-POFs in ionizing radiation monitoring can be in principle twofold. First, fiber core doped with scintillating compound acts as a transducer converting the ionization deposited by high-energy radiation or particles to optical photons. Second, thanks to the fiber cladding, the generated optical signal can then be guided along the fiber toward the read-out optoelectronics. As scintillation light is normally isotropic, addition of a cladding layer around a scintillating core increases the light collection capability of the system and enables efficient extraction of the scintillating signal from extended lengths of the fiber. Secondary cladding can be added to further increase the light collection efficiency.

Due to its advantageous optical and thermal properties, polystyrene is the most common base material for cores of modern S-POFs. Upon irradiation, PS itself produces broadband RIL with maximum around 305 nm (**Figure 13.3a**) [65]. However, the quantum efficiency of this emission is rather low, typically around 3% [66]. In addition, transparency of PS at UV wavelengths is very poor due to the electronic absorption and Rayleigh scattering (**Figure 13.3b**) [65]. Consequently, scintillation of pure polystyrene is not suitable for radiation monitoring applications.

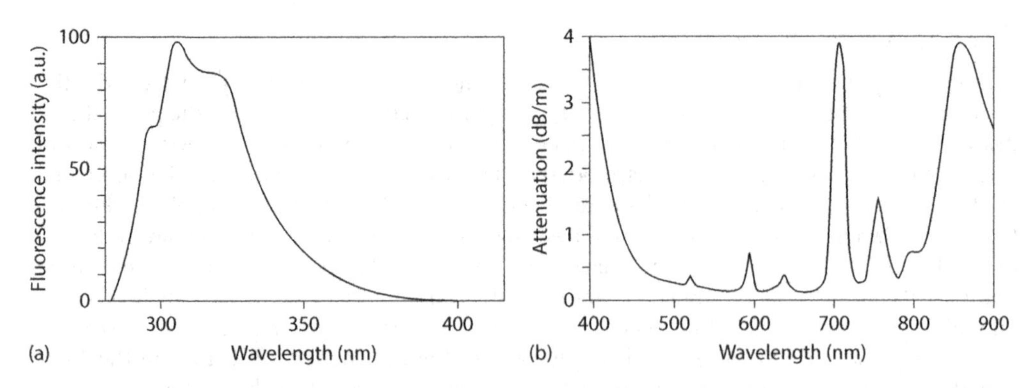

FIGURE 13.3 (a) Emission and (b) absorption spectrum of pure (undoped) polystyrene. (Reprinted from *Nucl. Instr. Meth. Phys. Res. A*, 427, Rebourgeard, P.H., et al., Fabrication and measurement of plastic scintillating fibers, 543–567, Copyright 1999, with permission from Elsevier.)

In order to increase the scintillation efficiency of the fiber, the core is doped with supplementary scintillating compounds (fluors). The aim of the doping is to address both aforementioned issues, i.e. to increase the quantum efficiency of the scintillation and to shift the emission to polystyrene's transparent region ($\lambda > 400$ nm). At the basic level, two- or three-component systems can be distinguished. The former consist of the core base material and primary scintillating dopant (scintillator). In the case of the latter, additional secondary dopant (wavelength shifter) is added to facilitate further emission shift to longer wavelengths. Several different families of compounds suitable for S-POF doping have been investigated. These include hydroxyflavons, hydroxybenzothiazoles, and hydroxybenzoxazoles [67]. Examples of some common dopants for two- and three-component systems can be found in **Table 13.2**. Excitation of the fiber base material is transferred to the molecules of primary dopant via non-radiative dipole-dipole coupling known as Förster resonance energy transfer [17,68]. Förster transfer is a fast process typically occurring on timescales below 1 ns. The efficient energy coupling requires presence of fluor molecules in the immediate vicinity of the excited site. Therefore, relatively high dopant concentrations, typically ≥1 mol% [67], are required to achieve efficient energy coupling to the scintillating fluor. Two-component formulation with fluors exhibiting large Stoke's shift are preferred in the majority of modern scintillating POF. Large Stoke's shift ensures that the scintillation falls into PS transparent region and also minimizes self-absorption by the dopant [69].

For some other dopants, however, the scintillation emission is still at wavelengths where PS exhibits large attenuation. In this case, additional wavelength-shifting (WS) dopant needs to be added to the composition. The secondary dopant should have the absorption band overlapping the emission band of the primary fluor and high quantum efficiency. The energy transfer between primary and secondary dopant is of radiative nature, therefore, much lower concentrations of wavelength shifter, typically <1 mol% [67], are sufficient for the efficient transfer. Nevertheless, fiber geometrical parameters need to be considered. Concentration of the secondary dopant is related to the mean transfer length, or in other words, to the mean free path of scintillation photons in the fiber. If this characteristic length gets comparable or larger than the fiber diameter, the efficiency of the radiative energy transfer will drop considerably as the large portion of the photons generated by the primary scintillator will escape the fiber core without interacting with the wavelength-shifting fluor. The smaller the fiber diameter needs to be, the higher concentration of the wavelength shifter is required. On the other hand, increased WS concentration might raise additional problems with the light self-absorption, especially for molecules with lower Stoke's shift [70]. This is associated with shorter attenuation length of the fiber and signal cross-talk issues in multi-channel systems. As a result of extensive research on the topic, dopant concentration has been already optimized for all commercial S-POFs.

Table 13.2 Properties of Selected Dopants for Scintillating POF

Compound	Maxima		Molar Mass (g/mol)	Decay Time (ns)	Fluor. Yield	Stokes Shift (eV)
	Abs. (nm)	Emiss. (nm)				
Two Component Scintillator						
p-terphenyl (in Cyclohexane)	275	340	230	0.95	0.93	0.86
PBD (in Cyclohexane)	300	360	354	1.0	0.83	0.69
Wavelength Shifter						
POPOP (in Polystyrene)	360	420	364	1.5	0.93	0.49
BBOT (in Cyclohexane)	370	425	430	1.1	0.74	0.43
TPBD (in Cyclohexane)	345	450	358	1.8	0.6	0.84
BDB (in Polystyrene)	350	415	414			0.55
One Component						
3HF (in Polystyrene)	340	530	238	8.0	0.4	1.27
PMP 420 (in Polystyrene)	300	415	264	3.0	0.88	1.11
PMP 450 (in Polystyrene)	305	435	294	3.6	0.74	1.24
PBBO (in Polystyrene)	330	395	347	2.1	0.79	0.62
R39 (in Toluene)	~330	480	346	~7.0		~1.17
R45 (in Toluene)	~330	490	364	~7.0		~1.22

Source: Reprinted from *Nucl. Instrum. Meth. Phys. Res. A*, 364, Leutz, H., Scintillating fibres, 422–448, Copyright 1995, with permission from Elsevier.

Selection of the most suitable fiber (fiber composition) depends largely on the nature of whole sensor system and application. Two-component systems with large Stoke's shift can minimize light self-absorption issues. They also typically operate on longer wavelengths, thus offering lower inherent attenuation of scintillation in the fiber. In addition, radiation tolerance of POFs generally increases with growing wavelength [54,70], which is an important factor for applications where larger overall doses are expected. On the other hand, according to Einstein's relations, emission on the longer visible wavelengths is typically associated with longer decay times [66]. Moreover, spectral sensitivity of system's light detector also needs to be considered for optimal choice of the scintillation wavelength.

Emission of scintillation light in OPS is isotropic. In order to improve efficiency of its collection and transfer toward the read-out optoelectronics, a thin cladding layer is added around the doped core material. This gives rise to scintillating polymer optical fibers. Diameter of S-POFs can range from tens of microns to several millimeters. Round geometry is most common, although square fibers are also available. Thickness of the cladding is usually 3%–4% of the overall fiber diameter. For a typical 1 mm thick S-POF with PS-based core and PMMA cladding, a minimum ionizing particle produces around 1500–2000 scintillation photons. Only about 6.64% of them will be trapped at the core-cladding interface and continue to propagate along the fiber [65]. Moreover, almost a half of the trapped light (3.2%) corresponds to skew (leaky) modes. In principle, light trapping at the cladding-air interface is possible as well. Due to higher refractive index difference, trapping efficiency of such a process is higher (23.83%) [67], giving rise to so-called cladding light. However, the attenuation of the cladding light is relatively high and its contribution to the overall guided light is significant only in the first few centimeters of the fiber [71]. Coating the fiber with extra-mural absorber (EMA) can be used to further minimize the influence of the cladding light. Alternatively, secondary cladding layer of fluorinated MMA can be added to increase the light trapping efficiency by up to 60% [72].

The number of photons reaching the read-out optoelectronics is further reduced due to light attenuation in the fiber. Attenuation of S-POFs is commonly characterized by attenuation length. This is the fiber length at which the intensity of produced scintillation light drops to 1/e of its original value. Attenuation length of commercial S-POFs is typically around 2–4 m. However, the light attenuation in S-POFs cannot be modeled by the usual simple exponential decay function of propagation distance. The effect is a result of combination of several factors including scintillation light reabsorption and contribution from leaky modes. Therefore, characterization of a fiber by a single effective attenuation length can be misleading. In addition, RIA incurred by the fiber throughout sensor lifetime cycle needs to be considered when designing the system [52–56].

Due to the low amount of photons that eventually reach the read-out end of the fiber, high requirements on the efficiency of the employed light detection system are imposed. Detectors with single-photon counting ability are typically required. Historically, photomultiplier tubes are the most common detectors used with scintillating fibers. Later on, also other technologies including solid state photomultipliers, CCD cameras with image intensifiers, and PIN or avalanche photodiodes have been employed. Photodetectors indisputably represent one of the key elements of scintillation-based radiation sensors. Nevertheless, the focus of this book is on POF technology, and the optoelectronic part of the sensors will not be discussed here in detail. General information and application-specific guidelines on selection of suitable photodetectors can be found in ref. [70,73,74]. More details on S-POF technology and related theory can be found in a number of topical and review publications, such as ref. [17,65,67,70,75].

13.3.2 Applications in Particle Physics

High-energy particle physics remains the most traditional application field for modern S-POFs. Two main implementations of S-POF detectors in high-energy physics are particle trackers (hodoscopes) and calorimeters. S-POF-based systems have been used for detection and tracking of particles in fixed-target, colliding-beam as well as astrophysics experiments.

Particle trackers are sub-detectors of complex monitoring systems around interaction (collision) point of particle accelerators. Their main task is to help reconstruct paths and vertex positions of charged particles generated in accelerator beam collisions. The main building block of a scintillating fiber (SciFi) particle trackers is a so-called fiber ribbon which is constructed by stacking several single layers of scintillating fibers. The layers are most commonly arranged in a simple staggered geometry where all fibers lie parallel to each other (**Figure 13.4a**). Alternatively, an arrangement with stereo layers placed at a small alternating angle with respect to the each other is often used as well (**Figure 13.4b**). This arrangement allows also derivation the axial coordinate of the particle track. The output ends of the fibers are typically bundled together and fed into a read-out optoelectronic system. Auxiliary clear (undoped) extension fibers are often used to guide scintillation signals out of high radiation environments which are detrimental to the optoelectronics. The other ends of the scintillating fibers are usually treated to reflect the light back toward the read-out system and increase the system's light collection yield. For systems using small diameter fibers, EMA coatings/paintings are often used to suppress inter-fiber cross talk caused by the unabsorbed primary UV photons from the neighboring fibers.

Particles passing through the fiber ribbon produce a scintillation signal in the fibers that they cross. The light signals from one or more fiber bundles are recorded by a space-resolved detection system, e.g. image intensified CCD camera. The detected "light track" (**Figure 13.5**) can be then used to derive the particle path through the fiber ribbon. Optimization of the tracker design typically requires balancing the contradicting requirements for high spatial resolution

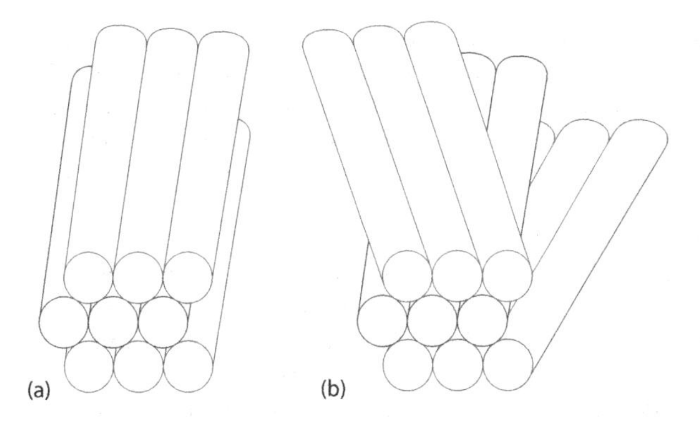

FIGURE 13.4 Schematic illustration of scintillating fiber ribbon with (a) simple staggered geometry and (b) stereo layers arrangement.

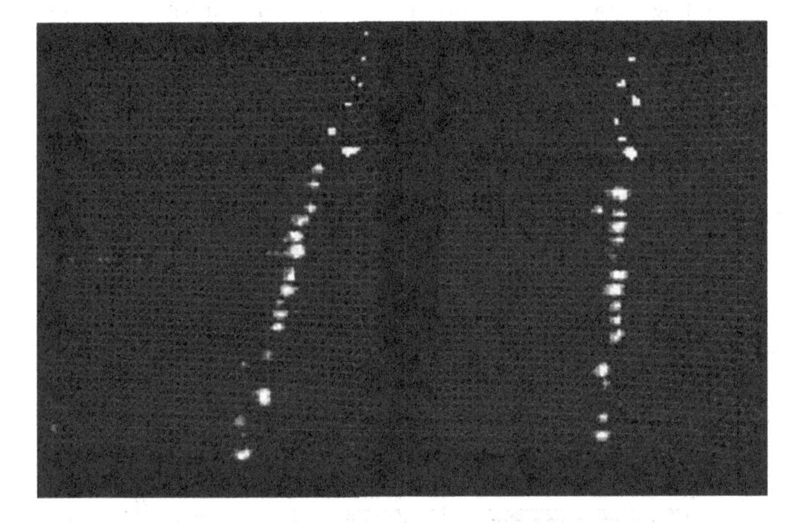

FIGURE 13.5 Examples of a raw CCD display "light track" generated by a muon passing through a scintillating fiber ribbon. The circles represent the calculated positions of the fibers on the CCD image. (Reprinted from *Nucl. Instrum. Meth. Phys. Res. A*, 265, Ansorge, R.E. et al., Performance of a scintillating fibre detector for the UA2 upgrade, 33–49, Copyright 1988, with permission from Elsevier.)

and high hit (detection) efficiency. While the former requires fibers with smaller diameters, the latter calls for thicker fibers. Designing is, therefore, a complex problem involving selection of suitable fibers, geometrical arrangement, and read-out system.

The first major implementation of S-POF particle detector technology was for the upgrade of CERN UA2 detector in the mid-1980s [76,77]. A 2.1 m long cylindrical detector consisting of roughly 60 thousand fibers arranged in 24 layers was constructed. The detector used 1 mm single-cladding fibers with PS core doped with butyl-PBD and POPOP producing scintillating light at around 440 nm. The fibers were grouped into triplet stereo layers and monitored by complex three-stage image intensified CCDs with 10 ms read-out cycle. The detector was used for particle tracking as well as pre-shower measurement. A single hit efficiency and resolution of >90% and 0.35 mm was achieved, respectively. Particle track resolution was estimated to be better than 0.2 mm.

One of the most extensive SciFi particle trackers was installed in CERN for the CHORUS experiment in the early 1990s [78,79]. Around 1.2 million 2.2 m long 0.5 mm S-POFs were arranged into multiple 7-layer plane fiber ribbons. Staggered geometry where consecutive fiber layers are mutually shifted by a half fiber diameter was used. TiO_2-based paint was used both as an EMA to avoid inter-fiber cross-talk and as a glue to fix the fibers together. A complex system of 58 image intensified CCDs was used to read-out the scintillation signal. The hit density between 5 and 7 hits per ribbon and the two-track resolution of about 0.54 mm was achieved.

A tracking detector consisting of roughly 275 thousand S-POFs was commissioned in the late 1990s as a part of the near detector for the K2K long baseline neutrino oscillation experiment [80]. The tracker consists of 20 layers of 2.6 × 2.6 m modules interleaved with water target tanks. Each of the modules consists of horizontal and vertical S-POF double-ribbon read out by image intensified CCD cameras. Kuraray double-cladding S-POF with 0.7 mm diameter was used for the detector. Detector position resolution and hit efficiency of around 0.8 mm and 92% was estimated, respectively.

Due to the good timing resolution, SciFi detectors are of importance for the high-rate experiments, such as CERN's DIRAC or COMPASS experiments [81,82]. Here, multiple staggered layers of 0.5 mm double-cladding Kuraray S-POFs were used in combination with position-sensitive photomultipliers. The system demonstrated high time resolution down to 300 ps, detection efficiency above 98%, and spatial resolution of about 125 μm. In addition to particle tracking, scintillating fiber detectors have been considered as fast triggers for these experiments [83].

A modern SciFi system was installed for the central fiber tracker (CFT) of the DØ spectrometer at the Fermi National Accelerator Laboratory (IL, USA) [84]. Around 80,000 double-cladding S-POF with 835 μm diameter were arranged on 8 concentric support cylinders placed around the beam (**Figure 13.6**). The radial distance of the cylinders from the beam ranges from 20 to 52 cm. Two innermost and six other cylinders were 1.66 and 2.52 m long, respectively.

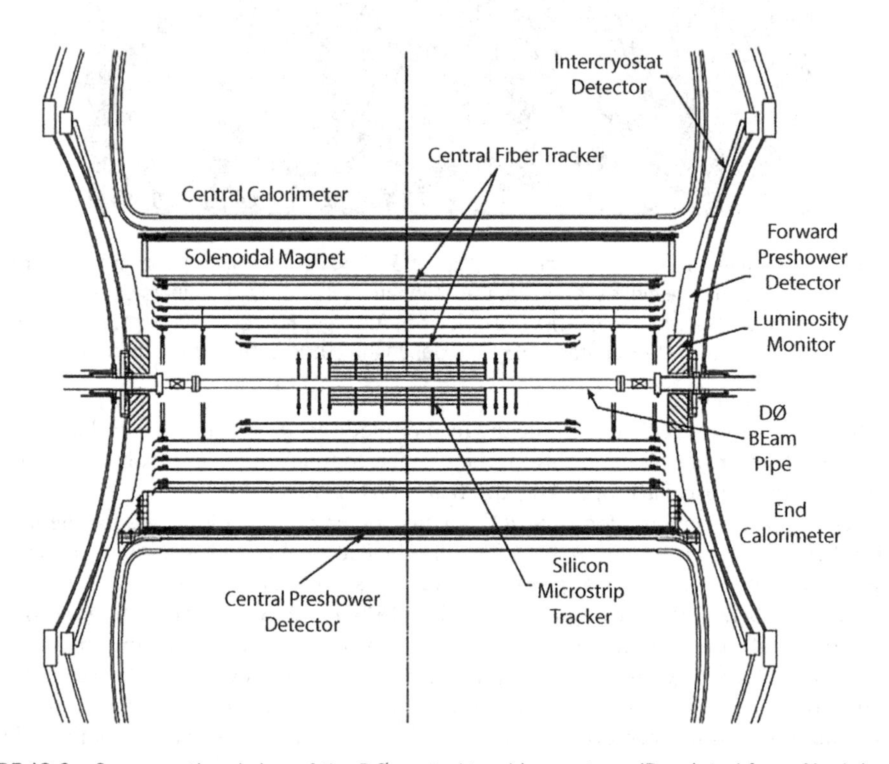

FIGURE 13.6 Cross-sectional view of the DØ central tracking system. (Reprinted from *Nucl. Instrum. Meth. Phys. Res. A*, 565., Abazov, V. M. et al., The upgraded DØ detector, 463–537, Copyright 2006, with permission from Elsevier.)

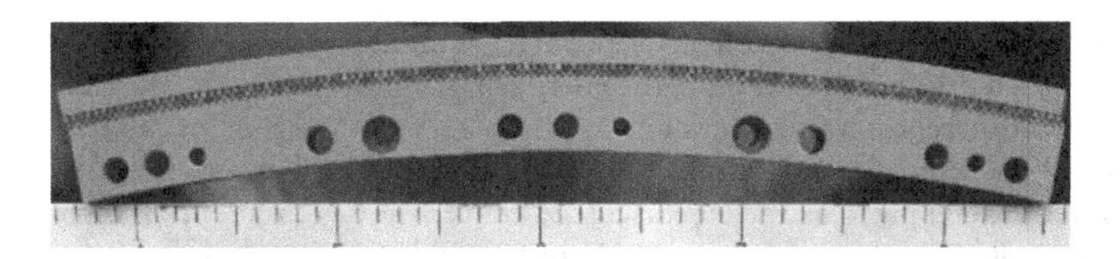

FIGURE 13.7 Illustration photo of S-POF double-ribbon used in the DØ central fiber tracker. (Reprinted from Nucl. Instrum. Meth. Phys. Res. A 565, Abazov, V.M. et al., The upgraded DØ detector, 463–537, Copyright 2006, with permission from Elsevier.)

Kuraray PS-based S-POF doped with p-terphenyl and 3HF emitting at 530 nm were used. Double-ribbons of 256 fibers were made by inserting the fibers into grooves machined in acrylic plates (**Figure 13.7**). Fiber ribbons were placed on both sides of the support cylinders. The fibers on the lower surface of the cylinders were mounted in the axial direction, i.e. parallel to the beam. The fibers on the upper surface were mounted alternatively at +3° or −3°. Auxiliary clear fibers were used to guide the light to a read-out system based on visible light photon counters (VLPCs). A hit efficiency of 99.5% and a doublet hit resolution of 100 μm was reported for the system.

A fairly new S-POF system is also installed at Rutherford Appleton Laboratory (UK) for particle tracing in the muon ionization cooling experiment MICE [85,86]. The tracker consists of two solenoidal spectrometers, each of which contains five planar scintillating fiber stations (**Figure 13.8**). The stations are composed of three staggered S-POF double-ribbons and have a circular active area with 30 cm diameter. Groups of seven 350 μm S-POF, doped with p-terphenyl and 3-hydroxyflavone, are coupled to 1.05 mm clear POF light which is read out by an optoelectronic system similar to that used in DØ CFT based on VLPCs. The detector reached space-point efficiency above 99% and channel resolution of 470 μm.

FIGURE 13.8 Photograph of a partially constructed MICE tracker module showing the five scintillating fiber stations. (Reprinted from *Nucl. Instrum. Meth. Phys. Res. A*, 732, Overton, E., Progress with the MICE scintillating fiber trackers, 412–414, Copyright 2013, with permission from Elsevier.)

Another common application of S-POFs in particle physics is calorimetry. Calorimeters are sub-detectors used primarily to measure the energy of incoming particles. They allow discrimination between different particle types and also can serve as triggers for selecting particular events. Operation principles and construction of scintillating fiber calorimeters is in many aspects similar to SciFi particle trackers discussed above. The key difference is introduction of a dense absorber interleaving individual S-POFs or S-POF layers. Substances with high effective atomic number Z_{eff} (Cu, Fe, Pb, W, U) are usually used as the passive absorbers for the particle calorimeters. Energy of the particle passing through the SciFi calorimeter is deposited predominantly in these passive absorber layers where showers of secondary electrons with progressively decreasing energy are created. The role of the active S-POF layers is to sample the energy deposited by the shower particles throughout the calorimeter. The detected scintillation intensity serves as a measure of the incident particle energy.

A typical example of a SciFi calorimeter is the electromagnetic calorimeter of the KLOE experiment at the DAΦNE collider (Italy) [87]. A total of 15000 km of fiber was used on the barrel-shaped calorimeter. The calorimeter was constructed from more than 50 individual modules which consisted of thin Pb plates with round grooves for 1 mm S-POFs. The modules were stacked into staggered lead-fiber matrix. Different types of S-POF were used for the inner and outer half of the calorimeter in order to balance the performance and the cost requirements. The modules were read out via auxiliary light guides coupled to the fine-mesh photomultiplier detector system. The calorimeter covered 90% of the total solid angle and achieved detection efficiency larger than 98% for gamma radiation with energy above 100 MeV.

Electromagnetic calorimeters based on the Pb/S-POF design as described above belong to the most compact and fastest calorimeters with high radiation resistance. Consequently, they have been investigated and employed for different experiments, e.g. for the Small Angle Tagger of the DELPHI experiment at CERN [88], for the JETSET detector at the CERN Low Energy Antiproton Ring [89], or for the muon g-2 experiment at Brookhaven National Laboratory (NY, USA) [90]. More recent developments explore utilization of W/S-POF combination and alternative fabrication techniques to produce more compact and cost-effective calorimeters [91,92].

The list of discussed application examples is non-exhaustive and some other cases can be found in ref. [74,93,94]. S-POF technology and SciFi particle detectors remain a subject of an active R&D in high energy physics. S-POF-based particle detectors are being developed for a number of experiments including LHCb at CERN [95,96], STAR and PHENIX experiments at Relativistic Heavy Ion Collider of Brookhaven National Laboratory (NY, USA) [92], and MUSE [97] or MEG II [98] experiments at the Paul Scherrer Institute (Switzerland). In addition to particle accelerator experiments, SciFi systems can be also used for particle detection in astrophysics and space applications [99,100].

13.3.3 Applications in the Medical Industry

Along with high energy physics, the second major application area of S-POFs is in the medical industry. Use of S-POFs has been explored in different medical areas including quality assurance for radiotherapy, *in vivo* dosimetry, and radiology. Real qualitative radiation dose measurement are required for a majority of the applications. General advantages of OFRS technology have been well recognized, and optical fiber dosimeters (OFDs) based on different principles have been developed over the years [1,2,101,102]. Many of these advantages, such as small dimensions or possibility of real-time measurement, remain highly relevant for medical applications. However, compared to other technologies, use of OPS for medical dosimetry offers significant advantages in terms of their water equivalence, linearity, dose-rate, and energy independence in the MeV energy range [101]. Intensity of the scintillator emission is linearly proportional to the energy deposited by incoming particles and is a measure of irradiation dose rate. Continuous monitoring of scintillation intensity over time can then be used to determine the overall received dose. Scintillation-based dosimeters can thus provide simultaneous dose and dose rate measurement.

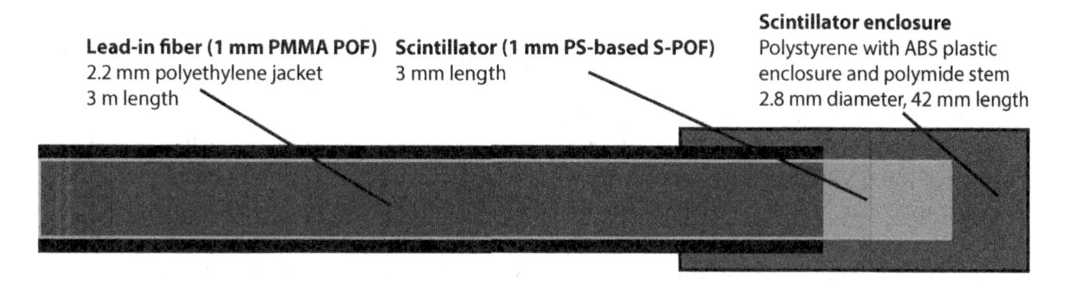

FIGURE 13.9 Schematic illustration of design of a commercial Exradin W1 scintillation fiber dosimeter from Standard Imaging Inc. (From Standard Imaging Inc., Exradin W1 Scintillator (product datasheet), https://d3udwuy5vw6moa.cloudfront.net/files/ExradinW1_DS_1335-20web.pdf.)

A typical scintillation fiber dosimeter (SFD) consists of a small scintillation element coupled to an auxiliary optical fiber guide which brings the generated scintillation light to the detector (**Figure 13.9**). The auxiliary fiber guide can be either silica fiber or POF. Nevertheless, over the years, polymer optical fibers proved to be the superior choice due to their lower cost, water equivalence, and increased robustness. The scintillation element can be in principle also an inorganic scintillator [103,104], or a bulk OPS [105,106]. However, here we will focus on the intrinsic plastic optical fiber dosimeters, where the S-POF itself is used as the scintillation element. Some examples of extrinsic SFDs incorporating polymer optical fibers will be discussed in Section 13.4.2. Compared to typical inorganic scintillators, scintillation of OPS is energy independent for photons and electrons in the Compton dominated energy range (above ~ 125 keV) [17]. This is the key advantage for the dosimetric applications as a single calibration factor is usually sufficient to characterize the sensor for a reliable operation in a large energy range. In addition, used plastic materials are close to water equivalent, i.e. they have properties closely compatible with biological tissue. This is beneficial for medical applications as POF-based sensors do not interfere with the energy deposition in the human body or reference medium as much as sensors containing inorganic materials with higher Z_{eff}. Finally, compared to the fiber dosimeters with bulk OPS, using S-POFs as the scintillation element was shown to increase the light collection efficiency due to the total internal reflection at the core-cladding boundary [107]. Nevertheless, one may still dispute the intrinsic nature of such fiber sensors as the lengths of used S-POFs are very short.

The most common dosimeter configuration consists of a short (few mm) piece of 1 mm diameter BCF-60 S-POF (Saint-Gobain Crystals) emitting in the green spectral region coupled to a 1 mm auxiliary PMMA POF light guide. However, different configurations also can be found throughout the literature. Compared to other S-POFs emitting at shorter wavelengths, green scintillation of BCF-60 provides better spectral separation from parasitic Cerenkov light produced in the optical light guides under irradiation [108]. Intensity of Cerenkov light emitted per volume is typically up to two orders of magnitude lower than the scintillation intensity [20]. However, its overall impact is scaled up by the extended length of the clear fiber guide in radiation field compared to the short sensing S-POF element. Therefore, generation of Cerenkov light is the main stem effect compromising the operation of SFDs. Different approaches including two-fiber subtraction, spectral, and time filtering techniques have been tested for minimizing the Cerenkov light impact on the sensor performance [109,110]. Two-fiber subtraction is the most traditional technique that has become a reference benchmark for other approaches. It is based on the utilization of secondary reference fiber without scintillation element to measure and subtract contribution of the Cerenkov light [105,111–113]. Spectral methods based on simple chromatic filtering of scintillation light [114,115] or more complex multi-spectral removal techniques [109,116–120] have been demonstrated. The latter represent the most favored method of Cerenkov light removal in modern systems. Alternative approaches based on signal temporal filtering [121] or utilization of hollow-core fibers [112,122] have been investigated as well. Fluorescence of the POF material can also contribute to the overall stem effect, especially at lower energies below the Cerenkov emission threshold. However, fiber fluorescence yield is considerably lower than for scintillation light and can be neglected in most of the cases [123].

Another potential issue of SFDs is the temperature dependence of the scintillation yield in the case of some S-POFs. More specifically, small decrease of scintillation intensity with rising temperature was observed for BCF-60 and BCF-12 S-POF from Saint-Gobain Crystals [124,125]. The effect can be compensated by using a multi-spectral detection approach, which is already being used for Cerenkov light filtering [126]. It is also important to note that the scintillation of OPS is energy independent only above certain threshold energy. This threshold is at the level of 100 keV for photons and electrons and can be considerably larger for heavier particles such as protons [105,106]. The exact threshold level is strongly material dependent [127]. Below the threshold, the light production suffers from quenching by the linear energy transfer, which could represent a problem for applications with proton beams [128] or in radiology [127,129].

One of the most common application areas of SFDs is quality assurance (QA) for external beam radiotherapy. This includes tasks such as measurement of depth-dose curves and profiles of photon and electron radiotherapy beams [109,120,131]. Due to the small size, SFDs can rival and even outperform standardly used ionization chambers. Moreover, they hold large potential for small field dosimetry required for modern highly-targeted radiotherapy techniques, such as stereotactic radiosurgery, volumetric modulated arc therapy (VMAT), and intensity-modulated radiation therapy (IMRT) [113,132–135]. As a result of extensive research in this area, the first commercial SFD for radiotherapy QA tasks Exradin W1 from Standard Imaging (WI, USA) was recently introduced [130]. Since then, a number of studies investigated its performance and compared it to other small-field dosimeters [136–142]. The sensor design is based on the SFD system described earlier by Therriault-Proulx et al. [118]. A 3 mm long PS-based S-POF with 1 mm diameter is coupled to a 3 m long PMMA-based lead-in POF (**Figure 13.9**). The system uses two-wavelength (blue-green) detection scheme to facilitate the removal of parasitic Cerenkov light [116]. The studies with CyberKnife systems demonstrated that the SFD can provide accurate measurement of the depth-dose curves, tissue-maximum ratios, off-axis ratios, and beam profiles in small-field conditions [136,137,142].

In addition to single-point sensors that need to be scanned through the beam field, 1D or 2D arrays of multiple SFDs can be constructed to facilitate simpler and faster beam characterization [131,133,135,143,144]. Lacroix et al. constructed a linear array of 29 SFDs and used it to measure depth-dose curves in a solid water phantom at simultaneous beam-profile measurement [131]. The system was based on a blue-emitting BCF-12 S-POFs from Saint-Gobain Crystals read out by a color CCD detector for chromatic removal of Cerenkov light. In another study, a crosshair array of 49 SFDs based on BCF-60 fibers read out by CCD camera was constructed (**Figure 13.10**) [135]. The sensor performance was compared with other dosimetry technologies including photon diodes, dosimetry film, and micro-ionization chamber. The results showed that the SFD array can provide precise real-time beam shape measurement in agreement with other standard techniques. The concept of 2D SFD arrays was taken one step further in a study by Guillot et al. who constructed large 26 × 26 cm^2 array containing 781 individual SFDs based on BCF-12 S-POF [143]. The SFDs were spaced equally 1 cm apart in a square shaped mesh, except for the central crosshair lines where 5 mm spacing was used. Two CCD detectors with a dichroic beam splitter and optical filters were used to implement two-wavelength chromatic removal of Cerenkov light. The system was shown to provide excellent precision and accuracy for QA tasks for small-field radiotherapy methods.

An alternative approach was proposed by Goulet et al. who constructed an array of longer S-POFs acting as a 1D dose integrators. They demonstrated that by using suitable signal processing, the system can be used as a real-time fluence monitor that can be inserted directly in a LINAC beam path [145,146]. In a later study, the concept was taken further by rotating the planar array of S-POFs, inserted in the beam, around the beam axis and using tomographic reconstruction algorithm to obtain a high-resolution 2D dose measurement [147]. The prototype

FIGURE 13.10 Photo of the SFD crosshair array prototype. (Reproduced from Gagnon, J.-C. et al.: Dosimetric performance and array assessment of plastic scintillation detectors for stereotactic radiosurgery quality assurance. *Medical Physics*. 39. 429–436. 2012, © 2012 *Am. Assoc. Phys. Med.*, with permission from John Wiley & Sons.)

consisted of 50 parallel BCF-60 S-POFs mounted next to each other on a 30 cm circular acrylic slab. The spacing between the fibers was 3.19 and 6.38 mm for 40 central and 10 outer fibers, respectively. Both ends of the scintillating fibers were connected to clear PMMA fibers which guided the scintillation signal toward the read-out CCD camera. The central axis of the detection disk was aligned with the LINAC beam direction and multiple measurements were taken at different disk rotational positions. A computational technique similar to CT image reconstruction was used to map the 2D distribution of the delivered dose. The technique was called tomodosimetry. Further extension for the 3D volumetric dosimetry was suggested [148,149], but requirements on complex rotational movements made the design rather impractical.

Another area where characteristics of SFDs are of big advantage is *in vivo* dosimetry [101,150]. For example, Klein et al. used BCF-60-based SFD mounted onto an endorectal balloon to verify dose delivered in IMRT and VMAT plans (**Figure 13.11**) [151]. The measurement was performed

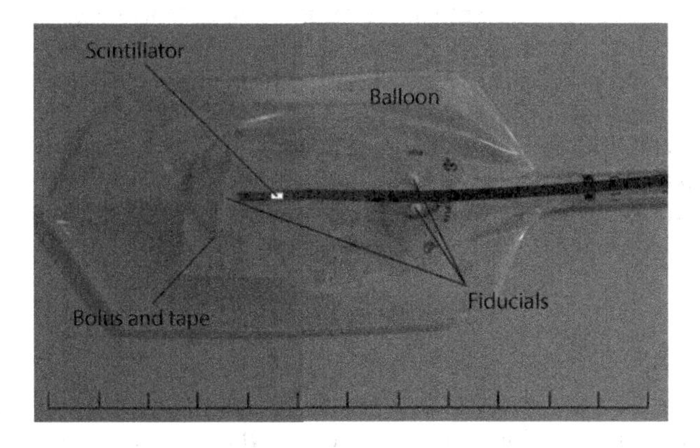

FIGURE 13.11 Photo of a partially inflated endorectal balloon with an SFD attached to it. (Reprinted from *Radiat. Meas.*, 47, Klein, D. et al., In-phantom dose verification of prostate IMRT and VMAT deliveries using plastic scintillation detectors, 921–929, Copyright 2012, with permission from Elsevier.)

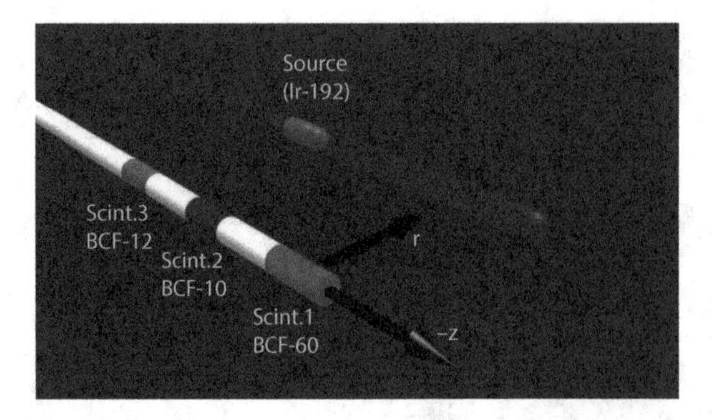

FIGURE 13.12 Schematic illustration of mSFD used for irradiation source position measurement. (Reprinted from Therriault-Proulx, F. et al., On the use of a single-fiber multipoint plastic scintillation detector for 192Ir high-dose-rate brachytherapy, *Medical Physics*, 2013, 40, 062101-n/a; © 2013 *Am. Assoc. Phys. Med.*, with permission from John Wiley & Sons.)

both in a deformable anthropomorphic and a rigid IMRT quality assurance phantom. The SFD measurement was in agreement with reference ionization chamber and within 1% of the treatment planning system dose. The SFD system was later successfully tested for rectal wall dose monitoring in patients undergoing IMRT for prostate cancer [152]. The research lead to the further development of the first commercial SFD for *in vivo* dosimetry [153]: the OARtrac system now distributed by Radiadyne (TX, USA) [154].

Utilization of S-POF-based SFDs for brachytherapy applications has been explored as well [118,119,155,156]. Therriault-Proulx et al. investigated the use of SFD based on BCF-60 S-POF for verification of dose delivery for an ^{192}Ir high-dose-rate (HDR) brachytherapy within a customized water phantom [118,119]. The dose and dose rates measured by SFD were compared to the doses from the treatment planning system and sample doses to the organs at risk. In later works, they developed a novel multipoint SFD (mSFD) based on hyperspectral filtering approach [157], and tested its performance in an analogical study for ^{192}Ir HDR brachytherapy [3]. The tested mSFD consisted of three different S-POFs emitting at different wavelengths connected in a series with a 2.2 cm long clear fiber between the first and second S-POF and a 2.3 cm long piece between the second and the third S-POF (**Figure 13.12**). The first and the second S-POF were 2 mm long, while the third one was slightly longer (3 mm) to compensate for higher attenuation caused by longer light path and multiple interfaces. The dose measurement accuracy was investigated for different source-to-detector distances and the results exhibited good agreement for all the parameters of interest. The sensor proved to be very effective for irradiation source position measurement and has high potential for clinical applications as it increases amount of information attainable from a single sensor line.

Scintillation-based OFRS for medical applications remain a very active research field with large number of publications produced every year. Complete review of the literature on the topic goes beyond the scope of this work and more information on OFDs and their application examples can be found in numerous topical reviews, e.g. [20,101,102,150,158,159].

13.4 POFs in Extrinsic Radiation Sensors

In addition to intrinsic OFRSs reviewed in previous sections, POFs play an important role also in a number of extrinsic radiation sensors, i.e. those where POFs are not used directly as a transducer for ionizing radiation. It is not surprising that the most notable examples of extrinsic OFRSs can be found in the two application areas already discussed: particle detection in high energy physics and dosimetry for medical applications. With regard to particle physics, we will briefly discuss the role of so-called wavelength-shifting POFs (WLS-POFs). Concerning applications in the medical industry, some examples of optical fiber dosimeters where POFs are used only as scintillation light guides will be addressed.

13.4.1 Wavelength-Shifting POFs

Wavelength-shifting fibers can be considered siblings of scintillating POFs sharing a majority of their characteristics. The production, materials, geometry, and even application tasks of WLS-POFs are almost the same as for scintillating fibers. However, unlike S-POFs, WLS-POFs do not require the presence of primary scintillation dopant and typically contain only secondary wavelength-shifting dopant, hence the name. Their main function is not direct production of scintillation, but rather collection (absorption) and re-emission of primary scintillation light from external bulk scintillators. Similar to S-POFs, WLS-POFs can be found in particle trackers, pre-shower detectors, and calorimeters. WLS-POFs are usually inserted in the grooves on the surface or incorporated inside of bulk scintillation strips or tiles. Scintillation strips and tiles can be prepared from OPSs in a variety of shapes, and different WLS-POF incorporation strategies can be found. The surface of the strip/tile is usually treated with a reflective material (e.g. TiO_2, MgO, aluminized Mylar or TYVEK) in order to increase the light collection efficiency by the WLS-POF. The scintillation light generated in the bulk scintillator eventually hits the WLS-POF where it is absorbed, re-emitted at longer wavelengths, and guided toward the read-out optoelectronics. Analogically to S-POF-based detectors discussed in Section 13.3.2, complex detector structures are created by stacking of multiple scintillating strips or tiles. Compared to S-POF based detectors, using larger bulk scintillator interrogated by a WLS-POF effectively increases active detection volume pertaining to a single read-out line. This is beneficial especially for large-area detectors as it helps to limit the costs associated with more expensive S-POFs and optoelectronics.

The first large-scale implementation of detector based on PS-based scintillation strips (doped with PPO and POPOP) interrogated by Kuraray Y11 WLS-POF was realized in the frame of MINOS experiment at Fermilab (IL, USA) [160]. The experiment aimed at studying neutrino oscillations and used a calorimeter comprising about 100 thousand $4.1 \times 1 \times 800$ cm (W×H×L) scintillation strips with a groove for integration of WS-POF. The strips covered the total are of around 28 000 m^2. Another Fermilab's experiment targeting neutrino studies, MINERvA, also adopted a similar technology [161]. However, this time, scintillation tiles of triangular cross-section with centrally integrated WLS-POFs were used. WS-POFs can be found in a number of other experiments, e.g. OPERA detector at CERN [162], electromagnetic calorimeter of the T2K experiment [163], pre-shower detectors at Fermilab's DØ [164] and CDF [165] experiments, calorimeter systems of CERN's LHCb [166], ATLAS [167], and CMS [168] experiments. More information and examples can be found in [74,94].

It is also worth noting that, whether it is detector based on S-POFs or WLS-POFs, additional clear POFs are almost always present as extension light guides bringing the scintillation signal to the read-out optoelectronics.

13.4.2 POFs in Extrinsic Optical Fiber Dosimeters

As it was already mentioned in Section 13.3.3, the sensing element in scintillation-based OFDs does not necessarily need to be an S-POF. Many examples of OFDs relying on bulk OPS or even inorganic scintillator element can be found throughout literature. Nevertheless, in the majority of these cases, polymer optical fibers still play an important role as they became the preferred option for auxiliary light guides that bring scintillation signal to the detection optoelectronics. Large-core POFs are cheap, robust, but still flexible and offer better biocompatibility than silica-based fibers. In addition, they do not fail in a brittle manner like silica fibers, hence, they are more acceptable even for *in vivo* applications. In recent years, increasing numbers of studies have been published on the topic of OFDs for medical applications, relying both on intrinsic and extrinsic sensor realization. Comprehensive overview of these studies goes beyond the scope of the chapter. Readers are advised to consult ref. [20,102,158] and citations therein for more detailed discussion of the topic.

Efficient coupling of scintillation light into the optical fiber guide is one of the main challenges of extrinsic OFDs. Different approaches for this task have been investigated. The most common sensor arrangement uses the simple butt-coupling of a bulk scintillator element to a lead-out POF guide [169–171]. This approach has been explored for example at University of Sydney (Australia) for the development of BrachyFOD™ system for *in vivo* brachytherapy dosimetry. The system uses commercial BC-400 OPS from Saint-Gobain Crystals, butt-coupled to a PMMA fiber with photomultiplier tube read-out. The performance of two different sensor variants based on 0.5 and 1 mm POF and scintillator geometry were first evaluated in phantom study [169]. In the next step, the system was tested in a clinical trial with 24 patients undergoing brachytherapy for prostate cancer [172]. The results indicated a maximum departure between measured and calculated dose of 9%, which further emphasizes the need for *in vivo* dose delivery assurance.

An alternative approach was explored at University of Limerick (Ireland), where a mixture of inorganic scintillation phosphor Gd_2O_2S:Tb and epoxy glue was used to cover the exposed end of the polymer optical fiber to form a low-dose X-ray dosimeter [173–175]. A short length of fiber jacket and cladding was removed from the distal end of the POF and replaced by a cylindrical mould containing scintillator-epoxy mixture. It was also shown that light collection efficiency can be improved by modification of the fiber tip before applying scintillation material [176,178]. Higher scintillation yield of inorganic scintillator allowed the system to be read-out by a scientific-grade spectrometer. The initial characterization in phantom studies demonstrated that the sensor has high sensitivity, good repeatability, and linear response from cGy to 16 Gy dose range. On the other hand, due to higher Z_{eff} of the scintillator, the sensor suffered from an energy dependence causing over-response at varying filed size and depth-dose measurements.

Later on, the same group developed different coupling technique relying on micro-machining a small hole at the POF tip which is then filled with scintillator-epoxy mixture [178–180]. This approach should enhance scintillation light coupling into the POF, yield more compact sensor, and improve manufacturing reproducibility. Sensors with three different scintillation materials, CsI:Tl, Gd_2O_2S:Tb and La_2O_2S:Eu, were tested with ^{125}I brachytherapy seeds. The sensor using Gd_2O_2S:Tb was found to provide highest scintillation yield [180]. The sensor was further tested both with ^{125}I source for brachytherapy applications [178], and with 6 MV photon beam from LINAC for radiotherapy applications [179].

In addition to scintillation-based sensors, POFs can be used as the signal light guides also in extrinsic OFDs relying on other sensing mechanism, for example using optically-stimulated luminescence [181,182] or radiochromic films [183,184].

References

1. A. L. Huston, B. L. Justus, P. L. Falkstein *et al.*, "Remote optical fiber dosimetry," *Nucl. Instr. Meth. Phys. B,* 184, 55–67 (2001).
2. S. O'Keeffe, C. Fitzpatrick, E. Lewis *et al.*, "A review of optical fibre radiation dosimeters," *Sens. Rev.*, 28(2), 136–142 (2008).
3. F. Therriault-Proulx, S. Beddar, and L. Beaulieu, "On the use of a single-fiber multipoint plastic scintillation detector for ^{192}Ir high-dose-rate brachytherapy," *Med. Phys.*, 40(6), 062101 (2013).
4. H. Henschel, M. Körfer, J. Kuhnhenn *et al.*, "Fibre optic radiation sensor systems for particle accelerators," *Nucl. Instrum. Meth. Phys. Res. A,* 526(3), 537–550 (2004).
5. S. Girard, J. Kuhnhenn, A. Gusarov *et al.*, "Radiation effects on silica-based optical fibers: Recent advances and future challenges," *IEEE Trans. Nucl. Sci.*, 60(3), 2015–2036 (2013).

6. E. J. Friebele, G. C. Askins, M. E. Gingerich *et al.*, "Optical fiber waveguides in radiation environments, II," *Nucl. Instrum. Meth. Phys. Res. B,* 1(2–3), 355–369 (1984).
7. E. J. Friebele, "Optical fiber waveguides in radiation environments," *Opt. Eng.,* 18, 552–561 (1979).
8. B. Brichard, A. Fernandez-Fernandez, H. Ooms *et al.*, "Radiation-hardening techniques of dedicated optical fibres used in plasma diagnostic systems in ITER," *J. Nucl. Mater.,* 329, 1456–1460 (2004).
9. S. Girard, J. Kuerinck, A. Boukanter *et al.*, "Gamma-rays and pulsed X-ray radiation responses of nitrogen-, germanium-doped and pure silica core optical fibers," *Nucl. Instrum. Meth. Phys. Res. B,* 215, 187–195 (2004).
10. S. Girard, Y. Querdane, C. Marcandella *et al.*, "Feasibility of radiation dosimetry with phosphorus-doped optical fibers in the ultraviolet and visible domain," *J. Non-Cryst. Solids,* 357, 1871–1874 (2011).
11. H. Bauker, F. W. Haesing, S. Nicolai *et al.*, "Fiber optic radiation dosimetry for medical applications," *Opt. Fibers Med.,* 1201, 419–430 (1990).
12. H. Bauker, and F. W. Haesing, "Fiber optic radiation sensors," *Proc. SPIE,* 2425, 106 (1994).
13. B. D. Evans, G. H. Sigel, J. B. Langworthy *et al.*, "The fiber optic dosimeter on the navigational technology satellite 2," *IEEE Trans. Nucl. Sci.,* 25(6), 1619–1624 (1978).
14. H. Bauker, F. W. Haesing, F. Pfeiffer *et al.*, "A fiber-optic twin sensor for dose measurements in radiation therapy," *Opt. Rev.,* 4(1A), 130–132 (1996).
15. M. K. Barnoski, M. D. Rourke, S. M. Jensen *et al.*, "Optical time domain reflectometer" *Appl. Opt.,* 16(9), 2375–2379 (1977).
16. I. Toccafondo, Y. E. Martin, E. Guillermain *et al.*, "Distributed optical fiber radiation sensing in a mixed-field radiation environment at CERN," *J. Lightwave Technol.,* 35(16), 3303–3310 (2017).
17. J. B. Birks, *The Theory and Practice of Scintillation Counting.* Pergamon press, Oxford, UK (1964).
18. G. F. Knoll, *Radiation Detection and Measurement.* John Wiley & Sons, New York (2010).
19. P. A. Cerenkov, "Visible radiation produced by electrons moving in a medium with velocities exceeding that of light," *Phys. Rev.,* 52, 378–379 (1937).
20. L. Beaulieu, and S. Beddar, "Review of plastic and liquid scintillation dosimetry for photon, electron, and proton therapy," *Phys. Med. Biol.,* 61, R305–R343 (2016).
21. S. Beddar, "Plastic scintillation dosimetry and its application to radiotherapy," *Radiat. Meas.,* 41, S124–S33 (2006).
22. B. Brichard, A. F. Fernandez, H. Ooms *et al.*, "Fibre-optic gamma-flux monitoring in a fission reactor by means of Cerenkov radiation," *Meas. Sci. Technol.,* 18, 3257–3262 (2007).
23. P. Gorodetzky, D. Lazic, G. Anzivino *et al.*, "Quartz fiber calorimetry," *Nucl. Instrum. Meth. Phys. Res. A,* 361, 167–179 (1995).
24. K. W. Jang, W. K. Yoo, H. S. Sang *et al.*, "Fiber-optic Cerenkov radiation sensor for proton therapy dosimetry," *Opt. Express,* 20(13), 13907–13914 (2012).
25. J. Son, M. Kim, D. Shin *et al.*, "Development of a novel proton dosimetry system using an array of fiber-optic Cerenkov radiation sensors," *Radiother. Oncol.,* 117, 501–504 (2015).
26. A. Darafsheh, R. Taleei, A. Kassaee *et al.*, "Development of a novel proton dosimetry system using an array of fiber-optic Cerenkov radiation sensors," *Med. Phys.,* 43, 5973–5980 (2016).
27. A. Darafsheh, R. Taleei, A. Kassaee *et al.*, "On the origin of the visible light responsible for proton dose measurement using plastic optical fibers," *Proc. SPIE,* 10056, 100560V-1 (2017).
28. W. Primark, "Fast-neutron-induced changes in quartz and vitreous silica," *Phys. Rev. B,* 110(6), 1240–1254 (1958).
29. E. Lell, N. J. Hensler, and J. R. Hensler, "Radiation effects in quartz, silica and glasses," *Progr. Ceramic Sci.,* 4, 3–93 (1966).
30. R. D. L. Kronig, "On the theory of the dispersion of X-rays," *J. Opt. Soc. Am.,* 12, 547–557 (1926).
31. H. A. Kramers, "La diffusion de la lumière par les atomes." *Atti Congr. Int. Fis.,* 2, 545–557 (1927).
32. H. A. Lorentz, "Über die Beziehungzwischen der Fortpflanzungsgeschwindigkeit des Lichtes der Körperdichte," *Ann. Phys.,* 9, 641–665 (1880).
33. L. Lorenz, "Über die refractionsconstante," *Ann. Phys.,* 11, 70–103 (1880).
34. A. Gusarov, and S. K. Hoeffgen, "Radiation effects on fiber gratings," *IEEE Trans. Nucl. Sci.,* 60(3), 2037–2053 (2013).
35. F. Berghmans, and A. Gusarov, Fiber Bragg Grating Sensors in Nuclear Environments, in *Fiber Bragg Grating Sensors: Recent Advancements, Industrial Applications and Market Exploitation,* A. Cusano, A Cutolo and J. Albert (eds.), Bentham Science Publisher, Sharjah, United Arab Emirates (2011).

36. K. Krebber, H. Henschel, and U. Weinand, "Fiber Bragg gratings as high dose radiation sensors?" *Meas. Sci. Technol.*, 17, 1095–1102 (2006).
37. J. Witt, M. Schukar, M. Breithaupt *et al.*, "MPOF LPGs as Radiation Sensors," *20th International conference on plastic optical fibers (POF 2011)*, 14-16 September 2011, Bilbao, Spain.
38. T. A. Hamdalla, and S. S. Nafee, "Bragg wavelength shift for irradiated polymer fiber Bragg grating," *Opt. Laser Technol.*, 74, 167–172 (2015).
39. S. O'Keeffe, A. Fernandez-Fernandez, C. Fitzpatrick *et al.*, "Real-time gamma dosimetry using PMMA optical fibres for applications in the sterilization industry," *Meas. Sci. Technol.*, 18, 3171–3176 (2007).
40. S. O'Keeffe, and E. Lewis, "Polymer optical fibre for in situ monitoring of gamma radiation processes," *Int. J. Smart Sens. Intel. Sys.*, 2(3), 490–502 (2009).
41. S. O'Keeffe, E. Lewis, A. Santhanam *et al.*, "Low Dose Plastic Optical Fibre Radiation Dosimeter for Clinical Dosimetry Applications," *The Eight IEEE Conference on Sensors (IEEE SENSORS Conference 2009)*, 25-28 October 2009, Christchurch, New Zealand, pp. 1689–1692.
42. S. O'Keeffe, E. Lewis, A. Santhanam *et al.*, "Variable sensitivity online optical fibre radiation dosimeter." *The Eighth IEEE Conference on Sensors (IEEE Sensors Conference)*, 25-28 October 2009, Christchurch, New Zealand, pp. 787–790.
43. P. Stajanca, L. Mihai, A. Sporea *et al.*, "Impacts of Gamma Irradiation on Cytop Plastic Optical Fibres," *25th International Conference on Plastic Optical Fibres (POF 2016)*, 13-15 September 2016, Birmingham, UK, pp. 114–117.
44. P. Stajanca, and K. Krebber, "Radiation Induced Attenuation in Perfluorinated Polymer Optical Fibers for Dosimetry Applications," *25th International Conference on Plastic Optical Fibres (POF 2016)*, 13-15 September 2016, Birmingham, UK, pp. 167–170.
45. P. Stajanca, L. Mihai, D. Sporea *et al.*, "Effects of gamma radiation on perfluorinated polymer optical fibers," *Opt. Mater.*, 58, 226–233 (2016).
46. P. Stajanca, and K. Krebber, "Towards on-line radiation monitoring with perfluorinated polymer optical fibers," *Proc. SPIE*, 10323, 103231O (2017).
47. P. Stajanca, and K. Krebber, "Radiation-induced attenuation of perfluorinated polymer optical fibers for radiation monitoring," *Sensors*, 17, 1959 (2017).
48. P. Stajanca, and K. Krebber, "POF as Radiation Sensors?," *26th International Conference on Plastic Optical Fibres (POF 2017)*, 13-15 September 2017, Aveiro, Portugal, paper 5.
49. C. Yan, S. H. Law, N. Suchowerska *et al.*, "Radiation damage to polymer optical fibres." In *Fifth Australasian Congress on Applied Mechanics* (*ACAM 2007*), 10–12 December 2007, Brisbane, Australia, Vol. 1, pp. 830–834.
50. M. S. Kovacevic, A. Djordjevich, S. Savovaic *et al.*, "Measurement of ^{60}Co gamma radiation induced attenuation in multimode step-index POF at 530 nm," *Nucl. Technol. Radiat. Protection*, 28(2), 158–162 (2013).
51. M. S. Kovacevic, S. Savovic, A. Djordjevich *et al.*, "Measurements of growth and decay of radiation induced attenuation during the irradiation and recovery of plastic optical fibres," *Opt. Laser Technol.*, 47, 148–151 (2013).
52. K. F. Johnson, *Present Status of Radiation Damage Effects in Plastic Scintillators and Fibers*, IEEE Nuclear Science Symposium and Medical Imaging Conference (NSS/MIC'92), Orlando, FL (1992).
53. Y. M. Protopopov, and V. G. Vasilchenko, "Radiation damage in plastic scintillators and optical fibers," *Nucl. Instrum. Meth. Phys. Res. B*, 95(4), 496–500 (1995).
54. K. Hara, K. Hata, S. Kim *et al.*, "Radiation hardness and mechanical durability of Kuraray optical fibers," *Nucl. Instrum. Meth. Phys. Res. A*, 411, 31–40 (1998).
55. E. C. Aschenauer, J. Bähr, R. Nahnhauer *et al.*, "Measurements of the radiation hardness of selected scintillating and light guide fiber materials," *DESY 99-079*, 99, 078 (1999).
56. K. Wick, and T. Zoufal, "Unexpected behaviour of polystyrene-based scintillating fibers during irradiation at low doses and low dose rates," *Nucl. Instrum. Meth. Phys. Res. B*, 185, 341–345 (2001).
57. C. Lethien, C. Loyez, J.-P. Vilcot *et al.*, "Exploit the bandwidth capacities of the perfluorinated graded index polymer optical fiber for multi-services distribution," *Polymers*, 3, 1006–1028 (2011).
58. Y. Koike, *Fundamentals of Plastic Optical Fibers*. John Wiley & Sons, Weinheim, Germany (2015).
59. G. T. Reynolds, and P. E. Condon, "Filament scintillation counter," *Rev. Sci. Instrum.*, 28, 1098–1099 (1957).
60. S. R. Borenstein, and R. C. Strand, "Scintillating optical fibers for fine grained hodoscopes," *IEEE Trans. Nucl. Sci.*, 29(1), 402–404 (1982).

61. S. R. Borenstein, R. B. Palmer, and R. C. Strand, "Optical fibers and avalanche photodiodes for scintillator counters," *Phys. Scripta,* 23, 550–555 (1981).
62. S. R. Borenstein, R. B. Palmer, and R. C. Strand, "A fine grained scintillating optical fiber hodoscope for use at ISABELLE," *IEEE Trans. Nucl. Sci.*, 28(1), 458–460 (1981).
63. L. R. Allemand, J. Calver, J. C. Cavan *et al.*, "Optical scintillating fibres for particle detectors," *Nucl. Instrum. Meth. Phys. Res.*, 225, 522–524 (1984).
64. H. Blumenfeld, M. Bourdinaud, and J. C. Thevenin, "Scintillating plastic fibres for calorimetry and tracking devices," *IEEE Trans. Nucl. Sci.*, 33(1), 54–56 (1986).
65. P. H. Rebourgeard, F. Rondeaux, J. P. Baton *et al.*, "Fabrication and measurement of plastic scintillating fibers," *Nucl. Instr. Meth. Phys. Res. A*, 427, 543–567 (1999).
66. I. B. Berlman, *Handbook of Fluorescence Spectra of Aromatic Molecules.* Academic Press, New York (1971).
67. C. P. Achenbach, *Active Optical Fibres in Modern Particle Physics Experiments.* Nova Science Publisher, New York (2004).
68. T. Förster, "Zwischenmolekulare energiewanderung und fluoreszenz," *Ann. Phys.*, 2, 55–75 (1948).
69. C. L. Renschler, and L. A. Harrah, "Reduction of reabsorption effects in scintillators by employing solutes with large stokes shift," *Nucl. Instrum. Meth. Phys. Res. A*, 235, 41–45 (1985).
70. H. Leutz, "Scintillating fibres," *Nucl. Instrum. Meth. Phys. Res. A*, 364, 422–448 (1995).
71. C. M. Hawkes, M. Kuhlen, B. Milliken *et al.*, "Decay time and light yield measurements for plastic scintillating fibers," *Nucl. Instrum. Meth. Phys. Res. A*, 292(2), 329–336 (1990).
72. S.-G. Crystals, "Scintillating Optical Fibers" (product brochure), https://www.crystals.saint-gobain.com/sites/imdf.crystals.com/files/documents/sgc-scintillation-fiber_0.pdf (accessed on October 1, 2019).
73. J. Boivin, "Systematic evaluation of photodetector performance for plastic scintillation dosimetry," *Med. Phys.*, 42(11), 6211–6220 (2015).
74. Y. N. Kharzheev, "Scintillation counters in modern high-energy physics experiments," *Phys. Part. Nuclei*, 46(4), 678–728 (2015).
75. T. O. White, "Scintillating fibres," *Nucl. Instrum. Meth. Phys. Res. A*, 273, 820–825 (1988).
76. R. E. Ansorge, C. Aurouet, P. Bareyre *et al.*, "Performance of a scintillating fibre detector for the UA2 upgrade," *Nucl. Instrum. Meth. Phys. Res. A*, 265, 33–49 (1988).
77. J. Alitti, A. Baracat, P. Bareyre *et al.*, "The design and construction of a scintillating fiber tracking detector," *Nucl. Instrum. Meth. Phys. Res. A*, 273, 135–144 (1988).
78. P. Annis, S. Aoki, G. Brooijmans *et al.*, "Performance and calibration of the CHORUS scintillating fiber tracker and opto-electronics readout system," *Nucl. Instrum. Meth. Phys. Res. A*, 367, 367–371 (1995).
79. S. Aoki, J. Dupont, J. Dupraz *et al.*, "Scintillating fiber trackers with optoelectronic readout for the CHORUS neutrino experiment," *Nucl. Instrum. Meth. Phys. Res. Sec. A* 344(1), 143–148 (1994).
80. A. Suzuki, H. Park, S. Aoki *et al.*, "Design, construction, and operation of SciFi tracking detector for K2K experiment," *Nucl. Instrum. Meth. Phys. Res. Sec. A*, 453(1), 165–176 (2000).
81. A. Gorin, S. Horikawa, K.-I. Kuroda *et al.*, "Scintillating fiber hodoscopes for DIRAC and COMPASS experiments," *Czech. J. Phys.*, 49(2), 173–182 (1999).
82. S. Horikawa, I. Daito, N. Doshita *et al.*, "A scintillating fiber tracker with high time resolution for high-rate experiments," *IEEE Trans. Nucl. Sci.*, 49(3), 950–956 (2002).
83. S. Horikawa, T. Toeda, I. Daito *et al.*, "Time resolution of a scintillating fiber detector," *Nucl. Instrum. Meth. Phys. Res. Sec. A*, 431(1), 177–184 (1999).
84. V. M. Abazov, B. Abbott, M. Abolins *et al.*, "The upgraded DØ detector," *Nucl. Instrum. Meth. Phys. Res. A*, 565(2), 463–537 (2006).
85. M. Ellis, P. R. Hobson, P. Kyberd *et al.*, "The design, construction and performance of the MICE scintillating fibre trackers," *Nucl. Instrum. Meth. Phys. Res. Sec. A*, 659(1), 136–153 (2011).
86. E. Overton, "Progress with the MICE scintillating fiber trackers," *Nucl. Instrum. Meth. Phys. Res. Sec. A*, 732(Supplement C), 412–414 (2013).
87. M. Adinolfi, F. Ambrosino, A. Antonelli *et al.*, "The KLOE electromagnetic calorimeter," *Nucl. Instrum. Meth. Phys. Res. Sec.* A, 482(1), 364–386 (2002).
88. P. Sonderegger, "Fibre calorimeters: Dense, fast, radiation resistant," *Nucl. Instrum. Meth. Phys. Res. Sec. A*, 257(3), 523–527 (1987).
89. D. W. Hertzog, P. T. Debevec, R. A. Eisenstein *et al.*, "A high-resolution lead /scintillating fiber electromagnetic calorimeter," *Nucl. Instrum. Meth. Phys. Res. Sec. A*, 294(3), 446–458 (1990).

90. S. A. Sedykh, J. R. Blackburn, B. D. Bunker *et al.*, "Electromagnetic calorimeters for the BNL muon (g–2) experiment," *Nucl. Instrum. Meth. Phys. Res. Sec. A*, 455(2), 346–360 (2000).
91. R. McNabb, J. Blackburn, J. D. Crnkovic *et al.*, "A tungsten/scintillating fiber electromagnetic calorimeter prototype for a high-rate muon (g–2) experiment," *Nucl. Instrum. Meth. Phys. Res. Sec. A*, 602(2), 396–402 (2009).
92. O. D. Tsai, L. E. Dunkelberger, C. A. Gagliardi *et al.*, "Results of R&D on a new construction technique for W/ScFi Calorimeters," *J. Phys.: Conf. Ser.*, 404(1), 012023 (2012).
93. R. C. Ruchti, "The use of scintillating fibers for charged-particle tracking," *Annu. Rev. Nucl. Part. Sci.*, 46, 281–319 (1996).
94. Y. Kharzheev, "Scintillation detectors in modern high energy physics experiments and prospect of their use in future experiments," *J. Laser Opt. Phot.*, 4(1), 1000148 (2017).
95. C. Joram, G. Haefeli, and B. Leverington, "Scintillating fibre tracking at high luminosity colliders," *J. Instrum.*, 10(8), C08005 (2015).
96. R. Greim, "A large scintillating fibre tracker for LHCb," *J. Instrum.*, 12(2), C02053 (2017).
97. E. O. Cohen, E. Piasetzky, Y. Shamai *et al.*, "Development of a scintillating-fiber beam detector for the MUSE experiment," *Nucl. Instrum. Meth. Phys. Res. Sec. A*, 815(Supplement C), 75–82 (2016).
98. A. Papa, F. Barchetti, F. Gray *et al.*, "A multi-purposed detector with silicon photomultiplier readout of scintillating fibers," *Nucl. Instrum. Meth. Phys. Res. Sec. A*, 787(Supplement C), 130–133 (2015).
99. D. J. Lawrence, L. M. Barbier, J. J. Beatty *et al.*, "Large-area scintillating-fiber time-of-flight/hodoscope detectors for particle astrophysics experiments," *Nucl. Instrum. Meth. Phys. Res. Sec. A*, 420(3), 402–415 (1999).
100. R. S. Miller, J. R. Macri, M. L. McConnell *et al.*, "SONTRAC: An imaging spectrometer for MeV neutrons," *Nucl. Instrum. Meth. Phys. Res. Sec. A*, 505(1), 36–40 (2003).
101. K. Tanderup, S. Beddar, C. E. Andersen *et al.*, "*In vivo* dosimetry in brachytherapy," *Med. Phys.*, 40(7), 070902-n/a (2013).
102. S. O'Keeffe, D. McCarthy, P. Woulfe *et al.*, "A review of recent advances in optical fibre sensors for *in vivo* dosimetry during radiotherapy," *Br. J. Radiol.*, 88(1050), 20140702 (2015).
103. K. W. Jang, D. H. Cho, W. J. Yoo *et al.*, "Fiber-optic radiation sensor for detection of tritium," *Nucl. Instrum. Meth. Phys. Res. Sec. A*, 652(1), 928–931 (2011).
104. M. W. Seo, J. K. Kim, and J. W. Park, "Test of a fiber optic scintillation dosimeter with BGO tip in a ^{60}Co irradiation chamber," *Prog. Nucl. Sci. Technol.*, 1, 186–189 (2011).
105. A. S. Beddar, T. R. Mackie, and F. H. Attix, "Water-equivalent plastic scintillation detectors for high-energy beam dosimetry: I. Physical characteristics and theoretical considerations," *Phys. Med. Biol.*, 37(10), 1883–1990 (1992).
106. A. S. Beddar, T. R. Mackie, and F. H. Attix, "Water-equivalent plastic scintillation detectors for high-energy beam dosimetry: II. Properties and measurements," *Phys. Med. Biol.*, 37(10), 1901–1913 (1992).
107. L. Archambault, J. Arsenault, L. Gingras *et al.*, "Plastic scintillation dosimetry: Optimal selection of scintillating fibers and scintillators," *Med. Phys.*, 32(7Part1), 2271–2278 (2005).
108. A. S. Beddar, T. R. Mackie, and F. H. Attix, "Cerenkov light generated in optical fibres and other light pipes irradiated by electron beams," *Phys. Med. Biol.*, 37(4), 925–935 (1992).
109. L. Archambault, A. S. Beddar, L. Gingras *et al.*, "Measurement accuracy and Cerenkov removal for high performance, high spatial resolution scintillation dosimetry," *Med. Phys.*, 33(1), 128–135 (2006).
110. P. Z. Y. Liu, N. Suchowerska, J. Lambert *et al.*, "Plastic scintillation dosimetry: Comparison of three solutions for the Cerenkov challenge," *Phys. Med. Biol.*, 56(18), 5805–5821 (2011).
111. B. Lee, K. W. Jang, D. H. Cho *et al.*, "Measurements and elimination of Cherenkov light in fiber-optic scintillating detector for electron beam therapy dosimetry," *Nucl. Instrum. Meth. Phys. Res. Sec. A*, 579(1), 344-348 (2007).
112. J. Lambert, Y. Yin, D. R. McKenzie *et al.*, "A prototype scintillation dosimeter customized for small and dynamic megavoltage radiation fields," *Phys. Med. Biol.*, 55(4), 1115–1126 (2010).
113. D. Létourneau, J. Pouliot, and R. Roy, "Miniature scintillating detector for small field radiation therapy," *Med. Phys.*, 26(12), 2555–2561 (1999).
114. S. F. de Boer, A. S. Beddar, and J. A. Rawlinson, "Optical filtering and spectral measurements of radiation-induced light in plastic scintillation dosimetry," *Phys. Med. Biol.*, 38(7), 945–958 (1993).

115. M. A. Clift, R. A. Sutton, and D. V. Webb, "Dealing with Cerenkov radiation generated in organic scintillator dosimeters by bremsstrahlung beams," *Phys. Med. Biol.*, 45(5), 1165–1182 (2000).
116. J. M. Fontbonne, G. Iltis, G. Ban *et al.*, "Scintillating fiber dosimeter for radiation therapy accelerator," *IEEE Trans. Nucl. Sci.*, 49(5), 2223–2227 (2002).
117. A. Darafsheh, R. Zhang, S. C. Kanick *et al.*, "Spectroscopic separation of Čerenkov radiation in high-resolution radiation fiber dosimeters," *J. Biomed. Opt.*, 20, 095001 (2015).
118. F. Therriault-Proulx, T. M. Briere, F. Mourtada *et al.*, "A phantom study of an in vivo dosimetry system using plastic scintillation detectors for real-time verification of ^{192}Ir HDR brachytherapy," *Med. Phys.*, 38(5), 2542–2551 (2011).
119. F. Therriault-Proulx, S. Beddar, T. M. Briere *et al.*, "Technical note: Removing the stem effect when performing Ir-192 HDR brachytherapy *in vivo* dosimetry using plastic scintillation detectors: A relevant and necessary step," *Med. Phys.*, 38(4), 2176–2179 (2011).
120. A. M. Frelin, J. M. Fontbonne, G. Ban *et al.*, "Spectral discrimination of Čerenkov radiation in scintillating dosimeters," *Med. Phys.*, 32(9), 3000–3006 (2005).
121. M. A. Clift, P. N. Johnston, and D. V. Webb, "A temporal method of avoiding the Cerenkov radiation generated in organic scintillator dosimeters by pulsed mega-voltage electron and photon beams," *Phys. Med. Biol.*, 47(8), 1421–1433 (2002).
122. J. Lambert, Y. Yin, D. R. McKenzie *et al.*, "Cerenkov-free scintillation dosimetry in external beam radiotherapy with an air core light guide," *Phys. Med. Biol.*, 53(11), 3071–3080 (2008).
123. F. Therriault-Proulx, L. Beaulieu, L. Archambault *et al.*, "On the nature of the light produced within PMMA optical light guides in scintillation fiber-optic dosimetry," *Phys. Med. Biol.*, 58(7), 2073–2084 (2013).
124. L. Wootton, and S. Beddar, "Temperature dependence of BCF plastic scintillation detectors," *Phys. Med. Biol.*, 58(9), 2955 (2013).
125. S. Buranurak, C. E. Andersen, A. R. Beierholm *et al.*, "Temperature variations as a source of uncertainty in medical fiber-coupled organic plastic scintillator dosimetry," *Radiat. Meas.*, 56(Supplement C), 307–311 (2013).
126. F. Therriault-Proulx, L. Wootton, and S. Beddar, "A method to correct for temperature dependence and measure simultaneously dose and temperature using a plastic scintillation detector," *Phys. Med. Biol.*, 60(20), 7927 (2015).
127. A. M. Frelin, J. M. Fontbonne, G. Ban *et al.*, "Comparative study of plastic scintillators for dosimetric applications," *IEEE Trans. Nucl. Sci.*, 55(5), 2749–2756 (2008).
128. L. Archambault, J. Polf, C., L. Beaulieu *et al.*, "Characterizing the response of miniature scintillation detectors when irradiated with proton beams," *Phys. Med. Biol.*, 53(7), 1865 (2008).
129. J. F. Williamson, J. F. Dempsey, A. S. Kirov *et al.*, "Plastic scintillator response to low-energy photons," *Phys. Med. Biol.*, 44(4), 857 (1999).
130. Standard Imaging Inc., "Exradin W1 Scintillator" (product datasheet), https://d3udwuy5vw6moa.cloudfront.net/files/ExradinW1_DS_1335-20web.pdf (accessed on October 1, 2019).
131. F. Lacroix, L. Archambault, L. Gingras *et al.*, "Clinical prototype of a plastic water-equivalent scintillating fiber dosimeter array for QA applicationsa," *Med. Phys.*, 35(8), 3682–3690 (2008).
132. J. Morin, D. Béliveau-Nadeau, E. Chung *et al.*, "A comparative study of small field total scatter factors and dose profiles using plastic scintillation detectors and other stereotactic dosimeters: The case of the CyberKnife," *Med. Phys.*, 40(1), 011719-n/a (2013).
133. L. Archambault, A. S. Beddar, L. Gingras *et al.*, "Water-equivalent dosimeter array for small-field external beam radiotherapy," *Med. Phys.*, 34(5), 1583–1592 (2007).
134. D. M. Klein, R. C. Tailor, L. Archambault *et al.*, "Measuring output factors of small fields formed by collimator jaws and multileaf collimator using plastic scintillation detectors," *Med. Phys.*, 37(10), 5541–5549 (2010).
135. J.-C. Gagnon, D. Thériault, M. Guillot *et al.*, "Dosimetric performance and array assessment of plastic scintillation detectors for stereotactic radiosurgery quality assurance," *Med. Phys.*, 39(1), 429–436 (2012).
136. P. Francescon, S. Beddar, N. Satariano *et al.*, "Variation of kQclin, Qmsrfclin, fmsr for the small-field dosimetric parameters percentage depth dose, tissue-maximum ratio, and off-axis ratio," *Med. Phys.*, 41(10), 101708-n/a (2014).
137. T. S. A. Underwood, B. C. Rowland, R. Ferrand *et al.*, "Application of the Exradin W1 scintillator to determine Ediode 60017 and microDiamond 60019 correction factors for relative dosimetry within small MV and FFF fields," *Phys. Med. Biol.*, 60(17), 6669 (2015).

138. A. R. Beierholm, C. F. Behrens, and C. E. Andersen, "Dosimetric characterization of the Exradin W1 plastic scintillator detector through comparison with an in-house developed scintillator system," *Radiat. Meas.*, 69(Supplement C), 50–56 (2014).
139. A. R. Beierholm, C. F. Behrens, and C. E. Andersen, "Comment on "Characterization of the Exradin W1 scintillator for use in radiotherapy" *Med. Phys.* 42, 297–304 (2015); *Med. Phys.*, 42(7), 4414–4416 (2015).
140. P. Carrasco, N. Jornet, O. Jordi *et al.*, "Response to "Comment on 'Characterization of the Exradin W1 scintillator for use in radiotherapy'" *Med. Phys.* 42, 297–304 (2015); *Med. Phys.*, 42(7), 4417–4418 (2015).
141. P. Carrasco, N. Jornet, O. Jordi *et al.*, "Characterization of the Exradin W1 scintillator for use in radiotherapy," *Med. Phys.*, 42(1), 297–304 (2015).
142. P. Francescon, W. Kilby, and N. Satariano, "Monte Carlo simulated correction factors for output factor measurement with the CyberKnife system—Results for new detectors and correction factor dependence on measurement distance and detector orientation," *Phys. Med. Biol.*, 59(6), N11 (2014).
143. M. Guillot, L. Beaulieu, L. Archambault *et al.*, "A new water-equivalent 2D plastic scintillation detectors array for the dosimetry of megavoltage energy photon beams in radiation therapy," *Med. Phys.*, 38(12), 6763–6774 (2011).
144. W. J. Yoo, J. Moon, K. W. Jang *et al.*, "Integral T-Shaped phantom-dosimeter system to measure transverse and longitudinal dose distributions simultaneously for stereotactic radiosurgery dosimetry," *Sensors*, 12(5), 6404 (2012).
145. M. Goulet, L. Gingras, and L. Beaulieu, "Real-time verification of multileaf collimator-driven radiotherapy using a novel optical attenuation-based fluence monitor," *Med. Phys.*, 38(3), 1459–1467 (2011).
146. M. Goulet, L. Gingras, and L. Beaulieu, "SU-GG-T-326: Experimental validation of a scintillating fiber detector for real-time quality control of MLC-driven radiotherapy treatment," *Med. Phys.*, 37(6Part20), 3261–3261 (2010).
147. M. Goulet, L. Archambault, L. Beaulieu *et al.*, "High resolution 2D dose measurement device based on a few long scintillating fibers and tomographic reconstruction," *Med. Phys.*, 39(8), 4840–4849 (2012).
148. M. Goulet, L. Gingras, L. Beaulieu *et al.*, "3D tomodosimetry using scintillating fibers: Proof-of-concept," *J. Phys.: Conf. Ser.*, 444(1), 012023 (2013).
149. M. Goulet, L. Archambault, L. Beaulieu *et al.*, "3D tomodosimetry using long scintillating fibers: A feasibility study," *Med. Phys.*, 40(10), 101703-n/a (2013).
150. B. Mijnheer, S. Beddar, J. Izewska *et al.*, "*In vivo* dosimetry in external beam radiotherapy," *Med. Phys.*, 40(7), 070903-n/a (2013).
151. D. Klein, T. M. Briere, R. Kudchadker *et al.*, "In-phantom dose verification of prostate IMRT and VMAT deliveries using plastic scintillation detectors," *Radiat. Meas.*, 47(10), 921–929 (2012).
152. L. Wootton, R. Kudchadker, A. Lee *et al.*, "Real-time in vivo rectal wall dosimetry using plastic scintillation detectors for patients with prostate cancer," *Phys. Med. Biol.*, 59(3), 647 (2014).
153. S. J. Klawikowski, C. Zeringue, L. S. Wootton *et al.*, "Preliminary evaluation of the dosimetric accuracy of the in vivo plastic scintillation detector OARtrac system for prostate cancer treatments," *Phys. Med. Biol.*, 59(9), N27 (2014).
154. Radiadyne, OARtrac®" (product website), http://radiadyne.com/oartrac.php (accessed on October 1, 2019).
155. L. M. Moutinho, I. F. Castro, H. Freitas *et al.*, "Scintillating fiber optic dosimeters for breast and prostate brachytherapy." *Proc. SPIE, 10058, 100580C* (2017).
156. B. Lee, W. J. Yoo, K. W. Jang *et al.*, "Measurements of relative depth doses using fiber-optic radiation sensor and EBT film for brachytherapy dosimetry," *IEEE Trans. Nucl. Sci.*, 57(3), 1496–1501 (2010).
157. F. Therriault-Proulx, L. Archambault, L. Beaulieu *et al.*, "Development of a novel multi-point plastic scintillation detector with a single optical transmission line for radiation dose measurement," *Phys. Med. Biol.*, 57(21), 7147 (2012).
158. P. Woulfe, F. J. Sullivan, and S. O'Keeffe, "Optical fibre sensors: Their role in in vivo dosimetry for prostate cancer radiotherapy," *Cancer Nanotechnol.*, 7(1), 7 (2016).
159. S. Beddar, and L. Beaulieu, *Scintillation Dosimetry*. CRC Press, Boca Raton, FL (2016).
160. MINOS Collaboration, The MINOS detectors technical design report. Fermilab, Batavia, IL (1998).

161. MINERvA Collaboration, "Design, calibration, and performance of the MINERvA detector," *Nucl. Instrum. Meth. Phys. Res. Sec. A*, 743(Supplement C), 130–159 (2014).
162. OPERA Collaboration, "The OPERA experiment in the CERN to Gran Sasso neutrino beam," *J. Instrum.*, 4(04), P04018 (2009).
163. D. Allan, C. Andreopoulos, C. Angelsen *et al.*, "The electromagnetic calorimeter for the T2K near detector ND280," *J. Instrum.*, 8(10), P10019 (2013).
164. P. Baringer, A. Bross, V. Buescher *et al.*, "Cosmic-ray tests of the DØ preshower detector," *Nucl. Instrum. Meth. Phys. Res. Sec. A*, 469(3), 295–310 (2001).
165. M. Gallinaro, A. Artikov, C. Bromberg *et al.*, "A new scintillator tile/fiber preshower detector for the CDF central calorimeter." *IEEE Trans. Nucl. Sci.*, 52, 879–883 (2005).
166. LHCb Collaboration, "The LHCb detector at the LHC," *J. Instrum.*, 3(08), S08005 (2008).
167. Atlas Collaboration, "The ATLAS experiment at the CERN large hadron collider," *J. Instrum.*, 3(08), S08003 (2008).
168. CMS Collaboration, "The CMS experiment at the CERN LHC," *J. Instrum.*, 3, S08004 (2008).
169. J. Lambert, D. R. McKenzie, S. Law *et al.*, "A plastic scintillation dosimeter for high dose rate brachytherapy," *Phys. Med. Biol.*, 51(21), 5505–5516 (2006).
170. J. Archer, E. Li, M. Petasecca *et al.*, "High-resolution fiber-optic dosimeters for microbeam radiation therapy," *Med. Phys.*, 44(5), 1965–1968 (2017).
171. W. J. Yoo, S. H. Shin, D. E. Lee *et al.*, "Development of a small-sized, flexible, and insertable fiber-optic radiation sensor for gamma-ray spectroscopy," *Sensors*, 15(9), 21265–21279 (2015).
172. N. Suchowerska, "Clinical trials of a urethral dose measurement system in brachytherapy using scintillation detectors," *Int. J. Radiat. Oncol. Biol. Phys.*, 79, (2011), 609–615.
173. D. McCarthy, S. O' Keeffe, E. Lewis *et al.*, "Optical fibre X-ray radiation dosimeter sensor for low dose applications." 121–124.
174. S. O'Keeffe, M. Grattan, A. Hounsell *et al.*, "Radiotherapy dosimetry based on plastic optical fibre sensors," *Proc. of SPIE*, 8794, 879418 (2013).
175. D. McCarthy, S. O' Keeffe, E. Lewis *et al.*, "Radiation dosimeter using an extrinsic fiber optic sensor," *IEEE Sens. J.*, 14(3), 673–685 (2014).
176. A. I. de Andrés, S. O'Keeffe, L. Chen *et al.*, "Highly sensitive extrinsic X-Ray polymer optical fiber sensors based on fiber tip modification," *IEEE Sens. J.*, 17(16), 5112–5117 (2017).
177. A. I. de Andrés, Ó. Esteban, and M. Embid, "Improved extrinsic polymer optical fiber sensors for gamma-ray monitoring in radioprotection applications," *Opt. Laser Technol.*, 93(Supplement C), 201–207 (2017).
178. P. Woulfe, S. O'Keeffe, and F. J. Sullivan, *Optical Fibre Luminescence Sensor for Real-time LDR Brachytherapy Dosimetry, Proc. SPIE*, 9916, 99160T, (2016).
179. S. O'Keeffe, W. Zhao, W. Sun *et al.*, "An optical fibre-based sensor for real-time monitoring of clinical linear accelerator radiotherapy delivery," *IEEE J. Sel. Top. Quantum Electron.*, 22(3), 35–42 (2016).
180. P. Woulfe, F. J. Sullivan, E. Lewis *et al.*, "Plastic optical fibre sensor for *in-vivo* radiation monitoring during brachytherapy," *Proc. SPIE*, 9634, 963421 (2015).
181. C. J. Marckmann, C. E. Andersen, M. C. Aznar *et al.*, "Optical fibre dosimeter systems for clinical applications based on radioluminescence and optically stimulated luminescence from Al_2O_3:C," *Radiat. Prot. Dosim.*, 120(1–4), 28–32 (2006).
182. C. E. Andersen, S. K. Nielsen, S. Greilich *et al.*, "Characterization of a fiber-coupled Al_2O_3:C luminescence dosimetry system for online *in vivo* dose verification during ^{192}Ir brachytherapy," *Med. Phys.*, 36(3), 708–718 (2009).
183. A. Croteau, S. Caron, A. Rink *et al.*, "Fabrication and characterization of a real-time optical fiber dosimeter probe," *Proc. SPIE*, 8090, 80900G (2011).
184. A. Croteau, S. Caron, A. Rink *et al.*, "Real-time optical fiber dosimeter probe," *Proc. SPIE*, 7894, 789406 (2011).

14

Microstructured POFs

Maryanne Large and Marcelo Martins Werneck

Contents

14.1 An Introduction to Microstructured Optical Fibers

Guidance of light requires a variation in refractive index, which in conventional optical fibers is achieved either by chemical doping or by the use of more than one material. Microstructured optical fibers [MOFs], also known as "photonic crystal" [PCF] or "holey" fibers, achieve light guidance through a patterning of tiny holes which run along the entire length of the fiber, as shown in **Figure 14.1**. In the simplest case, where the core of the fiber is solid, MOFs can be thought of as using holes to "dope" the material with air, and thus lower its effective refractive index. Using the holes to modify the material in this way, it is straightforward to produce what is effectively a step index fiber, or indeed any other refractive index profile. This might suggest that MOFs are not very different from conventional fibers, but the reality is much more interesting. The hole structure is intrinsically dispersive, and the size and placement of the holes can dramatically affect the optical performance. Indeed, the microstructure can be used as a modal "sieve" which confines the fundamental mode much more effectively than higher order modes (Russell 2003). This effect allows MOFs to remain single-moded over a very large frequency range (Knight et al. 1996, Birks et al. 1997), and to be single-mode with a large mode area (Knight et al. 1999).

Modifications of the microstructure have allowed an increasing number of specialty applications to be realized. For example by making the arrangement of holes, or the holes themselves have different profiles in the *x* and *y* directions, it is possible to make highly birefringent or polarization maintaining fibers (e.g., Ortigosa-Blanch 2000, Steel et al. 2001, Issa 2004a). The dispersion properties of MOFs have also attracted attention (e.g., Knight et al. 1999, Reeves et al. 2002, Saitoh et al. 2003), particularly because of the possibilities they offer for dispersion compensation. By changing the core size of the fibers, it is also possible to make them have either very low or very high optical non-linearity (Broderick et al. 1999), and an appealing use of this property was the development of a fiber system for supercontinuum generation (Ranka et al. 2000, Wadsworth et al. 2004a). By making the bridges between the holes extremely thin, it is possible to make fibers that are essentially "air clad," and such fibers can have numerical apertures greater than 0.9 (Wadsworth et al. 2004b).

FIGURE 14.1 The original "endlessly single mode" fiber. (From Knight et al., *Opt. Lett.*, 21, 1547–1549, 1996; Courtesy of University of Bath.)

A particularly significant feature of MOFs is that they offer the ability to guide light in air or other low index materials through the photonic bandgap effect. Photonic bandgap effects are most simply understood in one-dimensional structures such as multi-layer stacks, which can be designed to reflect particular wavelengths. These wavelengths are described as lying within the "band gap" of the structure—they cannot be transmitted, and so are reflected. If we imagine this multi-layer stack to be rolled up into a cylinder, we obtain a "Bragg fiber," in which the wavelengths with the bandgap are transmitted along the hollow core. Exactly this approach has been used to produce a "swiss roll" fiber, in which a two material multi-layer is rolled up to produce a hollow core fiber (Fink et al. 1999, Kuriki et al. 2004). A simpler approach, which requires a single material, is to make a two-dimensional microstructure. Microstructured fibers analogous to Bragg fibers can be produced using ring structures (Argyros et al. 2001, 2002), or more commonly a two-dimensional array of holes, as shown in **Figure 14.2**, is used to produce the bandgap. This not only allows wavelengths to be guided that previously could not because of material absorption, it also allows guidance in materials of low refractive index, such as liquids or gases (**Figure 14.3**).

These diverse properties are not only valuable in isolation, they can also be combined in new ways, to make, for example, large core fibers with very high numerical apertures for high power laser applications (Limpert et al. 2003, Wadsworth et al. 2003), or fibers in which the holes not only produce light guidance, but can also introduce other materials: gases for liquid or gas sensing or low threshold Raman effects (Benabid et al. 2002b, 2005, Fini 2004,

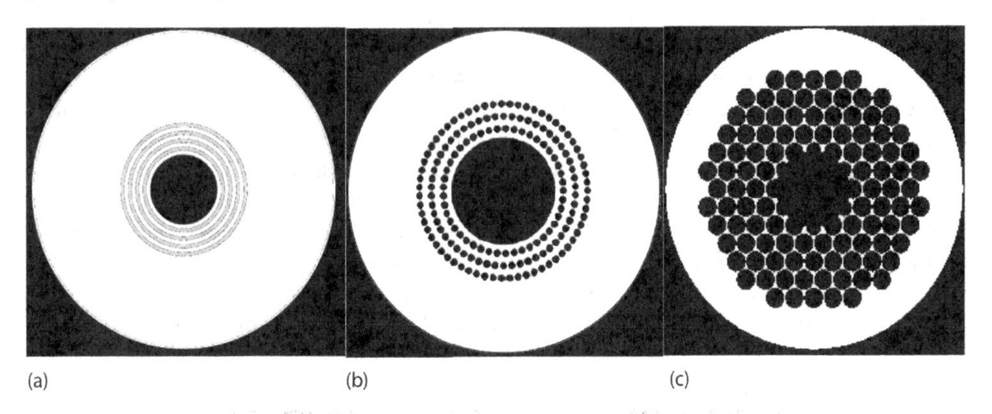

FIGURE 14.2 Schematic diagrams of the cross-section of a hollow-core: (a) Bragg fibered, (b) ring-structured Bragg fiber, and (c) a photonic bandgap fiber. Air regions are shown in black. (Courtesy of Alexander Argyros, OFTC.)

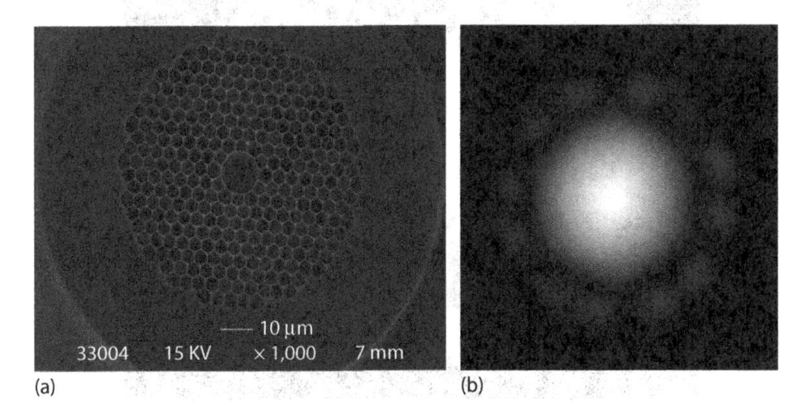

FIGURE 14.3 An electron micrograph of an air guiding "photonic band gap" fiber (a) and guiding in the hollow core (b). This fiber was made by Blaze Phontics Ltd. (Courtesy of Tim Birks.)

Ritari et al. 2004), or to produce tunable devices (Mach et al. 2002). Other novel applications of MOFs include multicore fibers for imaging and interconnects (van Eijkelenborg 2004c) and the guidance of particles (Benabid et al. 2002a), and potentially atoms, in hollow core fibers, using light guided by the fiber to prevent the particles from hitting the walls (Dall et al. 2002).

The significance of MOFs is not merely that they allow the optical properties to be tuned in unusual ways. By separating light guidance from the material properties, they have expanded the range of materials that can be used. In fact the use of photonic band gaps means that we are not even limited to using transparent material. Many fiber systems suffer from issues associated with the interfaces of different materials (Monro et al. 2003), or the presence of dopants. These can adversely affect the mechanical properties of the fiber, as they do in the soft glasses, and historically this has prevented the widespread use of some of these materials, even though they have desirable transmission characteristics, particularly in the mid-infrared region. In polymer optical fibers, dopant diffusion is one of the limiting factors in the temperature stability. Indeed, temperature stability may make MOFs a superior solution even in applications where conventional fibers work well, such as birefringent fibers. The temperature variation in the birefringence is far less in MOFs than it is in conventional HiBi fibers (Issa 2004b).

Most work in microstructured fibers to date has been in been in silica, but increasingly, other materials are being explored, including the soft glasses (Kiang et al. 2002, Kumar et al. 2002, 2003, Monro et al. 2000, 2003), as well as polymers (van Eijkelenborg 2001, Choi et al. 2001). These new materials will expand further the rich field of MOFs.

In my own area, microstructured polymer optical fibers [mPOF] are already having a significant impact on the field. Using mPOFs, it is possible to make large core graded index fibers that are temperature stable with a single material (**Figure 14.4**) without using complex chemistry or requiring sensitive process control. The ease with which single mode POF can be fabricated by this technique is also significant, and presents many possibilities in applications such as sensing. The recent fabrication of UV written gratings (Dobbs et al. 2005), as well as mechanically produced long period gratings in mPOF (van Eijkelenborg 2004b), further enhances these possibilities. Indeed, in some ways the prospects of microstructured fibers in polymer represent a more fundamental shift for polymers than for silica fibers, where the performance of conventional fibers has been optimized over many years, and is exceptionally good. By contrast, the performance of polymer optical fibers

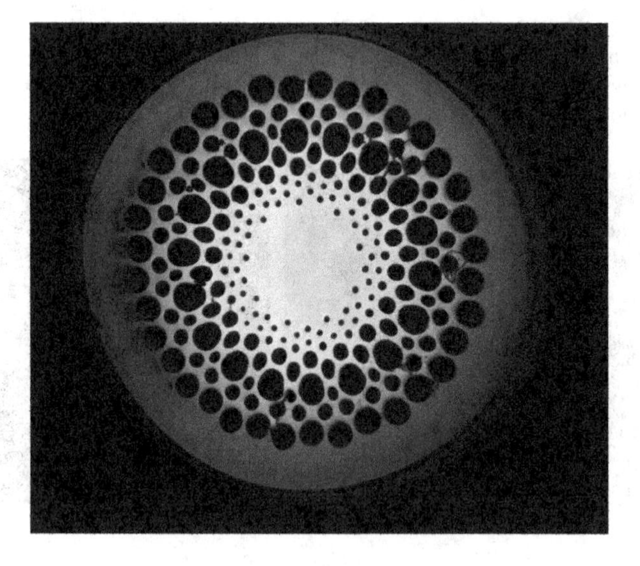

FIGURE 14.4 A graded index mPOF. The outer diameter of the fiber is 250 microns. (Courtesy of the OFTC.)

has been limited by losses and the lack of a generic process to produce arbitrary refractive index profiles: both issues are addressed by using microstructures. It is already feasible that mPOF will surpass conventional POF in terms of loss, by reducing scattering due to the core cladding interface and the presence of dopants. In the longer term, air guiding mPOF (Argyros et al. 2005c) could lower the transmission losses below that of any conventional POF. Currently, the best conventional non-fluorinated POF has a loss minimum of 150 dB/km at 650 nm (Daum et al. 2002) while the best mPOF, made of the same material, has a loss of 192 dB/km (Lwin 2005a). MPOF have also been made in the extremely low loss fluorinated material, CYTOP (Kondo et al. 2004), an exciting development because not only does this material have very low loss, its transmission window extends to 1300 nm. No loss figure, however, is quoted for this fiber.

While the losses of MOFs in silica remain above those of conventional fibers, their performance is improving rapidly. For solid core fibers the best loss figure now stands at 0.3 dB/km, and for hollow core fibers at 1.2 dB/km (Roberts et al. 2005). For telecommunications transmission, the extent to which these figures fall further will determine the application of these fibers. For many other applications, including devices, their performance already makes them a viable solution to many problems.

14.2 Fabrication of Microstructured Fibers

14.2.1 Preform Production

A major challenge in making microstructured fibers in any material is to make the preform. Indeed, the ability to make suitable preforms at low cost, and with appropriate quality, is one of the factors that will determine the uptake of this technology. For silica fibers, preforms are usually made by hand, using capillary stacking, a technique used in the earliest fibers (Knight et al. 1996, Birks et al. 1997). Capillaries are carefully drawn in a conventional draw process to give good uniformity in diameter, then cut into lengths and stacked in the desired pattern. The difficulty in working with very thin capillaries means that diameters of larger than 0.5 mm are preferred (Knight et al. 2001). In materials that are more flexible than silica, such as polymer, capillaries may need to be significantly thicker to produce a uniform stack (Huang et al. 2004; Kondo et al. 2004).

The capillary stack, similar to the one shown in **Figure 14.5**, can then be drawn into a fiber. For solid core fibers, the core is formed by the inclusion of a solid rod in the preform (see **Figure 14.5**) while in air-core fibers, one or more capillaries are removed from the stack and replaced with a larger capillary to preserve the structural integrity of the preform. The integrity of the capillary stack requires the packing to be very tight, in order to avoid later deformation, and may also be improved by pressurizing and/or collapsing various sections of the preform. Capillary stacking has also been used for a variety of non-silica materials. A tellurite fiber drawn from a capillary stack is shown in **Figure 14.6**.

FIGURE 14.5 Preforms can be produced by stacking capillaries. In this case the core is formed by replacing the central capillary with a solid rod. (Courtesy of the OFTC.)

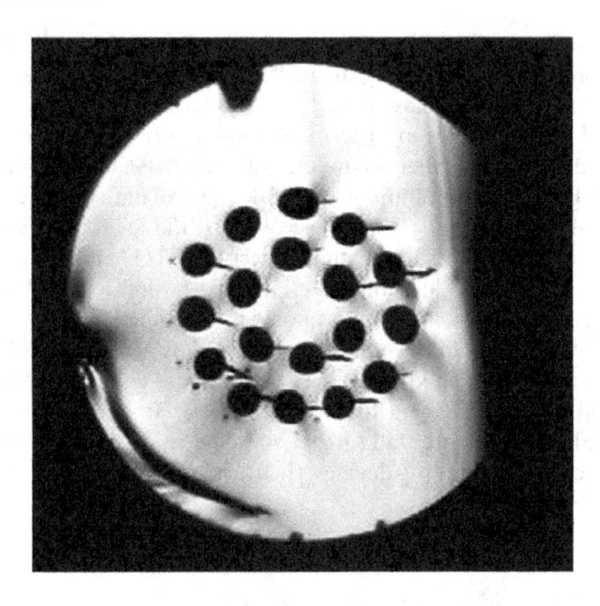

FIGURE 14.6 A tellurite microstructured fiber drawn from a capillary stack. (Courtesy of Cristiano Cordeiro, UNICAMP, Brazil.)

The structure can be modified by including rods rather than capillaries, and further variation can be produced by using different internal and external capillary diameters. Capillaries or rods of other materials, for example those including a dopant, can also be incorporated, making this an extremely versatile, if time consuming technique. Indeed, variations of capillary stacking have been used to produce extremely novel structures and even devices. By stacking and drawing capillaries down to an intermediate size, not much larger than a fiber, it is possible to create a "scaffold" structure into which can be placed single or multimode fibers. After drawing down, the core of a single mode fiber becomes negligibly small, and the whole fiber essentially becomes an undoped core of the microstructured fiber. This technique was used recently to produce a solid photonic bandgap fiber (Argyros et al. 2005a), shown in **Figure 14.7**. It has also been used to produce an efficient tapered transition from a conventional single mode fiber to the core of a microstructured fiber (Leon-Saval et al. 2005).

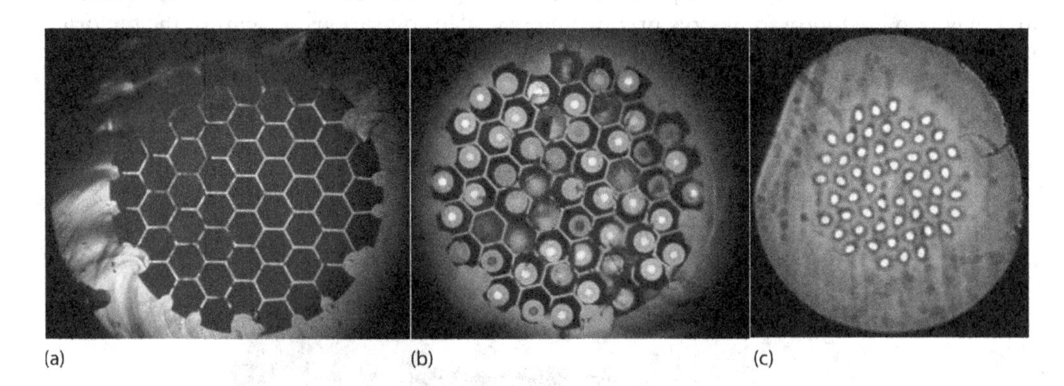

FIGURE 14.7 Stacked capillaries are drawn down to an intermediate stage to form a scaffold structure (a). This can then be filled as desired with conventional fibers, either single or multi-mode (b), and the whole structure drawn down to a fiber (c). Note that these images are not to scale. (From Argyros et al., *Opt. Express*, 12, 2688–2698, 2004.)

A variety of other methods have been used to produce preforms, particularly for non-silica materials. Motivations include the ability to produce structures that cannot be easily made by capillary stacking (a process that usually results in fibers having hexagonal symmetry), or a desire to make preform production easier, faster, and more automated. Most of these techniques result in a monolithic preform, in which the entire hole structure is produced as a single larger entity. This has many advantages but makes it more difficult to incorporate other materials, which may be required for applications such as fiber lasers. In some cases, this has led to fabrication techniques being combined, most often through the use of an intermediate stage, to produce the final fiber.

One of the most widely used methods to produce preforms in non-silica materials is extrusion (Allan et al. 2001, Kiang et al. 2002), in which heated bulk material is forced through a die under pressure (see **Figure 14.8**). The method is attractive because it allows arbitrary structures to be produced, and because it genuinely offers a route to scaling up manufacturing of preforms. A disadvantage of extrusion is that it adds and extra thermal cycle to the processing, which causes material difficulties. At this stage, fibers made using extrusion have more problems with surface quality, and perhaps contamination, than those produced by capillary stacking, though this may merely reflect the need for further development. Co-extrusion has also been used to produce preforms of more than one material, as shown in **Figure 14.9**.

FIGURE 14.8 Examples of extruded soft glass preforms. (Courtesy of the Optical Research Centre, University of Southampton, and the University of Adelaide.)

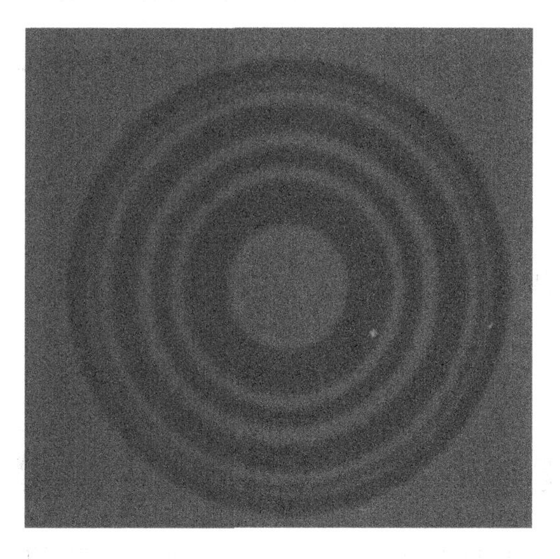

FIGURE 14.9 Two co-extruded soft glass materials. (Courtesy of the Optical Research Centre, University of Southampton.)

Microstructured polymer optical fibers (mPOF) probably have the widest range of possible fabrication methods, bring as they do the diversity of polymer processing techniques, including casting, molding, and extrusion. As with other materials, the problem is less trivial than it may seem, particularly when there is the need for very uniform and complex hole structures, with a diversity of wall thicknesses between the holes and extremely large aspect ratios. These difficulties are considerably reduced if the preforms can be made larger, though this requires a significantly larger draw ratio to be used. Casting has been used to make microstructured fibers in polymer, and probably offers the best long-term solution in terms of material purity as the reaction can be sealed from contamination. It may also give better surface quality than other techniques. So far, however, the best results have been obtained simply drilling the preforms, a technique that has also been used in glasses (see **Figure 14.10**), where the brittleness of the material makes it distinctly more difficult.

In polymers, this process can be carried out accurately and in an automated fashion using a computer-controlled drill, and it allows a very large diversity of structures to be rapidly prototyped, making it an ideal research technique. It can readily produce structures that are not easily produced by capillary stacking, for example the core structure seen in **Figure 14.11**.

FIGURE 14.10 A drilled fluoride glass preform. (Courtesy of Pam MacNamara, OFTC.)

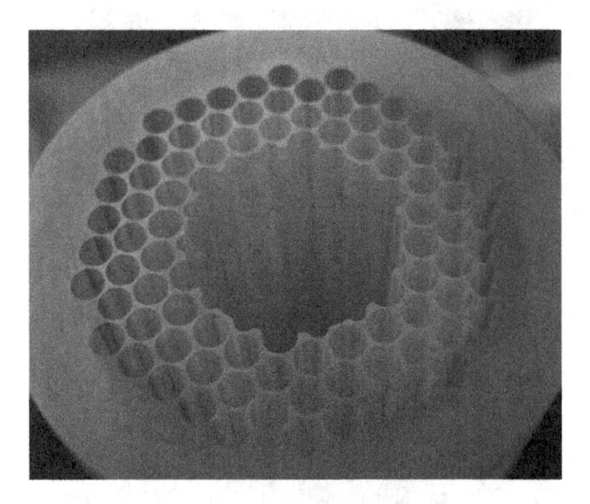

FIGURE 14.11 A drilled polymer preform for making an air-guiding photonic band gap fiber. Note that by using drilling it is possible to obtain different core shapes from that obtained by capillary stacking. This has some advantages in terms of optical performance. (Courtesy of the OFTC.)

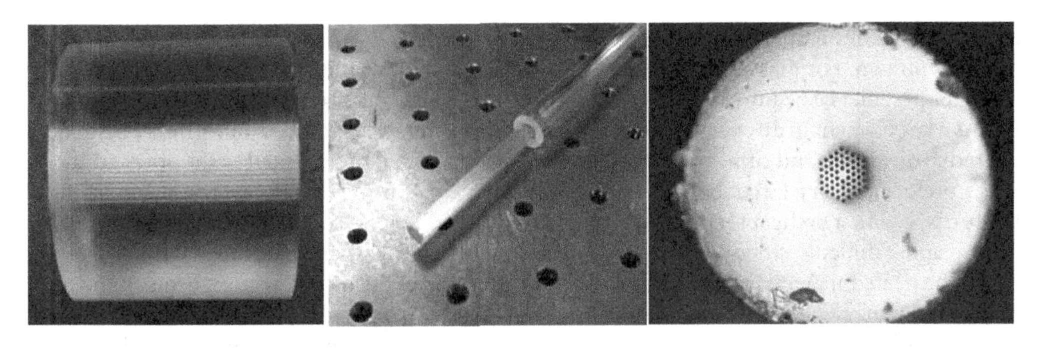

FIGURE 14.12 Preforms in the polymer fabrication process can be made very short (here 7 cm long) and fat, requiring a two stage draw process. The preform is drawn to an intermediate or "cane" stage of about 50–100 mm in diameter, and may be sleeved, as here, before being drawn to fiber. The fiber is 500 microns in diameter. (Courtesy of the OFTC.)

Its disadvantages include the introduction of contaminants required as cutting fluids (van Eijkelenborg 2004a), which increase the loss, and the inability to scale up production. The preforms that have been produced by this method are unusually short and fat (typically having a 7 cm length and diameter) requiring a two-stage draw process in which the first stage is noncontinuous (Barton et al. 2004), as shown in **Figure 14.12**.

An obvious feature of microstructured fibers and their preforms is their large surface area, and it has proved possible to exploit this to solution-dope monolithic performs of a single material (Large et al. 2004). Using appropriate choice of solvents, doped solutions can be introduced through the holes of the microstructure. The solvent acts as a plasticizer which increases the diffusion of the dopant dramatically. When the solvent is removed through heating, the dopant remains in the polymer matrix.

14.2.2 The Draw Process

The draw process for microstructured fibers is essentially the same as that used for other optical fibers. A schematic of a draw-tower, together with a photograph of a commercial draw tower (this one designed for drawing polymer fibers), is shown in **Figure 14.13**. The preform is fed into a

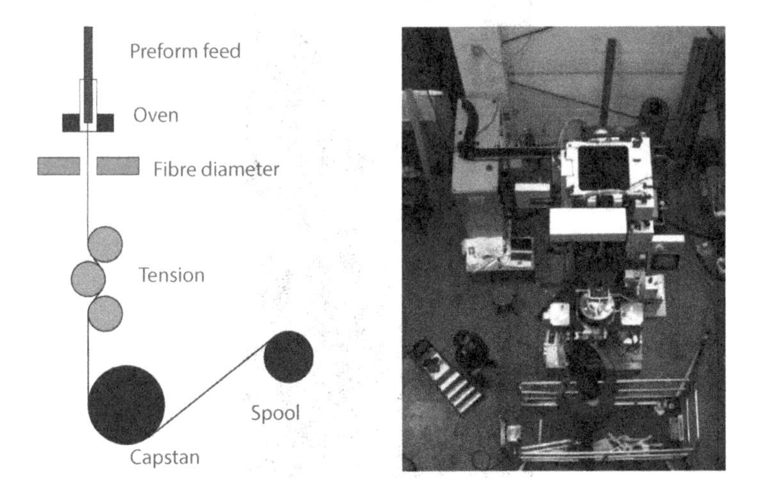

FIGURE 14.13 A schematic of a draw-tower, together with a photograph of a tower designed for drawing polymer fibers. (Courtesy of the OFTC.)

furnace to raise its temperature to the point where the viscosity is sufficiently lowered for the material to be drawn. For glasses, this requires a temperature above the "softening" temperature T_s, while for polymers it requires a temperature above the "glass transition" temperature T_g. The furnace is a key element in any such system, because the heating needs to be as uniform as possible in the radial direction, and sufficiently rapid to allow new material to be heated as the fiber is drawn. If the preform is very large, or the heating very inefficient, it may be necessary to supplement the furnace with a preheat section. The presence of a large air fraction in the preform may make heating more difficult, since air is a very good insulator (Lyytikäinen et al. 2004).

As the heated material is pulled down, a characteristic deformation region forms, known as the "neckdown" region, where a rapid change in radius can be observed. An example is shown in **Figure 14.14**. The length of this neckdown region is related to both to the length of the furnace and thermal properties of the material, principally, how rapidly the viscosity changes with temperature. As it turns out, the geometry of the neckdown region (for example, how long it is, and its slope) is critically important in drawing microstructured fibers. A number of systems may be used to control the draw process, but the most important is a feedback system, which monitors the fiber diameter and uses this to control the fiber draw rate. Other parameters are also often monitored during the draw, including the draw tension and the temperature.

In some cases, where a very large reduction in radius is required, more than one draw process may be needed. For example, in the case of mPOF, a two-stage process is generally used to reduce the preform from a diameter of about 7 cm to a fiber diameter of between 120 and 500 microns (see **Figure 14.12**). In this case the normal fiber draw process is preceded by another draw process, which is essentially a non-continuous "stretching" (Barton et al. 2004), which reduces the diameter to an intermediate size (a "cane," see **Figure 14.12**).

The draw process for microstructured optical fibers does pose some challenges. The major difficulties center on obtaining the desired microstructure in the fiber from the preform.

FIGURE 14.14 A "neckdown" region for a microstructured polymer optical fiber. This is the deformation region where the preform changes dimension rapidly. (Courtesy of the OFTC.)

Although the structure in the fiber can be close to a scaled down version of preform, it is rarely identical: holes may increase or decrease their relative size, and even change shape significantly. The degree of size and shape change depends both on the material parameters and the draw conditions on the particular draw tower used. An additional difficulty in surveying the experience of groups around the world has been a desire to keep sensitive fabrication details in-house.

Most people working in the area seem to have determined the operating conditions that produce the best results empirically, an approach that can be quite effective, particularly for structures having a single hole size. Knight and others (Knight et al. 2001, Eom 2002) have noted the importance of controlling surface tension during the draw process of silica MOFs. In fibers, where the slope of the neck down region is small, surface tension is simply inversely proportional to the hole radius. Its effect is to cause hole collapse, and this has led to MOFs being drawn at comparatively lower temperatures than conventional fibers because this yields a higher viscosity which limits the collapse. Generally, silica MOFs are drawn at a temperature of about 1900°C (Bjarklev et al. 2003), about 200 degrees colder than conventional silica fibers. The lower bound of operating temperatures is determined by the mechanical strength of the fiber, which will snap if drawn at too high a viscosity. Tuning the draw conditions, particularly the draw temperature, has a positive side-allowing different fiber structures to be drawn from the same preform. As the temperature is increased, the relative hole size is reduced.

The competition between surface tension and viscosity has become something of a theme for workers in the field. An early attempt to obtain a more rigorous understanding of the process was provided by Fitt and co-workers (Fitt et al. 2001, 2002). They considered the behavior of a hollow glass fiber, but did not consider more complex cases in which the holes were not centrally located, or the interactions between holes. Another early study by Deflandre (2002) considered thermal effects on the periodicity and hole size during the process. In fact, modeling the draw process for MOFs with arbitrary structures is extremely demanding because of the number of substantially deformed three-dimensional free surfaces.

By far the most complete study of drawing MOFs has been performed by Xue et al. (2005a), who used both analytical and numerical approaches. Xue modeled both the steady state and transient draw processes for isothermal conditions and Newtonian materials, using a wide variety of hole structures. The transient draw process has been used to describe the initial "stretching" on a tower of a short preform while the steady state process applies to the draw to fiber. He has shown how the sizes and shapes and holes can be changed during the draw process. The most powerful aspect of Xue's work is that it allows generic conclusions to be drawn about the recommended draw parameters. He has achieved this by the use of scaling analysis and dimensionless quantities, which allow the effect of different materials and draw conditions to be directly compared.

The three most important parameters he identifies are the capillary number C_a, draw ratio D_r and the aspect ratio ε. The capillary number is defined by the properties of the material being drawn, and defines the relative significance of surface tension and viscosity. The draw ratio and the aspect ratio are determined by the draw conditions—how fast the fiber is drawn, and the length of the neckdown region. By varying these three parameters it is largely possible to adjust the structure in the fiber.

The mathematical definitions of the parameters are as follows:

The capillary number:

$$C_a \equiv \eta V_i / \sigma$$

where η and σ are the viscosity and the surface tension coefficients of the material and V_i is the preform feed rate; the draw ratio D_r:

$$D_r \equiv V_f / V_i$$

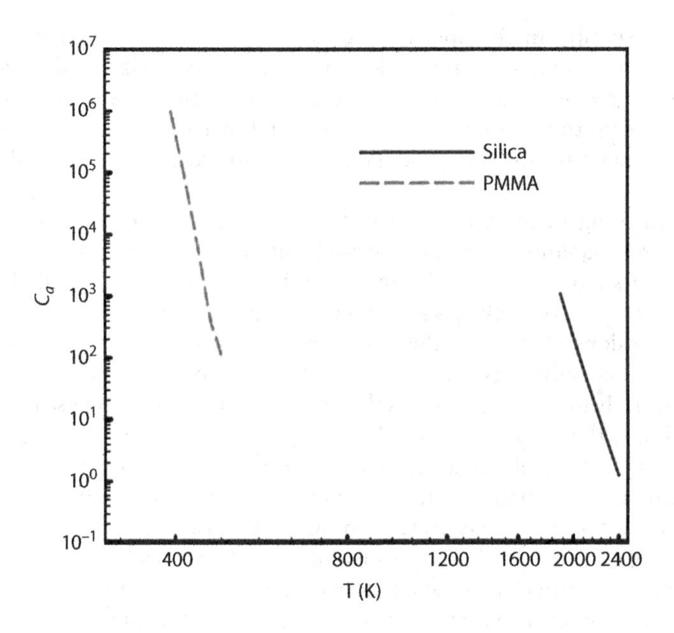

FIGURE 14.15 The values of the capillary number C_a in the operating temperature range for silica and PMMA. (Courtesy of Shicheng Xue, University of Sydney.)

where V_f is the draw speed of the fiber and the aspect ratio ε:

$$\varepsilon \equiv R_i / L$$

where R_i is the initial preform radius and L is the length of the neckdown region.

This analysis immediately makes it clear why very different results can be expected when drawing MOFs from different materials. **Figure 14.15** shows the value of the capillary number for the two most commonly used MOF materials, silica and the polymer, polymethylmethacrylate [PMMA], over their respective temperature regimes, and at typical feeding speed (V_i = 2.5 mm/min). For PMMA, Ca varies from 106 to 102; for silica, however, it varies from 103 to 100. It is this difference which is at the heart of the very different types of hole deformation that occur in these materials. When the capillary number is large, surface tension can actually be neglected, and this is often the case with PMMA, while hole collapse caused by surface tension is almost always important for silica.

The effects of changing the aspect ratio and draw ratio are most easily seen in structures where holes of quite different sizes are placed in close proximity. In this case the holes tend to interact strongly because both important forces, the surface tension and the viscous force, have a $1/r$ dependence (r is the radius of the hole).

The interaction of differently sized holes has actually been used to good effect, for example to produce elliptical holes for birefringent effects (Issa 2004a). It may, however, be problematic in a particularly important case—that of the air core fiber, in which there are normally two sizes of holes: those defining the cladding structure, and the core itself. **Figures 14.16** and **14.17** illustrate the effect of changing the draw ratio and aspect ratio respectively. Hole deformation is most pronounced when the draw ratio and aspect ratio are both large, corresponding to a rapid draw and a short neckdown region. These are also the criteria which promote hole expansion.

There is a clear tendency for the holes to match their curvatures at their point of proximity. As might be expected, this effect is successively reduced with distance, as shown in **Figure 14.18**.

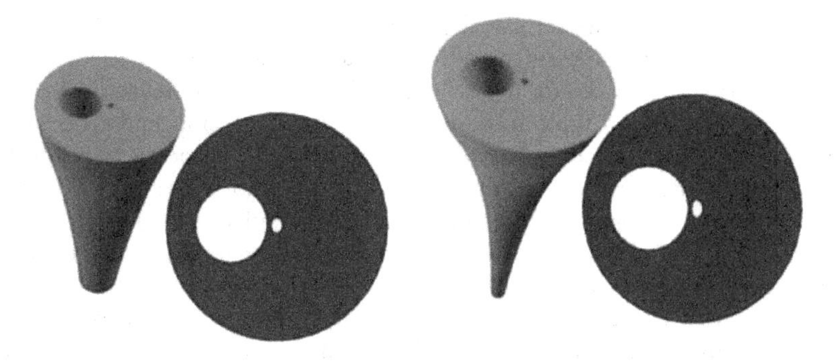

FIGURE 14.16 The effect of draw ratio: the neck-down shapes and the fiber cross-section for draw ratios $D_r = 10$ (left) and $D_r = 100$ (right). The material properties of PMMA have been used in this simulation, and an aspect ratio of 0.25. (Courtesy of Shicheng Xue, University of Sydney.)

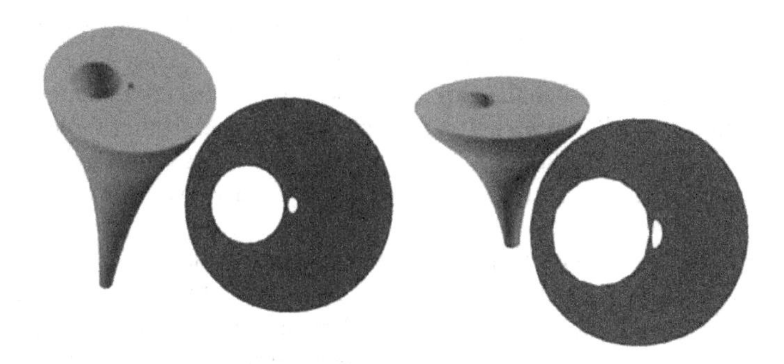

FIGURE 14.17 The effect of changing the aspect ratio: the neck-down shapes and the fiber cross-section for $\varepsilon = 0.25$ (left) and 0.5 (right). The draw ratio used is 100 in both cases, and the material properties of PMMA have been used. (Courtesy of Shicheng Xue, University of Sydney.)

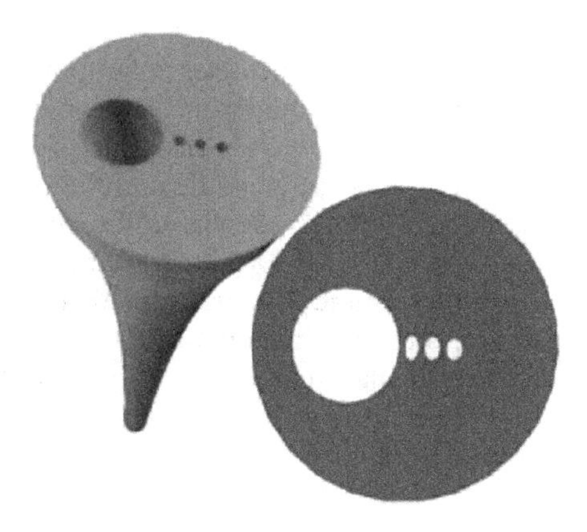

FIGURE 14.18 Deformations in hole shape decrease with increasing distance from the larger hole. (Courtesy of Shicheng Xue, University of Sydney.)

These deformations in shape are actually driven by deformations in size. In this simulation, the capillary number and draw parameters cause the holes to expand in relative terms. As they do so, they compress the material between the holes. Quite different effects, however, can be expected for a silica draw, where the system is typically dominated by surface tension, and hence hole collapse. In this case, there will also be an elongation of the holes, but in the perpendicular direction.

It is very clear from these results that the presence of other holes can exert a dramatic influence on the type of hole deformation that occurs if they are placed in close proximity to one another. The deformation behavior of a hole is determined by the force balance around the hole, which is, in turn, determined by the relative size variations of the hole as well as all its neighboring holes. Compressive force around a hole will be produced when hole expansion is dominant; while a stretching force around a hole will be produced when hole collapse is dominant. Therefore, in principle, hole structure can be manipulated by pressurizing a collapsing hole and depressurizing an expanding hole. However, in practice, it is difficult to apply the technique. With pressurization, the size of the hole increases exponentially with the pressure applied, so it is certainly effective in keeping holes open. On the other hand, the holes may explode due to over pressurization, and it is easy to introduce asymmetries in the fiber, particularly if the seals on the holes are not perfect.

Like surface tension, pressure has a strong dependence on the area of the holes and varies along the axial position (due to temperature gradient along the draw direction, which causes a variation in viscosity). Also, it is much more effective at keeping large holes open than small ones. Indeed, it is conceivable that the expansion of larger holes could squeeze neighboring small holes and result in smaller holes being deformed and closed. If the whole perform is to be pressurized, this implies that the larger holes should be made relatively smaller in the perform, with the expectation that they will expand to the correct dimensions. Alternatively, it is possible to pressurize only some of the holes.

The relationship between the perform and fiber structures is not a simple one, and depends on many factors. It is, however, both possible to develop rules of thumb based on the material and draw parameters, and to do accurate, though demanding, numerical modeling. The relationships between the material and process parameters described here are of course not a substitute to experimental testing, but they do suggest which may be the most productive directions to pursue. In the long run, it will be feasible to run these numerical simulations backwards to reverse engineer the appropriate perform for a particular fiber. Such a process is particularly powerful when coupled to a perform fabrication technique that can produce arbitrary structures.

14.3 Modeling of Microstructured Fibers

As with other optical fibers, modeling MOFs involves solving Maxwell's equations for the appropriate fiber geometry, and with correct boundary conditions. This process is carried out numerically, and yields the modes of the fiber, from which optical properties such as the birefringence or non-linearity can be deduced. Ray optics, while useful for highly multimode conventional fibers, have never been used for MOFs because the additional complexities they introduce does not make this a meaningful approach. A wide variety of modeling techniques have been applied to MOFs, but each has advantages and disadvantages that make it hard to identify any as a generically best technique. Depending on the application of interest, the reliability of the technique, its ability to calculate a particular function, the ease with which it can be automated, and robustness may motivate the choice of a particular technique. This chapter does not aim to do more than give an overview of some of the important issues and approaches to the problem. More complete reviews of this area can be found elsewhere (Peyrilloux et al. 2002, Bjarklev et al. 2003).

Some useful insights can be obtained by simply modeling MOFs as a kind of traditional optical fiber, in which the presence of holes was used to modify the refractive index of the cladding. Thus, the fiber shown in **Figure 14.1** could be considered a step index fiber. This suggests that the properties of MOFs could be modeled by considering them as conventional fibers, with the cladding index being determined by an appropriate homogenization of the high and low index materials, and indeed this was the approach used in the initial publications (Birks et al. 1997). The "effective index" approach can yield important optical information about the behavior of the fibers. For fibers using the conventional hexagonal hole structure, recent work has identified parameters equivalent to the well-known fiber parameters V (normalized frequency) and W (normalized transverse attenuation constant) (Mortensen et al. 2003, Nielsen et al. 2003, Saitoh et al. 2005) and effective area A (Mortensen 2002). Such approaches are useful for gaining insights into how MOFs can be designed, and how they compare to conventional fibers. However, by definition, they cannot be used to determine the most interesting properties of MOFs, in which they exhibit very different behavior to that of conventional fibers. These properties relate, for example, to polarization, dispersion, and loss.

Most conventional fibers use a refractive index contrast between core and cladding which is relatively small. In silica step index fibers the refractive index of core and cladding are about 1.48 and 1.46 respectively. In polymer fibers the contrast is somewhat larger, with the values being of order 1.49 and 1.41. In both cases, though, the values are small enough to allow the "weak guidance" approximation to be used (Gloge 1971). This approximation, perhaps more accurately a "low contrast" approximation, allows a considerable simplification to be made in the modeling. In the most general case, solving Maxwell's equations requires that six equations be solved, respectively relating to the three components of E and H. The "weak guidance" approximation reduces this to a single equation: the scalar wave equation (Snyder et al. 1983). While again, some features of MOFs can be deduced from this approach, it is not strictly appropriate because the refractive index contrast between air and silica or polymer is so large that the weak guidance approximation is not valid.

The most important property that is affected by this is the polarization, which is strongly affected by the nature of the interface. Thus, if the polarization properties of the modes are required, a vector rather than scalar treatment has to be used.

Fortunately, this is not quite six times harder than the scalar case, as might be supposed. In the fiber geometry, it means that two equations rather than one need to be solved.

Another difference to conventional fibers is in the dispersion. Clearly, the hole structure is inherently dispersive because the effects of hole size are strongly dependent on wavelength. This allows the dispersion properties of MOFs to be varied in ways that are well beyond what is possible in conventional fibers, but also necessitates a correct modeling of the hole structure if the dispersion is to be correctly predicted.

The subtlest of problem in modeling MOFs, however, relates to the fact that modes are not confined in the sense that they are in traditional fibers. Two types of leakage can occur in microstructured fibers. The light can tunnel through the air holes, in a process akin to quantum mechanical tunneling, or it can be lost through the bridges between the holes. This means that the propagation constant β of the modes in the fiber is a complex rather than a real quantity, as it is in conventional fibers. Indeed, this leakage is responsible for some of the unique properties of MOFs, because the degree to which a mode is confined depends on how its geometry compares to that of the bridges. In other words, the hole structure can act as a "sieve" confining modes which are too large to pass through the bridges far more than those with smaller lobes.

The leakage of light through the microstructure is shown schematically in **Figure 14.19**.

For all modes, confinement loss will depend on the number of rings of holes and the air fraction of each ring. Modeling this confinement loss has proved to be one of the most difficult aspects of modeling MOFS. Unless this "leakage" of light through bridges is properly

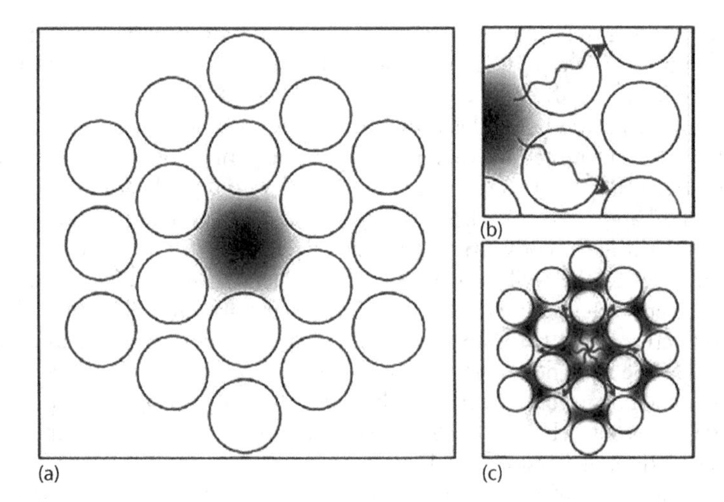

FIGURE 14.19 The fundamental mode in a MOF is well confined by the hole structure (a), though some light tunnel through the air holes (b), and through the bridges between the holes (c). The latter process strongly affects the confinement of higher order modes, whose smaller features are less confined by the hole structure. (Courtesy of Steven Manos, Optical Fibre Technology Centre, University of Sydney.)

considered, some features of modeling will be incorrect. In particular, if not properly accounted for, the light lost through leakage is likely to reappear in the calculations as spurious reflections. To a large extent, the diversity of methods available for modeling MOFs reflects the variety of ways in which this problem is treated.

Some of these approaches are summarized in **Table 14.1**. A number of techniques add an "artificial" layer beyond the hole structure to emulate the leaky nature of the microstructure. This layer can variously be absorbing, transparent, or "perfectly matched." In the latter case the

Table 14.1 Approaches Used to Address the "Leaky" Boundary Conditions in MOFs

Implementation	Advantages	Disadvantages
Absorbing layer	Automated mode finding. Efficient for multi-mode fibers.	Suffers from spurious modes. Requires very much larger computational domain to minimize back reflection.
Transparent boundary conditions	Highly automated mode finding.	Requires iteration for each mode. Approximate.
Perfectly matched layer	Automated mode finding. Efficient for multi-mode fibers.	Requires skillful choice of parameters for each application. Can suffer from spurious modes.
Multipole expansion	Most efficient for circular holes. No artificial boundary required.	Slow (sometimes manual) mode searches in the complex plane. Hole shapes currently restricted.
Adjustable boundary conditions	Highly automated mode finding.	Requires iteration for each mode.

Source: Table courtesy of Nader Issa.

artificial layer absorbs all the incident light without giving back reflection. These artificial layer methods can work very well, but require skill in their application.

The approach to the boundary conditions is of course only a part of the formalism needed for modeling. Boundary conditions need to be used in the context of a generic modeling technique such as finite element or plane wave decomposition methods. These methods have been well developed for modeling conventional fibers and in some cases are available as commercial software (see, for instance, RSOFT: https://www.synopsys.com/optical-solutions/rsoft.html and FEMLAB: https://www.comsol.com/products).

Finite element methods use a numerical approach to the problem which involves dividing the computational domain (a physical region which includes all of the microstructure) into uniform sub-regions that are sufficiently small that the properties are uniform within the element. Maxwell's equations are then applied to each element, with continuity conditions applied at each boundary. This is extremely powerful in that it allows the modal properties of arbitrary structures to be calculated, but is relatively computationally intensive. An issue that can arise in finite element approaches is the role of the symmetry of the mesh used to divide up the computational domain. Clearly, the results should be independent of the type of mesh, but in some cases the relationship of the mesh and fiber symmetry can cause anomalous results. An implementation of the finite element approach is commercially available (https://www.comsol.com/products) and widely used.

Another class of techniques expresses the solution to Maxwell's equation as a superposition of known functions. A particularly widely used approach is the plane wave decomposition (Broeng et al. 1999). Although this method can be used for arbitrary hole structures, it is most appropriate for truly periodic structures, where Bloch theory allows the solution of the wave equation to be expressed as a plane wave modulated by the periodicity of the crystal. Incorporating non-periodic features, which in the fiber case necessarily includes the core, requires a "supercell" to be defined, which includes the core, and is periodically repeated. A large number of terms are required to accurately represent these features, making the method computationally intensive. It has, however, been widely used by the MOF community.

Another technique, which is commercially available, is the beam propagation technique (www.rsoftdesign.com). This approach was initially intended to model complex beam trajectories, not those of fibers, where there is no variation in structure along the propagation direction. The features of the structure through which the light is traveling are divided into thin discrete layers, and the effects of diffraction and refraction are alternately considered. This is, again, a computationally intensive approach, which is probably most suited to fiber devices or tapers, where the features vary along the length.

Another approach that has been used is the multipole technique (White et al. 2001b). The multipole method is similar to other expansion methods but uses many expansions, one based on each of the holes of the structure. Modes are found by combining the expansions of each element and adjusting the expansion coefficients to meet the boundary conditions. The nature of the expansion does not require the artificial periodicity of the other techniques and so can allow confinement losses to be calculated. As microstructures are inherently leaky (although increasing the number of rings can make the losses arbitrarily small), this is a real requirement.

Computationally this can be an extremely efficient method, and is the most analytically rigorous of all the methods described. Being a more analytical approach than some of the others, it can sometimes yield more profound physical insights, allowing, for example, the effect of a single feature to be isolated from those of the environment, or studying the effects of finite periodic structures.

One method that has been developed specifically to allow arbitrary structures to be modeled is the adjustable boundary condition method (Poladian et al. 2002, Issa et al. 2003a, 2005), in which an iterative regime is used to match the fields inside and outside the computational domain. Complex hole structures can be "constructed" from annular sectors, and if a large number of sectors are used, the desired hole structures can be closely matched. The final calculation uses the fact that the sectors each share a common origin, and the expansion is done in terms of functions chosen for their integrability. For reasons of computational efficiency, the method uses a hybrid approach to solving the differential equations in the waveguide cross-section. A Fourier decomposition method is used in the angular direction, while a finite difference method is used in the radial direction.

With all approaches, particularly numerical ones, appropriate care needs to be taken to ensure that the results are free from artifacts such as resolution errors. The convergence of the solutions should preferably be checked at several resolutions, particularly when the confinement loss is low. Another issue that needs to be approached with care is correct mode tracking—where modes have very similar propagation constants, it is easy to inadvertently "track" the wrong mode, or switch between modes. Spurious modes can also be problematic in some cases. For these reasons, repeated checking of the results is advisable, including checking different methods against each other.

14.4 Effective Index Guiding Microstructured Fibers

Effective index guiding fibers are the most easily understood microstructured fibers because the way they guide light is essentially the same as that of a conventional fiber. The microstructure is simply used as a kind of "doping" to change the effective refractive index of the cladding, and total internal reflection acts to confine the light within the higher refractive index core. This simple explanation, however, belies the unusual scope of the properties and applications they have enabled. Indeed, it is not feasible in this chapter to give a comprehensive review of applications. Rather, what follows is a discussion of some examples, which demonstrate how the use of a microstructured fiber has redefined what is possible in these areas. It is hoped that this combination of applications and physical principles will at least give a flavor of this rapidly expanding field.

14.4.1 Single Mode Fibers

The first MOF ever produced (Knight et al. 1996) illustrated both similarities and differences with conventional fiber. The fiber, shown in **Figure 14.1**, had a relatively low air fraction, and, as would be the case with a conventional step index fiber with the same core/cladding refractive indices, was single mode. What was less obvious was that it is was endlessly single mode; in other words, there was no cut-off wavelength at which it became multi-mode. The single-mode property can be understood in several ways. In the last section we saw that the microstructure acts as a "sieve," which only effectively confines the fundamental mode (see **Figure 14.19**). But it is also possible just to see the fiber as a conventional fiber with a core/cladding index contrast and core size that only permits a single mode. The number of modes an optical fiber supports can be related to the V parameter (the normalized frequency) given by:

$$V = \left(\frac{2\pi a}{\lambda}\right)\left(n_{\text{core}}^2 - n_{\text{cladding}}^2\right)^{1/2}$$

where a is the core radius, λ is the free space wavelength, and n_{core} and n_{cladding} are the core and cladding refractive indices respectively. If V is less than 2.405 the fiber is single-mode. In conventional fibers, the dependence of V on wavelength also defines the cut-off wavelength—when the wavelength is sufficiently short, an additional mode will "fit" in the core and the fiber will become multi-mode. In MOFs, however, this need not happen. The cut-off condition may never be reached. In both conventional fibers and MOFs, diffraction causes the modes at longer wavelengths to be less tightly confined in the core than they are at short wavelengths. Thus, at longer wavelengths, modes "see" more of the cladding, and hence experience a lower effective index. In MOFs, however, the situation is complicated by the fact that the cladding in MOFs

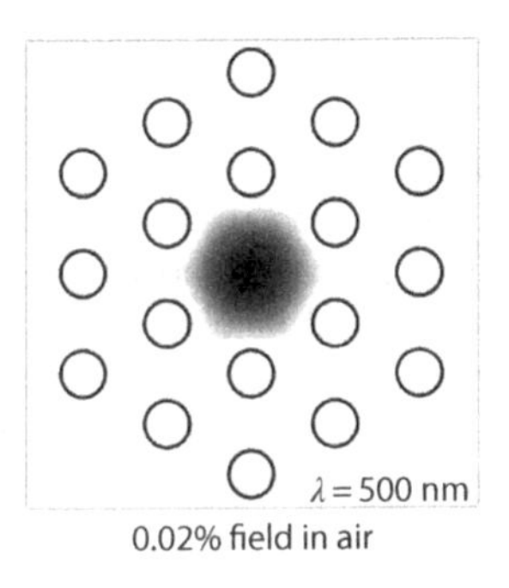

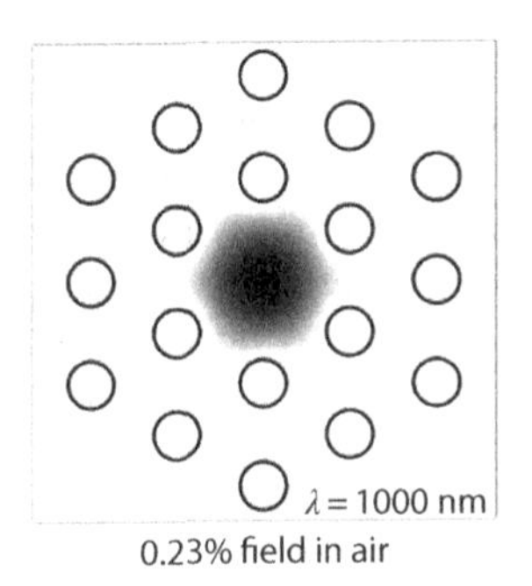

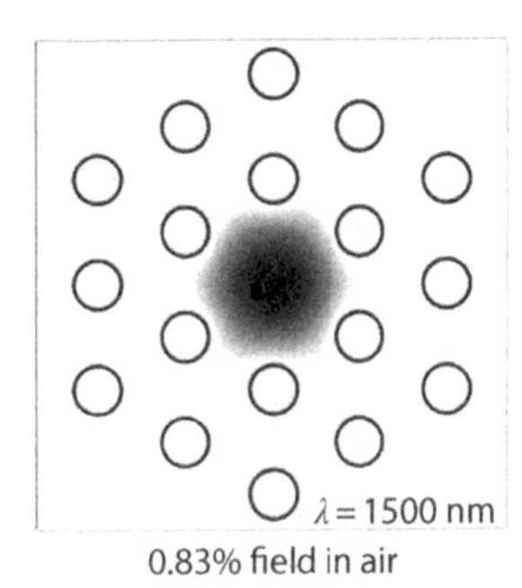

FIGURE 14.20 Microstructured fibers can be made "endlessly single mode." This is possible because of the unique dispersion properties of the fiber. The effective index of the cladding is reduced at longer wavelengths due to the field increasingly occupying the air regions.

is intrinsically highly dispersive, not just because the of the relationship of hole sizes to wavelength, but also because of the varying degrees to which a mode can occupy the high and low index regions of the cladding. The effect of this is to make $n_{cladding}$ dependent on wavelength, counteracting the dependence on wavelength in the V parameter so that the cut-off condition may never be reached (**Figure 14.20**).

Obtaining endlessly single mode behavior requires the appropriate choice of hole structure. For the most common, triangular lattice MOFs (shown, for example, in **Figures 14.1** and **14.3**), the condition for can be related to diameter of the holes d, and their spacing, Λ. When $d/\Lambda < 0.406$ the fiber will be single-mode for all core sizes (Kuhlmey et al. 2002). As implied by the V parameter, fibers with higher air fractions may also be single-mode, if the core is sufficiently small. In still other cases, the fiber may be considered "effectively" single-mode; higher order modes have a sufficiently high loss. The fiber may then be considered single-mode over lengths that are long enough for the higher order modes to have decayed away (Argyros 2002). The "endlessly single mode" behavior is one of the most attractive features of MOFs, and has been widely exploited. In the future it may be even more significant in telecommunication networks, a prospect that looks more likely as the loss of these MOFs drops to levels that are competitive with the best conventional fibers. Single-mode fibers with large core areas were highlighted initially as an important benefit of MOFs (Knight et al. 1999), though it has subsequently become clear that large core single mode fibers are extremely sensitive to bend losses.

In conventional polymer optical fibers, making low loss single mode fibers has always been problematic, limiting their uses for many applications, particularly sensing, for which they would otherwise be well suited. Thus, the development mPOFs (microstructured polymer optical fiber) has the potential to redefine POF applications.

14.4.2 Multimode Fibers

While the telecom demands on silica fibers are almost exclusively for single mode fibers, there are several important applications for which large core, multi-mode fibers are likely to be required. These typically involve short distance, high data transmission rate communications such as local area networks (LAN), fiber to the home, and even chip-to-chip communication. In such cases the key performance indicators are the ease with which the fibers can be connected and the bandwidth. The most important aspect for connectivity is the size of the core. Single mode fibers are difficult and expensive to join because their cores are very small and need to be aligned to a precision of less than 1 micron. Fibers with much larger cores are difficult to make in glass, while polymer fibers can easily be made with cores 1 mm in diameter while still being flexible. For this reason, most people believe that short distance, high-speed connections will probably be made using polymer optical fibers.

Large core fibers, however, are multi-mode and can suffer badly from modal dispersion, which causes signals to become "smeared," limiting the rate at which data can be transmitted without ambiguity. This effect can be greatly reduced by using the appropriate refractive index profile across the fiber. Using conventional polymer fiber technology, it is difficult to control the refractive index profile of a fiber, as there is no generic technique, analogous to chemical vapor deposition in silica fibers, for producing the profile of choice. Microstructured polymer optical fibers (mPOFs) have provided one possible solution to this problem. Using a microstructure, it is relatively easy to produce a refractive index profile of choice (**Figure 14.21**). These graded index fibers have so far been shown to have bandwidth properties competitive with other fibers (van Eijkelenborg 2004c), with the most recent experimental results (Large et al. 2005) indicating bandwidths of 4.4 Gbits/s over 100 m.

These results are, in some ways, very surprising, given the very large refractive index contrast used. The fractional index change Δ is given by:

$$\Delta = \left(n_1 - n_2\right) / n_1$$

where n_1 and n_2 are the maximum (i.e. the core) and minimum refractive indices respectively. In conventional graded index fibers, this strongly determines the modal pulse spread in the fiber (Palais 1992):

$$\Delta\left(\tau / L\right) = \frac{n_1 \Delta^2}{2c}$$

The value of Δ in the graded index mPOF is nearly an order of magnitude larger than those used in conventional fibers, yet the experimentally observed bandwidths show that this has not caused a correspondingly large pulse spreading.

A number of issues may contribute to the unusual behavior of the graded index MOF. Conventional fibers are circularly symmetric and have a well-known set of modal degeneracies that cause the coupling between higher order modes to be stronger than between lower order modes. MOFs are not circularly symmetric, so that these degeneracies no longer exist.

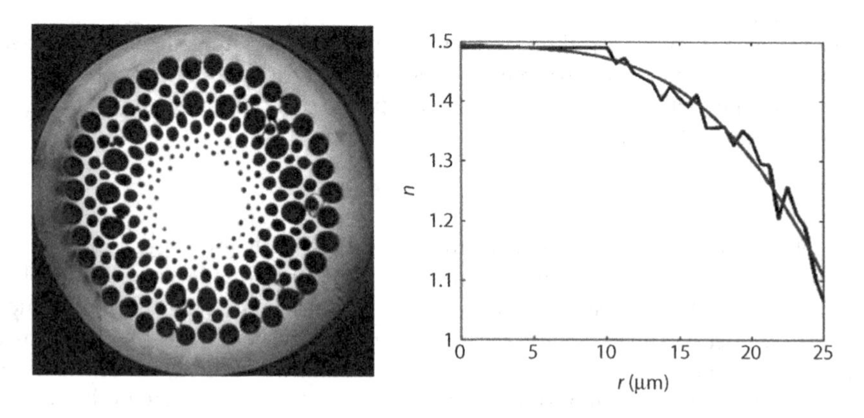

FIGURE 14.21 A graded index MOF fabricated in polymer, and its effective refractive index profile. The desired profile, in this case a quadratic change in refractive index, is shown in red. (Courtesy of the Optical Fibre Technology Centre, University of Sydney.)

A different pattern of near-degeneracies or mode distributions will change the pattern of coupling between high order modes. In fact, it may be possible to design MOFs to produce specific patterns of coupling.

Another factor that may be significant in determining the bandwidth performance is fact that the differential mode attenuation in MOFs is far larger than in conventional fibers, and may vary by many orders of magnitude. This could have the effect skewing of the stable mode distribution strongly toward the lower order modes (Barton et al. 2003). Thus, it seems likely that the graded index MOF may achieve its high bandwidth performance by favoring transmission within a relatively narrow range of low order modes, rather than the very wide range of modes that in principle could be supported by the fiber.

Clearly, more work needs to be done to understand these issues, but it is already clear that despite the similarities to conventional fibers, this is yet another application where the behavior of MOFs falls outside our usual intuition.

14.4.3 High Numerical Aperture Fibers

One area where MOFs should provide an obvious advantage over conventional fibers is those applications which require efficient light capture. For a conventional step index fiber, the light capturing ability of a fiber, or numerical aperture (*NA*), is determined solely by the refractive index contrast between core and cladding:

$$NA = \left(n_{\text{core}}^2 - n_{\text{cladding}}^2\right)^{1/2} = \sin\alpha$$

where α is the acceptance angle of the fiber. Clearly, MOFs, which allow the cladding index to be very close to 1, offer the best possible prospects for making very high *NA* fibers. Such fibers, known as "air-clad" fibers, have indeed been produced (Bouwmans et al. 2003, Limpert et al. 2003; Wadsworth et al. 2004b), and have proved invaluable, particularly for fiber laser applications for which efficient use of the pump light is critical. There are a variety of other applications in fields as diverse as spectroscopy, astronomy (Mediavilla et al. 1998), and the detection of charged particle and ionizing radiation (Achenbach et al. 2003) which also require high *NA* fibers.

The complications arise because of course MOFs are not actually "air-clad," but have bridges which suspend the core. To understand how they work and predict their properties, we need to be able to relate the numerical aperture to the geometrical parameters of the fiber such as the bridge thicknesses, lengths, and the number of holes. These parameters are identified in **Figure 14.22**.

The first studies relating the numerical aperture to these parameters in multimode fibers (Bouwmans et al. 2003, Wadsworth et al. 2004b) modeled the bridges as slab wave-guides that could couple light out of the core when their effective indices were close. Thus, the effective index of the fundamental bridge mode, rather than the cladding mode as previously, provided the cut-off condition for guidance in the core:

$$\overline{NA} = \left(n_{\text{core}}^2 - n_{\text{bridge}}^2\right)^{1/2}$$

This relatively simple approach provided good agreement with experiment, but did not provide any insights about how the number of bridges, bridge length, length of the fiber, or the presence of additional rings impacted on the *NA*. These questions were covered in an analysis by Issa (2004c). The results indicated that the major parameter determining the *NA* was the scaled bridge thickness, with a weak dependence on the number of bridges. The length of the fiber also plays a minor role, because the "leaky" nature of MOFs means that some modes will be effectively lost over sufficiently long lengths (**Figure 14.23**).

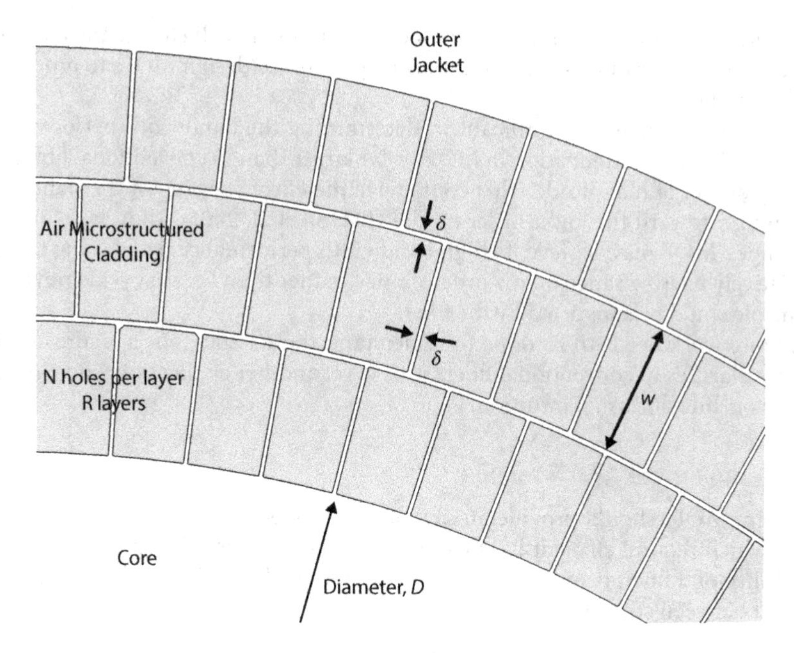

FIGURE 14.22 A schematic of an "air-clad" MOF, identifying the key parameters. (From Issa, *Opt. Lett.*, 29(12), 1336–1338, 2004a.)

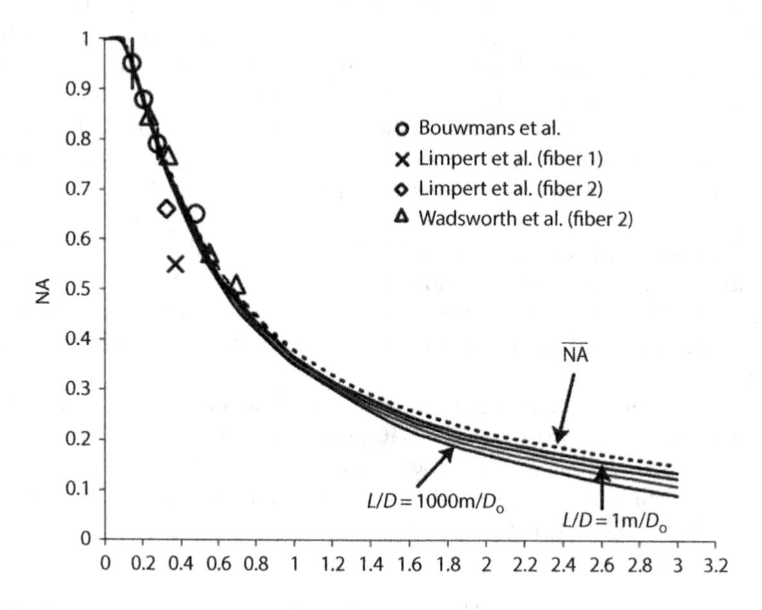

FIGURE 14.23 The calculated dependence of the numerical aperture on the scaled bridge thickness, δ/λ. The results are shown for four different fiber lengths (1 m, 10 m, 100 m, and 1000 m) for a core diameter D_0 of 150 microns. Also shown is the NA curve obtained using the fundamental bridge mode and agreement with the experimental values taken from literature. (From Issa, *Opt. Lett.*, 29(12), 1336–1338, 2004a.)

The effect of additional layers of microstructure proved to be more significant. The efficiency of coupling between the core and bridge modes depends not only on close matching of the effective indices of the modes, but also on the matching of their polarizations. When an additional ring of holes is added, a series of "T" intersections are formed (as shown in **Figure 14.23**). Like the radial bridges, these intersections can be considered as slab waveguides, but, critically, with a perpendicular orientation. Thus, light that has coupled out of the core into the radial bridge will be unlikely to be coupled out further into the "T" bridge, as the polarizations of the bridge modes mirror their perpendicular orientation. Light that couples effectively to the radial bridges will have the wrong polarization to couple to the "T" bridges. At a practical level, the addition of extra rings can give an improvement in NA of up to 20%, and also allows the bridge thicknesses to be greatly increased, while maintaining very high NAs. A caveat that should be added is that deviations from the assumed rectangular bridge shape can reduce the NA (Lwin 2005b).

14.4.4 Highly Birefringent Fibers

Highly birefringent [HiBi] or polarization maintaining fibers are needed for many applications, precisely because a small degree of birefringence is easily introduced accidentally. Slight asymmetries in the fabrication process, such as the core not being perfectly circular or even non-uniform heating of the fiber in the ground, can cause a measurable degree of birefringence. This is problematic because the fundamental mode of optical fibers is polarization degenerate, and will readily split into two modes. HiBi fibers are widely used to introduce polarization stability, by separating the effective indices of the modes to a degree where they do not interact. They are also widely used in applications such as interferometric and polarimetric sensors.

In conventional fibers, a high degree of birefringence is introduced by using material stress. The most common "PANDA" or "Bow tie" fibers use inclusions outside the guiding region with slightly different thermal properties to the glass, so that when the fiber cools after drawing a high degree of stress is frozen in. These fibers produce a birefringence that is largely independent of wavelength, but strongly temperature sensitive. The birefringence B is defined as the difference between the effective indices in the x and y directions. It is sometimes quoted in terms of the beat length L_B, the length for the two modes to differ in phase by 2π, resulting in the polarization state identical to that of the input:

$$B(\lambda) = n_x - n_y = \frac{\lambda}{L_B(\lambda)}$$

The values of B for conventional HiBi fibers vary from about 1×10^{-4} (Corning PMF28) to 6×10^{-4} (Fibercore HB800).

MOFs were identified early on as potential candidates for HiBi fibers, because of their unparalleled ability to exploit waveguide or "form" birefringence. They can readily be made with structures for which the x and y orientations are not equivalent, for example by having a non-circular core, and their large index contrast between air and glass or polymer ensures that the interface interactions will have a strong influence on the polarization properties. This was one of the first applications of MOFs investigated in detail, with the world-record birefringence being claimed by a MOF in 2000 (Ortigosa-Blanch 2000) for a fiber with a beat length of 0.4 mm, and a birefringence of 3.7×10^{-3} at 1540 nm. There have since been wide range MOFs for high birefringence either made or investigated. A fabricated example with elliptical holes and core

FIGURE 14.24 MOFs can be made highly birefringent by using holes structures that differ in the *x* and *y* directions. In this case, the holes and the core of a conventional triangular array have been made elliptical. This has been done by the inclusion of large external holes, which distort the structure. (From Issa, *Opt. Lett.*, 29(12), 1336–1338, 2004a.)

is shown in **Figure 14.24**. In general, the form birefringence of MOFs is larger than the stress birefringence of conventional HiBi fibers, but has other slightly different properties. The birefringence is strongly dependent on wavelength, but very weakly dependent on temperature. One novel feature of MOFs has been exploited to make the birefringence tunable. By selectively filling some of the holes with polymer, a high degree of birefringence can be introduced, with tenability being provided by temperature-induced changes to the refractive index of the polymer (Kerbage et al. 2002).

14.4.5 Non-linear Fibers

Most of the applications discussed so far do not strongly depend on the material used to make the fiber, provided that it is transparent enough to allow transmission. But material properties do play a critical role in some optical processes. When light interacts with matter, the electric field of the light induces a polarization in the material, resulting in the generation of another field—that of the light traveling in the material. At low intensities the polarization induced in the material is linearly dependent on the incident electric field, but at higher intensities the polarization can become non-linear. Physically, this is associated with multiple photon interactions, or the incident light interacting with an external electric field. These interactions transiently change the optical properties of the material, causing processes such as second harmonic generation, in which the energy of two identical photons is combined to form a photon of twice the energy, stimulated scattering processes such as Raman and Brillouin scattering and pulse compression, due to an intensity dependent change in the refractive index.

In fact, non-linear optical effects occur in all optical systems, but without very high intensities they are generally too weak to be observed. In optical fibers they are particularly important because their low loss and long interaction lengths make it easy to observe small effects. In transmission fibers non-linear effects are generally unwanted, but there are a number of applications for which they are very desirable. Such applications include the transmission of solitons, pulse shaping, wavelength conversion, optical, data regeneration, optical demultiplexing, and Raman amplification.

The non-linearity of an optical fiber is characterized by its effective nonlinearity γ:

$$\gamma = \frac{2\pi n_2}{\lambda A_{eff}}$$

where n_2 is the nonlinear refractive index of the material and A_{eff} is the effective mode area. The simplest way of controlling non-linear effects is therefore by changing the confinement in the core. By increasing the air fraction in the cladding and using very thin bridges, MOFs can ensure that light is very tightly confined in the core. If the core is made very small, this can cause a dramatic enhancement of the fiber non-linearity. Fibers with tiny cores have been made in pure silica with nonlinearities some 50x that of a standard telecommunications fiber (Monro et al. 2003). Much larger effects can be obtained when a material with a higher n_2 is used to make the fiber such as chalcogenides, lead silicates, and heavy metal oxides such as tellurite.

The world record nonlinearity in an optical fiber is now held by a lead silicate glass MOF, with a value of γ nearly 2000x that of a standard communications fiber (Petropoulos et al. 2005).

High confinement is, however, not the only advantage that MOFs bring to this field. They also bring unprecedented control of the dispersion properties. For example, it is straightforward to engineer the position of the zero group velocity dispersion so that it coincides with a particular wavelength. Short pulses at this wavelength can then generate strong non-linear effects because of the phase-matching of different frequencies. One of the most important applications of MOFs to date has used exactly this effect to produce what is in effect a white light laser: a cascading series of nonlinear effects results in a dramatic spectral broadening known as "supercontinuum generation."

Supercontinuum generation is not a new phenomenon, having been first observed in 1970 (Alfano et al. 1970), but MOFs have proved to be particularly effective in exploiting it because of their unusual dispersion properties. After the discovery of supercontinuum generation in MOFs (Ranka et al. 2000), considerable research interest focused on optimizing the experimental parameters to the point where supercontinuum sources have become a readily available, and invaluable, research tool that can easily be assembled in a lab (Wadsworth et al. 2004a). The output of one such system is shown in **Figure 14.25**, together with the fiber used. While non-linear optics offer many attractive possibilities for using MOFs, there is no doubt that supercontinuum generation currently represents one of its so many applications.

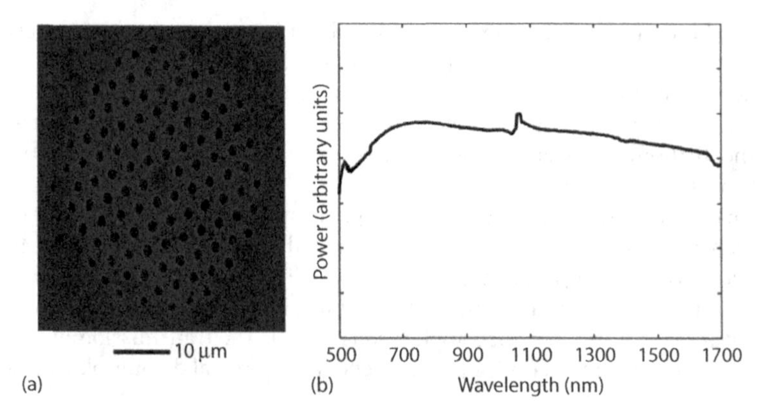

FIGURE 14.25 An electron micrograph of fiber used for supercontinuum generation (a) and the output of the fiber (b). The exact details of the experiment are given by Wadsworth et al. (2004a). This fiber was made by Blaze Photonics Ltd. (Courtesy of Tim Birks.)

14.5 Photonic Bandgap Guiding Fibers

14.5.1 The Mechanism of Photonic Bandgap Guidance

While effective index guiding in MOFs are "similar but different" to conventional fibers, bandgap guiding fibers are just different, and require us to re-think many of our perceptions about how fibers behave. The key practical distinction between effective index guiding and photonic bandgap guiding is that the latter allows light to be guided in low index materials such as air, something that has only been previously possible using metal-coated hollow waveguides.

In fact, though we are accustomed to thinking that it is impossible to guide light in air without using a microstructure, some limited guidance in air is possible even in a simple capillary. Just as it is possible to "skim" stones along the surface of the water, and to see a reflection in a transparent slide if viewed at the correct angle, so it is possible for light at glancing incidence to be guided by a transparent capillary. At these angles the Fresnel reflection coefficient is close to 1, allowing modes to propagate in a hollow core. These modes are of course very leaky, and do not have the discrete spectral features associated with bandgap guidance (Issa et al. 2003b).

The modes of a capillary are very leaky because the cladding of the capillary supports a continuum of modes to which the light can couple and escape the core. Introducing microstructure into the cladding breaks the continuum into discrete bands. The bands can be separated and regions between the bands contain no modes—these are band gaps. If light in the core of a fiber has a wavevector that corresponds to a bandgap, there are no cladding modes it can couple to and it remains guided in the core. Outside a bandgap the wavevector will correspond to a mode of the structure and so can couple to it and leak out of the core.

The exact properties of the fiber depend on the microstructure, which is periodic in either one or two dimensions. These are respectively known as "Bragg fibers" and photonic bandgap fibers, though in fact both types of fibers are bandgap fibers. Indeed, it is also possible to make a "Bragg" fiber using rings of holes (Argyros et al. 2003, 2005c, Vienne et al. 2004). These structures, illustrated together with the more conventional designs, are shown in **Figure 14.2**. The Bragg fiber case can be understood most easily. The periodic arrangement of high and low index regions reflects a range of wavelengths. This is simply a multilayer stack, whose properties are very well known and do not change when the stack is rolled into a cylindrical geometry to form the cladding region of a fiber. This cladding reflects exactly the same range of wavelengths

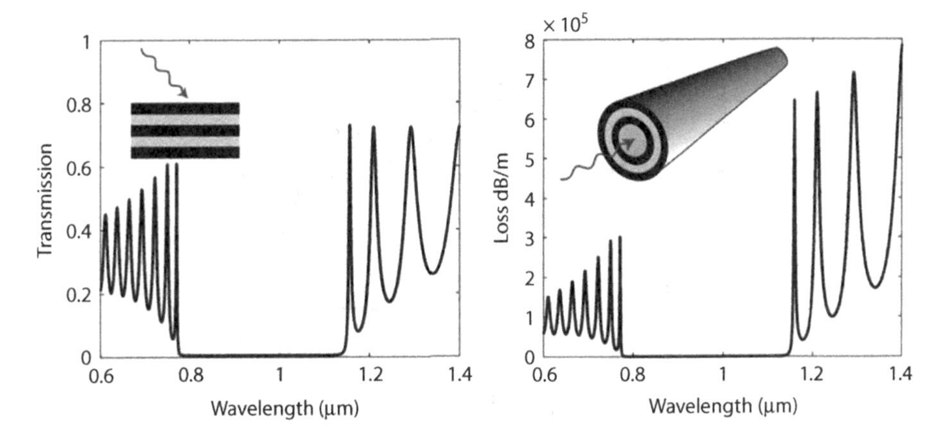

FIGURE 14.26 The transmission of a multilayer stack (left) depends on the periodicity of the stack and its refractive index contrast. Some wavelengths are strongly reflected and have almost no transmission through the stack. If the same stack is rolled into a tube (right), the wavelengths that are strongly reflected will be confined to the core and will propagate along its length. Note that the calculations used in this diagram were based on a 16 layer stack. (Courtesy of Steven Manos, Optical Fibre Technology Centre, University of Sydney.)

giving low loss transmission in the hollow core (Argyros 2002; **Figure 14.26**). Outside this bandgap region, however, light in the core can leak out through coupling to the modes of cladding structure.[1]

Two-dimensional photonic bandgap structures can be viewed as an array of tiny rods, suspended by small connecting bridges (**Figure 14.27**). Each rod acts like an optical fiber, which can support a number of modes defined by the V parameter. The presence of an array of rods broadens these discrete modes into bands (**Figure 14.28**). Light is well confined in the core region when the effective index of the modes is very different to that of the core mode, preventing any coupling between them which would cause light to leak into the cladding. Thus, the edges of the bandgaps correspond to the cut-off frequencies of the modes.

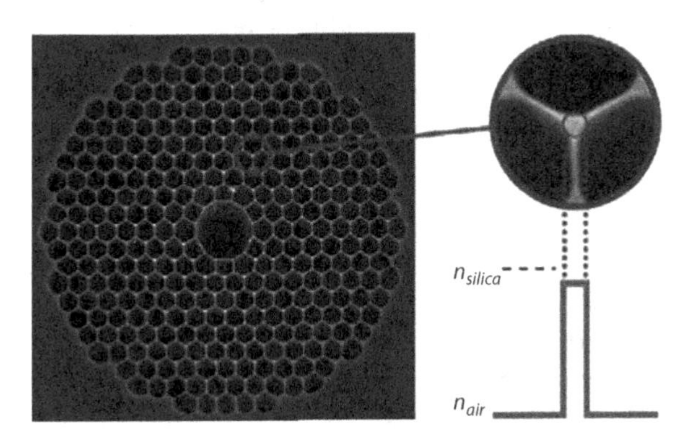

FIGURE 14.27 The cladding structure of a photonic bandgap fiber can be thought of as a two-dimensional array of tiny rods suspended in air, which can support modes. (Courtesy of Steven Manos, Optical Fibre Technology Centre, University of Sydney.)

[1] An animation illustrating this for a ring structured Bragg fiber can be downloaded at: http://www.opticsexpress.org/abstract.cfm?URI=OPEX-12-12-2688

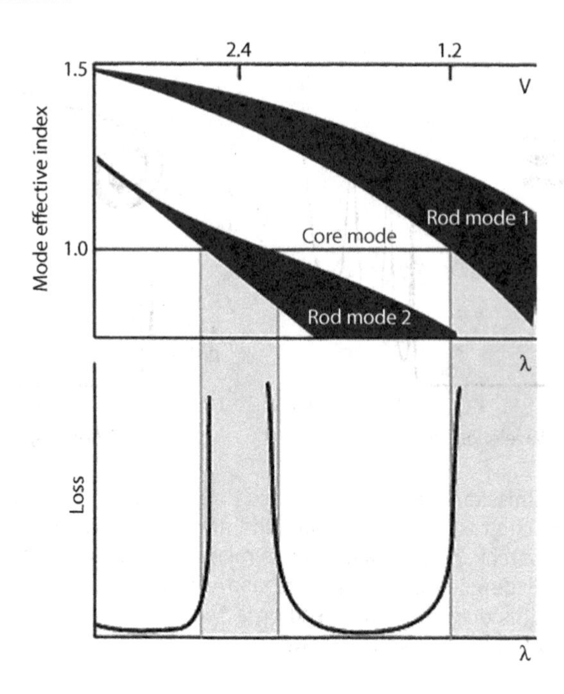

FIGURE 14.28 The presence of many linked guiding rods broadens the dispersion curves into bands. When the rod modes have very different effective indices to that of the core mode, they do not couple and the core mode can propagate with low loss. In the case shown, the core is air, corresponding to an effective index of 1. When the rods and the core modes have very similar effective indices (at the mode cutoffs) and some spatial overlap, they couple easily and the core mode becomes very lossy. (Based on Argyros et al. 2005c.)

Most photonic bandgap fibers are made using a very high index contrast (glass or polymer and air), but it is also possible to make photonic bandgap fibers with a very low index contrast (Argyros et al. 2004), for example by embedding an array of doped silica rods in a pure silica background. Such structures can readily be made by the capillary stacking process (see **Figure 14.7**). Indeed, fibers produced by this process are in some ways more "ideal" structures, because the high index rods are not joined by bridges. They also make it possible to clearly visualize the rod modes and core modes (**Figure 14.29**).

14.5.2 Issues Affecting the Performance of Bandgap Fibers

Photonic bandgap fibers have generated enormous interest since they were first proposed but have proved difficult to make, with very few groups in the world successfully achieving guidance in air. Understanding the physical processes underlying photonic bandgap guidance has been essential in elucidating why they are so hard to make. When these two-dimensional photonic bandgap structures were first fabricated there was considerable interest in the degree of periodicity that was required to allow hollow core guidance. Most believed that the structures needed to be very close to perfectly periodic in order to allow coherent scattering, and in experimental groups there was often a degree of bewilderment when extremely regular structures did not work.

The nature of this problem has become clearer with the realization that the bandgaps can be closely associated with the resonant behavior of the high index regions, cylinders for the Bragg

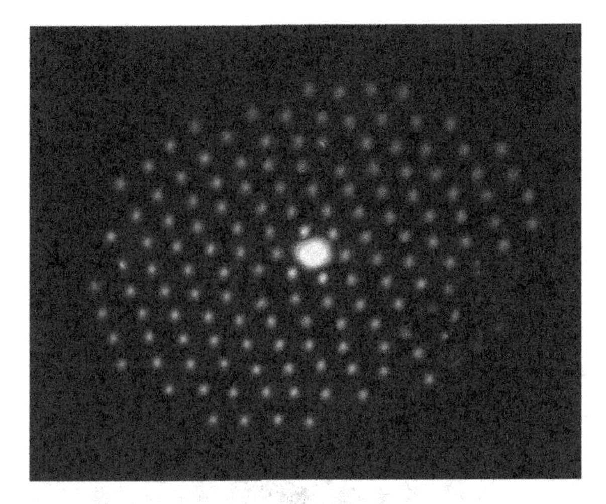

FIGURE 14.29 A low contrast photonic bandgap fiber produced using pure and doped silica (see Figure 14.7) illustrates the interaction of the rod and core modes. Although the intensity in the core has saturated the photograph, it is clear that the rods guide a different range of wavelengths to the core mode. The wavelengths of light in the core correspond to bandgaps in the cladding structure, formed by an array of rods. The edges of the bandgap correspond to the cut-off frequencies of the rod modes. (Courtesy of Alexander Argyros and the University of Bath.)

fibers, and rods for the 2-dimenisonal structures. These high index regions can be considered as small Fabry-Perot resonators (Abeeluck et al. 2002, Litchinitser et al. 2002), and destructive inference causes some wavelengths to be strongly reflected by them. These "anti-resonant" wavelengths define the bandgaps of the microstructure, so it is the regularity of these regions that is most significant in producing air guidance. By contrast, the exact spacing of the rods is much less important for short wavelengths, although it becomes more significant at longer wavelengths (Abeeluck et al. 2002). Experiments with low index bandgap fibers (Argyros et al. 2004) have clearly demonstrated that bandgap guidance can be achieved in fibers where the structure is quite deformed.

Much more important than the exact periodicity of the array of rods is the width of the bridges connecting them. These bridges determine the coupling between the rod modes, and hence the width of the bands in the cladding structure. When the bands broaden too much, the bandgaps become shrink and may even disappear. From a fabrication perspective, the need for very thin connecting bridges turns out to be the major challenge, particularly for silica fibers where the capillary number tends to promote hole collapse during the draw.

The other major challenge in making bandgap fibers is to control the interaction with surface modes (**Figure 14.30**). Modes are known to be associated with the termination of periodic structures, and occur even when the structure is "perfect." Perturbations to the perfect structure, such as a thickening of the ring around the core, may increase the number of surface modes. They are problematic in bandgap fibers because their location means that they overlap spatially with the core mode. If the effective index of the core mode is close to that of the surface mode, they will allow light to be coupled out of the core (**Figure 14.31**). Surface modes can also indirectly couple light to highly lossy radiation modes, which would normally not interact with the core due to their spatial separation (West et al. 2004).

Surface modes are considered to be the main loss limitation to bandgap fibers (Roberts et al. 2005). While it is impossible to remove surface modes, it is possible to modify the

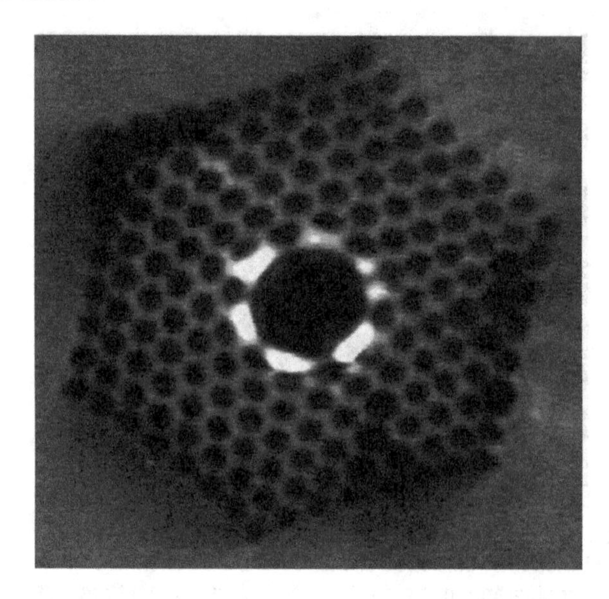

FIGURE 14.30 The region surrounding the core of a bandgap fiber can determine many of its properties. The solid region surrounding the core can support modes, as shown here, which can couple light out of the core. The impact of surface modes can be deduced by appropriate design of the region around the core. (Courtesy of the Optical Fiber Technology Centre, University of Sydney.)

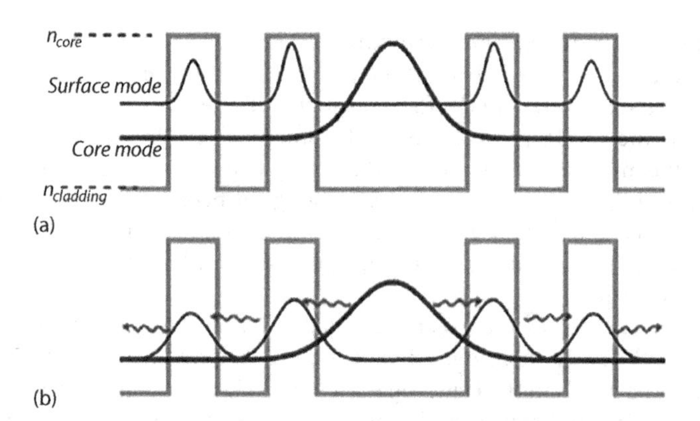

FIGURE 14.31 The high index regions around a low index core may support modes that have a strong impact on the fiber transmission. If the surface modes and core modes have very different effective indices (a) they will not interact. If however they have a similar effective index to the core modes, and a degree of spatial overlap due to their evanescent tails (b), they will cause coupling from the core to the cladding. Surface modes can also indirectly couple light to highly lossy radiation modes, which would normally not interact with the core due to their spatial separation. (Courtesy of Steven Manos, Optical Fiber Technology Centre, University of Sydney.)

design of the core/cladding interface to reduce their effects, and to ensure that they are outside the bandgap (West et al. 2004). The core/cladding interface shown in the preform in **Figure 14.11** is regarded as more optimal that the cladding termination formed by capillary stacking (see for example **Figure 14.3**) because material has effectively been removed from the interface.

14.5.3 Applications of Photonic Band-Gap Fibers

To some extent, the impact and application of bandgap fibers will be determined by their loss performance. By guiding light in air, these fibers offer what is in principle the perfect fiber. Air transmission of light is free of the limitations that occur in all other optical systems—air does not absorb light, it has non-linear response, and no dispersion. The decrease in loss in bandgap fibers that has occurred since their invention is impressive, but the figures are still considerably above those of conventional optical fibers. If the loss could be reduced below these levels, bandgap fibers could transform telecommunications. Even without this, however, bandgap fibers have opened up completely new areas. While silica has excellent transmission properties in the telecommunications window, other wavelengths of interest are absorbed. In fact, there are whole wavelength ranges for which it is impossible to produce low loss optical fibers. The mid-infrared and UV regions are both fall into this category. Hollow core guidance allows transmission within the absorption bands of the material, so it is possible that this will offer the best solution to many applications. Similarly, high intensity transmission is required in a number of industrial applications. Even when the absorption is very small, this can still cause problems if the intensity is sufficiently high, so guidance in air would be distinctly preferable. In a similar context, a particularly significant development has been the development of hollow core fibers in polymer (Argyros et al. 2005c). In many ways polymer is a more attractive material to work with than silica—it is cheaper, easier to process, more flexible, and lighter. But as an optical material, it has always been constrained by its higher absorption. Photonic bandgap guidance offers a possible solution to this.

Of course, transmission in air is not the only thing that is enabled by this technology. It has become possible to guide light in any low refractive index material, such as a gas or liquid. Some of the most exciting work being carried out in hollow core fibers recently has been in the area of gas based non-linear optics (Benabid et al. 2002b, 2005), where hollow, core fibers offer an unusual combination of long interaction lengths, good beam quality and high intensity transmission at low power that is close to ideal (Russell 2003). Other liquids too can be guided. From a sensing perspective, most samples of biological interest are aqueous, and maintaining a strong interaction between the solution and light has always been difficult. Again, bandgap fibers offer a possible solution (Fini 2004, Large et al. 2005).

14.5.3.1 A Final Thought...

Microstructured optical fibers are still a new field. The ultimate performance limits of these fibers are still unclear, and it remains to be seen how revolutionary they will be, either in a fundamental physics context or as commercial successes. Indeed, even the fundamental mechanisms of loss and transmission are still the subject of research. While this makes the field less defined than other areas of optics, that is in many ways their most appealing feature. It is a time when all things are still possible, and major contributions are still to be made.

14.5.3.2 Acknowledgments from Maryanne Large

Over the years that I have been working in microstructured fibers, I have learned much from my interaction with other members of the mPOF group at the University of Sydney, and it is a pleasure to acknowledge the contribution of the entire group to this work. I would also like to thank a number of people more specifically for their help in writing this chapter. My discussions with Alex Argryos and Leon Poladian clarified my understanding of many of the issues raised here, and their careful reading of the text improved it enormously. Steven Manos prepared a number of the images and was extremely patient in dealing with the numerous changes I requested. In Brazil, I would like to thank Cristiano Cordeiro for his comments on the text, and more generally the enthusiasm he brings to all discussions of this area. Finally, thank you to Marcelo Werneck, who kindly invited me to contribute the chapter, little realizing how slow I would be in delivering it!

I would like to dedicate this chapter to my long-term friend and colleague, Leon Poladian, who died in 2018.

14.5.3.3 Acknowledgments from Marcelo Werneck

The text above was prepared by Maryanne in 2005 as a chapter for a book I was writing about optical fibers. I had intended to publish it, but, for a number of reasons, it did not come to press. The manuscript was kept on a CD for more than 10 years, and I did not see Maryanne again in any POF conference. However, by fall 2018, when I was finishing this book, I found the CD buried into an archaeological layer in my office and I decided to include it in the book. For a piece of good luck, I was able to contact Maryanne and she agreed to my using the material for the present book. So I thank to her for writing this manuscript that is finally going to press, with my excuses for the 14 years she was probably waiting to see the book on a library shelf.

The rest of this chapter was completed by myself showing different kinds of sensors that apply microstructured optical fiber.

14.6 mPOF Sensors

14.6.1 Introduction

As it has been seen throughout this chapter, microstructured polymer optical fibers (mPOF) guide light using a pattern of holes running longitudinally along the fiber. These guiding mechanisms are identical to those of silica photonic crystal fibers (PCF); however, by using PMMA many advantages are achieved. The combination of PMMA and microstructure allows many sensor applications to be explored. One of these advantages is the possibility of producing single-mode (SM) mPOF for use in the visible spectrum. This capacity allows for instance the possibility of inscribing fiber Brag grating (FBG) and long period grating (LPG) in such fibers.

Even with a loss of ~1 dB/m in a core diameter of 8 μm an SM mPOF for a wavelength range of 500–700 nm can be advantageous. This is because the Young's modulus is significantly lower than that of silica, allowing much larger strains to be reached. The elastic limit of PMMA fibers is ~4%, and the breaking strain is around 25% as compared with 0.03 $\mu\varepsilon$ for the rupture limit of a standard optical fiber SMF-28 from Corning (Antunes et al. 2012). This rupture limit corresponds to about 0.68 kg for an SMF-28.

Although strain sensing using silica fibers is a well-established technology, polymer fibers would clearly be used in a much higher range as compared with silica.

It has been shown that an mPOF FBG presents a linear response of approximately −11 nm per % strain and even a strain as low as 0.05% can be resolved, presenting a shift of ~0.5 nm.

14.6.2 LPG in mPOF

Comparing FBG with LPG we first recall that in FBGs the resonance frequency is a function of the refractive index and the grating periodicity:

$$\lambda_B = 2n_{eff}\Lambda \quad (14.1)$$

where λ_B is the Bragg wavelength, η_{eff} is the effective refractive index of the fiber, and Λ is the periodicity of the grating (Werneck et al. 2017).

In long period grating (LPG), unlike FBGs in which counter-directional coupling occurs in the core, codirectional coupling is between a core and a cladding mode, and this fundamental physical difference presents different effects. Notice that **Equation 14.1** gives the FBG resonance frequency which depends on the core refractive index. On the other hand, the LPG resonance frequency is given by:

$$\left(n_{eff}^{\text{core}} - n_{eff}^{\text{clad}}\right)\Lambda_{LPG} = \lambda_{LPG} \quad (14.2)$$

Where n_{eff}^{core} and n_{eff}^{clad} are effective refractive indexes of the core and cladding modes, respectively, Λ is the *LPG* period, and λ_{LPG} is the resonance coupling wavelength.

Notice that the *LPG* resonance frequency also depends on the cladding effective refractive index. This means that it is possible to use an *LPG* as a refractive index based chemical or biological sensor.

LPGs are sensitive not only to temperature and strain as are FBGs, but also to a bending, to hydrostatic pressure, to torsion, and to ambient refractive index changes and, as such, are an excellent core element for specialized ultra-sensitive industrial sensors.

The sensitivity to the refractive index of the surrounding external medium makes LPG very useful as the core element for building chemical and bio-medical sensors. The change in the ambient index changes the effective index of the cladding mode and this in turn leads to wavelength shifts of the resonance dips in the LPG transmission spectrum that can be monitored and processed by test and measurement equipment.

The period of such a structure is of the order of a fraction of a millimeter. In contrast to the fiber Bragg gratings, LPGs couple copropagating modes with close propagation constants; therefore, the period of such a grating can considerably exceed the wavelength of radiation propagating in the fiber. Because the period of an LPG is much larger than the wavelength, LPGs are relatively simple to manufacture. Since LPGs couple copropagating modes, their resonances can only be observed in transmission spectra. The transmission spectrum has dips at the wavelengths corresponding to resonances with various cladding modes (in a single-mode fiber).

It is possible to inscribe an FBG in POF using conventional methods utilized for silica. In this case, some investigators use benzyl dimethyl ketal (BDK) dopant that acts as a photoinitiator, which triggers a photopolymerization process in the PMMA when irradiated with UV light and therefore, the photosensitivity of the fiber is increased.

14.6.3 Strain Measurement

Min et al. (2018) used the following protocol for inscribing LPG in an mPOF: A krypton fluoride (KrF) excimer laser system operating at 248 nm wavelength was employed with a pulse duration of 15 ns with about 3 mJ of energy. A plano-convex, 20 cm focal length cylindrical lens was used to focus the laser beam into the fiber core. A translation stage with 10 μm pression shifted the laser beam applying a point-by-point technique. The grating transmission spectrum was monitored by using a super luminescent diode and an OSA. Each of the 25 inscribed lines possessed 0.2 mm in width and shifted 1 mm from the other lines, producing an LPG with 25 mm in length. From these parameters the resonance appears in the range of 800 nm to 900 nm, presenting a sensitivity of about -2.21 ± 0.05 nm/mε.

Now, recalling that a silica FBG presents a sensitivity of about 1.2 nm/mε at 1550 nm (Werneck et al. 2017), we notice that an mPOF LPG almost doubles this sensitivity.

Another interesting feature of mPOF is that either FBG or LPG's sensitivity to strain is negative, that is, the resonance frequency decreases with an increasing strain, opposite to FBG/LPG in silica.

To explain this behavior, we recall the classical equation of the FBG sensitivity to strain (Werneck et al. 2017):

$$\frac{\Delta\lambda_B}{\lambda_B} = (1 - \rho_e)\varepsilon_z \tag{14.3}$$

where ρ_e is the photo-elastic coefficient of the fiber, that is, the variation of the refractive index of the fiber material with respect to the applied strain. In some solids, depending on the Poisson ratio of the material, this effect is negative, that is, when one expands a transparent medium, as an optical fiber for instance, the index of refraction decreases due to the decrease of density of the material, making ρ_e negative. Then, when an extension is applied to the fiber, the two terms between parentheses in **Equation 14.3** produce opposite effects, one by increasing the distance between gratings and thus augmenting the Bragg wavelength and the other by decreasing the effective refractive index and thus decreasing the Bragg wavelength. What controls the sign of the FBG/LPG sensitivity is the combined effect of both phenomena. In case of silica, $\rho_e = -0.22$, and therefore the effect is a positive behavior, whereas in most polymers, ρ_e is greater than unit, producing a negative sensitivity.

14.6.4 Temperature Measurement

Since silica FBG is largely used to temperature measurement, it is worth to compare silica and mPOF sensitivity to temperature.

Let's start recalling that the FBG sensitivity to temperature follows the classic equation:

$$\frac{\Delta\lambda_B}{\lambda_B} = (\alpha + \eta)\Delta T \tag{14.4}$$

where α is the thermal expansion of fiber material and η is the thermo-optic coefficient, representing the temperature dependence of the refractive index (d_n/d_T) of the fiber material.

The parameters in **Equation 14.4** have the following values for a silica fiber with a germanium-doped core:

$$\alpha = 0.55 \times 10^{-6}/°\text{C and } \eta = 8.6 \times 10^{-6}/°\text{C}.$$

Thus, the sensitivity of the grating to temperature at a wavelength range of 1550 nm is:

$$\frac{\Delta\lambda_B}{\Delta T} = 14.18 \text{ pm}/°\text{C} \tag{14.5}$$

Now, considering that most polymers present a thermo expansion coefficient much larger than that of silica, we must expect a larger sensitivity of an FBG/LPG in mPOF to temperature.

Woyessa et al. (2017) fabricated and characterized a polycarbonate (PC) mPOF capable to operate at a temperature up to 100°C with the following characteristics:

- Fiber propagation loss: ~4.06 dB/m at 819 nm
- Fiber drawn temperature at 170°C and drawing stress at 10.5 MPa
- Core and cladding diameter are 10 μm and 125 μm, respectively
- Holes diameter and the pitch size are 2.5 μm and 6.25 μm, respectively

Then a fiber Bragg grating was inscribed in the PC mPOF applying a phase mask technique and a HeCd CW UV laser at 325 nm and 5 mW output power. The grating had a period of 572.4 nm and 2 mm in length, and its resonance frequency was located at 892.24 nm with reflection strength of 30 dB and a full width half maximum (FWHM) of 0.92 nm.

The fiber was then inserted into a temperature chamber that could also control the relative humidity and presented a temperature sensitivity of about 26 pm/°C at both 50%–90% RH.

This result confirms the expected higher sensitivity of a mPOF FBG to temperature as compared to about 14 pm/°C at 1550 nm for a silica FBG, as reported by Werneck et al. (2017).

14.6.5 Relative Humidity Measurement

Some polymers are humidity sensitive and strongly absorb water. PMMA, for instance, can absorb up to 2.1% at 23°C whereas PC absorbs about 0.3%. Other polymers, such as Topas and Zeonex, have been reported Woyessa et al. (2017) to be insensitive to humidity. In their work they inform that the moisture absorption leads to a change in the refractive index and size of the fiber. It is easy to conclude that a change in these parameters affect the Bragg wavelength, according to **Equation 14.1**.

The authors reported in the work a relative humidity (RH) sensitivity of 7.31 ± 0.13 pm/% RH in the range 10–90% RH at 100°C.

14.6.6 Birefringence Measurement

Different from POFs, single-mode mPOFs exhibit polarization properties that make them potentially interesting for their use in the design and development of polarimetric-based sensors.

Considering that SM mPOF behaves as a linear birefringent system with defined optical axes, it is reasonable to consider that externally induced forces such as bending, twisting, and pressure would induce retardations or light rotation as predicted theoretically.

This is the work of Durana et al. (2017) in which they studied the effect of external forces on the mPOF birefringence. Bend-induced light retardation in one of the axes varies linearly with the inverse square of the bending radius of the fiber and asymmetrical lateral stress or pressure, inducing light retardation with the applied force. As to twist-induced rotation, the electric field rotates linearly with the angle through which the fiber is twisted.

The effect can be explained by the birefringence theory. When polarized light propagates through an optical fiber, induced stress mechanisms (both intrinsic and external) affect to the state of polarization of light. This effect normally is avoided as it causes instability problems in the polarization state of the propagating light. However, this very effect can be used for development of useful devices such as sensors.

The setup was designed to measure birefringence by the use of a linearly polarized light from a He-Ne laser at a wavelength of 632.8 nm making a known angle to the one of the axes of the fiber (Durana et al. 2017).

The laser light was injected into a 1 m-length of an SM mPOF using a microscope objective. The light emerging from the output end, elliptically polarized, was collimated by another microscope objective and was made to pass through a quarter-wave plate and a second polarizer before reaching a photodetector.

In this way, any bending, pressure, or stress experienced by the fiber will change the light polarization and consequently the output power injected into the photodetector. Three parameters have been measured in this study: bend, twist, and transverse stress.

14.6.7 Fiber Bending Measurement

For the bend-induced birefringence we recall that when one bends a fiber, a stress distribution is created inside the fiber and as a consequence, the photoelastic effect produces changes of refractive index within the fiber that yields the bend-induced birefringence.

When plotted, the birefringence per unit length versus the inverse square of the bending radius produced a linear relationship with $R_2 = 0.99964$ (Durana et al. 2017).

14.6.8 Twist Measurement

When a fiber is twisted, there is a coupling between $\pi/2$-dephased fiber modes inducing a rotation of the output polarization. It has been shown that the phase rotation is proportional to the twist angle in a range between 0 and 150 (Durana et al. 2017).

14.6.9 Transverse Stress

When the fiber is pressed into an angled V-groove, linear birefringence is induced due to photo-elastic effect. The experiments have shown that the birefringence is proportional to the applied force in a range between zero and 6 N (Durana et al. 2017).

References

Achenbach et al., "Computational studies of light acceptance and propagation in straight and curved multimode active fibers," *J. Opt. A* 5, 239–249, 2003.

Abeeluck et al., "Analysis of spectral characteristics of photonic bandgap waveguides," *Opt. Express* 10(23), 1320–1333, 2002.

Alfano et al., "Emission in the region 4000 to 7000 Å via four-photon coupling in glass," *Phys. Rev. Lett.* 24, 584, 1970.

Allan et al., in "Photonic crystal fibers: Effective-index and band-gap guidance," *Photonic Crystal and Light Localisation in the 21st Century*, C. M. Soukoulis, Ed., Kluwer Academic, Dordrecht, the Netherlands, pp. 305–320, 2001.

Antunes et al., "Mechanical properties of optical fibers," in *Selected Topics on Optical Fiber Technology*, M. Yasin, Ed., Intech Press, 2012, doi:10.5772/2429.

Argyros et al., "Ring structures in microstructured polymer optical fibers," *Opt. Express* 9(13), 813–820, 2001.

Argyros, "Guided modes and loss in Bragg fibers," *Opt. Express* 10(24), 1411–1417, 2002.

Argyros et al., "Counting modes in optical fibers with leaky modes," *Symposium on Optical Fiber Measurements*, Boulder, CO, 2002.

Argyros et al., "Analysis of ring-structured Bragg fibers for TE mode guidance," *Opt. Express* 12, 2688–2698, 2004.

Argyros et al., "Photonic bandgap with an index step of one percent," *Opt. Express* 13, 3014, 2005a.

Argyros et al., "Guidance properties of low-contrast photonic bandgap fibers," *Opt. Express* 13, 2503–2511, 2005b.

Argyros et al., "Hollow-core microstructured polymer optical fiber," *Submitted to the Polymer Optical Fiber Conference*, Hong Kong, 2005c.

Barton et al., "Characteristics of multimode microstructured POF performance," *Proceedings of the 12th International Plastic Optical Fibers Conference*, Seattle, WA, pp. 81–84, 2003.

Barton et al., "Fabrication of microstructured polymer optical fibers," *Opt. Fiber Technol.* 10, 325–335, 2004.

Benabid et al., "Particle levitation and guidance in hollow-core photonic crystal fiber," *Opt. Express* 10, 1195–1203, 2002a.

Benabid et al., "Stimulated Raman scattering in hydrogen-filled hollow-core photonic crystal fiber," *Science* 298, 399, 2002b.

Benabid et al., "Compact, stable and efficient all-fiber gas cells using hollow-core photonic crystal fibers," *Nature* 434, 488–491, 2005.

Birks et al., "Endlessly single-mode photonics crystal fiber," *Opt. Lett.* 22, 961, 1997.

Bjarklev et al., *Photonic Crystal Fibers*, Springer Science+Business Media, Dordrecht, 2003.

Bouwmans et al., "High-power Er:Yb fiber laser with very high numerical aperture pump–cladding waveguide," *Appl. Phys. Lett.* 83, 817–818, 2003.

Broderick et al., "*Nonlinearity in holey optical fibers: Measurement and future opportunities*," *Opt. Lett.* 24(20), 1395–1397, 1999.

Broeng et al., "Photonic crystal fibers: A new class of optical waveguides," *Opt. Fiber Technol.* 5, 305–330, 1999.

Choi et al., "Fabrication of properties of polymer photonic crystal fibers," *Proceedings of the Plastic Optical Fibers Conference*, Vrije Universiteit, Amsterdam, the Netherlands, pp. 355–360, 2001.

Dall et al., "Single mode hollow optical fibers for atom guiding," *Appl. Phys. B* 74, 11–18, 2002.

Daum et al., *POF—Polymer Optical Fibers for Data Communication*, Springer, 2002.

Deflandre, "Modeling the manufacturing of complex optical fibers: The case of the holey fibers," *Proceedings 2nd International Colloquium*, Valenciennes, France, pp. 150–156, 2002.

Durana et al., "Study of the influence of various stress-based mechanisms on polarization of an SM mPOF for the development of useful devices," *J. Light. Technol.* 35(14), 3035–3041, 2017.
Eggleton et al., "Grating resonances in air–silica microstructured optical fibers," *Opt. Lett.* 24, 1460–1462, 1999.
Eom et al., "Optical properties measurement of several photonic crystal fibers," *SPIE, Photonics West*, San Jose, 2002.
Dobb et al., "Continuous wave ultraviolet light-induced fiber Bragg gratings in few- and single-mode microstructured polymer optical fibers," *Optics Letters* 30, 3296–3298, 2005.
Fink et al., "An Omni directional reflector," *Science* 282, 1679–1682, 1999.
Fini, "Microstructure fibers for optical sensing in gases and liquids," *Meas. Sci. Technol.* 15, 1120–1128, 2004.
Fitt et al., "Modeling the fabrication of hollow fibers: Capillary drawing," *J. Lightwave Technol.* 19(12), 1924–1931, 2001.
Fitt et al., "The mathematical modeling of capillary for holey fiber manufacture," *J. Eng. Math.* 43, 210–227, 2002.
Gloge, "Weakly guiding fibers," *Appl. Opt.* 10, 2252, 1971.
Huang et al., *Design, Fabrication and Characterization of Microstructured Polymer Optical Fibers*, session TU3C (16) 8 1242, Photonics West San Jose, 2004.
Issa et al., "Vector wave expansion method for leaky modes of microstructured optical fibers," *J. Light. Technol.* 24(4), 1005–1012, 2003a.
Issa et al., "Identifying hollow waveguide guidance in air-cored microstructured optical fibers," *Opt. Express* 11(9), 996–1001, 2003b.
Issa, "Fabrication and study of microstructured optical fibers with elliptical holes," *Opt. Lett.* 29(12), 1336–1338, 2004a.
Issa, "High numerical aperture in multimode microstructured optical fibers," *Appl. Opt.* 43(33), 6191–6197, 2004b.
Issa, "Light acceptance properties of multimode microstructured optical fibers: Impact of multiple layers," *Opt. Express* 12(14), 3224–3235, 2004c.
Issa, et al., "Modes and propagation in microstructured optical fibers," PhD thesis, University of Sydney, 2005.
Kerbage et al., "Highly tunable birefringent microstructured optical fiber," *Opt. Lett.* 27(10), 842–844, 2002.
Kiang et al., "Extruded single mode non-silica glass holey optical fibers," *Electron. Lett.* 38(12), 546–547, 2002.
Knight et al., "All silica single-mode optical fiber with photonic crystal cladding," *Opt. Lett.* 21(19), 1547–1549, 1996.
Knight et al., "Large mode area photonic crystal fiber," *Electron. Lett.* 34, 1347, 1999.
Knight et al., "'Holey' silica fibers" in *Optics of Nanostructured Materials*, V. Markel and T. George, Eds., John Wiley & Sons, 2001.
Kondo et al., "Fabrication of polymer photonic crystal fiber," paper B-7, *10th Microoptics Conference*, Jena, Germany, 2004.
Kumar et al., "Tellurite photonic crystal fiber," *Opt. Express* 11(20), 2641–2645, 2003.
Kumar et al., "Extruded soft glass photonic crystal fiber for ultrabroad supercontinuum generation," *Opt. Express* 10, 1520–1525, 2002.
Kuhlmey et al., "Modal 'cutoff' in microstructured optical fibers," *Opt. Lett.* 27, 1684–1686, 2002.
Kuriki et al., "Hollow multilayer photonic bandgap fibers for NIR applications," *Opt. Express* 12, 1510–1517, 2004.
Large et al., "Solution doping of microstructured polymer optical fibers," *Opt. Express* 12(9), 1966–1971, 2004.
Large et al., "Microstructured POF and applications," *14th International Conference on Polymer Optical Fiber*, Hong Kong, China, 2005.
Leon-Saval et al., "Splice-free interfacing of photonic crystal fibers," *Opt. Lett.* 30(13), 1629, 2005.
Limpert et al., "High-power air-clad large-mode-area photonic crystal fiber laser," *Opt. Express* 11, 818–823, 2003.
Litchinitser et al., "Antiresonant reflecting photonic crystal optical waveguides," *Opt. Lett.* 27(18), 1592–1594, 2002.

Lwin, "Suspended core microstructured polymer optical fiber: Connecting to reality," *30th Australian Conference on Optical Fiber Technology (BGPP/ACOFT)* Star City, Sydney, Australia, 2005a.
Lwin, "Progress on low loss of microstructured polymer optical fibers," *14th Polymer Optical Fiber Conference*, Hong Kong, 2005b.
Lyytikäinen et al., "Heat transfer within a microstructured polymer optical fiber preform," in SPECIAL ISSUE: Recent developments in modelling and simulation in polymers and composites processing, *Modelling Simul. Mater. Sci. Eng.* 14, 1–11, 2004.
Mach et al., "Tunable microfluidic optical fiber," *Appl. Phys. Lett.* 80, 4294–4296, 2002.
Mediavilla et al., *Fiber Optics in Astronom*, Vol. 3. Astronomical Society of the Pacific, San Francisco, CA, 1998.
Min et al., "Microstructured PMMA POF chirped Bragg gratings for strain sensing," *Opt. Fiber Technol.* 45, 330–335, 2018.
Monro et al., "Chalcogenide holey fibers," *Electron. Lett.* 36, 1998–2000, 2000.
Monro et al., "Holey optical fibers: Fundamental properties and device applications," *C. R. Physique* 4, 175–186, 2003.
Mortensen et al., "Modal cutoff and the V parameter in photonic crystal fibers," *Opt. Lett.* 28, 1879–1881, 2003.
Mortensen, "Effective area of photonic crystal fibers," *Opt. Express* 10(7), 341–343, 2002.
Nielsen et al., "Photonic crystal fiber design based on the V-parameter," *Opt. Express* 11, 2762–2768, 2003.
Ortigosa-Blanch, "Highly birefringent photonic crystal fibers," *Opt. Lett.* 25(18), 1325, 2000.
Palais, *Fiber Optic Communications*, Prentice Hall, NJ, 1992.
Petropoulos et al., "High nonlinearity holey fibers: Design, fabrication and applications," paper CFI2-4-INV, IQEC/CLEO-PR2005, Tokyo, 2005.
Peyrilloux et al., "Comparison between the finite element method, the localized function method and a novel equivalent averaged index method for modelling photonic crystal fibers," *J. Opt. A: Pure Appl. Opt.* 4, 257–262, 2002.
Poladian et al., "Fourier decomposition algorithm for leaky modes of fibers with arbitrary geometry," *Opt. Express* 10(10), 449–454, 2002.
Ranka et al., "Visible continuum generation in air-silica microstructure optical fibers with anomalous dispersion at 800 nm," *Opt. Lett.* 25, 25, 2000.
Reeves et al., "Demonstration of ultra-flattened dispersion in photonic crystal fibers," *Opt. Express* 10, 609–613, 2002.
Ritari et al., "Gas sensing using air-guiding photonic bandgap fibers," *Opt. Express* 12, 4080–4087, 2004.
Roberts et al., "Ultimate low loss of hollow-core photonic crystal fibers," *Opt. Express* 13, 236–244, 2005.
Russell, "Photonic crystal fibers," *Science* 358–362, 299, 2003.
Saitoh et al., "Chromatic dispersion control in photonic crystal fibers: Application to ultra-flattened dispersion," *Opt. Express* 11, 843–852, 2003.
Saitoh et al., "Empirical relations for simple design of photonic crystal fibers," *Opt. Express* 13(1), 267–274, 2005.
Snyder et al., *Optical Waveguide Theory*, Chapman & Hall, London, UK, 1983.
Steel et al., "Elliptical-hole photonic crystal fibers," *Opt. Lett.* 26, 229–231, 2001.
van Eijkelenborg et al., "Microstructured polymer optical fiber," *Opt. Express* 9(7), 319–327, 2001.
van Eijkelenborg, "Imaging with microstructured polymer fiber," *Opt. Express* 12(2), 342–346, 2004c.
van Eijkelenborg, "Bandwidth and loss measurements of graded-index microstructured polymer optical fiber," *Electron. Lett.* 40(10), 592–593, 2004a.
van Eijkelenborg, "Mechanically induced long-period gratings in microstructured polymer fiber," *Opt. Commun.* 236, 75–78, 2004b.
Vienne et al., "Ultra-large bandwidth hollow-core guiding in all-silica Bragg fibers with nanosupports," *Opt. Express* 12, 3500–3508, 2004.
Wadsworth et al., "High power air-clad photonic crystal fiber laser," *Opt. Express* 11, 48–53, 2003.
Wadsworth et al., "Supercontinuum and four-wave mixing with Q-switched pulses in endlessly single mode photonic crystal fibers," *Opt. Express* 12(2), 299–309, 2004a.
Wadsworth et al., "Very high numerical aperture fibers," *Photonic. Technol. Lett. IEEE* 16(3), 843–845, 2004b.
West et al., "Surface modes in air-core photonic band-gap fibers," *Opt. Exp.* 12(8), 1485–1496, 2004.
White et al., "Confinement losses in microstructured optical fibers," *Opt. Lett.* 26(21), 1660–1662, 2001a.

White et al., "Calculations of air-guided modes in photonic crystal fibers using the multipole method," *Opt. Express* 11, 721–732, 2001b.
Werneck et al., "Fiber Bragg gratings: Theory, fabrication, and applications," *Tutorial Texts in Optical Engineering*, TT114, 256, SPIE PRESS, Bellingham, WA (softcover), Published 5th September 2017.
Woyessa et al., "Low loss polycarbonate polymer optical fiber for high temperature FBG humidity sensing," *IEEE Photonic. Technol. Lett.* 29(7), 575–578. doi:10.1109/LPT.2017.2668524.
Xue et al., "Fabrication of microstructured optical fibers, Part I: Problem formulation and numerical modelling of transient draw process," *J. Light. Technol.* 23(7), 2245–2254, 2005a.
Xue et al., "Fabrication of microstructured optical fibers, Part II: Numerical modelling of steady-state draw process," *J. Light. Technol.* 23(7), 2255–2266, 2005b.
Xue et al., "The role of material properties and drawing conditions in the fabrication of microstructured optical fibers," *J. Light. Technol.* 24(2), 853, 2005c.

15

POF Applications

Marcelo Martins Werneck

Contents

15.1 Introduction

This chapter aims to report some POF applications that have been used by the Photonics and Instrumentation Laboratory (LIF) in specific areas and have not been published so far as conventional papers. They are specific applications designed to solve specific problems in the oil and gas industries and electric power sector. The idea of presenting these field applications is that they might serve as initiators to catalyze readers' new ideas for general POF applications.

Some of these ideas are patented but can be improved by the reader and possibly can generate new patents.

We start with a displacement sensor used to monitor contact blades of a substation switch, also known as a disconnector. The second application comes from the first one, applying the same sensor technology to monitor opening and closing of tanker truck valves. Then, we go on to describe an oil leaking sensor employed to monitor oil transportation by floating hoses in an offshore oil production platform.

The last application described here is a solar tracker with a Fresnel lens and a POF bundle used to ambience illumination and microalgae cultivation. We decided to include this application in this book because, although not being a sensor as this book centers on, it deals with another foot of the famous POF tripod applications: telecom, sensor, and illumination.

As a final remark, notice that the applications described in this chapter were developed not as scientific research with an academic point of view, but as engineering solutions aimed at specific problems. Therefore, this chapter will not be presented as a scientific paper but rather, as a condensed report.

15.2 Sensor for Monitoring High-Voltage Switch

15.2.1 Introduction

A switch or disconnector is power substation maneuvering equipment, a component of the electric power system, that has the function of establishing the electric connection between generators, transformers, consumers, and transmission lines and disconnecting them, according to the requirements of the service.

Disconnectors are mechanical maneuvering devices which, in the open position, ensure an insulation distance and, in the closed position, maintain the continuity of the electric circuit under specified conditions. The disconnector can only be operated either to open or to close when a negligible current is passing through it and when there are no significant or sudden variations of voltage between its terminals. It is also capable of conducting currents under normal

FIGURE 15.1 A 720-kV disconnector opening with electric arcs following the contact blades.

circuit conditions and, for specified times, currents under abnormal conditions, such as short circuit currents. **Figure 15.1** shows a 720-kV disconnector opening with an electric arc following its contact blades.

The disconnector is composed of a long aluminum arm that closes or opens the circuit between two insulated contacts, each one on the top of a ceramic insulator. The contact resistance must be in the order of a few mΩ in order to dissipate the minimum power. Suppose, for instance, a current of 8,000 A across a disconnector with a contact resistance of 10 mΩ. The dissipated power will be: $P = IR^2 = 8{,}000\ A \times (10\ m\Omega)^2 \cong 1\ W$. If the contact resistance is too large, the dissipation power increases and, depending on the situation, the temperature can reach high values, capable to destroy the disconnector contacts.

The disconnector is driven by a motor that can be operated at a distance to move the contact rods to open or close (see **Figure 15.1**). To monitor the closing or opening operation, this equipment contains microswitches mechanically connected to its main axle so that they can inform the operator of the actual state of the disconnector. The microswitches can indicate the complete opening or the complete closing of the disconnector.

However, it is a general consensus among technicians and operators to not rely on this method, since even if the disconnector does not close completely, the indicators can indicate total closure if there is a slip between the motor shaft and the control rod. This also occurs when the indicators may show a full aperture, but the disconnector may not be fully open. In view of such a problem, all technicians agree that the best method is visual inspection. However, when a technician is moved to the disconnector during its operation, to verify that the key has opened or closed completely, it generates an additional cost involving operators and time.

15.2.2 Project Description

This project is composed of fiber optic sensors installed next to the electrical contacts of the disconnector, capable of monitoring its closure and opening during routine operations. The system takes advantage of the movement of the disconnector electrical contact during its closing to trigger an optical sensor.

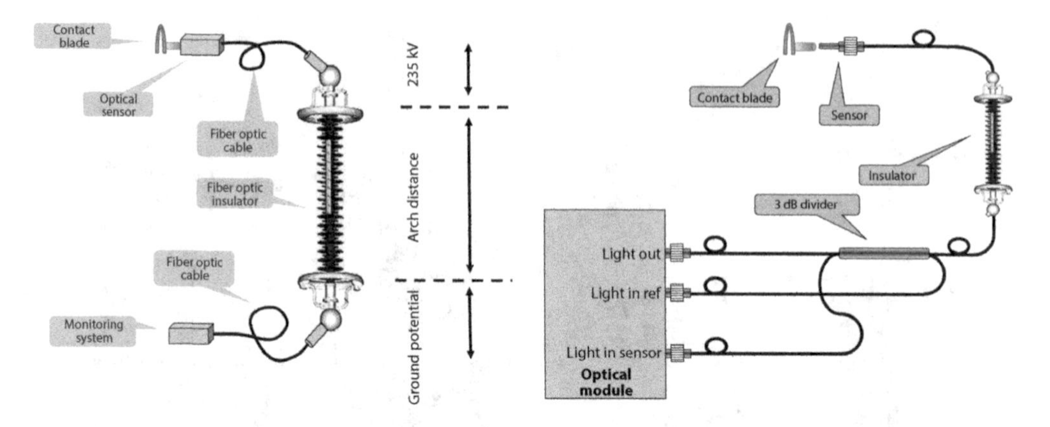

FIGURE 15.2 Left: The sensing system; Right: The optical setup.

Light from a light source (LED or LASER) is generated in the optoelectronic interrogation system and travels through a plastic optical fiber (POF) to the electrical contact of the disconnector where the transducer is located. Inside the transducer is the sensor, moved by the disconnector contact and is composed of a micro-mirror that reflects the light coming from the optical fiber back to the optoelectronic system when the electrical contact of the switch closes. The reflected light travels the optical fiber back to the optoelectronic system where it is measured by a photodetector. The optical power measured by the photodetector indicates the position of the electric contact. An alarm signal will be sent remotely to the control center if the disconnector does not close completely, thus preventing an operational accident.

The proposed method was the measurement of the movement of the contact blade. Every time the disconnector arm locks into the contact head, the contact blades move a few millimeters. Therefore, the sensor has to be a high precision displacement sensor. This was performed using a reflective power amplitude system in which the micro-mirror, driven by the contact blade, approaches or moves away from the POF tip. **Figure 15.2** (left) shows the idea and at the right the block diagram of the optical setup is shown. The sensor is a 1-mm SMA connector appropriate for a 1-mm PMMA POF.

Figure 15.3 shows a drawing the device, completely waterproofed and adequate to the harsh environment existent in outdoor power substations. The total contact blade displacement is about 300 µm with a resolution in the order of 10 µm. The top figure shows the main parts in a 2-D view whereas the lower figure shows a 3-D view.

15.2.3 Prototype Tests

Figure 15.4 shows the calibration graph of the output power versus blade distance. Notice the provision of an uncertainty range between open and close, within which the system does not show either open or closed status.

Figure 15.5 (left) shows the prototype installed on a real contact head and at right is a picture of the switch in closed state, with the movable arm fit inside the contact head. Notice the reddish contact blades at either internal side of the contact head.

15.2.4 Field Installation

After laboratory tests, three sensors were installed on an operational disconnector, in Brazil's southwest Coxipó substation, located in the State of Mato Grosso. **Figure 15.6** shows a picture

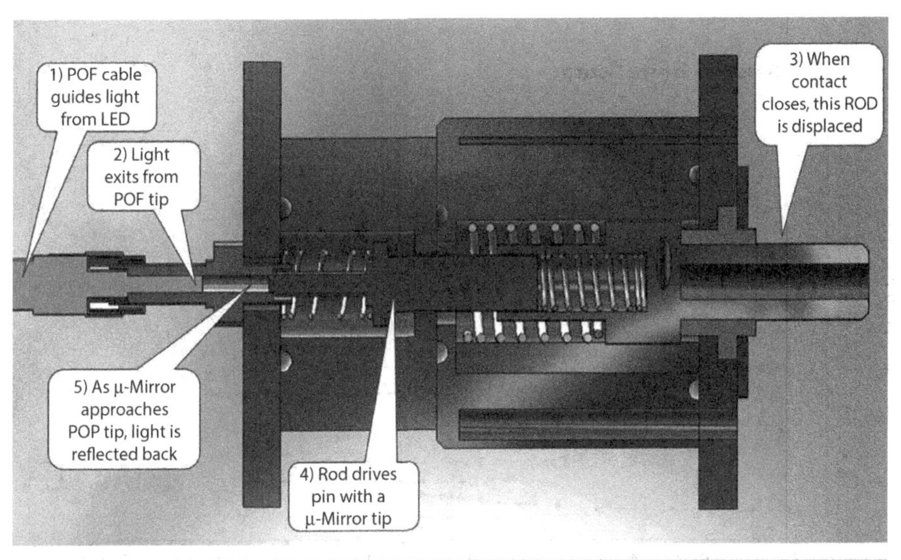

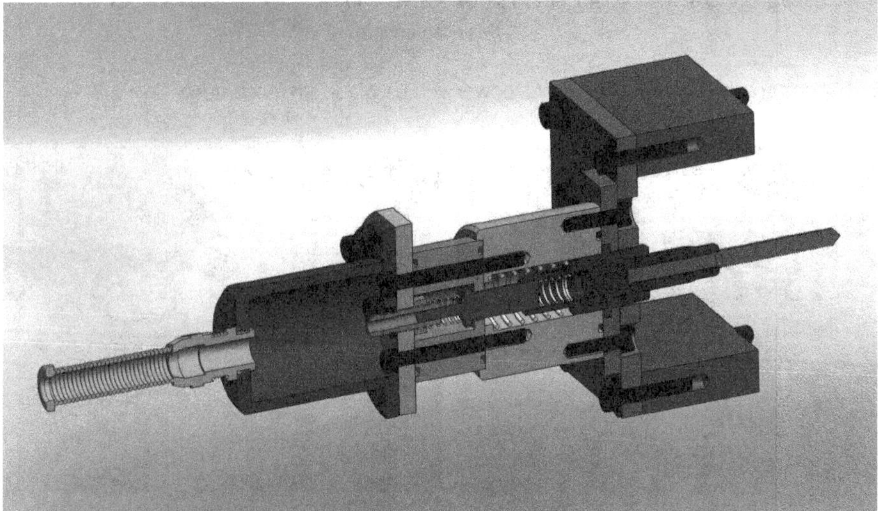

FIGURE 15.3 The transducer. Top: Main parts; Bottom: 3-D view.

of the installation stage at left and, at right, one of the three transducers installed at the contact head of a disconnector.

Figure 15.7 shows three industrial prototypes developed by a spinoff company that are to be installed in an operating disconnector for commissioning tests.

15.2.5 Conclusions

The system was tested in field and proved to be operational. The Photonics and Instrumentation Laboratory filed a patent (Werneck et al. 2012) and is now preparing a new prototype to be installed in an operational disconnector, aiming to be commissioned soon.

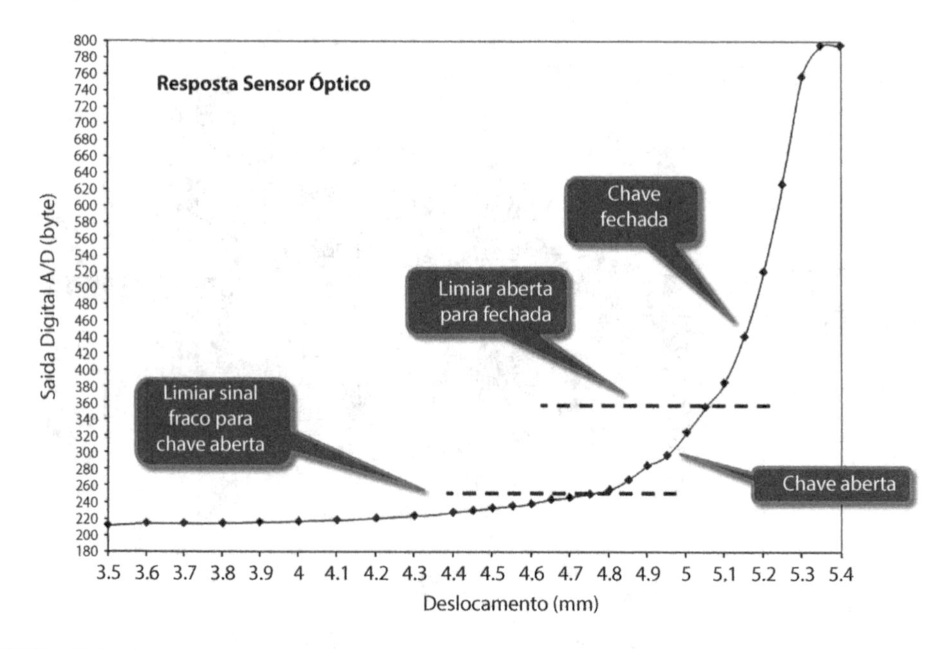

FIGURE 15.4 A graph showing the output power versus blade displacement.

FIGURE 15.5 Left: The prototype installed on a real contact head; Right: Picture of the switch in closed state, with the movable arm fit inside the contact head. The sensor is fixed at the left of the contact head with the POF cable connected to the SMA connector.

FIGURE 15.6 Left: The installation of a prototype in a real switch; Right: The transducer installed at the contact head of the disconnector.

FIGURE 15.7 Three industrial prototypes of the transducer.

15.3 Monitoring Oil Trucker Valves

15.3.1 Abstract

Constant news of theft and adulteration of fuels has been conveyed by the media. In Manaus, State of Amazonas in Brazil, in 2015, 26 gang members were arrested for diverting around 500,000 liters of fuel per month. The system to be described here was developed by the Photonics and Instrumentation Laboratory (LIF) at the Universidade Federal do Rio de Janeiro (UFRJ) in partnership with a private company. It is composed by a fiber optic sensor, located behind the valve cover that allows the fuel inlet or outlet of the tank of a tanker truck. From there, an optical fiber proceeds inside the tank to its top, when it enters an electronic enclosure sealed on the outside with no internal access, and with only a small waterproof antenna, for interpretation and transmission of data to a cell phone or tablet. The data refer to valve openings and closures, which could only occur when the truck departs and arrives at its destination. The sensor does not prevent theft of fuel, but records the date and time of the occurrence, enabling verification and taking of measures by the companies. Fiber optic technology has the advantage of not involving electricity, as opposed to conventional sensors which, in order to be used in explosive environments, undergo several tests from accredited laboratories to meet the intrinsic safety standards.

15.3.2 Introduction

Brazil is ranked by FreightWatch International, one of the largest cargo transportation consultants, as the country with the highest risk of cargo theft in the world. Thus, this project came about after we came across with various reports of cargo thefts across the country, especially in fuel theft. This has motivated LIF scientists, who have been working with fiber optic technologies for years, to create and develop a product that could generate and guarantee safety for the transport of cargo, especially fuel.

In the first stage of the project we developed a prototype of the equipment that works in the laboratory thus proving the proof of concept, showing that the technology works.

The fuel transport segment is migrating to a type of valve known as bottom-load, that is, the fuel load in the truck tank is made by the lower valve, the same used for discharge. There are a number of advantages, such as the filling time of a tanker truck with these valves is 25% of the time of trucks with top-loading valves or safety because the fuel does not come in contact with the air, unlike the top-loading So all new trucks are leaving the factory with this system but there are still tank trucks in operation with top-loading type valves.

From a valve of one of the models used by tanker trucks we developed the prototype. On the laboratory bench the valve is opened and closed several times at different times and intervals. The system stores a digital log by counting valve status (open or closed) and the time and date of the occurrence of each event. This data is stored in the device memory and can be downloaded at any time by a mobile application or a tablet.

The operation of the prototype allows the construction of the future product, which will provide operational safety by minimizing the risk of fuel fraud, as well as minimizing the risks of explosion, fire, intoxication, contamination, and spills, ensuring confidence for all involved stakeholders.

The long-term goal is to develop equipment based on fiber-optic technology that is capable of stopping fuel theft in tank trucks, providing operational and environmental safety, and minimizing the risks of falling performance and accidents.

After the successful completion of this project, it is intended to develop applications for the transportation of other loads, residences, etc.

The technology was developed to be applied, in the first moment, in bottom-loading valves, in the near future, top-loading valves, and, with later developments, to transport vehicles of other loads.

The system developed is compact, robust, safe, simple, and inexpensive, capable of sending valve opening and closing information through a wireless communication system. In addition, there are no equipment, sensors, or devices on the market capable of monitoring these valves despite the existence of various equipment available to prevent fuel theft. The most used ones monitor the truck, not the load. The few that monitor the cargo are imported and expensive, and therefore, they are not yet implanted in Brazil.

15.3.3 Description of the Sensor

The safety sensor (SenSeg) consists of an Arduino UNO microcontroller with an SD memory card module, a real-time clock module and a Bluetooth module. An optical sensor is installed inside the valve in contact with its main axle. When the valve is operated to open, its axle moves internally, triggering the sensor. An optical fiber carries the valve opening information, from inside the valve through the fuel tank up to its top where the control box is located, outside the tank but attached to the tank wall. In this way the opening and closing movement of the valve is monitored by the microcontroller that encrypts and stores all events with date and time in the SD card memory.

A tablet or smartphone-type mobile device in the immediate vicinity of the truck may request the sending of the stored data on the microcontroller's SD card via Bluetooth technology. The data arrives encrypted and is transferred to the memory of the mobile device where it can be analyzed by means of a password and transferred to a central computer, if desired.

Figure 15.8 (top) shows the position of the sensors inside the oil-truck. All valves including the top hatch may contain a sensor in order to monitor opening and closing events.

Figure 15.8 (bottom) shows the schematic diagram that includes electronic, mechanical, and optical hardware. The computer shown in the figure is for calibration and modification or software installation. The system is self-sustaining and independent, requiring only 12 V power from the vehicle's electrical system.

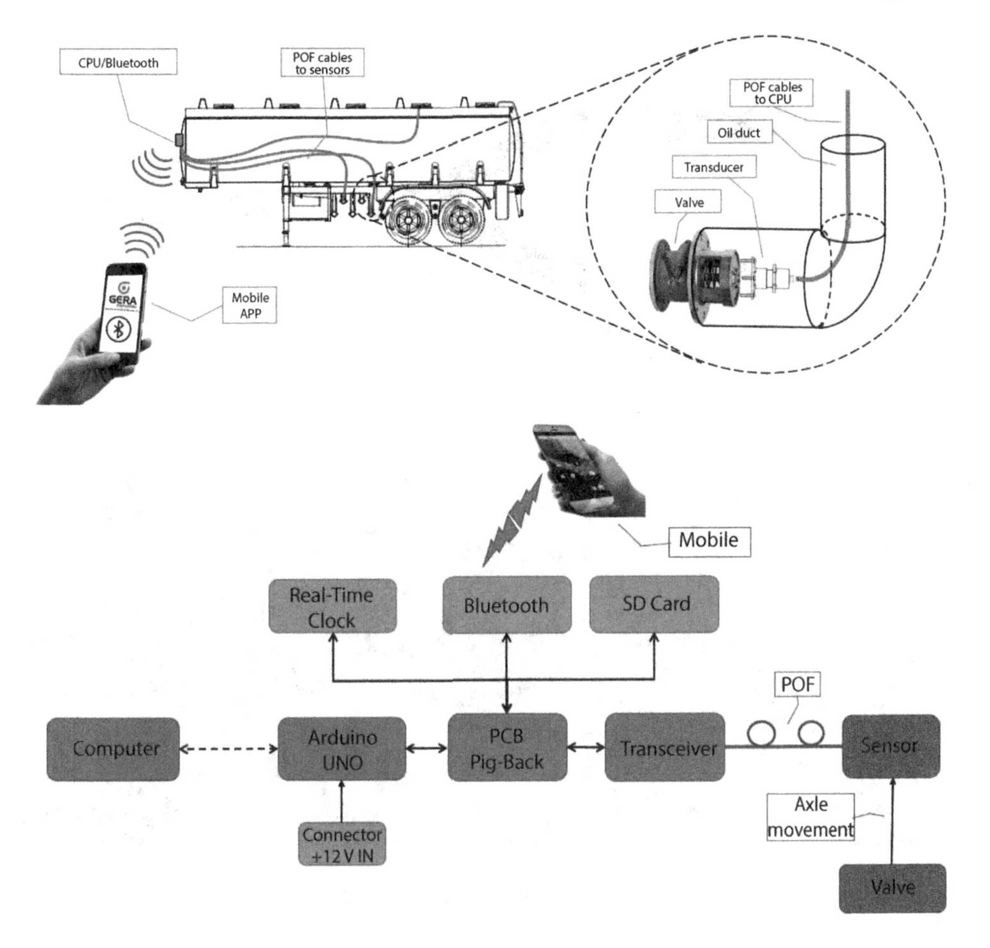

FIGURE 15.8 Top: The position of the sensors inside the oil-truck. All valves including the top hatch may contain sensors in order to monitor opening and closing events. Bottom: Block diagram of the complete system including optical, mechanical, and electronic parts.

The system is composed of:

- Arduino Microcontroller UNO: Controls the operation of the sensor; saves the data in an SD card;
- Real-time clock module (RTC): Electronic peripheral that stores and calculates time and date;
- SD card module: Electronic reading and writing drive of SD memory cards; the control of its operation is performed by the Arduino UNO;
- Transceiver: Electronic peripheral that is part of the optical sensor; its operation is controlled by the Arduino UNO.

15.3.4 Optical Setup

The microprocessor controls the LED that injects light into a plastic optical fiber. The light emitted by the LED is guided to the sensor where it is reflected back by the micro-mirror. A percentage of the light that is reflected returns by the same fiber, divides at the 3-dB coupler, and reaches the photodetector (**Figure 15.9**).

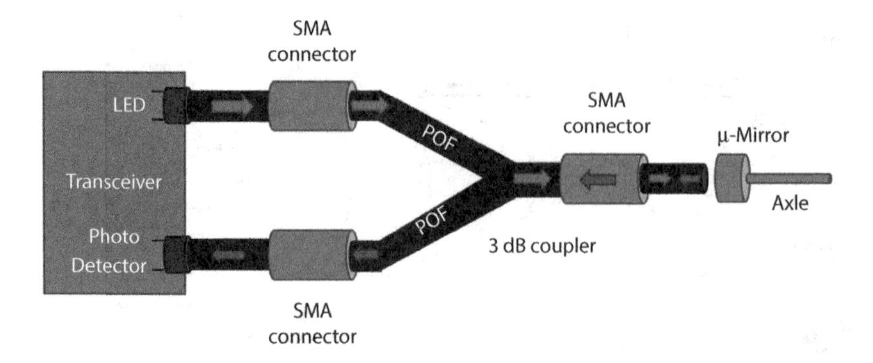

FIGURE 15.9 Schematic diagram of the optical path. All connectors are SMA type for 1-mm POF.

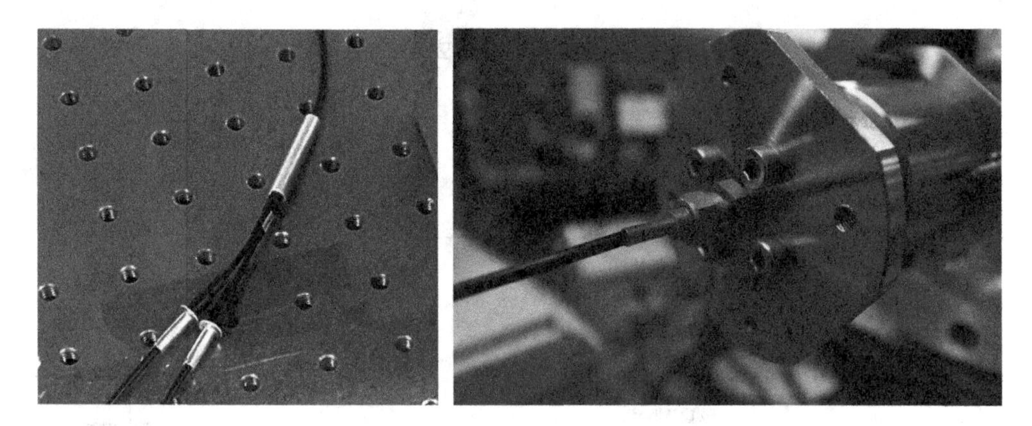

FIGURE 15.10 Left: 3-dB coupler connector; Right: Front view of sensor showing the POF SMA optical connector.

The photodetector's output voltage is proportional to the distance between the micro-mirror and the tip of the fiber.

Figure 15.10 (left) shows the 3-dB coupler connectorized to the POF cables (black), and **Figure 15.10** (right) shows the front view of the sensor with the SMA optical connector.

15.3.5 Position Sensor

Figure 15.11 illustrates the sensor operation mechanism. The valve lever moves the axle forward and backward when the valve is operated. When the valve is closed, the micro-mirror does not touch the fiber end and little light returns. When the valve opens, the axle approaches the mirror to the fiber end increasing the amount of light arriving at the photodetector.

The following sequence is followed by the algorithm in the firmware of the microprocessor:

1. The microprocessor triggers the LED and collects reflected light from the photodetector.
2. The microprocessor interprets the power of the reflected light and decides the state of the valve (open-closed-indefinite).
3. The process is repeated indefinitely at a programmable frequency.
4. The microprocessor stores these data at a programmable interval by creating a log consisting of the valve status and the date and time read.

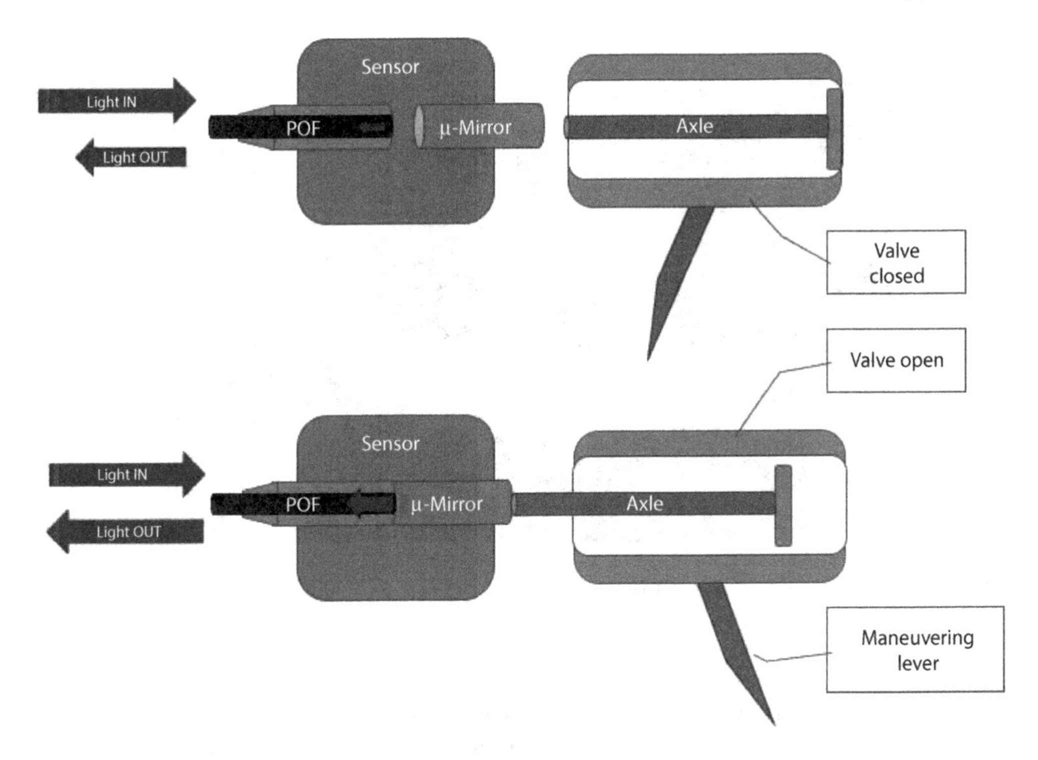

FIGURE 15.11 The working principle of the sensor coupled to the valve. Top: The valve is in closed position and the micro-mirror is far away from the fiber tip, therefore little light is returned by the fiber. Bottom: The valve is in open position and the micro-mirror is touching the fiber end forcing light to return to the photodetector.

5. When a laptop, smartphone, or tablet mobile device matched the system via Bluetooth technology, the application allows to:
   a. Check and set the date and time of the system clock, if necessary
   b. Download the log stored in memory
   c. Set the program in terms of reading frequency
   d. Create a calibration table that relates the optical power measured by the photodetector to the state of the valve
   e. Modify the program when necessary

Figure 15.12 shows the housing of the sensor in IP66 class, appropriated to work immerse into the fluid.

15.3.6 Electronic Setup

Recalling back to **Figure 15.8**, the following functions are performed by the microprocessor:

- Measurement of output voltage of the transceiver to verify whether the valve is open or closed.
- Measurement and control of the current in the transceiver LED.
- Save the valve status and date in the SD card.
- Display the operation and calibration interface of the system in an external computer.

FIGURE 15.12 Sensor housing in IP66 class.

A transceiver (Avago Technologies, USA) was used in conjunction with the POF sensor. The transceiver contains the LED and the photodetector within a small package that also contains fiber connectors. The transceiver requires a source of electrical power that is supplied by the Arduino. The Arduino generates a PWM signal with 490 Hz and amplitude from zero to 5 V, which is filtered by a low pass filter that extracts the average level of the PWM. An operational amplifier acts as a buffer, avoiding impedance mismatch between the load and the filter. The average level of the PWM is fed into the transmitter LED.

A real-time clock module (RTC) was used to control the date and time (see **Figure 15.8**). The Arduino checks the time and date on the RTC and stores it on the SD card when the valve is opened or closed. An LC STUDIO SD card read/write module was used for data storage.

15.3.7 The Software

The software for Arduino operation allows an interface of interaction between the sensor and the user as well as the control of all operations of storage and transmission of the data. **Figure 15.13** shows the program menu. The "Debug" is enabled to check the output voltage and current in the transceiver LED for calibration, with possible adjustments to the sensor and voltage limits that define the opening and closing of the valve.

15.3.8 Sensor Installation on the Valve

The tank trucker valve, provided by the customer, has an internal axle that performs a translatory movement when the valve opens and closes. This axle was used to control the position of the sensor rod (see **Figure 15.11**). The optical sensor was used to measure the movement of the central valve axle.

Figure 15.14 (left) shows the valve as seen from its front. The flange at the valve bottom is connected to the tank and by moving the lever, the operator can open or close the valve. **Figure 15.14** (right) shows the valve bottom where one can see the central axle at its lowermost position.

Figure 15.15 shows the sensor installed at the back of the valve. At left, the valve in closed position, the axle is retreated. At right the valve is open and the axle is at its outermost position, pushing the sensor rod to inside and thus triggering the system.

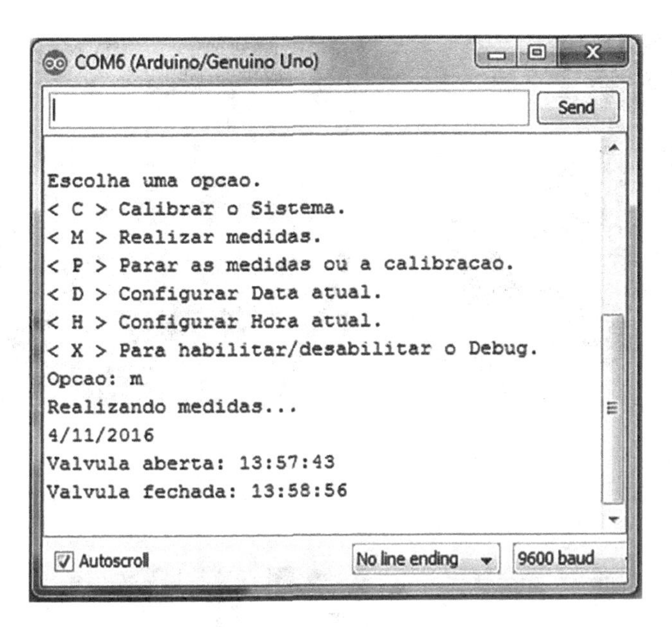

FIGURE 15.13 The smartphone interface menu.

FIGURE 15.14 Left: The valve as seen from its front. The flange at the valve bottom is connected to the tank and by moving the lever, the operator can open or close the valve; Right: The valve bottom where one can see the central axle at its lowermost position.

Figure 15.16 shows the complete prototype on the bench, before and after being inserted into its protection box. In the picture at left it is possible to see the transceiver with the two optical fiber cables and the 3-dB coupler.

In operating the valve, the system automatically stores the event together with the date and hour. All logs are stored in the SD card, the same as used in a photographic camera. From a computer or any mobile device with Bluetooth it is possible to download the log as shown in **Figure 15.17**.

FIGURE 15.15 Sensor installed at the back of the valve. Left: The valve in closed position with its axle retreated; Right: The valve is open and the axle is at its outermost position, pushing the sensor rod to inside and thus triggering the system.

FIGURE 15.16 Left: The prototype on the bench, before being inserted into the protection box. Notice the transceiver with the two optical fiber cables and the 3-dB coupler; Right: Prototype operating together with the sensor installed on the back of the valve.

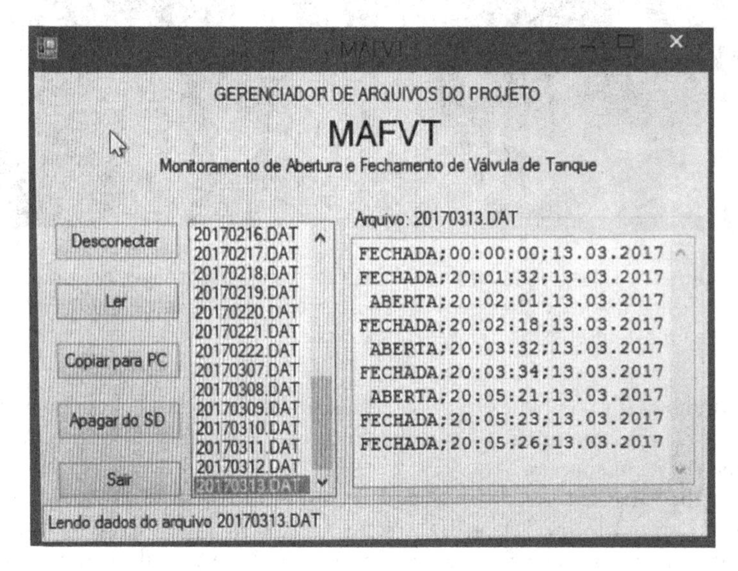

FIGURE 15.17 The screen of a paired mobile device showing the logged events.

15.3.9 Conclusions

It was concluded from the results that the prototype has all the conditions to be taken to the field for tests on tanker trucks. The system as described may be an important tool for preventing fraud and, in particular, minimizing the risk of theft and adulteration of fuel. In addition, a patent was filed in order to protect the invention (Werneck et al. 2017).

The system when installed and in operation, will allow:

- Increased operational safety;
- Minimizing the risk of a fall in performance due to fuel fraud;
- Minimization of socio-environmental risks associated with explosion, fire, intoxication, contamination, and spills;
- Minimizing the institutional risk of being linked to accidents.

15.4 Oil Leaking Sensor for Offshore Platforms

15.4.1 Introduction

The double carcass offshore hoses: The hose is supplied with special electronic sensors at each end of the hose. They are placed at diametrically opposed positions so that one unit is always visible to the operator. The sensor detects the presence of oil if the primary carcass has failed during operations. This detection system allows the operator to continue loading or discharging the oil at the terminal. The operator can then change the hose at a more convenient operational window without incurring any downtime or demurrage charges. The sensor is an optical autonomous unit that detects the presence of oil at the interface of the primary and secondary carcasses of the double carcass hose. If oil is detected at the unit, it will activate a series of specially designed pulsating LEDs.

15.4.2 The Hose

In offshore oil load and discharge operations, there is a risk of oil leakage to the sea from floating or submarine hoses. Hose failure may result from overpressure of the system, a puncture from outside, tensile break of the hose, defects in the manufacture, or construction or design of the hose. Because of the risk of failure inherent in single carcass hose construction, a double carcass hose is used in most applications. A double carcass hose construction utilizes an outer hose carcass and an inner hose carcass. The outer hose functions to hold the oil that could leak through the inner hose carcass.

A buffer space is designed between the carcass layers to retain oil that leaks from the inner carcass. The hoses are about 10 m long, 50 cm in diameter, and weigh about 500 kg. When in use, each hose is connected to another in an end-to-end basis to form a hose line for transporting oil under pressure. **Figure 15.18** shows a depiction of the hose, outside view and inside view. The hose carcasses are made from rubber, the same rubber used for automobile tires. The hose ends are flanges made of steel that are used to connect one hose to the other, in order to make a long hose line, capable of connecting a ship, for instance, to the oil platform.

An example of the use of a double carcass hose is depicted in **Figure 15.19** where a ship is funded close to an oil collector buoy that joins together several oil wells.

In any application, either floating or submarine, leakage from the hose results in undesirable consequences. In order to minimize the damage resulting from an undetected leak, various leak detection systems have been proposed and adopted. Leak detection devices present various configurations, operable under varying principles, mounted at the nipple region of an underwater hose connection as shown in **Figure 15.18**.

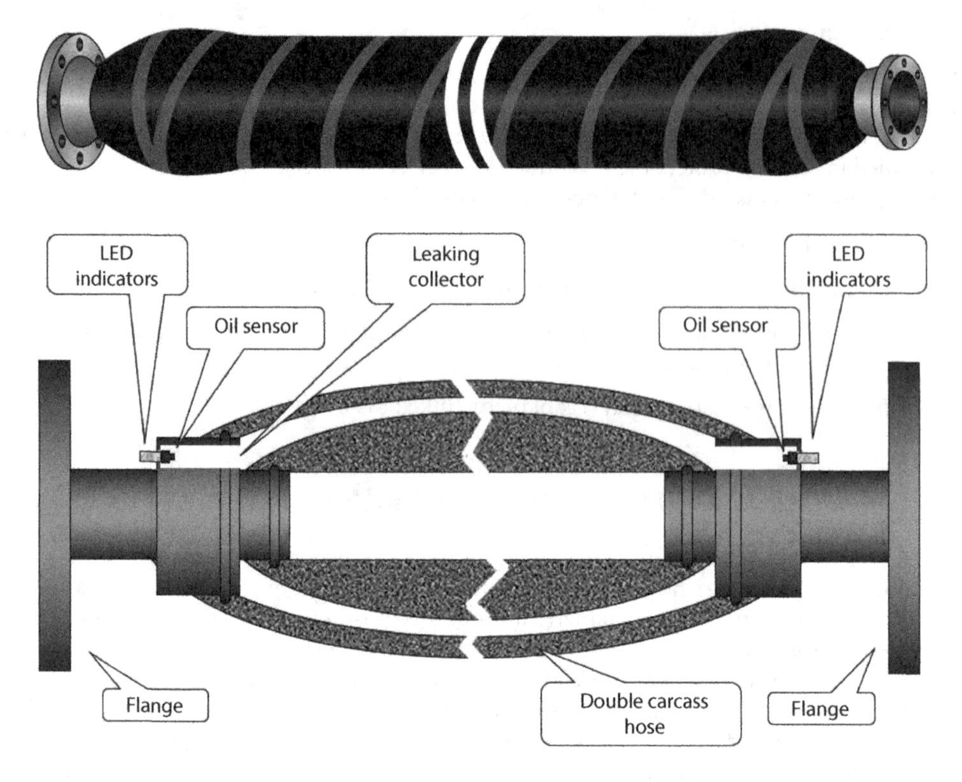

FIGURE 15.18 A depiction of a double carcass hose. Top: Outside view (Courtesy of Goodyear Seawing–Brazil). Bottom: Inside view of hose. The carcasses are made of rubber and the flanges are made of steel.

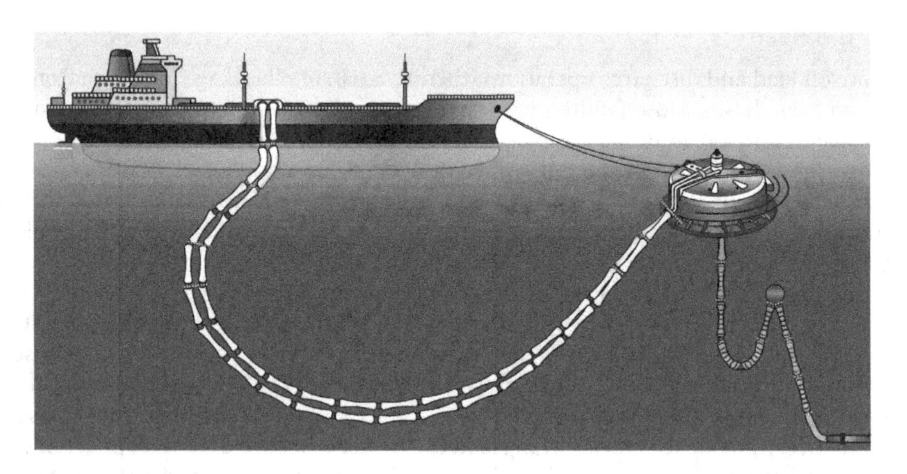

FIGURE 15.19 An application of a long line of double-carcass oil hose. A ship is funded close to an oil collector buoy that joins together several oil wells view. (Courtesy of Goodyear Seawing – Brazil.)

One method of leak detection is based on the contact between a sensing medium and the oil (US Patent No. 5,654,499). Another method is based on a pressure sensitive switch that measures the pressure of leakage fluid between carcass layers (US Patent No. 4,465,105). Finally, another method is an electro optical sensor that utilizes an infrared beam that senses the fluid level (US Patent No. 5,714,681).

In general, in offshore oil transfer operations, there is a constant physical check of hose and leak detectors since most leak detectors produce visual signals when triggered. For this reason, a continuous monitoring of each hose line is required in an activity done by the oil company or by an independent service provider.

However, local sea conditions may make it difficult for the monitoring of all the sensors located on the hoses due to the fact that the sensor may be instantly hidden by the sea or the sea itself may be high. Further, there could be a substantial danger to operational personnel and also a risk of incorrect data being logged. Finally, hose lines move as a result of seawater waves and wind conditions that can cause sensor position at a particularly hidden location or cause erroneous data collection.

15.4.3 The Sensor

The sensor developed in this study is comprised of a steel housing that is mountable to the flange of a hose line segment and having part inside the hose collection space and part outside the flange where indication LEDs are visible from outside. **Figure 15.18** shows the sensor locations, installed at both hose flanges.

Figure 15.20 shows an exploded view schematic of the complete transducer in which all elements can be seen. It is comprised of the main parts, the sensor, the control electronics containing IR LED and photodiode and indicating LEDs, and the transmitters. A transmitter continuously sends a radio beep containing the sensor ID and its state, if alarmed or not. But because the transducer can be immersed into the water or outside the water, depending on the position of the hose when floating on the sea, two transmitters were incorporated into the

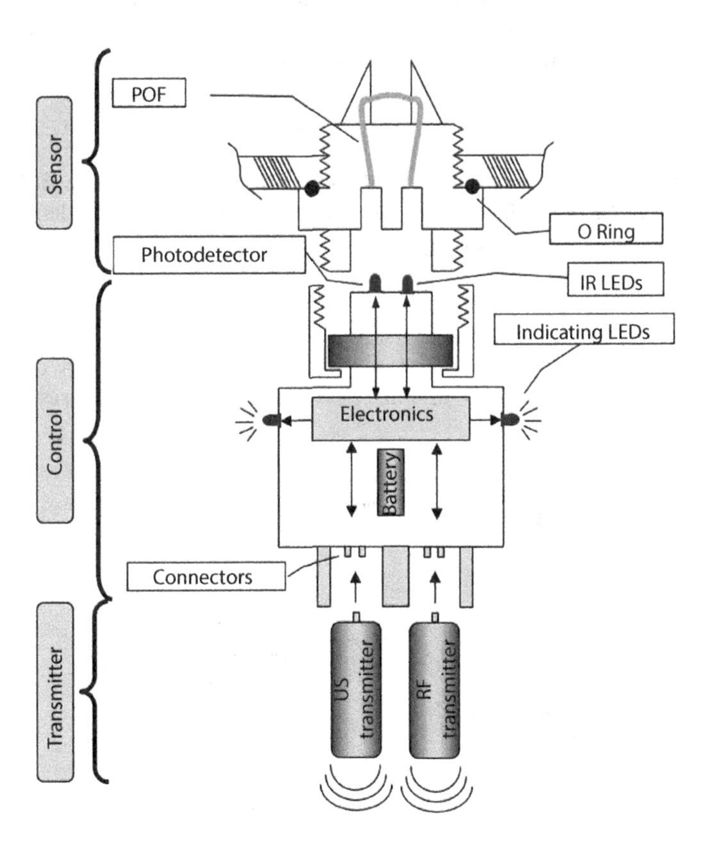

FIGURE 15.20 An exploded view schematic of the complete transducer.

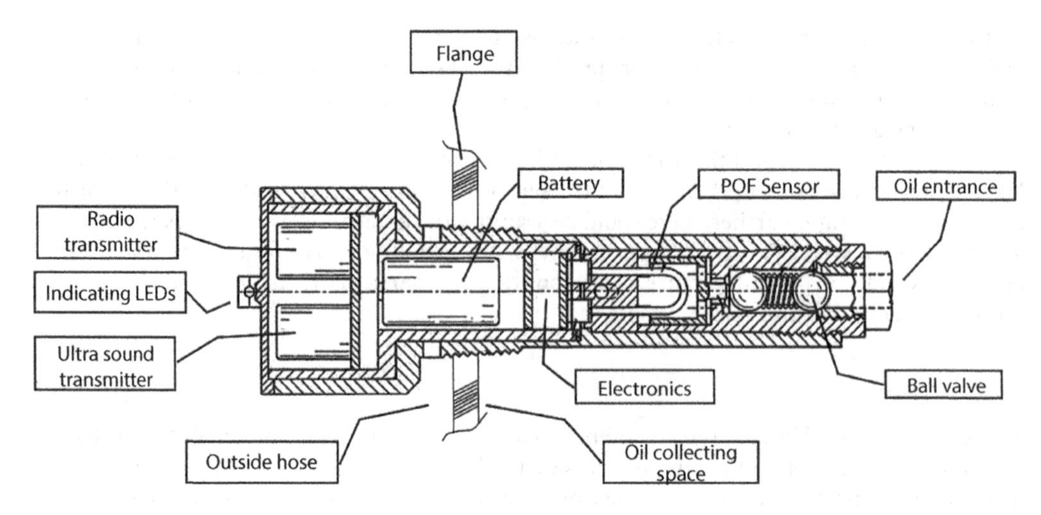

FIGURE 15.21 Schematic diagram of the transducer with the main parts. (Modified from US Patent No. US 7,387,012 B2.)

system. In case the transducer is outside the water, a radio transmitter will transmit the beep whereas if the transducer is immersed in water, un ultrasound transmitter will send the beep.

Figure 15.21 shows a schematic diagram of the sensor housing, as built.

Notice the ball valve that allows oil to flow inside the transducer. The coil retains a small amount of oil; however, in case of leaking, the pressure is high enough to press the ball over the coil allowing the oil to enter the transducer and be sensed by the POF sensor.

15.4.4 The Sensing Principle

The sensing principle is by the evanescent field. With air around the U-shape POF sensor, the light from the infrared LED is guided by the curved fiber reaching the photodetector at the other end. **Figure 15.22** shows a schematic diagram of the physical principle of sensing.

If any substance, including water, oil, gasoline, etc. touches the U-shape sensor, the refractive index (RI) outside the fiber changes from RI = 1 for air to something greater. This modifies the guiding conditions of the fiber decreasing the number of guided modes so that more light will be lost to the surroundings. The leakage sensor was designed to be capable to sense not only petrol but also water, alcohol, gasoline, and diesel. If the inner carcass ruptures, the fluid will be

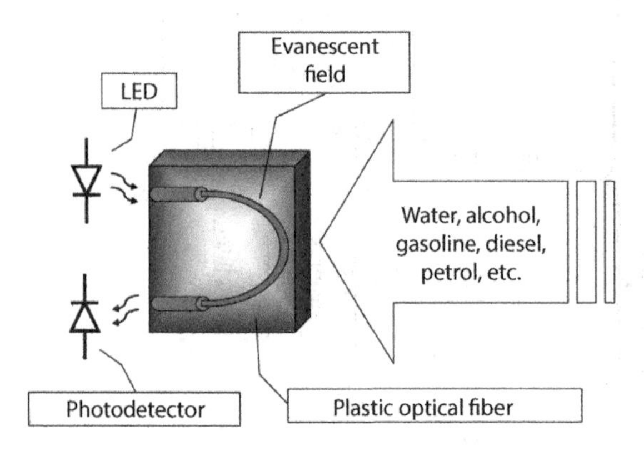

FIGURE 15.22 The physical principle of oil sensing.

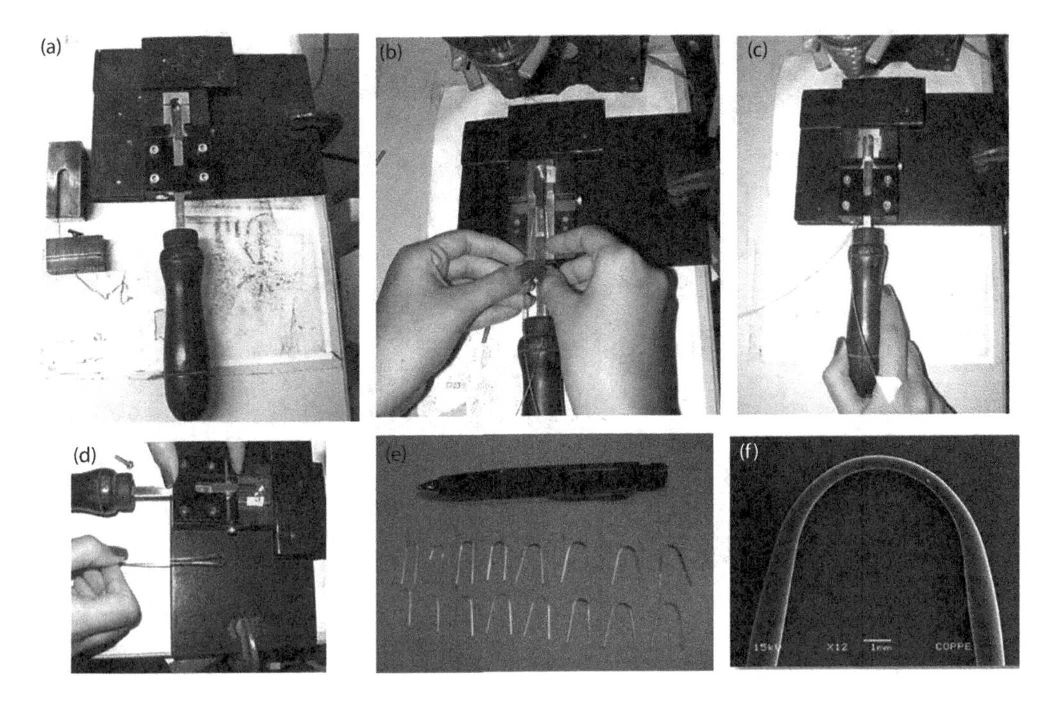

FIGURE 15.23 Several steps of the sensor fabrication. (a) Mold; (b) fiber inserted into the mold; (c) heat blow on (top of picture) to soft the fiber close to the melting point and handle being pressed, forcing the fiber to follow the inner curve of the mold; (d) fiber obtained after freezing; (e) sensors cut appropriated length and polished; and (f) picture of fiber at SEM.

in contact with the sensor. On the other hand, if the outer carcass ruptures, then water will leak inside the collecting space and the sensor will also detect this kind of failure.

The sensor was produced from a 1-mm diameter PMMA POF (Mitsubishi Rayon, Japan). **Figure 15.23** shows the steps we used to fabricate the sensor. A specially designed mold with a 1-cm diameter curve was fabricated in our workshop (A). Next, a 10-cm length of pristine fiber is cut and inserted into the mold (B). Then a heat blower heats the fiber to about 60°C while the handle presses the fiber against the mold curve (C). The result is a perfectly reproducible curved fiber (D) which is cut into the appropriated length and is polished at both ends (E). The resulted U-shaped sensor is shown under a scanning electron microscope (SEM). Notice that there is a taper at the curved part of the sensor. The taper increases the sensitivity because the evanescent field is augmented due to the decrease of the V-number of the wave guide.

The next step in the fabrication is the calibration of the sensor. For this, we developed a simple electronic circuit that produces an output voltage according to the RI of the surrounding material. We used as standard pure distilled water and a water with dissolved sucrose in it to obtain several different RIs. **Figure 15.24** shows the calibration bench with water and a sample of petroleum.

With this procedure we obtained the calibration curve shown in **Figure 15.25**, describing the output signal of the embarked electronics inside the sensor versus the RI of the leaking fluid.

Notice that the output signal keeps decreasing as the fluid RI increases. The lower limit is for pure petroleum, which is completely opaque and attenuates the evanescent wave in such a way that no detectable light reaches the photodiode.

In this way the indicating LEDs can produce different signals according to the detected RI, so that the technicians can tell whether it is the outer carcass (water leaking, RI = 1.33) or the inner carcass (fluid leaking, RI from 1.36 to 1.47) that failed.

FIGURE 15.24 The calibration bench.

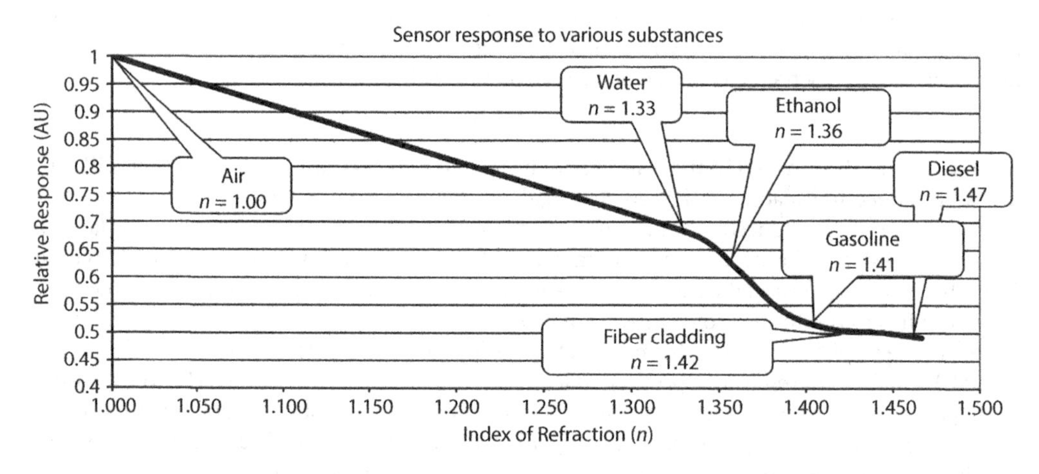

FIGURE 15.25 Sensor response for various substances that can be transported by the hoses.

In **Figure 15.26** it is shown a disassembled transducer and its various parts: the housing, the two transmitters, the electronics case, and battery.

15.4.5 Field Commissioning

With the sensor ready and tested in laboratory, the commissioning phase begins. The test was made at the Ocean Tank, an oceanic facility located at the Universidade Federal do Rio de Janeiro (UFRJ) campus. In **Figure 15.27** we can see a picture of a hose with an installed sensor. The first test was supposed to check the waterproof capability of the transducer and the maximum distance from transmitter to receiver capable to produce reliable information. **Figure 15.28** shows the hose being hauled by a crane to be submerged into the Ocean Tank.

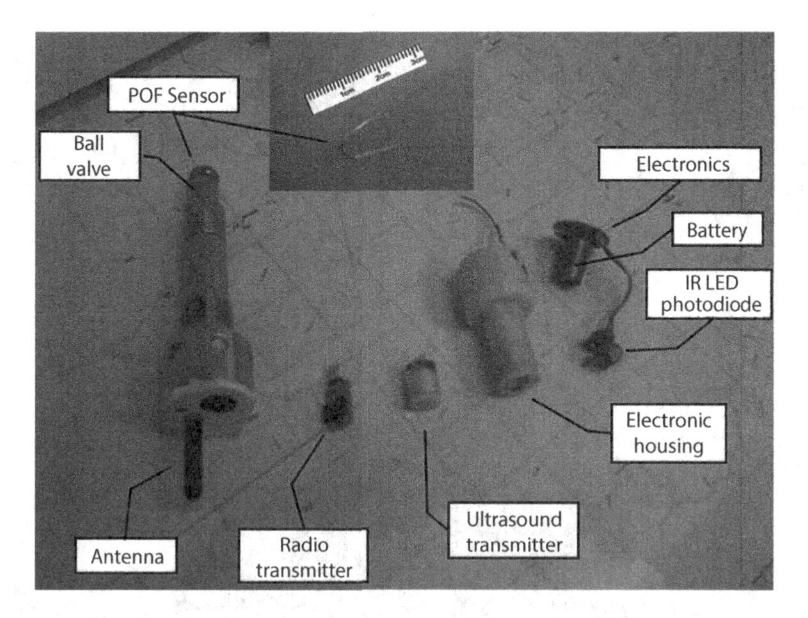

FIGURE 15.26 A disassembled transducer and its various parts: the housing, the two transmitters, the electronics case, and battery.

FIGURE 15.27 A picture of a hose with an installed sensor.

15.4.6 Satellite Telemetry System

To meet the operators' need of being able to monitor hoses continually at offshore terminal locations, we have developed a remote monitoring system as an option to the LED leak detection system. This allows the terminal operator to check the condition and status of each single hose

FIGURE 15.28 The hose being hauled by a crane to be submerged into the Ocean Tank at the Federal University of Rio de Janeiro.

in a hose line during load from the control room or any location while still in service. The system uses the sensor described above equipped with radio and ultrasonic microtransmitters.

The system checks continuously for the presence of oil and or petroleum products between the primary and secondary carcasses of the hose. The transducer then sends a signal to a receiving station (normally at the platform or a buoy). The receiving unit processes and stores the information from all the sensors of all individual hoses. This information is then relayed via INMARSAT satellite to a terrestrial station. The transmitted information is then available through a website 24 hours a day 365 days a year from any global point with Internet access. This allows the terminal operators as well as supervisors and managers to check the system even while away from the terminal location. **Figure 15.29** shows a schematic diagram of the network.

15.4.7 Field Installation

After the commissioning made in the Ocean Tank at UFRJ, the transducer system was put to work in a real operational platform. In this case, a long 300-m hose line was instrumented with hundreds of transducers and the hose was used to load pre-processed petroleum from a production platform to a tanker. **Figure 15.30** shows an illustration of the situation. It is important that the tanker be anchored at a secure distance from the platform to avoid collision, normally several hundred meters.

The hose is stored in a reel at the platform deck. The platform reel is just like a garden hose reel just in a different scale, as seen in **Figure 15.31**. In this reel are stored several hundred meters of the hose line. Each hose contains two transducers, one at each hose end. It is possible to see two of those sensors at the center of the image, fixed to the flanges of two hoses connected together. **Figure 15.32** shows a detail of the reel in which it is possible to see four sensors. They are protected by a steel cage in order to avoid the flange of one hose touch the flange of another, while floating free on the sea.

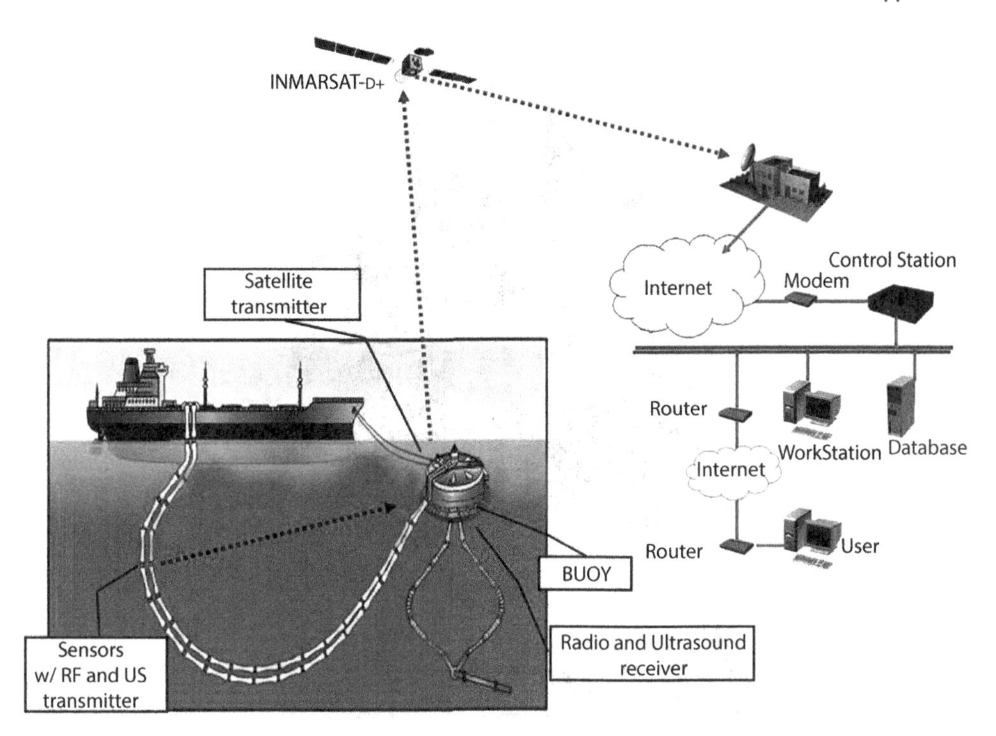

FIGURE 15.29 Schematic diagram of the network designed to allow operators access data collected by the sensors.

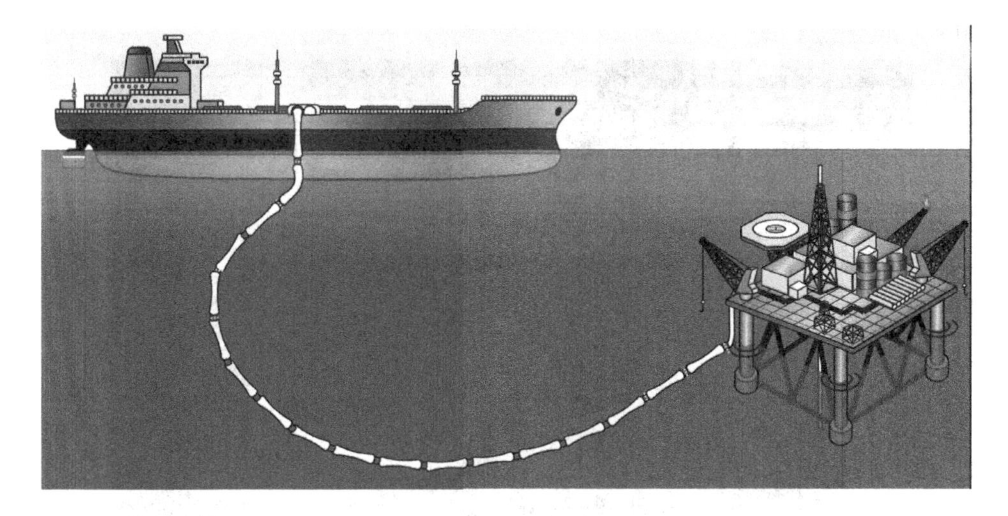

FIGURE 15.30 Loading pre-processed petroleum to a tanker anchored close to a production platform. (Courtesy of Goodyear Seawing–Brazil.)

The deployment of the hose line needs the support of the tanker personnel. A motor dinghy is launched from the tanker while the reel is unrolled. When the tip of the hose line touches the ocean, the hose starts floating and the operators in the boat lace the hose and tow the hose line to the tanker where it is connected and the loading starts. **Figure 15.33** shows the beginning of this operation when the reel has been launched to the sea and the operator is waiting for the motor dinghy to start the towing.

FIGURE 15.31 The hose reel on the deck of the platform. Notice two of sensors at the center of the image, fixed to the flanges of two hoses connected together.

FIGURE 15.32 A detail of the reel in which it is possible to see four sensors installed at the flange of four hoses. The sensors are protected by a steel cage in order to avoid the flange of one hose touch the flange of another.

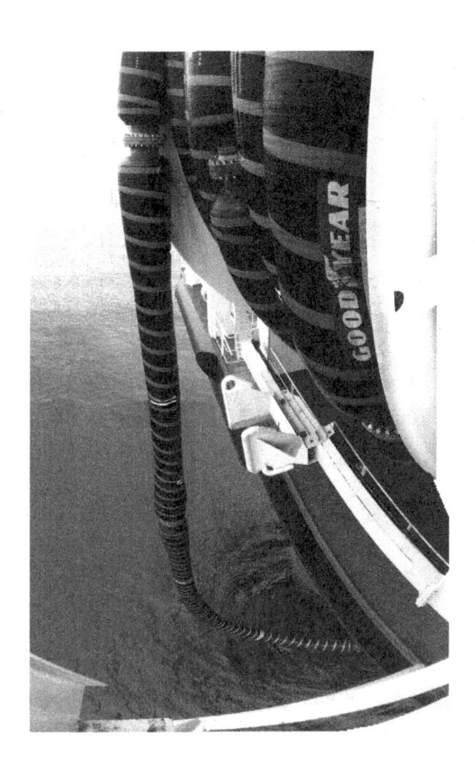

FIGURE 15.33 A line hose is unrolled into the ocean to start the loading operation.

15.4.8 Conclusions

The sensor developed in this project was commercialized by Goodyer together with their hoses all over the world. To protect the invention, Goodyear company deposited four patents under its property, but mentioning the name of the inventors (Werneck et al. 2008a, 2008b, 2009a, 2009b). The Photonic and Instrumentation Laboratory participated in the installation and replacement of sensors in many countries such as Thailand, Singapore, Venezuela, Brazil, etc.

It is important to note how such an insignificantly small sensor, a U-shape POF with 2 cm in length, could involve so many people, make part of so important procedures, and capable to stop a loading operation that costs US$100,000 a day, just because a suspected leaking was alarmed by one single sensor. At the end it was configured that the leaking was not from the inner fluid being transported by the hose, but rather due to the sea water leaking inside a defective hose.

One year after the end of this project, this very same sensor with the same sensing principle was successfully used for another measurand: bacteria. In Chapter 9, Dr. Regina Allil describes the application of U-shaped POF sensor for *Escherichia coli* detection.

15.5 A Sun Tracker for Ambience Illumination

Alexandre Silva Allil

15.5.1 Introduction

Brazil belongs to the group of countries with the highest solar irradiation index in the world, but it underuses its natural power. Although Brazil's hydropower makes up 68.5% of total energy produced; in 2018 we have only 5.3% in Eolic production and a meager 0.01% in photovoltaics.

Considering the solar irradiation all over the word, Brazil is among the regions with the highest solar irradiation, according to Solargis (2018). **Table 15.1** shows the average annual solar energy for some countries with high irradiation on the planet.

Table 15.1 Average Yearly Solar Irradiation of Some Countries	
Country	**Average Yearly Energy (kWh/m²)**
Australia	3,000
Chile	2,740
Sub-Saharan Africa	2,560
Saudi Arabia	2,400
Brazil	2,300
South Africa	2,270
USA	2,200
China	2,120
Spain	1,800
Germany	1,200

Source: Solargis, Weather data and software for solar power investments–Global horizontal irradiation, www.solargis.com/products/maps-and-gis-data/free/download/world, 2017.

In 2015, solar power was only about 1% of the global electricity demand in the world and additionally, illumination consumes about 20%, catalyzing the research in solar energy for illumination.

Inspired in these facts, the objective of this work was to develop a solar illumination system by the use a solar tracker that follows the Sun from sunrise to sunset. The objective was to apply the system as an alternative light generation in environments such as study rooms, museums, warehouses, underground parking, and commercial buildings, for instance.

The solar tracker was totally developed at the Photonics and Instrumentation Laboratory and consists of an opto-electromechanical system, using a motor, an electronic set up for control and an optical system comprised of a Fresnel lens and a plastic optical fiber (POF) bundle. In addition, some sensors are employed to monitor the process such as the light intensity captured by the optical fibers, ascension and declination angles by the use of an accelerometer, and the overall measurement of the solar radiation with a pyranometer.

15.5.2 Plastic Optical Fibers

Plastic optical fiber was used in this work rather than silica fiber for the several advantages POF offers. One of the best characteristics is their diameter and numerical aperture that allows the capture of more light than the silica fibers. On the other hand, one must mention the greatest disadvantage of POF that is the larger attenuation in the visible spectrum, as compared with silica fibers. **Table 15.2** shows the relevant characteristics of POF as compared to silica fibers.

Notice that we choose unconventional diameters for silica and POF. In reality there are many choices for each fiber type, but these two diameters are cited as examples for lightening applications. The greater the fiber diameter, the greater the amount of light captured; however, the greater the diameter, the greater the losses when making a bundle as the empty spaces between fibers increase.

Attenuation is a serious problem when large distances between light source and illumination site have to be covered. For silica fiber, the attenuation is much smaller than that of POF; however, as the silica density is higher, the bundle will be heavier and also much less flexible than a bundle made of POF.

Table 15.2 Comparison of Different Characteristics between POF and Silica Fiber

Parameter	POF	Silica Fiber
Material	PMMA[a]	Silica[b]
Diameter	2 mm	0.20 mm
Attenuation at visible spectrum	100 dB/km	8 dB/km
Numerical aperture	0.5	0.22
Price per meter	1 US$/m	8 US$/m
Maximum temperature	80°C	Up to 900°C
Specific mass	1 g/cm³	2.5 g/cm³
Bundle flexibility	High	Low

Source: Allil, R.C. et al., *Solar Energy*, 174, 648–659, 2018.
[a] Poly(methyl Methacrylate).
[b] SiO_2.

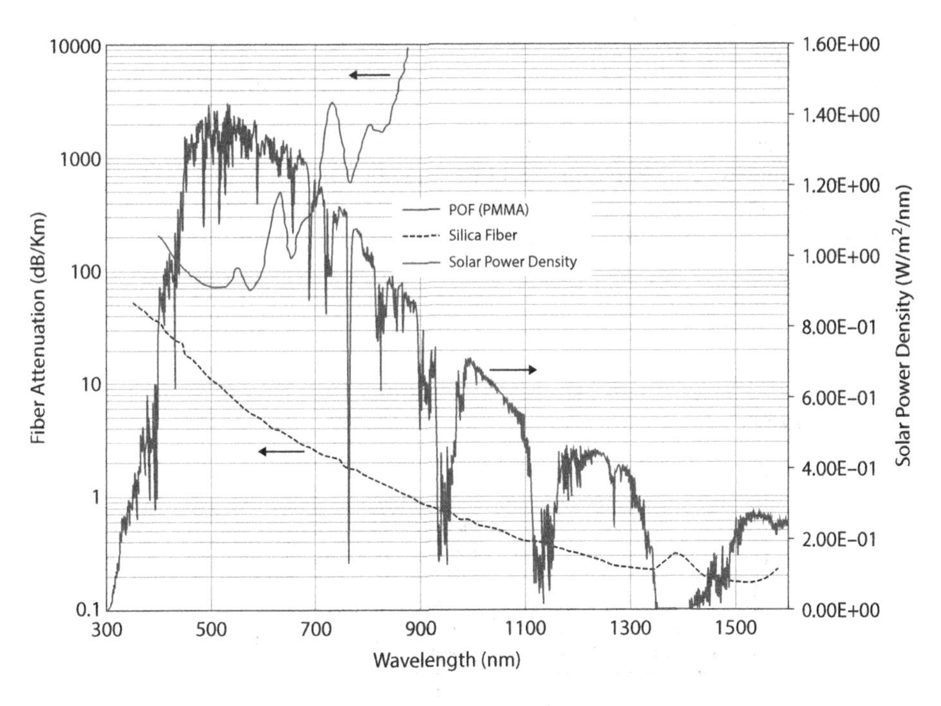

FIGURE 15.34 Comparison of solar irradiance spectrum with the attenuation of silica fiber and POF. (Solar spectrum adapted from Wikimedia Commons, https://en.wikipedia.org/wiki/Sunlight, 2018.)

Temperature is another important factor, because every incident photon from the Sun that is not guided by the fiber will be converted into heat. **Figure 15.34** shows the attenuation of POF compared with silica fiber and the solar irradiance available at Earth surface.

Notice that POF does not transmit the near infrared portion of the solar spectrum well whereas silica does. Therefore, under sunlight POF will heat more than silica, which will make necessary the use of an infrared filter because of the limited operating temperature of PMMA. On the other hand, since POF does not transmit IR, the illumination provided by a POF system will be made only on the visible part of the spectrum.

15.5.3 The Solar Tracker

The system is able to follow the Sun from the sunrise to the sunset using an algorithm that calculates the Sun position based on the time of year, the longitude, and the latitude of the location. An algorithm was developed to calculate the time of sunrise and sunset at any day in the year, given the latitude and longitude coordinates of the location. From this data it is possible to calculate the duration of the day and then the initial and final right ascension angle values. Hence, with the sunrise and sunset hour, the whole tracking system works completely automated through the days along the year.

The algorithm was implemented in C++ and uploaded into an Arduino® microprocessor. The microprocessor controls a stepper motor driven by a power transistor bridge. **Figure 15.35** shows the block diagram of the system.

An Arduino microcontroller is responsible for the control of the tracker through a power driver and a stepper motor. The pyranometer informs the amount of available solar irradiation. The photodetector receives the light from one single fiber from the POF bundle and, together with the pyranometer information, it is possible to know the amount of light concentrated by the solar tracker. These two facts should be always proportional to each other, otherwise the tracker is missing the solar direction.

The stepper motor is controlled by the microcontroller from the information provided by the accelerometer that gives the tilt angle of the tracker.

Initially the system was developed and tested at the laboratory. **Figure 15.36** shows the solar tracker being assembled at the Photonics and Instrumentation Laboratory.

In **Figure 15.36**, the motor (1), driven by the control box (2), can turn the frame (3) following the Sun all over the sky from sunrise to sunset, according to the algorithm responsible by the

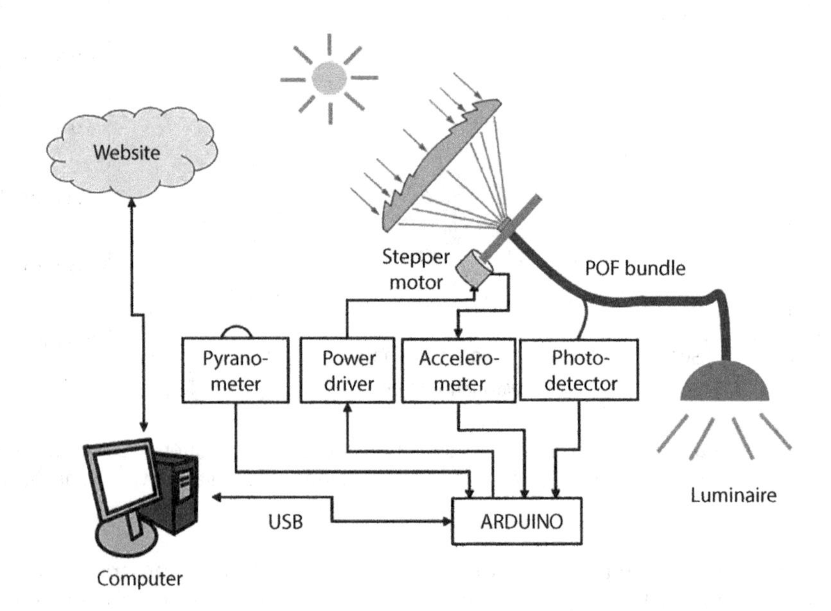

FIGURE 15.35 Block diagram of the system. Two POF bundles were built that can be inserted in the solar tracker, one for the luminaires and one for the photobioreactor. (Modified from Allil, R.C. et al., *Solar Energy*, 174, 648–659, 2018.)

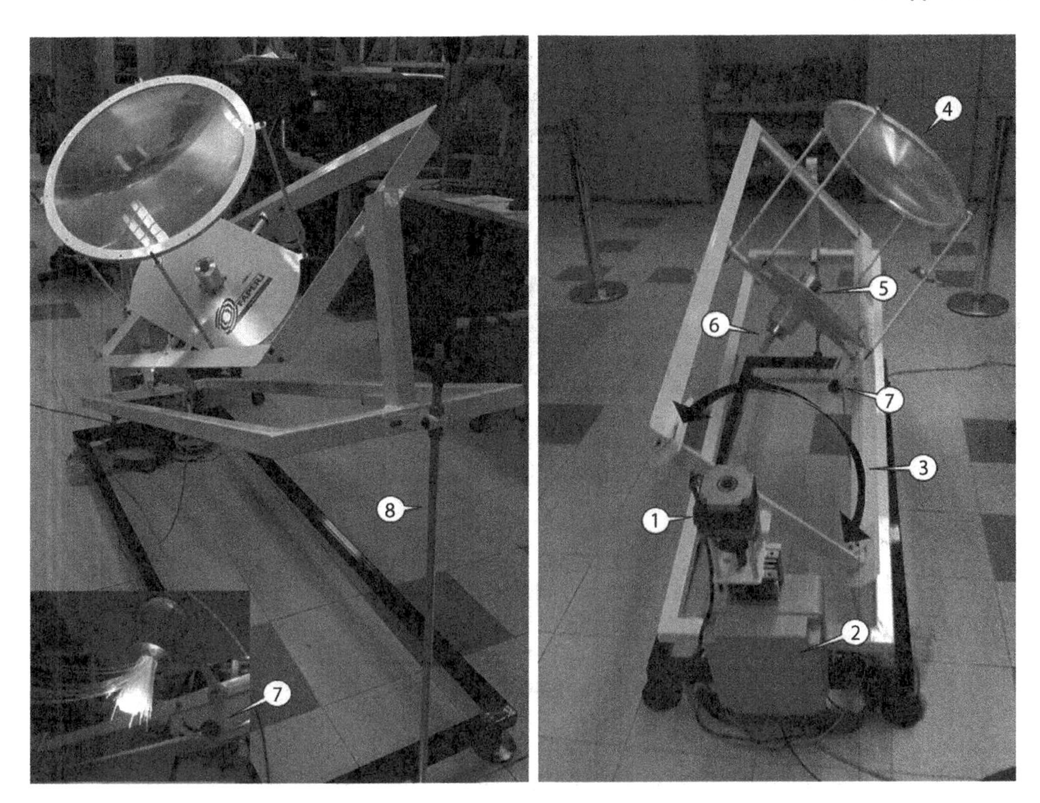

FIGURE 15.36 Two views of the solar tracker being assembled in laboratory. Lower left inset: POF bundle support in test with short fibers illuminated by an artificial light. (Reprinted from Allil, R.C. et al., *Solar Energy*, 174, 648–659, 2018. With permission.)

calculation of the sunrise time. The Fresnel lens (4) rotates together with the frame of the system focusing the light into the tip of the POF bundle (5), supported by a tube (6). In (7) there is an adjustment of the declination angle that will be carried out manually in a weekly schedule. Since the Sun rises every day at a different position, the Fresnel lens support can be turned manually to the specific direction on the horizon. The latitude setting (8) coincides with the inclination angle of the Earth axis and has to be adjusted according to geographical position of the installation and will be kept fixed thereafter.

15.5.4 Optical System

There are many challenges in producing a light guide illuminated by the Sun. First, there is a trade-off between the concentration factor and the temperature. The larger the concentration factor, the more light is conducted by the waveguide, but, because the POF does not transmit infrared (see **Figure 15.34**), this part of the spectrum will be converted in heat. Since the PMMA, as all polymers in fact, is a thermal insulator, it is not possible to dissipate this heat. The solution is to defocus the lens or decrease the lens diameter, with the decrease of guided light as a payback. Additionally, we filtered out the infrared part of the spectrum, since this part not only cannot be used for illumination, but also, it is not transmitted by the fiber. The filter used is a silica 12.5 mm diameter IR cutoff filter (Edmund Optics, 2018) with a cutoff wavelength at 750 nm.

The other difficult decision in constructing an optical fiber bundle is the trade-off between the fiber diameter, the bundle flexibility, the total length of the fibers, and the packing-factor as the larger the fiber diameter, the harder will be the bundle do curve.

FIGURE 15.37 Left: The tip of the bundle after the polishing process. Right: Tip of the bundle after compression under heat. (Reprinted from Allil, R.C. et al., *Solar Energy*, 174, 648–659, 2018. With permission.)

The fiber bundle has a diameter of 25 mm, which fits 120 fibers with 2 mm diameter. The challenge in making a bundle is the arrangement of its tip in order to keep the packing factor as tight as possible as any light that falls in the space between two fibers will not be guided.

After joining the fibers in the bundle, it is necessary to polish the tip in order for each fiber to capture the same amount of light. **Figure 15.37** (left) shows the tip of the bundle after the polishing process. By heating the bundle tip close to the melting point and compressing it inside an adjustable ring, one can eliminate the empty spaces between the fibers. **Figure 15.37** (right) shows the result, after the compression, where it is possible to notice the hexagonal fiber tip format.

15.5.5 Luminaires

Two types of luminaires were developed, one for diffuse lighting and another one for focused lighting. Two POF bundles with 120 fiber and 7 m in length were assembled, one for each luminaire.

For the spot luminaire, all fibers were inserted and glued into a 50-mm diameter PMMA disk with 120 equally spaced holes. After gluing, the disk face was polished so all fibers illuminate equally and homogeneously. Then, the disk with the fixed fibers was inserted into a commercial parabolic reflector. The focus divergence can be calculated from the POF numerical aperture shown in **Table 15.2**, NA = 0.5, and the divergence angle 60°.

For making the diffuse luminaire it was necessary to modify the fiber tips so that the light leaves the fiber laterally at several angles. As the goal is to spread the light as much as possible, the part of the fiber to be modified must extend a few centimeters from the end of the fiber.

Several ways of modifying the fiber in order to allow the escape of light from the sides were studied, but the better solution was the transversal grooves that were made with the use of a die to carve a continuous thread along the fiber end. **Figure 15.38** shows a microscopic picture of one fiber tip, after the process. We used a commercial fluorescent light reflector to case the fibers. As this luminaire is 1.2 m in length, we used six 20-cm sections with 20 fibers each to fill the entire space of the luminaire.

15.5.6 Results

Figure 15.39 shows the solar tracker installed in field, with the tracker positioned to the Sun on the beginning of the day at the roof of the Technology Center at the Universidade Federal do Rio de Janeiro. The local coordinates are: Latitude: 22°51′40.37″S, Longitude: 43°13′49.46″O.

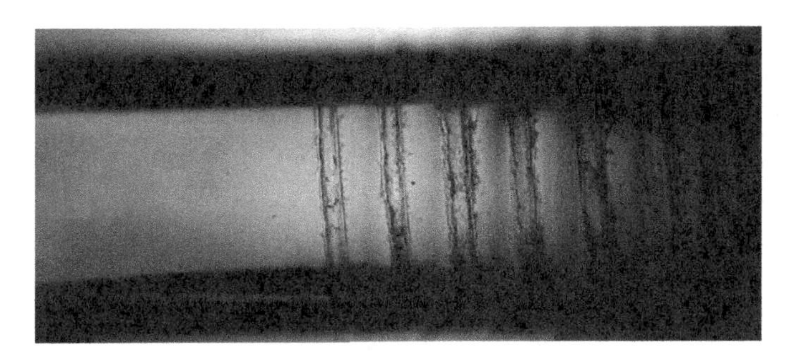

FIGURE 15.38 A microscopic picture of a fiber tip, after the threading process.

FIGURE 15.39 Top: Solar tracker system installed on the roof of the Technology Center at the Universidade Federal do Rio de Janeiro. Bottom left: Details of the solar tracker, showing the stepper motor at center left, the microprocessor control box at bottom left, the Fresnel lens, and the IR filter at the lens focus. Bottom right: Adjustment of the declination angle which allows the tracker to be directed to the sunrise direction. (Reprinted from Allil, R.C. et al., *Solar Energy*, 174, 648–659, 2018. With permission.)

Due to the POF attenuation of about 0.1 dB/m at the center of visible spectrum, we will experience an attenuation of 50% per each 30 m of fiber bundle. A fiber length of 7 m would provide an attenuation of 0.7 dB or about 15%. Aiming at an evaluation of the color distortion due to the PMMA attenuation spectrum, we measured the spectra of the Sun and that of the light guided by the POF bundle using an optical spectrum analyzer (Model HR4000 Ocean Optics, Florida, USA).

Figure 15.40 shows the solar spectrum (dotted line) superimposed over the fiber attenuation spectrum (solid line) and the measured spectrum at the end of the bundle (dashed line). Notice that the spectrum at the end of the bundle coincides with that of the Sun, except on a wavelength of about 620 nm (orange region of the visible spectrum) where the PMMA presents a strong attenuation.

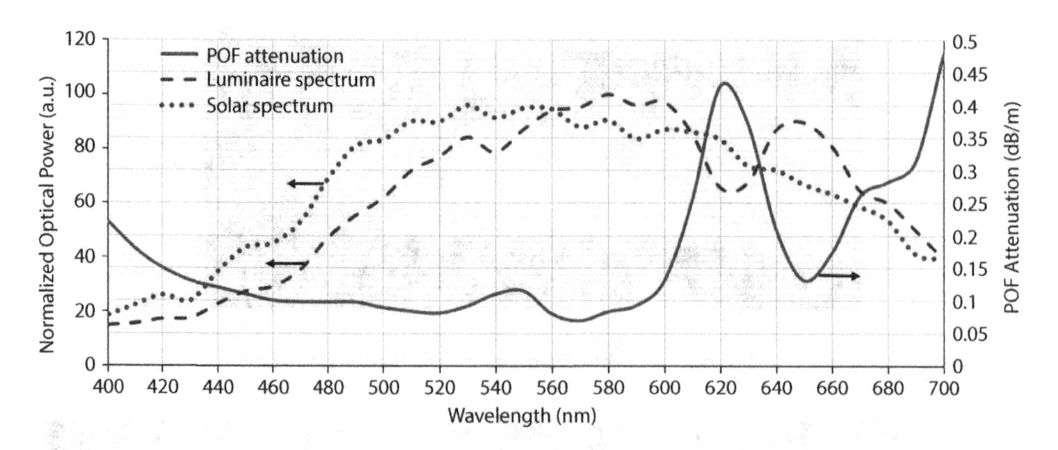

FIGURE 15.40 The solar spectrum (dotted line) superimposed on the POF attenuation spectrum (solid line) and the measured spectrum at the end of the POF bundle (dashed line). (Reprinted from Allil, R.C. et al., *Solar Energy*, 174, 648–659, 2018. With permission.)

After that point, however, there is another transmission window at 650 nm (red region) where the POF bundle spectrum presents a red peak. For a relatively small length such as the one used here, there will not be much color modification on the spectrum, but for longer lengths there will be less red and blue, which will distort the spectrum from white to a greenish tone.

As detailed above, two luminaries were built, one spot and one diffuse. The illumination provided by these two luminaires are show in **Figure 15.41**, at left the spot luminaire and at right the diffuse luminaire.

To evaluate the distribution of light in the illuminated areas by the POF luminaires, illuminances were taken at various points in the room using a lux meter (Model Extech HD450, New Hampshire, USA). Measurements were taken approximately 80 cm from the floor level. At the moment of making these measurements we did not have an IR filter; therefore, we used a lower concentration for controlling the POF temperature. **Figure 15.42** shows the measurements at table level.

In order to maintain the temperature at the tip of the bundle below 70°C, we controlled the system concentration by adjusting the focus position of the Fresnel lens. Experimental measurements confirmed that the best concentration was 74x. In a normal day with an illuminance of 100,000 lux and with a concentration of 74x, all fibers receive 7.4×10^6 lux. Since optical fibers present a diameter of 2 mm and therefore an area of 3.14×10^{-6} m^2, the number of lumens received per fiber is 23.25 lumens.

Since the fiber bundle has 120 fibers, a total of 2,790 lumens are injected into the bundle. In the visible range, the average fiber attenuation is about 0.1 dB per meter or a total of 0.7 dB of light losses in 7 m of optical fiber.

Equation (15.1) gives us the relationship between the transmittance T in dB with the transmittance in percent:

$$T(dB) = 10\log_{10}\left(\frac{P_{in}}{P_{out}}\right) = 10\log_{10}(\text{Transmittance}) \tag{15.1}$$

where P_{in} and P_{out} are the input and output power in the bundle, respectively. Solving the equation, we get the value of 0.85 for the bundle transmittance, which presents therefore, a loss of 15%. Now we can calculate the number of lumens arriving at the end of the POF bundle available to illuminate the ambience: 2,371.5 lumens.

FIGURE 15.41 Luminaries developed and their illumination effect. Top left: Spot luminaire; Bottom left: Illumination obtained with the spot luminaire; Top right: Diffuse luminaire; and Bottom right: Illumination obtained with diffuse luminaire. (Reprinted from Allil, R.C. et al., *Solar Energy*, 174, 648–659, 2018. With permission.)

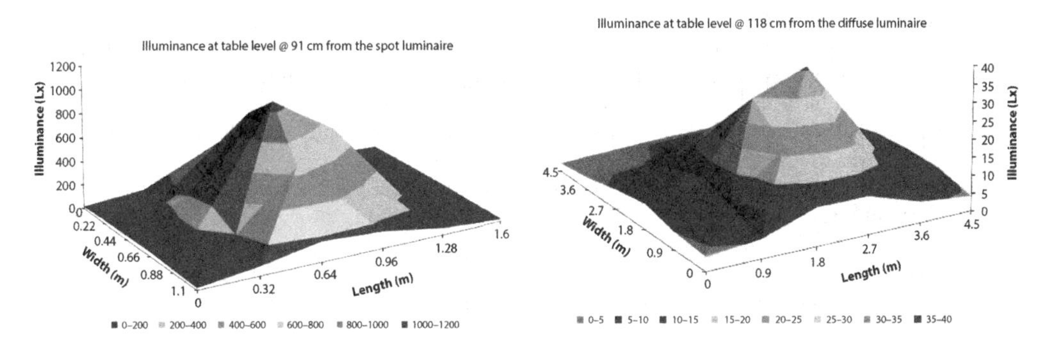

FIGURE 15.42 Results from measurements of illuminance in lux of spot luminaire (left) and diffuse luminaire (right). (Reprinted from Allil, R.C. et al., *Solar Energy*, 174, 648–659, 2018. With permission.)

According to DiLaura et al. (2011) an office room should have 500 lux, or 500 lumens/m^2. Therefore, to illuminate a 9 m^2 room we would need 4,500 lumens or about two solar trackers or one solar tracker with two Fresnel lenses and two POF bundles.

In order to increase illumination level, the natural solution would be to increase concentration, but that would increase temperature above the PMMA temperature limit. Another way to increase illumination, without compromising temperature, would be to keep the bundle tip with the same solar concentration but increase the number of fibers in the bundle, with a consequent increase in the bundle diameter and, of course, an increase in the fiber cost.

15.5.7 Conclusions

It is known that, for every innovation that will substitute conventional counterparts, there will be an initial investment which will take some time to be amortized, depending on the savings such innovation will allow. The cost to construct the tracker is very small with the exception of the optical fiber. The POF we used costs about US$1 per meter and since the bundle contains 120 fibers, a 10-m length bundle will cost US$1,200.00.

With respect to ambience illumination, the solar tracker proved to be able to substitute a 40 W fluorescent light bulb. Thus, considering the irrelevant energy costs of running a stepper motor, each system is capable of saving about 0.5 kWh per day, when substituting a light bulb for a 10-hour daily usage. The time that will take to amortize the initial implantation costs will depend on the electricity cost per kWh of each specific region.

References

Allil, R.C., Manchego, A., Allil, A., Rodrigues, I., Werneck, A., Diaz, G.C., Dino, F.T., Reyes, Y., Werneck, M. “Solar tracker development based on a POF bundle and Fresnel lens applied to environment illumination and microalgae cultivation”. *Solar Energy*, 174, 648–659, 2018. doi:10.1016/j.solener.2018.09.061.

DiLaura, D., Houser, K., Mistrick, R., Steffy, G., Eds., *The Lighting Handbook*, Illuminating Engineering Society of North America, New York, 2011.

Flexible leak detection system and method for double carcass hose, European Patent EP 1879008 A3. The Goodyear Tire & Rubber Company, Filing Date: July 13, 2007. Publication Date: May 21, 2008.

Flexible leak detection system and method for double carcass hose, European Patent EP 1879008 B1. Veyance Technologies, Inc, Filing Date: July 13, 2007. Publication Date: December 30, 2009.

Leak detection sensor system and method for double carcass hose, United States Patent US 7,387,012 B2. Veyance Technologies, Inc, Filing Date: July 14, 2007. Publication Date: June 17, 2008.

Solargis. Weather data and software for solar power investments - Global horizontal irradiation. www.solargis.com/products/maps-and-gis-data/free/download/world, 2017 (accessed in May 1, 2018).

Werneck, M.M., “Application of POF sensors in energy, oil and biotechnology”, *3rd International POF Modelling Workshop 2015*, September 21, 2015, Nurnberg, Germany, joint with the 24th International Conference on Plastic Optical Fibers “Bio-sensors with POF Technology”, held at Technische Hochschule Nurnberg Georg Simon Ohm Nuremberg, Germany, September 22–24, 2015.

Werneck et al., “Flexible leak detection system and method for double carcass hose”, European Patent EP 1879008 A3. The Goodyear Tire & Rubber Company, Filing Date: July 13, 2007. Publication Date: May 21, 2008a.

Werneck et al., “Leak detection sensor system and method for double carcass hose”, United States Patent US 7,387,012 B2. Veyance Technologies, Inc, Filing Date: July 14, 2007. Publication Date: June 17, 2008b.

Werneck et al., “Flexible leak detection system and method for double carcass hose”, European Patent EP 1879008 B1. Veyance Technologies, Inc, Filing Date: July 13, 2007. Publication Date: December 30, 2009a.

Werneck et al., “Flexible leak detection system and method for double carcass hose”, United States Patent US 7,509,841 B2. Veyance Technologies, Inc, Filing Date: July 14, 2006. Publication Date: March 31, 2009b.

Werneck, M.M. et al., Sistema de Monitoramento de Abertura e Fechamento de Chaves Seccionadoras. Patent deposited at the Instituto Nacional de Propriedade Industrial (INPI) under the number Protocolo 12120000050 in February 7, 2012. Software registered under number 12868-5.

Werneck, M.M., R.C.B.S. Allil, C.C. Carvalho, "Método e Aparato para Monitoramento da Abertura e do Fechamento de uma Válvula", patent deposited at the Instituto Nacional de Propriedade Industrial (INPI) under the number BR 10 2017 016783 6 in August 4, 2017, Protocol 870170055858.

Werneck, M.M., R.C.S.B. Allil, C.C. Carvalho, F.L. Maciel, D.M. Santos and F.V.B. Nazaré, "Applications of POF Sensors in the Electric Energy Sector" in *POF Simulation beyond Data Transmission: Summary of the 3rd International POF Modelling Workshop 2015*, 2015 C.A. Bunge, R. Kruglov, Herstellung und Verlag: BoD – Books on Demand, Norderstedt, pp. 105–122, December 2, 2015.

Index

Note: Page numbers in italic and bold refer to figures and tables, respectively.

F

G

H

I

K

L

M

Q

R

S

9781032337555